AN

ELEMENTARY TREATISE

ON

ASTRONOMY;

IN TWO PARTS.

THE FIRST CONTAINING

A CLEAR AND COMPENDIOUS VIEW OF THE THEORY;

THE SECOND,

A NUMBER OF PRACTICAL PROBLEMS.

TO WHICH ARE ADDED,

SOLAR, LUNAR, AND OTHER ASTRONOMICAL TABLES.

By JOHN GUMMERE, A.M.

MEMBER OF THE AMERICAN PHILOSOPHICAL SOCIETY, AND CORRESPONDING MEMBER OF THE ACADEMY OF NATURAL SCIENCES, PHILADELPHIA.

Sixth Edition.

REVISED AND ADAPTED TO THE PRESENT STATE OF THE SCIENCE,

By E. OTIS KENDALL, A.M.

PROFESSOR OF MATHEMATICS AND ASTRONOMY IN THE CENTRAL HIGH SCHOOL OF PHILADELPHIA.

PHILADELPHIA:
E. C. & J. BIDDLE, No. 39 SOUTH FOURTH STREET.
1854.

NOTE BY THE PUBLISHERS.

For the Sixth Edition of Gummere's Astronomy, the work has again been carefully revised by Professor E. O. Kendall, and such alterations and additions as were called for by the advance of the science have been made.

E. C. & J. BIDDLE.

Philadelphia, October, 1854.

STEREOTYPED BY L. JOHNSON AND CO.
PHILADELPHIA.
PRINTED BY T. K. AND P. G. COLLINS.

PREFACE TO THE FOURTH EDITION.

THE great and steadily increasing favour bestowed on this work, as modified by its author in 1842, would seem to indicate that no further changes were needed, or would be acceptable to teachers: the undersigned has therefore, in revising the third edition for the press, thought it best to confine himself principally to a correction of typographical, and other errors; and to such additions as the recent progress of the science demanded. Some additions have been made in the chapter on instruments; and, throughout the First Part, such changes and modifications of the formulæ and demonstrations introduced, as have been dictated by eight years' experience in the use of this work as a class book. The tables of the elements of the planetary orbits, have been arranged in a more convenient form, and extended so as to include those of the new planets, as far as they are at present known. These elements have been invariably derived from the most reliable sources. In the Second Part, very many inaccuracies have been corrected, and several problems and examples of a practical nature inserted. In connection with one of these problems, a table of reductions to the meridian has been given at the end of the book. With this exception, and that of an alteration in Table IX., adapting it to the present time, the tables are the same as those in the third edition.

E. O. KENDALL.

Central High School, Philadelphia, February 1, 1851.

PREFACE TO THE THIRD EDITION.

In preparing this edition, the greater part has been written anew and so modified as to increase the value of the work as a text book. Several of the more abstruse investigations of the First Part have been omitted, references being made to larger works, and others have been transferred to the Appendix. Many of the figures illustrating the text have been rendered more perspicuous by a change in the construction, and a number of new ones are added. The progressive state of the science has claimed attention, and notices of recent results have been introduced.

The Appendix, in addition to the other matter, contains Professor Bessel's late investigation of formulæ for computing Solar Eclipses, Occultations and Transits, reduced to a more elementary form. In the Second Part, the formulæ are applied in the computation of these phenomena.

The Tables are nearly the same as in the last edition. Instead, however, of some small ones, which have been omitted, a table of Logarithms, and one of Logarithmic Sines and Tangents to four decimal figures, have been inserted. These are convenient in many computations not requiring greater precision.

John Gummere.

Haverford School, 9*mo.* 1842.

CONTENTS.

PART I.

2

APPENDIX TO PART I.

PART II.

The following Alphabet is given in order to facilitate, to the student who is unacquainted with it, the reading of those parts in which the Greek letters are used.

Letters.	Names.	Letters.	Names.
Α α	Alpha	Ν ν	Nu
Β β ϐ	Bēta	Ξ ξ	Xi
Γ γ	Gamma	Ο ο	Omicron
Δ δ	Delta	Π ϖ π	Pi
Ε ε	Epsilon	Ρ ρ	Rho
Ζ ζ	Zēta	Σ σ ς	Sigma
Η η	Eta	Τ τ	Tau
Θ ϑ θ	Thēta	Υ υ	Upsilon
Ι ι	Iōta	Φ φ	Phi
Κ ϰ	Kappa	Χ χ	Chi
Λ λ	Lambda	Ψ ψ	Psi
Μ μ	Mu	Ω ω	Omega

AN

ELEMENTARY TREATISE

ON

ASTRONOMY.

PART I.

CHAPTER I.

GENERAL PHENOMENA OF THE HEAVENS—DEFINITIONS AND PRELIMINARY OBSERVATIONS.

1. *Astronomy*, or *Plane Astronomy*, is the science which treats of the motions, distances, magnitudes, and appearances of the heavenly bodies. *Physical Astronomy* applies the principles of mechanics to explain their motions.

2. *General Observations.* If, on a clear night, we fix our attention on the heavens, and continue to observe them at intervals for a few hours, we may, without the aid of any instruments, make some useful observations. It will be seen that the stars retain the same positions with regard to one another; but that their positions with respect to the earth are continually changing. Those in the eastern part of the heavens become more and more elevated, and others that were not at first visible, come into view, or *rise.* Those in the western part, descend lower and lower till they go out of view, or *set.* In the southern part, some will be observed to rise,

ascend to small elevations, and then descend and set, to the west of their places of rising.*

3. *Circumpolar Stars.* If we direct our attention towards the north, different phenomena are presented. In that part of the heavens, there are many stars which do not set. Those that are descending continue to do so till they arrive at certain lowest points, and then begin to ascend. They appear to revolve or describe circles about a certain star which seems to remain stationary. This star is called the *Pole Star.* All the stars that do not set are called *Circumpolar Stars.*

4 *North Pole.* When the pole star is more accurately observed by the aid of suitable instruments, it ceases to appear stationary. It is found to have an apparent motion in a small circle, round a certain point as a centre, or geometric pole, distant about $1\frac{1}{2}°$ from it. This point is called the *North Pole of the heavens*, or simply the *North Pole.*

5. *Diurnal Motion.* If our observations are repeated on successive evenings, we find the same stars moving in the same manner, and occupying, at any given time in the evening, very nearly the same positions with regard to the earth as at the same time the preceding evening. The stars, therefore, and indeed all the heavenly bodies, appear to revolve round the earth from *east* to *west*, in about twenty-four hours. This motion is called the *Diurnal Motion.*

6 *The Moon.* If the situations of the moon be observed on successive nights, it will be found that she changes her position among the stars, moving among them from *west* to *east:* that is, in a direction *contrary* to that of the diurnal motion. By this motion she makes a complete circuit of the heavens in about twenty-seven days.

7 *The Sun.* The sun also appears to have this motion from *west* to *east*, among the stars. This may be inferred from observing the position of different groups of stars after the sun has

---

* Here and in other parts of the work, unless the contrary is mentioned, the observer is supposed to be in the United States, or southern or middle parts of Europe.

set. If our observations are repeated at intervals for some weeks or months, we shall find that the sun appears continually to approach the stars to the eastward of him. He thus, in the course of a year, appears to make an entire circuit of the heavens.

Owing to this apparent annual motion of the sun, the groups of stars visible at a given hour in the evening, and their positions at that hour, are very different at different seasons of the year.

8. *Planets.* There are likewise several stars which have motions among the other stars, moving generally like the sun and moon, from *west* to *east;* though sometimes for short periods they appear to move in the *contrary* direction. These are called *Planets.*

Five of the planets, named *Mercury*, *Venus*, *Mars*, *Jupiter*, and *Saturn*, are visible to the naked eye, and were known to the ancients. Within the last seventy-five years, thirty-three others have been discovered with the aid of the telescope, without which they are invisible. Their names are *Uranus*, *Neptune*, *Flora*, *Melpomene*, *Clio*, *Euterpe*, *Vesta*, *Iris*, *Metis*, *Phocea*, *Massalia*, *Hebe*, *Lutetia*, *Parthenope*, *Fortuna*, *Thetis*, *Amphitrite*, *Astræa*, *Egeria*, *Irene*, *Thalia*, *Eunomia*, *Proserpina*, *Juno*, *Ceres*, *Pallas*, *Bellona*, *Calliope*, *Psyche*, *Hygeia*, *Themis;** these with the exception of Uranus and Neptune are called *Asteroids.*

9. *Satellites.* Observations with the telescope show that some of the planets are accompanied by one or more smaller bodies, whose positions are continually varying. These small bodies, are called *Moons*, *Satellites*, or *Secondary Planets;* those named in the preceding article being called *Primary Planets.*

Of the *Satellites* known at this time, four revolve around *Jupiter*, eight around *Saturn*, six around *Uranus*, and one around *Neptune.*

10. *Planetary Regions.* The sun, moon, and planets, with their satellites, move through nearly the same region of the heavens, the courses of the moon and planets, except *Pallas* and one or two other *Asteroids*, not differing greatly from that of the sun.

11. *Comets.* There is a class of bodies that appear occasionally in various parts of the heavens, moving in various directions among the fixed stars, and only continuing visible for a few weeks or months. These are called *Comets.*

A comet is not unfrequently accompanied by a faint brush of light, projecting from it on the side opposite the sun, and extending in some cases to a great distance. This is called the *tail* of the comet.

---

* The names of the two last discovered have not yet been announced.

12. *The Earth.* The earth is a body of a globular form. This may be inferred from the following well known facts. When persons on board a ship at sea, observe another ship receding from them, they first lose sight of the hull; then of the lower sails; afterwards of those that are higher; and lastly of the most lofty sails. To an observer at the mast head, the receding ship continues visible long after it has ceased to be seen by those on deck; the different parts eventually disappearing to him in the same order as to those below. This takes place in whatever direction the ship recedes, and in whatever part of the ocean the observations are made. Hence it follows that the surface of the ocean must be globular; and as the general level of the land does not greatly differ from that of the ocean, the whole earth may be regarded as a globular body.

By methods that will be noticed in a subsequent chapter, it has been ascertained that the earth is nearly, though not exactly, a perfect sphere, and that its diameter is about 7912 miles.

13. *Fixed Stars.* Those stars which do not sensibly change their positions with regard to one another are called *fixed stars.* They are at an immense distance from the earth. This may be inferred from the fact that the angular distance of any two of them is found to be the same at whatever part of the earth the observation is made. It has indeed been ascertained by means which may hereafter be understood, that the distance is so immensely great that the angle contained between two lines conceived to be drawn from one of them to opposite sides of the earth must be less than the ten thousandth part of a second. We may therefore regard the diameter of the earth as an insensible quantity in comparison with the distance of the stars.

In consequence of their immense distance, the fixed stars appear merely as luminous points, even when viewed with telescopes of high power; whereas the planets, when thus viewed, present sensible and measurable discs.

14. *Celestial Sphere.* It is not supposed that the fixed stars are all at the same distance from the earth. But since their distances are all so exceedingly great that no change of position on the earth produces any appreciable change in their positions with re-

gard to one another, we may regard them as placed in the concave surface of an immense hollow sphere, having its centre at the centre of the earth. This imaginary sphere is called the *Celestial Sphere.*

15. *Diurnal motions of the fixed stars.* Each star in its diurnal motion, moves uniformly, in a circle of which the north pole of the heavens is a geometric pole. The method by which the truth of this proposition is established, will be given in a subsequent chapter.

16. *Stars during the day.* The strong light of the sun overpowering the feebler light of the stars renders them invisible to the naked eye during the day time. But by the aid of a telescope the brighter stars, except those near the sun, may be distinctly seen, and observations may be made on them in the full light of day.

17. *Copernican System.* Copernicus, a celebrated Prussian astronomer, who flourished in the early part of the sixteenth century, formed a theory or system to account for the apparent motions of the heavenly bodies. According to this system, the apparent diurnal motion from *east* to *west* is produced by a *rotation* of the earth from *west* to *east*, about a line or axis passing through its centre and the north pole of the heavens: the apparent annual motion of the sun is produced by a real motion of the earth, round the sun at *rest;* the planets also revolve round the sun at different distances and in different times; and the moon revolves round the earth and with it round the sun; the revolutions all being from *west* to *east.* The truth of this system, called, from its author, the *Copernican System,* is confirmed by many astronomical facts; and no fact inconsistent with it is known to exist. Astronomers, therefore, adopt it as the true system. Some confirmations of its truth will be noticed in subsequent parts of the work.

18. *Order of the planets.* The order of the primary planets with regard to their distances from the sun, including the earth as one, and also those discovered since the time of Copernicus, is Mercury, Venus, the Earth, Mars, the Asteroids in the order given in Article 8, Jupiter, Saturn, Uranus, and Neptune.

Mercury and Venus, having their *orbits,* or paths which they describe round the sun, within that of the earth, are called *inferior* planets. The others, whose orbits are without that of the earth, are called *superior* planets.

19. *Characters.* The following characters, by which the sun, moon, and planets are sometimes designated, should be impressed on the memory of the student.

Sun	☉	Flora	⑧	Lutetia	㉑
Mercury	☿	Metis	⑨	Calliope	㉒
Venus	♀	Hygeia	⑩	Thalia	㉓
Earth	⊖ or ♁	Parthenope	⑪	Themis	㉔
Moon	☾	Clio or Victoria	⑫	Phocea	㉕
Mars	♂	Egeria	⑬	Proserpina	㉖
Ceres	⚳ or ①	Irene	⑭	Euterpe	㉗
Pallas	⚴ or ②	Eunomia	⑮	Bellona	㉘
Juno	⚵ or ③	Psyche	⑯	Amphitrite	㉙
Vesta	⚶ or ④	Thetis	⑰	Jupiter	♃
Astræa	⑤	Melpomene	⑱	Saturn	♄
Hebe	⑥	Fortuna	⑲	Uranus	♅ or ⛢
Iris	⑦	Massalia	⑳	Neptune	♆

In the above table the asteroids are arranged in the order of their discovery.

20. *Solar System.* This expression simply implies the sun and bodies connected with him, as the planets and satellites, the earth and moon included, and comets, without any reference to their arrangement.

21. *Attraction of Gravitation.* That force which causes a heavy body to descend to the earth, when left free to move, is called *gravity*, or the *attraction of gravitation.* Sir Isaac Newton assuming this force to decrease in intensity in the inverse ratio of the square of the distance from the earth's centre, found that, at the distance of the moon, it would be just sufficient to retain her in her orbit around the earth. Pursuing his investigations he found that the assumption of similar forces in the sun and planets, varying in the direct ratio of the mass of the body and the inverse ratio of the square of the distance, would account, on mechanical principles, for the motions of the latter, and for other known astronomical facts. He therefore inferred that attraction of gravitation is a universal property of matter, and that its intensity, or the force with which it acts, varies in the direct ratio of the mass, and the inverse ratio of the square of the distance.

This theory of universal gravitation is the foundation of Physical Astronomy; which, originating with Newton, has, through his labours and those of many eminent men since his time, as Clairaut, Euler, Lagrange, and especially Laplace, become a science of great extent.

DEFINITIONS OF TERMS.

22. The *Axis of the Heavens* is an imaginary straight line, joining the north pole of the heavens and centre of the earth, and extending to the southern part of the celestial sphere. It is the line about which the heavens appear to revolve. The point in which it meets the southern part of the celestial sphere is called the *South Pole of the Heavens*, or simply the *South Pole*. Thus, if C, *Fig.* 1, represent the centre of the earth, and P, the north pole of the heavens, then will PP′ be the axis of the heavens, and P′, the south pole.

23. The *North and South Poles of the Earth*, are the points in which the axis of the heavens intersects the surface of the earth. Thus p and p' are the north and south poles of the earth.

The line pp', terminated by p and p', is called the *axis of the earth.*

24. The *Celestial Equator*, or simply tho *Equator*, is the great circle in which a plane through the earth's centre, and perpendicular to the axis of the heavens, intersects the celestial sphere. The circle EWQO represents the celestial equator.

The north and south poles of the heavens, P and P′, are evidently the geometric poles of the equator.

25. The *Terrestrial Equator* is the circle in which the plane of the celestial equator intersects the earth's surface. The semicircle *ewq* represents one half of the terrestrial equator.*

26. A *Declination Circle* is any great circle which passes through

* The celestial equator is sometimes called the *Equinoctial.* More commonly, however, simply the term equator is applied both to the celestial and terrestrial equator; the context serving to designate which is intended. The same observation applies to the terms, pole of the heavens and terrestrial pole.

the poles of the celestial equator. It intersects the equator at right angles. The circle PQP′E is a declination circle; and PSP′, PS′P′ are arcs of declination circles.

27. The *Declination* of a heavenly body, or of any point in the celestial sphere, is the arc of a declination circle, intercepted between the equator and the centre of the body, or the point. The declination is reckoned *north* or *south*, according as the body is on the north or south side of the equator. Thus DS is the declination of a body at S; it is north.

The *Polar Distance* of a body is its distance from the *north* pole, measured on the arc of a declination circle. Thus PS is the polar distance of a body at S. The declination and polar distance are the complements of each other.

It is also evident that the angle pCS, contained between the earth's axis and a line joining the centres of the earth and body, expresses the polar distance of the body; for this angle is measured by PS.

28. The *Vertical Line*, at any place, is a straight line in the direction of gravity at that place: that is, in the direction of the plumb-line when a plummet freely suspended, is at rest. The line AZ represents the vertical line at the place A.

A plane that passes through the vertical line is called a *Vertical Plane;* A plane that is perpendicular to the vertical line is called a *Horizontal Plane.*

Regarding the earth as a sphere, a vertical line produced downwards passes through its centre.

29. The *Zenith* and *Nadir* are the points, *above* and *below*, in which the vertical line at a place meets the celestial sphere. Thus Z is the zenith and N the nadir, of the place A.

30. The *Horizon* of a place, or, as it is sometimes called, the *Sensible Horizon*, is the circle in which a plane passing through the place, and perpendicular to the vertical line at the place, cuts the celestial sphere. The plane itself is also frequently called the *horizon.*

The *Rational Horizon* is the circle in which a plane through the earth's centre, and parallel to the sensible horizon, cuts the celestial sphere. The circle HORW represents the rational horizon of the place A.*

With reference to the fixed stars, the sensible horizon may be regarded as being the same with the rational horizon (13).

31. The *Meridian* of a place is the declination circle which passes through the zenith of the place. It cuts the horizon at right angles in two opposite points, called the *north* and *south* points of the horizon. The circle HPZRN is the meridian of the place A; and H and R are the north and south points of the horizon.

32. A *Vertical Circle* is any great circle which passes through the zenith and nadir of a place. It cuts the horizon at right angles. The meridian ZRNH is a vertical circle, and ZSB, ZS′B′ and ZS″B″ are arcs of vertical circles.

33. The *Prime Vertical* is that vertical circle which is at right angles to the meridian of a place. It cuts the horizon in two opposite points, called the *east* and *west* points of the horizon. The straight line ZN represents the prime vertical, seen edgewise; and O and W, the east and west points of the horizon.

34. The *Altitude* of a heavenly body is the arc of a vertical circle, intercepted between the horizon and the centre of the body. Thus BS is the altitude of a body at S; and RM is the altitude of a body on the meridian at M. The latter is called the *meridian altitude.*

35. The *Zenith Distance* of a body is its distance from the zenith, and is equal to the complement of the altitude of the body. Thus ZS is the zenith distance of a body at S.

36. The *Azimuth* of a body is the arc of the horizon intercepted between the north or south point of the horizon, and a vertical circle through the centre of the body. Thus BR is the azimuth of a body at S; from the *south* towards the *west.*

The altitude and azimuth of a heavenly body are the co-ordinates that determine its position with reference to the horizon and meridian of a place.

* To avoid confusion in the figure, the sensible horizon is not represented

37. The *Culmination* of a body is the passage of its centre over the meridian of a place. This is also called the *Transit* of the body over the meridian.

The circumpolar stars pass the meridian twice in every diurnal revolution; once above, and once below the pole. These meridian passages are called respectively, *Upper* and *Lower Culminations.*

38. The *Hour Angle* of a body is the angle contained between the meridian and a declination circle through the centre of the body. Thus MPS is the hour angle of a body at S.

CHAPTER II.

ASTRONOMICAL INSTRUMENTS.

39. The *Astronomical Clock* is a clock constructed with great care and accuracy, and furnished with a *compensating pendulum:* that is, a pendulum with a rod so formed by a combination of materials, that its length is not sensibly affected by changes in the temperature of the air.

40. A *Chronometer* is a balance watch, constructed with various improvements and refinements of modern art, so as to insure great precision in its movement.

41. The *rate* of a clock or chronometer is its gain or loss in twenty-four hours. If it gains half a second in twenty-four hours, its rate is + 0.5 sec.; if it loses 1.4 sec. in that time, its rate is — 1.4 sec.

42. The *Vernier* is a divided arc or line, moveable along another graduated arc or line,* and serving to determine the values of fractional parts of the divisions of the latter. It is an appendage to various astronomical instruments, and to some others.

To explain the principle on which the vernier is constructed, let AB, *Fig.* 9, be an arc of a circle divided into degrees and subdivided into 20′ spaces, and let the vernier arc CD, be taken equal

---

* Sometimes the vernier has a fixed position and the graduated arc is moveable

in length to nineteen of these spaces and be divided into twenty equal parts. Each vernier space will then be $\frac{19}{20}$ of a space on the graduated arc. Hence if the line marked 0 on the vernier, called the *zero* of the vernier, coincide with a division line on the arc, as in the figure, it is evident the first division line of the vernier must fall behind the next division line of the arc, by $\frac{1}{20}$ part of a space on the arc, that is by 1′; the second division line of the vernier must fall behind the following one on the arc by 2′; and thus on. Consequently, if the vernier is moved forward till one of its division lines coincides with a division line on the arc, the zero must then be as many minutes forward of a division line on the arc, as is expressed by the number of the vernier division line. We may, therefore, for any position of the vernier, determine the place of the zero, which is the object required, by observing which of the vernier division lines coincides, or is the nearest to coincidence, with one on the arc, and adding the corresponding number of minutes to the degrees and minutes denoted by that division line on the arc, which next precedes the zero of the vernier. Thus in *Fig.* 10, the zero of the vernier stands forward of 15° 20′ on the graduated arc, and the eighth division line of the vernier coincides with a division line of the arc. Hence the arc indicated by the vernier is 15° 28′.

By making the vernier equal in length to fifty-nine divisions on the arc, instead of nineteen, and dividing it into sixty equal parts, it would evidently serve to *read off*, or indicate the fractional part of a division to the accuracy of $\frac{1}{60}$ of 20′, that is to 20″.

The *reading* of the vernier, that is, the precision with which it will indicate the arc, is varied according to the size of the instrument. In instruments of large size it is sometimes made to read to single seconds.

43. The *Reading Microscope* is an appendage frequently attached to instruments, instead of the vernier, and for the same object. It is commonly regarded as determining the arc with greater precision than the vernier.

In the body of the microscope a small frame is placed, across which are two spiders-lines intersecting each other in an acute angle. This frame with its spiders-lines is moveable by means of a screw having a graduated head. It may, therefore, by turning the screw, be moved till a division line of the graduated arc is seen to bisect the acute angle formed by the spiders-lines. When this is done, the number of whole turns of the screw gives the minutes, and the part of a turn as indicated by the graduated head gives the additional seconds, intercepted between the first position of the zero of the microscope and the division line.

44. A *Transit Instrument* is an instrument used for observing transits of the heavenly bodies over the meridian. It is made of various sizes, the length of the telescope forming a prominent part of it, varying from about twenty inches to ten feet. Those of the larger sizes are made to rest on stone piers, and are called *fixed* instruments. The smaller ones are placed on moveable stands, and are called *portable* instruments.

In *Fig.* 11, which represents a portable transit instrument, AB is a telescope firmly connected with an axis CD, which is at right angles to the optical axis of the telescope. The horizontal axis CD, terminates in two cylindrical pivots which rest in angular notches in pieces of metal called Y's. The Y's are attached to the upper ends of the upright pieces FF of the stand; one of them admits of a small lateral motion by means of a screw *a*, and the other, by means of a screw, not seen in the figure, admits of a small vertical motion. A graduated circle H is firmly fixed on the extremity of one of the pivots which extends beyond its Y for this purpose, and must, therefore, revolve as the telescope is turned to different altitudes. The double vernier index *e*, *e*, which may be placed in a horizontal position by means of a spirit level *f*, serves to direct the telescope to a given altitude.

The spirit level E, which rests on the pivots of the axis, is used in conjunction with the foot screw *b* of the stand, or with the screw that gives a vertical motion to one of the Y's, to place the axis in a horizontal position. When thus placed, the level is removed, and the telescope has then a free motion for all altitudes.

In the tube of the telescope near to the eye end A, a flat ring is placed, across the middle of which a spiders-line is fixed in a horizontal position. This is crossed at right angles by five equidistant parallel lines, as represented in *Fig.* 12. The ring is moveable by means of screws which connect it with the tube, and may be so adjusted, that the middle vertical line shall pass exactly through the optical axis of the telescope; the horizontal line at the same time passing through it, or very nearly so. To render the lines visible at night, a lantern I is placed opposite one end of the horizontal axis. The axis is hollow, and the light of a lamp in the lantern falls on a reflector placed in the tube of the telescope, at an angle of 45° with the axis, and is thence reflected so as to illumine the lines. The reflector has an aperture in its middle, sufficiently large to allow a free passage to the light from the body viewed.

When the direction of the meridian at any place is known, if the stand of the instrument be placed on some firm support, and turned till the optical axis of the telescope is nearly in that direction, it may be brought to be exactly so by means of the screw a, which moves one of the Y's, and the pivot of the horizontal axis resting in it. When this is done and the axis made truly horizontal, the middle vertical spiders-line will move in the plane of the meridian as the telescope is turned to different elevations; and consequently when a star or other heavenly body is bisected by that line, it must be on the meridian.

In transit observations it is usual to observe the time of the passage of the body over each of the vertical lines, and to take the *mean* of these times, as being generally more accurate than the observed time of the passage over the middle line. But if the observations are good, the difference between the mean and the middle time will never amount to a second. When the observed body is the sun, moon, or a planet, the times of passage of both the western and eastern limbs are observed; the mean of which gives the time of passage of the centre. If only one limb is visible, the time of passage of the centre is found by applying the computed interval of time occupied by the semi-diameter in its transit, to the observed time of passage of the visible limbs.

45. A *Transit Circle* or *Meridian Circle* is an instrument used for observing the meridian altitude, zenith distance, or polar distance of a heavenly body. It is an important instrument, differing from the simple transit instrument, in having, instead of the circle H, a much larger and very accurately graduated vertical circle

firmly connected with the telescope and axis, and consequently revolving as the telescope revolves. It is nearly represented in its principal parts by the upper part of *Fig.* 13, to which reference is made in the next article.

A *Mural Circle* is a meridian circle of large size, having its axis extending through a massive stone pier; the circle and telescope being fixed on one extremity of the axis, and a counterpoise on the other. In this, as in the transit instrument, provision is made for giving slight horizontal and vertical motions to one end of the axis. The angle is read off by stationary microscopes or verniers, usually six in number, attached to the pier.

The *Mural Quadrant* is a modification of the mural circle; a quadrant being substituted for the complete circle. This instrument is now rarely used, as the circle, though smaller, affords much more accurate results.

The *Zenith Sector* is an instrument used for measuring the meridian zenith distances of stars, that culminate within a few degrees of the zenith. In this instrument the graduated arc does not exceed 20°. It can, therefore, be made with a much larger radius than either the circle or quadrant, and admits of a more minute subdivision in the graduation of the arc.

The diameter of the mural circle at the National Observatory at Washington is 5 feet. The largest mural circles that have yet been constructed, are 8 feet in diameter. The celebrated zenith sector of Dr. Bradley, formerly at the Greenwich Observatory, and now at the Cape of Good Hope, has an arc of 12½ feet radius.

46. An *Altitude and Azimuth Instrument* is an instrument used for observing, at the same time, both the altitude and azimuth of a body in any part of the heavens.

In *Fig.* 13, which represents the instrument, the circular plate C has on it a graduated azimuth circle. This plate is attached to a tripod stand supported by three feet screws, two of which are shown at A and B. Firmly connected with the tripod, is a vertical axis, which passes through the centre of the azimuth plate and through two collars in the conical piece E, which projects upwards from the plate D. The whole of the instrument above the azimuth plate C is moveable about this vertical axis, and its position at any time is determined by a pointer, which gives the arc on the azimuth circle to five minutes, and two opposite reading microscopes, one of which is seen at F, which give the additional minutes and seconds.

The two connected vertical circles, K, K, are firmly attached to the telescope and horizontal axis, and, therefore, turn with the telescope as it is directed to different altitudes. That to the left is graduated, and the altitude or zenith distance of the body to which the telescope is directed is read off by the reading microscopes R, R. The spirit level, Q, Q, is used in making the vertical axis of the instrument truly vertical. The screws, whose heads are seen at P and P, serve to elevate or depress the Y's in which the pivots of the horizontal axis rest. By means of these and a striding level, such as is used with the transit instrument, the axis is made horizontal and placed at such a height that the zeros of the reading microscopes, R, R, shall be at opposite points of the graduated vertical circle. The plate T is a stand for a lantern to illuminate the spiders-lines, of which there are five horizontal as well as five vertical ones. When the ring to which these are attached is properly adjusted, by means of the screws which connect it with the tube, the intersection of the middle lines of the two sets is exactly in the optical axis of the telescope.

When the instrument is properly adjusted and placed so that, as the telescope revolves, its optical axis moves in the plane of the meridian, it may be used either as a transit instrument or transit circle

47. The *Equatorial* is an instrument consisting of the same essential parts as the altitude and azimuth instrument. It is so mounted that one of the axes is at right angles to the plane of the equator and the other parallel to it. The circles connected with these axes are called, respectively, the *Hour* and *Declination Circles*. The former is usually graduated into hours, and parts of an hour; and is so adapted to the axis that, when the telescope points toward a star on the meridian, the vernier will read 0 hrs., 0 min., 0 sec. If, then, the telescope be turned slowly westward, about the axis, at a rate corresponding with the diurnal motion of the heavens, it will be constantly directed towards the star, and the vernier will indicate, upon the hour circle, the star's hour-angle at

any moment. The declination circle is so adjusted that, when the optical axis of the telescope is in the plane of the equator, the vernier reads 0°; and hence, if the telescope be directed to any star, this vernier will indicate the star's declination.

The larger instruments of this class are generally provided with clock work, which communicates to the telescope a slow motion from east to west, causing it to follow a star for any length of time.

48. A *Sextant* is an instrument used for measuring the angular distance between two heavenly bodies or other objects. In Fig. 14, which represents a sextant, A, A, is a double frame connected by small pillars *a*, *a*, &c. The arc BC is usually graduated to 10′ and subdivided by a vernier E, to 10″. The degrees are numbered from 0° near B to about 130° near C; the construction of the instrument being such that half degrees on the arc correspond to whole degrees of the angle measured, they are for convenience regarded and numbered as whole degrees. The microscope H may be moved over the vernier, and aids in distinguishing the division line of the vernier that coincides, or is the nearest to coincidence with a division line on the arc. A glass reflector F, called the *index glass*, is attached perpendicularly to the index IE, which is moveable about the centre of the circular part I; this centre being also the centre of the graduated arc BC. Another glass G, called the *horizon glass*, is attached at right angles to the frame of the instrument, being parallel to the index glass when the index is at zero of the arc. The lower half of this glass is silvered so as to make it a reflector; the upper half is clear. A small telescope is placed in a ring L, and may be so adjusted by a screw M, that its optical axis shall be directed towards the division between the silvered and unsilvered parts of the horizon glass, or a little higher or lower, as may be desired. At K and N are sets of dark glasses of different colours, one or more of which may be interposed between the index and horizon glasses, or horizon glass and telescope, or both, to moderate the light and heat of the sun when that body is observed. The instrument, when in use, is held in the hand by a handle at O; or it is sometimes attached to a stand, called a centre of gravity stand, which admits its being placed at any inclination to the horizon.

In observing the angle contained between two bodies, the sextant is placed so that the telescope is directed to one of them, which is seen through the clear part of the horizon glass, and being then turned till its plane corresponds with the directions of both bodies, the index is moved till the second body appears to be in contact with the first; its light having been reflected from the index glass to the silvered part of the horizon glass, and thence through the telescope. The arc indicated by the vernier index, is then the measure of the angle contained between the nearest limbs of the two bodies.

The sextant, with the aid of an instrument called an *artificial horizon*, consisting of a small trough containing mercury, covered by a glass roof to protect the surface from agitation by the wind, serves to observe, with considerable accuracy, the altitude of a heavenly body. In doing this, the image of the body seen by reflection from the index and horizon glasses, is brought into contact with the image reflected from the surface of the mercury, and seen through the clear part of the horizon glass; the arc indicated by the index being then *double* the altitude of the lower limb of the body.

49. The *Reflecting Circle* is an instrument constructed on the same principle, and used for the same purposes as the sextant. In it, the arc BC extends to the whole circumference, and there are three vernier indices. It is regarded, on several accounts, as being a rather more accurate instrument than the sextant.

50. A *Micrometer* is an instrument or appendage, which, when attached to a telescope, serves to measure with great accuracy small angles, such as the apparent diameters of the sun, moon, or planets, or the angular distance between two bodies very near to each other.

This instrument is constructed in various ways. In Troughton's spiders-line micrometer, which may be substituted for the usual eye-piece of a telescope, there are two small frames placed side by side, at right angles to the optical axis, and moveable by screws with graduated heads. Across each frame a spiders-line is fixed at right angles to the direction in which it moves. The heads of the screws may be so adjusted, that they shall be at zero when the frames have such a position, that one of the lines is directly behind the other, the two appearing as one line. When thus adjusted, if the screws be turned till one of the lines appears to touch one edge or limb of a heavenly body, and the other to touch the opposite limb, the contained angle, or apparent diameter of the body, becomes known from the number of whole turns and parts of a turn of the screws, required to put them in those positions.

A *Heliometer* is a telescope fitted up with a peculiar kind of micrometer, for the especial purpose of measuring the apparent diameters of the heavenly bodies, or other small angular spaces.

A *Position Micrometer* is an instrument, which serves not only to measure the angular distance between two contiguous bodies, but also the angle contained between the arc of a great circle joining them, and a declination circle passing through one of them.*

CHAPTER III.

TO PLACE AN INSTRUMENT IN THE PLANE OF THE MERIDIAN.—SIDEREAL TIME.—TERRESTRIAL MERIDIAN.—LATITUDE AND LONGITUDE OF A PLACE.

51. *To place an altitude and azimuth instrument in the plane of the meridian.* Let ENWS, *Fig.* 2, be the horizon of a place A; Z the zenith, P the north pole, NZS the meridian, and S′ and S″ two positions of the same star when at equal altitudes B′S′ and B″S″, on opposite sides of the meridian. Then, since a star, in its

---

* From the preceding brief notices, the student may obtain a general view of the constructions and uses of the instruments mentioned, sufficient to enable him to comprehend the astronomical observations to which reference will be made in subsequent parts of the work. For full descriptions of these and various other astronomical instruments, with the methods of adjusting and using them, he may be referred to the second volume of Dr. Pearson's Treatise on Practical Astronomy. Those who have not access to the large work of Dr. Pearson, may obtain considerable information from a small work on the principal mathematical instruments used in Surveying, Levelling, and Astronomy, by Simms. An American edition, edited by J. W. Alexander, has been published in Baltimore.

apparent diurnal motion, describes a circle about the pole P (15), we have in the two spherical triangles ZPS′ and ZPS″, the side PS′ equal to PS″; and we have also ZS′ and ZS″ equal, being the complements of the equal altitudes B′S′ and B″S″, and PZ common. The angles PZS′ and PZS″ are therefore equal, and consequently their measures, the azimuths NB′ and NB″, are also equal.

Hence, the instrument being placed on a firm support, and properly adjusted and levelled, let the altitude of a star, when at a position S′ to the east of the meridian, be observed, and let the azimuth arc be also read off. Let the star be again observed, when, after having passed the meridian, it has arrived at a position S″, in which its altitude is the same as before, and let the azimuth arc be again read off. From these azimuth arcs, the azimuth arc B′B″ becomes known. Then, if the instrument be turned eastwardly through an arc equal to B″N, the half of B″B′, and be clamped in that position, the telescope, when turned about its horizontal axis, will, if the observations have been accurately made, move in the plane of the meridian NZS.

To ascertain whether the instrument is truly placed in the plane of the meridian, let several culminations of a circumpolar star, both above and below the pole, be observed, and the time, as shown by a good clock or chronometer, be noted. Then, as the diurnal motion of a star is uniform (15), and as the star must therefore be as long to the east of the meridian as to the west, it follows that if the interval during which the star appears to be to the east, is equal to that during which it appears to be to the west, the instrument is truly placed. If the intervals are unequal, the instrument deviates towards the side of the less interval, and should be slightly moved in a contrary direction. The observations and movement of the instrument should be repeated, till the intervals are found to be equal.

When the instrument is thus truly placed in the meridian, it may be used for observing the culminations and meridian altitudes of the heavenly bodies.

52. *To place a transit instrument in the meridian.* A transit instrument, or transit circle, may be placed in the plane of the meridian by first putting it by estimation nearly in that position,

and then proceeding as directed in the latter part of the last article.*

53. A *Sidereal Day* is the interval between two consecutive passages of a fixed star over the same meridian. It is about four minutes shorter than the common day.

Sidereal Time is time reckoned by sidereal days. A clock that is so adjusted as to move through 24 hours in a sidereal day is called a *sidereal clock*, or is said to be regulated to keep sidereal time.

The point of time at which the sidereal day commences according to the present usage of astronomers, and also a very slight change in the definition of a sidereal day, will be noticed in a subsequent chapter. For the present it may be regarded as commencing when any fixed star, selected at pleasure, passes the meridian of a place; the clock being regarded as being regulated to sidereal time when it is so adjusted as to mark and continue to mark 0 h. 0 m. 0 sec. at the instant that star passes the meridian. The numbering of the hours of the sidereal day, is continued from 0 hours to 23 hours.

54. The *Terrestrial Meridian* of a place is the intersection of the plane of the meridian of the place with the earth's surface. It is very nearly, though not exactly, a circle.

The terrestrial meridian of a place is usually considered as only extending from pole to pole. Thus $pAp'a$, *Fig.* 1, being the intersection of the plane of the meridian of the place A, with the earth's surface, the half pAp' is called the terrestrial meridian of the place A; the other half being called the *opposite* meridian.

55. The *Latitude of a place*, or as it is sometimes called the *Geographical Latitude*, is the arc of the meridian intercepted between the zenith of the place and the equator. It is said to be *north* or *south* according as the zenith is north or south of the equator. Thus ZQ, *Fig.* 1, is the latitude of the place A, to the north.

It follows, from the definition and article (27), that the latitude of a place is the same as the declination of the zenith of the place.

It is also evident that, regarding the earth as a sphere, and consequently a terrestrial meridian as a circle, the latitude of a place

* The details of this and other methods of adjusting these instruments to the plane of the meridian are given in treatises on practical astronomy.

is its distance from the terrestrial equator, measured in degrees and parts of a degree, on the terrestrial meridian through the place. For the arcs ZQ, and Aq, being measures of the same angle ZCQ, contain the same number of degrees.

56. A *Parallel of Latitude* is any small circle on the earth's surface, parallel to the terrestrial equator. Thus Ana is one half of a parallel of latitude through the place A.

57. *The Latitude of a place is equal to the altitude of the pole at that place.* For the sum of ZQ and ZP, is equal to the sum of PH and ZP, each sum being equal to a quadrant. Hence ZQ= PH. But ZQ is the latitude of the place A, and PH is the altitude of the pole at that place; this altitude, in consequence of the extreme minuteness of CA, in comparison with CH (13), being the same whether observed from A or C.

58. *The Latitude of a place is equal to the half sum of the greater and less meridian altitudes of a circumpolar star, at the place.*

Let the circle FGIK, meeting the meridian of the place A, in F and I, be the circle described by a circumpolar star in its diurnal motion. Then will PI = PF (15), and HI and HF will be the greater and less meridian altitudes of the star. Now,

$$HP = HI - PI,$$
$$\text{and } HP = HF + PF = HF + PI.$$

Hence by adding, $2\ HP = HI + HF$; or $HP = \frac{1}{2}(HI + HF)$. But (57), HP is equal to the latitude of the place.

Remark. This proposition assumes the two culminations to be on the same side of the zenith. If they are on different sides, the supplement of the greater altitude must evidently be substituted for the altitude itself. It may further be remarked that, in applying the proposition to find the latitude of a place, the *observed* altitudes require small corrections. These will be noticed in the chapter on refraction.

59. *First Meridian.* The *first meridian* is the meridian of some place arbitrarily selected, to which the positions of the meridians of other places are referred. The place selected for a first

meridian, is commonly that of some noted Astronomical Observatory.*

60. *Longitude.* The *Longitude* of a place is the angle contained between its meridian and the first meridian; it is measured by the arc of the equator intercepted between the two meridians. Longitude is reckoned *east* or *west*, according as the meridian of the place is east or west of the first meridian. Thus, assuming the meridian PSP′ of a place *s*, to be the first meridian, the angle DPQ, or the arc DQ, will be the longitude of the place A, to the east; and the angle DPD′ or the arc DD″ will be the longitude of a place *s*′, to the west.

As two terrestrial meridians form the same angle as their corresponding celestial meridians, the angle between the terrestrial meridian of a place and the terrestrial first meridian, expresses the longitude of the place.

61. *Selection of the first Meridian.* The selection of the first meridian being arbitrary, different first meridians are used in different countries. The English take for theirs, the meridian of their celebrated Observatory at Greenwich, near London, and the French, that of their public Observatory at Paris. In the United States, longitude is usually reckoned from the meridian of Greenwich.

62. *Position of a place.* The longitude and latitude of a place, determine its relative position on the earth's surface. For the longitude gives the position of the terrestrial meridian of the place, and the latitude gives the position of the place on that meridian.

When the longitude and latitude of a city, or place of considerable extent, are given, they are generally to be understood as applying to the central part, or to some prominent public edifice in the place.

63. *Difference of time under different Meridians.* As the diurnal motion of a fixed star, in the circle it appears to describe from east to west, is uniform (15), and the time of its complete

---

* An *Astronomical Observatory* is a building furnished with instruments for the especial purpose of making astronomical observations.

revolution is twenty-four sidereal hours, it follows that, in each sidereal hour, it must move through the twenty-fourth part of 360°, that is, through 15°; and in the same proportion for other times. Hence the star in its westwardly motion is on the meridian of a place in east longitude, earlier, and on that of a place in west longitude, later, than on the first meridian, by intervals of time at the rate of a sidereal hour for each 15° in the longitude. Thus, if LUMV, *Fig.* 1, be the circle described by the star in its diurnal motion, and PSP′, the meridian of a place *s*, be the first meridian; and if the longitudes of the places A and *s*′, whose meridians are PMP′ and PS′P′, be 30° east and 30° west; then will the star be at M two sidereal hours earlier, and at S′ two sidereal hours later, than at S. It therefore follows, that if we suppose sidereal clocks at the places A, *s*, and *s*′, to be adjusted to mark 0 h. 0 m. 0 sec., when this star is on their meridians respectively, then at the instant the star is at S on the meridian of *s*, and consequently the clock at that place marks 0 h. 0 m., the clock at the place A in 30° east longitude, must mark 2 h. 0 m., the star having been on its meridian two hours previously; and the clock at the place *s*′, in 30° west longitude, must mark 22 h. 0 m. of the preceding sidereal day, the star not arriving at the meridian of that place till two hours later. The same relation must exist among the times marked by the clocks at those places at any other instant of time. For instance, if when some other star is on the meridian of the place *s*, the clock at that place marks 7 h. 10 m. 30 sec., the clock at the place A, must at that instant mark 9 h. 10 m. 30 sec., and the clock at the place *s*′, 5 h. 10 m. 30 sec.

Hence the sidereal time reckoned at a given instant at a place in *east* longitude is *later*, and at a place in *west* longitude is *earlier*, than that reckoned at the first meridian.

It is the same with time reckoned in the usual way, that is, by the diurnal motion of the sun, when allowance is made for some little inequalities in that motion. Thus, when it is nine o'clock in the evening at Greenwich, it is only 59 m. 20 sec. past three o'clock, in the afternoon, at Philadelphia, the longitude of which is 75° 10′ west.

64. *Expression of longitude in time.* In consequence of the connection between the longitudes of places and the times reckoned

at them at the same instant, longitude is frequently expressed in time; one hour corresponding to 15°, one minute to 15′, and one second to 15″. Thus, long. 75° 10′ W., and long. 5 h. 0 m. 40 sec. W., are synonymous expressions.

65. *To find the longitude of a place by a chronometer.* Let a chronometer, which keeps time accurately, be carefully adjusted to the time at a place, the longitude of which is known. Then being carried to the place of which the longitude is required, let the time shown by it at any instant be compared with the correct time reckoned at the place at that instant, and let the difference be marked *east* or *west* according as the time at the place is *later* or *earlier* than that shown by the chronometer: that is, than the time, reckoned at the same instant, at the place of known longitude. Then (63) by *adding* this difference to the known longitude, expressed in time, if it is of the *same* name with that longitude, or *subtracting*, if it is of a *different* name; the longitude of the required place will be obtained.

It is not requisite that the chronometer should be so regulated as neither to gain nor lose any time. This would be difficult, if not impracticable. It is only requisite that its rate (41) should be well ascertained, as allowance can then be made for its gain or loss during the time of its transportation from one place to the other.

Other methods of finding the longitude of a place will be noticed in a subsequent chapter.

CHAPTER IV.

FIGURE AND DIMENSIONS OF THE EARTH—GEOCENTRIC LATITUDE OF A PLACE.

66. By *the figure of the earth* is meant the general form of its surface, supposing it to be uniform, or that it corresponds with the surface of the ocean. This excludes the consideration of the irregularities in its surface, proceeding from mountains and valleys; which are indeed very minute in comparison with its whole extent.

67. *The angle formed by the vertical lines at two places on the same terrestrial meridian, expresses the difference of the latitudes of the places.*

Let CZ and CZ′, *Fig.* 1, be the vertical lines at the two places A and A′, on the same meridian pAp'. Then Z and Z′ being the zeniths of these places, ZQ and Z′Q are the latitudes (55), and consequently ZZ′ is the difference of the latitudes. But ZZ′ is the measure of the angle ZCZ′ formed by the vertical lines.

This is still true if the vertical lines meet at some distance from the centre of the earth, as must be the case if the earth is not a perfect sphere. For, in consequence of the immense distance of the points Z and Z′, the angular distance between them is sensibly the same at a little distance from the centre as at the centre itself.

68. *Length of a degree of latitude.* The length of a degree of latitude or of a *degree of the meridian* is the distance, expressed in linear units, between two points on the same terrestrial meridian, the difference of whose latitudes is one degree.

The length of a degree of latitude may be obtained by finding the latitudes of two points on the same meridian, that do not differ in latitude more than a few degrees, measuring the distance between them, and then making the proportion; as the difference of the two latitudes is to one degree, so is the measured distance to the length of a degree. For supposing A and A′ to be the points, we have the proportion: as ang. ZCZ′ : 1° : : length of AA′ : length of a degree; in which the angle ZCZ′ is equal to the difference of the latitudes (67). The proportion is rigorously true on the supposition that the earth is a sphere, and consequently AA′ the arc of a circle; and for a small deviation in the form of the earth from a sphere, it is not sensibly erroneous, especially for the degree at the middle of the arc. Supposing the earth to be a sphere, the product of the length of a degree by 360, gives its circumference.

The difference of latitude between the two places A and A′, may be found without knowing the latitude of either. For if, at the two places, the meridian zenith distances ZM and Z′M of the same star, be observed and corrected for refraction,* we have ZZ′ = ZM — Z′M.

---

* See next chapter.

The distance between the two places is not found by direct measurement. This would be a very tedious operation, and would generally, from irregularities in the earth's surface, be deficient in accuracy. An extent of level ground is selected and a horizontal line BC, *Fig.* 3, of a few miles in length, called a *base* line, is measured with the utmost care and precision. Then, supposing A′, one of the places, to be visible from B and C, the horizontal angles of the triangle A′BC are carefully measured with a theodolite, or altitude and azimuth instrument. A station D, visible from B and C, being chosen, the angles of the triangle BCD are observed. Another station E, visible from C and D, being taken, the angles of the triangle CDE are observed. Proceeding thus, on both sides of the base line if requisite, the places A and A′ become connected by a series of triangles, in which the angles are all known, and also the side BC. From these data, other sides, and then the distance AA′, may be computed.

69. *The length of a degree of latitude increases from the equator to the pole.*

This may be inferred from inspection of the following table, which contains the length of a degree of latitude at several different latitudes, selected from measurements which have been made with great care, in various parts of the earth.

Country.	Arc measured.	Latitude of middle of the arc.	Mean length of a degree.	Observers.
			Miles.	
Peru	3° 7′ 3″	1° 31′ 0″	68.714	Condamine, &c.
India	15 57 40	16 8 22	68.759	Lambton, Everest.
Pennsylvania	1 28 45	39 12 0	68.899	Mason, Dixon.
France.....	12 22 13	44 51 2	69.041	Delambre, Mechain.
England....	3 57 13	52 35 45	69.123	Roy, Kater.
Sweden.....	1 37 19	66 20 10	69.277	Svanberg.

70. *A terrestrial meridian is an Ellipse, having the axis of the earth for its less axis and a diameter of the equator for its greater axis.*

The variation in the length of a degree of latitude proves that the meridian is not a circle; and the small amount of that variation shows that its deviation from a circle is not great. As the whole deviation is not great, a small portion of the meridian in any part may, without sensible error, be regarded as the arc of a circle; the radius of the circle to which the arc appertains, evidently increasing as the length of the degree of latitude increases, that is, from the equator to the pole. Now as the radius of an arc increases, its curvature decreases. The curvature of the meridian must therefore decrease in proceeding from the equator to the pole. This is the case with an ellipse in passing from the extremity of the major axis to that of the minor axis. Hence the form of the meridian corresponds in this respect with that of an ellipse, as $epqp'$, *Fig.* 4, in which pp', the axis of the earth, is the less axis, and eq, a diameter of the equator, is the greater axis.

Taking into view the actual lengths of a degree at different latitudes, it has been proved, by analytical investigations not adapted to the present work, that the meridians are really ellipses, or very nearly so; in which the less axis, or axis of the earth, is less than the greater, or a diameter of the equator, by about $\frac{1}{300}$ part of the latter.

71. *Figure and dimensions of the Earth.* From measurements which have been made at right angles to the meridian, it appears that the equator and parallels of latitude are circles, or nearly so. It therefore follows from the last article that the form of the earth is that of an *oblate spheroid;* that is, of a solid, such as would be generated by the revolution of a semi-ellipse pqp', about its minor axis pp'.

From computations made from the most accurate measurements, it has been found that the *equatorial* diameter of the earth is 7925 miles, and the *polar* diameter, or axis, is 7899 miles; the difference between them being 26 miles. Consequently the *mean* diameter is 7912 miles, and the mean circumference 24856 miles.

Hence the mean length of a degree of the meridian is $69\frac{1}{20}$ miles, the mean length of a minute is $1\frac{1}{6}$ miles, and the mean length of a second is 101 feet. It therefore follows, that in changing our position in a north or south direction, by only 101 feet, we make a change of one second in our latitude.

The length of a degree of the equator is $69\frac{1}{6}$ miles.

72. *Ellipticity or Oblateness of the Earth.* It is frequently

found convenient to denote the equatorial radius of the earth by a unit, or 1, and to express other large lengths and distances by means of this unit.

The fraction which expresses the difference between the equatorial and polar radii of the earth, when the equatorial radius is denoted by a unit, is called the *ellipticity* or *oblateness* of the earth. It is also sometimes called the *compression* of the earth. Hence (70), the ellipticity or oblateness is $\frac{1}{300}$.

73. *The ellipticity of the earth may be deduced from experiments with a pendulum.*

The number of oscillations made in any given time, as for instance in a sidereal day, by the same pendulum retaining the same length, is found to be different at different places on the earth's surface. It is least at the equator, and continually increases towards the poles. A pendulum oscillating sidereal seconds at the equator, and consequently making there 86400 oscillations in a sidereal day, would, on being transported to Philadelphia, make nearly 100 more in the same time. Now the motion of the pendulum depends on the force of gravity; and it is proved, in treatises on mechanics, that the number of oscillations made by the same pendulum in a given time, varies as the square root of that force. Hence it follows that the force of gravity increases from the equator to the poles, and that the law of this increase may be determined by experiments with a pendulum. This increase in the force of gravity, indicating a decrease in the distance from the earth's centre, is connected with the figure of the earth, and formulæ have been obtained which serve to determine the latter from the former. Computations, founded on numerous accurate experiments with a pendulum, made at various places, give for the ellipticity nearly the same value as that obtained from the measurement of degrees of the meridian.*

* Dr. Bowditch, in his excellent Translation of Laplace's Mécanique Céleste, with a Commentary, obtains, from a combination of several of the most accurately measured arcs of the meridian, a result a little less than $\frac{1}{300}$; and from a combination of many observations made with the pendulum, a result a little greater than $\frac{1}{300}$. Hence he infers that $\frac{1}{300}$ may be regarded as being very nearly the true value of the ellipticity or oblateness of the earth.

74. The *Eccentricity* of the earth is the distance between a focus of any of the elliptical meridians and the centre.

To *find the eccentricity*. Let f, *Fig.* 4, be one of the foci, and put $e = f\text{C}$ = the eccentricity. Then by conic sections, $fp = e\text{C} = 1$. Hence,

$$1 - e^2 = p\text{C}^2 = \frac{299^2}{300^2}.$$

From which we easily find,

$$e = \frac{\sqrt{599}}{300} = 0.08158.$$

75. The *Geocentric Zenith* of a place is the point in which a straight line from the earth's centre, passing through the place, meets the celestial sphere.

The *Geocentric Latitude* of a place, sometimes called the *reduced latitude*, is the arc of the meridian intercepted between the equator and the geocentric zenith of the place. The difference between the latitude and the geocentric latitude is called the *reduction of latitude.*

76. *The tangent of the latitude of a place is to the tangent of the geocentric latitude as the square of the equatorial radius of the earth is to the square of the polar radius.*

Let Z, *Fig.* 4, be the zenith of the place A, and z the geocentric zenith. Then ZGQ is the latitude of A, and zCQ is its geocentric latitude. Let AD be drawn perpendicular to eq. Put $\phi = \text{ZGQ}$ = the latitude, and $\phi' = z\text{CQ}$ = the geocentric latitude.

Then in the right angled triangle ACD, we have AD = CD tang ϕ', and in the right angled triangle AGD, we have AD = GD tang ϕ. Hence CD tang ϕ' = GD tang ϕ; or, CD : GD : : tang ϕ : tang ϕ'. But by conic sections, CD : GD : : $q\text{C}^2 : p\text{C}^2$. Consequently tang ϕ : tang ϕ' : : $q\text{C}^2 : p\text{C}^2$.

Cor. If $q\text{C} = 1$, and e = earth's eccentricity, we have (74), $p\text{C}^2 = 1 - e^2$. Hence, tang ϕ : tang ϕ' : : $1 : 1 - e^2$; or,

tang $\phi' = (1 - e^2)$ tang ϕ ....(A)

D

CHAPTER V.

ASTRONOMICAL REFRACTION.

77. *Earth's atmosphere.* From the science of pneumatics we learn that the earth's atmosphere is an elastic medium, the density of which continually decreases as the distance from the general surface of the earth increases. The law of decrease is such that, as the height *increases* in *arithmetical* progression, the density *decreases* nearly in *geometrical* progression. The actual decrease is such that, at the height of 3½ miles, the density is only about one half as great as at the earth's surface; at the height of 7 miles, about one fourth as great; at the height of 10½ miles, about one eighth as great; and thus on. The whole height of the atmosphere is not known. But from the preceding law, it follows that, at the height of 40 or 50 miles, its density must be extremely small, so as to be nearly or quite insensible. The density at any given place varies with the pressure, as indicated by a barometer, and with the temperature, as indicated by a thermometer; but this variation is not great.

If the density of the whole mass of the atmosphere was *uniform* throughout, and the same that it is at the earth's surface when the barometer stands at 30 inches, and Fahrenheit's thermometer at 50°, which is regarded as being nearly the *mean* density at the earth, the height would then be 5.13 miles.

78. *Refraction.* The science of optics teaches us that, when a ray of light passes *obliquely* from one medium into another of different density, it becomes bent, or *refracted;* the ray in the second medium, called the *refracted* ray, taking a different direction from that in the first, which is called the *incident* ray. Both rays lie in the same plane with a perpendicular to the common surface of the two mediums at the point of passage from one to the other. When the ray passes from a *rarer* to a *denser* medium, the refracted ray is bent towards the perpendicular to the common surface, making with it a less angle than that made with it by the incident ray. Thus an incident ray SA, *Fig.* 5, entering obliquely

a second medium of greater density at A, takes a direction AB, making the angle BAD, which is called the *angle of refraction*, less than the angle SAE, which is called the *angle of incidence*. The angle BAC, which expresses the difference between the directions SA and AB, of the incident and refracted rays, is called the *refraction*.

For the same two mediums, the amount of refraction changes with a change in the angle of incidence. The law of this change is such that the sine of the angle of incidence is to the sine of the angle of refraction in a constant ratio, which is called the *index of refraction*. Thus if I be the angle of incidence, R the angle of refraction, and m the index of refraction, the value of which for different mediums is determined by experiment, we have sin I : sin R : : m : 1; or, $\sin I = m \sin R$. For the passage of a ray of light from a vacuum into air of a mean density, or that which it has when the barometer stands at 30 inches, and the thermometer at 50°, the value of m is 1.000284.

When a ray passes through a medium composed of strata of different densities, bounded by *parallel planes*, the whole refraction is the same, as if the incident ray had at once entered the last stratum with its first angle of incidence; the direction of the ray in the last stratum being the same in either case. Thus, if a ray SA, *Fig*. 6, in passing through such a medium, takes the directions AB, BC, a ray S′A′ entering the last stratum at the same angle of incidence with SA, will take a direction A′C′, parallel to BC. When the strata are indefinitely thin and their number indefinitely great, or, which amounts to the same, when the density continually varies from A to C, the broken line ABC becomes a curve. The whole refraction is however still the same, provided the density at the surface C remains unchanged: that is, the whole refraction for a given angle of incidence depends entirely on the density at the second surface.

79. *Astronomical Refraction*. As the density of the earth's atmosphere continually increases from its upper surface to the earth (77), it follows, from the last article, that when a ray of light, from any of the heavenly bodies, enters the atmosphere obliquely, it becomes bent into a curve, concave towards the earth. The

density in the upper parts of the atmosphere being very small, the curve at first deviates very little from a straight line, but the deviation becomes greater as it approaches the earth. Both the straight and curved parts of the ray must necessarily lie in the same vertical plane; for, as the corresponding parts of the atmosphere on each side of a vertical plane may be regarded as of equal density, there is no cause for a deviation to either side. The whole change produced in the direction of the ray in traversing the atmosphere is called the *astronomical refraction.*

80. *Astronomical refraction increases the altitude of a heavenly body, but does not affect the azimuth.*

Let S*a*A, *Fig.* 7, be a ray which, proceeding from a body S, enters the atmosphere at *a*, and being bent by refraction, meets the earth's surface at A; and let AS′ be a tangent to the curve A*a* at A. Then will the ray enter the eye of an observer at A, in the direction S′A, and consequently the body S will appear to be in the more elevated position S′. As the tangent AS′ must be in the same vertical plane with the ray A*a*S, the azimuth of the body is not affected by refraction.

It follows that the altitude of a heavenly body is obtained by *subtracting* the refraction from the *observed* altitude, and the zenith distance, by *adding* the refraction to the *observed* zenith distance.

81. *At the zenith, the refraction is nothing.* For, in consequence of the corresponding density of the atmosphere on every side of a vertical line, there is no cause for a ray entering it in that direction to deviate from its rectilineal course.

82. *To obtain approximate formulæ for computing the refraction due to any altitude or zenith distance.*

As the upper and under surface of that portion of the atmosphere through which a ray of the heavenly bodies passes in its course to a place on the earth's surface, do not differ much from parallel planes, we may obtain approximate formulæ for the refraction, by assuming the density to be uniform throughout, and the same that it is at the earth's surface (78). Let *bd*, *Fig.* 8, be a part of the boundary of the atmosphere on this supposition, S*a* a ray from a body S, which being refracted at *a*, meets the earth's surface at A, and let C be the centre of the earth, and Z the zenith of the place A. Then, to an observer at A, the body will appear in the direction AS′, and the angle SaS′ will be the refraction corresponding to the apparent zenith distance ZAS′. Put,

ρ = CA = radius of the earth, assumed to be a sphere,
$h = ea$ = height of a uniform atmosphere,
Z = angle ZAS′ = observed zenith distance,
I = ,, Cac = angle of incidence,
R = ,, CaA = ,, refraction,
r = ,, Aac = ,, SaS′ = the refraction,

Then, since I = Cac = CaA + Aac = R + r, we have (78),

$$\sin (\mathrm{R} + r) = m \sin \mathrm{R} \ldots\ldots\ldots\ldots \text{(A)};$$

or, (App.* 13), sin R cos r + cos R sin $r = m$ sin R; or, dividing by cos R, we have,

$$\text{tang R} \cos r + \sin r = m \text{ tang R}.$$

But since m differs but little from a unit (78), it is evident from equat. (A), that R + r must differ but little from R, and consequently r must be a small angle. Taking therefore the angle instead of its sine (App. 51), and assuming cos r = 1, we have

$$\text{tang R} + \frac{r}{\omega} = m \text{ tang R};$$

$$\text{or,} \quad r = (m - 1)\, \omega \text{ tang R} \ldots\ldots\ldots\ldots \text{(B)}.$$

Now in the triangle CAa we have, Ca : CA : : sin CAa, or sin ZAS′ : sin CaA, or $\rho + h : \rho$: : sin Z : sin R. Hence,

$$\sin \mathrm{R} = \frac{\rho}{\rho + h} \sin \mathrm{Z} \ldots\ldots\ldots\ldots \text{(C)}.$$

Taking m = 1.000284 (78) ,and substituting for ω its value 206264″.8 (App. 51), we have, $(m—1).\ \omega$ = 58″.6. Hence, since ρ = 3956 (71) and h = 5.13 (77), the formulæ (C) and (B) become,

$$\left.\begin{aligned} \sin \mathrm{R} &= \frac{3956}{3961.13} \sin \mathrm{Z} \\ r &= 58''.6 \text{ tang R} \end{aligned}\right\} \ldots\ldots\ldots\ldots \text{(D)}.$$

The degree of accuracy of these formulæ may be tested by finding the latitude of a place from the observed upper and lower

* Appendix to part 1.

meridian altitudes of different circumpolar stars (58), using the formulæ in computing the refractions; which must be subtracted from the observed altitudes to obtain the correct altitudes. If the state of the air is the same or nearly the same as that assumed in finding the formulæ, and if no one of the lower altitudes of the stars employed is less than about 20°, the latitude as obtained from different stars will be sensibly the same. But if the lower altitude of any one of the stars is much under 20°, the latitude found from that star will be decidedly too great. Whence it follows, that, for a low altitude, the refraction computed by the formulæ is too small. It may thus be ascertained that, for altitudes of 20° and upwards, the refractions computed by the formulæ do not err to the amount of a second; but for lower altitudes the error becomes considerable, amounting at the horizon to several minutes.

83. *Tables of Refraction.* The complete investigation of astronomical refraction is a subject of great difficulty. It has claimed the attention of many eminent mathematicians,* and formulæ have been obtained which give the amount of the refraction with great precision, except for altitudes under 12° or 14°; and for these they give it very nearly. These formulæ take into view the changes in the density of the air at the earth's surface as indicated by the barometer and thermometer. From the formulæ, tables have been computed, from which the refraction corresponding to a given observed altitude is easily obtained. In these tables, the principal columns contain the refractions computed for a density of the air corresponding to some medium heights of the barometer and thermometer. These are called *mean refractions.* Other columns contain the corrections due to given changes in the states of these instruments.

84. *Refraction increases the visible continuance of the heavenly bodies above the horizon.*

As refraction increases the altitudes of the heavenly bodies, it must accelerate their rising and retard their setting, and thus render them longer visible. The refraction at the horizon is about

---

* Laplace, in the Mécanique Céleste; Prof. Bessel, in the *Fundamenta Astronomiæ;* Dr. Young, in the Transactions of the Royal Society of London for 1819 and 1824; Ivory, in the same Transactions for 1823; and various others.

34′, which is rather greater than the apparent diameter of the sun or moon. Either of these bodies may therefore be wholly visible when it is really below the horizon.

85. *Oval form of the discs of the sun and moon when near the horizon.* This is an effect of refraction. As R must be nearly equal to Z (82. D), and as the tangent of an angle increases rapidly when the angle approaches to 90°, it is evident from the expression for r (82. D), that the refraction must increase rapidly near the horizon. Hence the lower part of the disc, when in that situation, is considerably more elevated by refraction than the upper; and consequently the vertical diameter and chords parallel to it are shortened, while the horizontal diameter and its parallel chords are not sensibly affected. This necessarily causes the disc to assume an oval form. The apparent diminution of the vertical diameter amounts, at the horizon, to about $\frac{1}{8}$ of the whole diameter.

86. *Apparent enlargement of the discs of the sun and moon when near the horizon.* Although this is not an effect of refraction, it may properly be noticed here. It is an optical illusion of the same kind as that which makes a ball or other object appear larger when seen at a distance on the ground than when viewed, at the same distance from the eye, on the top of a high steeple. Our judgment of the magnitude of a distant object depends not only on the angle it subtends at the eye, but also on a concurring though sometimes very erroneous impression with regard to the distance; the same object, seen under the same angle, appearing larger as there is an impression of greater distance. Now in viewing the sun or moon when at or near the horizon, the various intervening objects near the line of sight, give the impression of its being more remote, than when seen in an elevated position. When the sun or moon is viewed through a smoked glass, which renders intervening objects invisible, the disc does not appear thus enlarged.

87. *Twinkling of the Stars.* From changes in the temperature, currents of air, and other causes, the atmosphere is continually more or less agitated. This agitation produces momentary condensations and dilatations in its constituent molecules, and thus occasions slight but sudden and continually repeated deviations in

the directions of the rays of light which traverse it. As the stars appear merely as luminous points, presenting scarcely any visible discs, these irregularities in the directions of their rays of light give to them the apparent tremulous motion called the *twinkling of the stars.*

The discs of the planets, though small, are much larger than those of the stars, as is shown by observations with the telescope. They are therefore less affected than the stars, and the twinkling is but little observable in them, except sometimes near the horizon, where the cause producing it usually acts with the greatest effect.

88. *Twilight or Crepusculum.* This depends on both reflections and refractions of the sun's rays in the atmosphere. When, in the evening, the sun has descended so far below the horizon as to cease to be visible by refraction (84), a portion of the lower part of the atmosphere ceases to receive his rays directly, and is only illumined by light diffused through it by reflection from the higher parts. As the sun continues to descend below the horizon, the part of the atmosphere that is not directly enlightened by his rays increases, and at the same time its illumination gradually diminishes, in consequence of the diminished portion of the atmosphere from which its light is received. This gradual diminution of the light continues till the sun has descended so far below the horizon as to cease to illuminate any sensible portion of the atmosphere above it. This takes place when he is about 18° below the horizon. The last appearance of twilight must evidently be in the western part of the heavens.

In the morning the twilight commences, or the first dawn of day is perceived in the eastern part of the heavens, when the sun has arrived within about 18° of the eastern horizon; and the light then increases in the same gradual manner as it diminishes in the evening.

CHAPTER VI.

APPARENT AND TRUE PLACES OF A BODY—PARALLAX—METHOD OF FINDING THE PARALLAX OF A BODY—PARALLAXES AND DISTANCES OF THE MOON AND SUN—THEIR APPARENT AND REAL DIAMETERS.

89. *Apparent and true places of a body.* The place which a planet or any other of the heavenly bodies, except the fixed stars, appears to occupy in the celestial sphere varies with a change in the position of the observer. Astronomers, therefore, in order to render their observations easily comparable, and for convenience in computations, reduce the place of a body as observed at any place on the surface of the earth, to that in which it would appear to be, if seen from the centre.

The place in the celestial sphere in which a body would appear to be as seen from any point on the earth's surface, if there were no refraction, is called the *apparent* place of the body; and that in which it would appear to be if seen from the centre of the earth, is called the *true* place. Thus, if C, *Fig.* 15, be the centre of the earth, A a place on its surface, Z the zenith of this place, B the place of a body, and b and c the points in which AB and CB produced meet the celestial sphere, then is b the apparent place and c the true place of the body.

90. The *Parallax* or *Parallax in altitude* of a body is the angle contained between two straight lines conceived to be drawn from the centre of the body, one to the centre of the earth and the other to a place on its surface. Thus, for the place A, the angle ABC is the parallax of a body at B.

The *Horizontal Parallax* is the parallax when the body is in the horizon, or, which is the same, when the apparent zenith distance is 90°. Thus, the angle AB′C is the horizontal parallax of the body.

The *Equatorial Parallax* of a body is its horizontal parallax for a place at the equator.

91. *The parallax of a body is equal to the difference between*

the apparent and true zenith distances of the body, or between the true and apparent altitudes.

For as ZAB is an exterior angle of the triangle ABC, we have ang. ZCB + ang. ABC = ang. ZAB; or ABC = ZAB — ZCB. But ABC is the parallax, ZAB the apparent zenith distance, and ZCB the true zenith distance. As the altitudes are the complements of the zenith distances, the difference between them must be the same.

Cor. It is evident that parallax *increases* the zenith distance, and consequently *diminishes* the altitude. Hence, to obtain the true zenith distance from the apparent, the parallax must be *subtracted;* and to obtain the true altitude from the apparent, it must be *added.*

92. *The sine of the parallax at any altitude is equal to the product of the sine of the horizontal parallax by the sine of the apparent zenith distance.*

Put, r = AC = radius of the earth,
D = CB = CB′ = distance of the body,
N = ang. ZAB = app. zenith distance,
p = „ ABC = the parallax,
P = „ AB′C = „ horizontal parallax.

Since the angles ZAB and CAB are supplements of each other, their sines are equal, and we have from the triangles CAB and CAB′,

$$D : r :: \sin N : \sin p,$$

and, $$D : r :: 1 : \sin P \qquad \text{(A)}.$$

Hence, $1 : \sin P :: \sin N : \sin p$,

or, $$\sin p = \sin P \sin N \qquad \text{(B)}.$$

As the parallax is always a small angle, that of the moon, which is much the greatest, being only about a degree, we may frequently take the parallax itself instead of its sine (App. 51). We then have,

$$p = P \sin N \qquad \text{(C)}.$$

When the *spheroidal* figure of the earth is taken into view, the zenith distance must be taken in reference to the geocentric zenith, and r must be the radius of the earth at the place of observation.

93. *Distance of a body in terms of the horizontal parallax and radius of the earth.*

From (92 A), we have,

$$D = r. \frac{1}{\sin P} = r \frac{\omega}{P} \text{ (App. 51.)} \dots\dots\dots\dots \text{(D)}$$

If R = the equatorial radius of the earth and π = the equatorial parallax, then

$$D = R \frac{1}{\sin \pi} = R \frac{\omega}{\pi} \dots\dots\dots\dots \text{(E)}$$

From these two expressions for D, we have the following relation between the equatorial parallax and the horizontal parallax at a place where r is the radius of the earth.

$$R : r :: \pi : P \dots\dots\dots\dots \text{(F)}$$

It also follows that the *parallaxes of different bodies, or of the same body at different distances, are inversely as the distances.*

For, let D and D′ be the distances of two bodies from the earth, and π and π' the corresponding equatorial parallaxes. Then,

$$D = R. \frac{\omega}{\pi} \text{ and } D' = R. \frac{\omega}{\pi'}$$

Hence, $\frac{D}{D'} = \frac{\pi'}{\pi}$, or $D : D' :: \pi' : \pi \dots\dots\dots\dots$ (G)

94. *To find the equatorial parallax of a body.*

Let B, *Fig.* 16, be the body, and A and A′ two places situated remote from each other on the same meridian. Let the meridian zenith distances ZAB and Z′A′B be observed at the same time by two observers at A and A′, and let them be corrected for refraction. Also let the meridian zenith distances ZAS and Z′A′S of a star which passes the meridian at nearly the same time with the body, be observed and corrected for refraction. [The student should here be reminded that the lines AS and A′S are sensibly parallel (13)]. Then BAS, which is the difference of the corrected values of ZAS and ZAB, is known; and also BA′S, the difference of Z′A′B and Z′A′S. Now, ALA′ being an exterior angle of the triangle ABL, ABA′ = ALA′ — BAS, but ALA′ = BA′S, and and hence, ABA′ = BA′S — BAS. If the zenith distance of the body is greater at each place than that of the star, as may sometimes occur when the zenith distances of the body and star are nearly the same, B falls between AS and A′S. In this case it will easily be seen that ABA′ = BA′S + BAS. Hence ABA′ = the difference or sum of the known angles BA′S and BAS, is known.

From the latitudes Zdq and $Z'd'q$ of the places A and A′, the geocentric latitudes zCq and $z'Cq$ may be found (76). The difference between Zdq and zCq gives the angle ZAz, and this angle taken from the zenith distance ZAB leaves the geocentric zenith distance zAB. In like manner we find the geocentric zenith distance z'A′B. Put,

N, N′ = the app. geocen. zen. distances zAB and z'A′B,
P, P′ = the horizontal parallaxes at A and A′,
p, p' = the parallaxes ABC and A′BC.
r, r' = the radii CA and CA′,

and let R and π be as in the last article.

Then since ABC + A′BC = ABA′, we have $p + p' = ABA'$.

But (92 C), $p = P \sin N = \pi . \frac{r \sin N}{R}$ (93 F)

and, $$p' = P' \sin N' = \pi . \frac{r' \sin N'}{R}$$

Hence, $$\pi . \frac{r \sin N + r' \sin N'}{R} = p + p' = ABA'$$

$$\text{or, } \pi = ABA' . \frac{R}{r \sin N + r' \sin N'}.$$

The values of r and r' may be found from the latitudes of the places A and A′ (App. 52). Hence the quantities in the expression for π, are all known.

It is not essential that the two observers should be on exactly the same meridian; for if the meridian zenith distances of the body be observed on several consecutive days, its change of meridian zenith distance in a given time will become known. Then if the difference of longitude of the two places is known, the zenith distance of the body, as observed at one of the meridians, may be reduced to what it would have been found to be if the observations had been made in the same latitude at the other meridian.

95. *Moon's parallax and distance.* In the year 1751, La Caille and La Lande, two French Astronomers, made corresponding observations on the moon; the former at the Cape of Good Hope

and the latter at Berlin. From these observations, others of a similar kind which have since been made, and from other methods, the moon's parallax has been ascertained with much greater precision than it was previously known. The parallax and consequently the distance (93) are found to vary considerably during a revolution of the moon round the earth. It is also ascertained that the least and greatest parallaxes, or greatest and least distances, in one revolution of the moon, differ materially from those in another. There is, however, a *mean* distance, a mean of the average greatest and least distances, that is not subject to this change. The parallax corresponding to this mean distance is called the *constant* of the parallax. The constant of the moon's equatorial parallax is found to be 57′ 4″. The equatorial parallax when least, is about 53′ 54″, and when greatest, 61′ 32″.

From tables that will be hereafter noticed, called *lunar* tables, the equatorial parallax of the moon may be obtained for any given time. The parallax computed from these is given in the Nautical Almanac* for every 12 hours throughout the year; whence it may easily be obtained for any intermediate time. From the equatorial parallax the horizontal parallax at a given place may be found by (93 F), or by a table computed for the purpose.

Taking the moon's parallax 57′ 4″, we have, (93 E),

$$D = R. \frac{\omega}{\pi} = R \frac{206265}{3424} = R \times 60.24 = 239{,}000 \text{ miles, nearly.}$$

Hence the moon's mean distance from the earth is about 60 times the equatorial radius of the earth or 239,000 miles nearly. The least distance is about 56 times the equatorial radius, and the greatest 64 times that radius.

96. *Sun's parallax and distance.* By the preceding method (94), the sun's parallax may be ascertained to be about 9″. By a

* The *Nautical Almanac* is an astronomical ephemeris, published annually at London and republished at New York. It contains a large amount of data of great importance to the mariner and also to the practical astronomer. It is usually published about three years prior to the year for which it is computed. The *Connaissance des Tems*, published at Paris, the *Astronomisches Jahrbuch*, published at Berlin, and the *Effemeridi Astronomiche*, published at Milan, are ephemerides of a similar character. The *American Ephemeris and Nautical Almanac*, a work

method that will be noticed in a subsequent chapter, his mean equatorial parallax has been found to be 8″.6. The parallax when least is about 8″.5, and when greatest about 8″.7. From the mean parallax the mean distance is found to be 23,984 times the equatorial radius of the earth, or 95,000,000 miles nearly (93 E).

97. *The apparent semidiameter or diameter of a body, seen at different distances, is inversely proportional to the distance.*

Let A and A′, *Fig.* 17, be two positions from which a body, whose centre is C, is viewed. Then AB and A′B′ being tangents to the body at B and B′, the angle CAB is the semidiameter of the body as seen from A, and CA′B as seen from A′. Put $\delta = \text{CAB}$, $\delta' = \text{CA'B}$, $\text{D} = \text{AC}$, and $\text{D}' = \text{A'C}$. Then $\text{D} \sin \delta = \text{CB} = \text{CB}' = \text{D}' \sin \delta'$, or $\text{D} \times \delta = \text{D}' \times \delta'$. Hence,

$$\text{D} : \text{D}' :: \delta' : \delta :: 2\,\delta' : 2\delta.$$

Regarding CB, *Fig.* 15, the distance of a body from the centre of the earth, as constant, the distance AB from a place on the surface must diminish as the altitude increases; and consequently the apparent semidiameter of the body, as seen from A, must increase. The apparent semidiameter, when the body is in the horizon, is sometimes called the *horizontal* semidiameter, and when it is elevated, the *augmented* semidiameter. When the expression, apparent semidiameter, is used without reference to the altitude of the body, it implies that of the body when in the horizon.

98. *The sine of the apparent zenith distance of a body is to the sine of the true zenith distance, as the apparent diameter of the body at that zenith distance is to the horizontal diameter.*

Let δ be the horizontal semidiameter of the body, and δ' the apparent diameter at B, *Fig.* 15. Then (97), $\text{AB}' : \text{AB} :: \delta' : \delta$, or, since AB′ may be regarded as sensibly equal to CB′ or CB, we have $\text{CB} : \text{AB} :: \delta' : \delta$. But $\text{CB} : \text{AB} :: \sin \text{ZAB} : \sin \text{ZCB}$. Hence,

$$\sin \text{ZAB} : \sin \text{ZCB} :: \delta' : \delta :: 2\delta' : 2\delta.$$

---

recently commenced by our government, is equal in value and in some particulars superior to any one published in Europe. The number for the year 1855 has already been published, and that for 1856 will shortly appear. It is under the superintendence of Commander Charles Henry Davis, of the U. S. Navy.

If the apparent diameter of a body be measured with a micrometer at any observed zenith distance, and the apparent and true zenith distances be obtained (80 and 91), the above proportion gives the horizontal diameter.

For the moon, the difference between the apparent diameters in the horizon and zenith, amounts to about half a minute. For other bodies, the difference is nearly or quite insensible.

99. *The sine of the equatorial parallax of a body is to the sine of the apparent semidiameter in a constant ratio.*

For if R = equatorial radius of the earth, R′ = radius of the body, and D = distance of the body from the earth, we have (93 E) $R = D \sin \pi$ and (97) $R' = D \sin \delta$. Hence $\sin \pi : \sin \delta :: R : R'$. Therefore, since R and R′ are constant quantities, the ratio of $\sin \pi : \sin \delta$, is constant. For the moon this ratio is ascertained to be, $\sin \pi : \sin \delta :: 1 : 0.27304$.

Cor. From the proportion we have $R' = R.\frac{\sin \delta}{\sin \pi} = R\frac{\delta}{\pi}$, or $2R' = 2R\frac{\delta}{\pi}$. Hence, putting d = equatorial diameter of the earth and d' = diameter of the body, we have $d' = d\frac{\delta}{\pi}$....................(H)

100. *Apparent and real diameters of the Sun and Moon.* The apparent diameter of the sun at his *mean* distance from the earth is 32′ 3″.6. When least, it is 31′ 32″.0, and when greatest, 32′ 36″.5.

The apparent diameter of the moon at her *mean* distance is 31′ 39″.6. When least, it is about 29′ 26″, and when greatest, 33′ 37″.

Taking the sun's apparent semidiameter at his mean distance, and the corresponding parallax (96), we find (99 H) the sun's real diameter to be nearly 112 times the equatorial diameter of the earth, or more than 880,000 miles. His bulk is therefore about fourteen hundred thousand times that of the earth.

In like manner we find the moon's diameter to be about $\frac{3}{11}$ of the equatorial diameter of the earth, or 2160 miles.

The moon's surface is therefore about $\frac{1}{13}$ of that of the earth, and her volume or bulk about $\frac{1}{49}$ of the earth's volume.

CHAPTER VII.

POLAR DISTANCE OF A BODY—APPARENT DIURNAL MOTIONS OF THE FIXED STARS UNIFORM—MOTION OF THE EARTH ON ITS AXIS.

101. *The polar distance of a body, when on the meridian, is equal to the sum or difference of the complement of the latitude of the place and the zenith distance of the body, according as it culminates to the south or north of the zenith.*

Let M, *Fig.* 1, be the point at which a body is when on the meridian of the place A. Then PM = PZ + ZM. But PM is the polar distance of the body, ZM its zenith distance, and PZ the complement of the latitude of the place. If the body be on the meridian at I, to the north of the zenith, we have PI = PZ — IZ; if at F, we have PF = FZ — PZ.

102. *To find the polar distance or declination of a body.* Let the meridian zenith distance of the body be observed at a place whose latitude is known, and be corrected for refraction and parallax. Then, by the last article, the polar distance becomes known. If the body is a fixed star, the zenith distance only requires correction for refraction, as the star has no sensible parallax. When the body has a sensible diameter, the apparent semidiameter *added* to, or *subtracted* from, the observed zenith distance of the *upper* or *lower* limb, when corrected for refraction and parallax, gives the true zenith distance of the centre.

The declination is evidently equal to the difference between the polar distance and 90°, and is *north* or *south*, according as the polar distance is *less* or *greater* than 90°. It therefore becomes known when the polar distance is known.

The polar distances or declinations of the heavenly bodies, are found to vary more or less from day to day, except those of the fixed stars, which continue sensibly the same for several days in succession; but after a longer interval, changes become also perceptible in them.

103. *The apparent diurnal motion of a fixed star is in a circle, of which the north pole of the heavens is a geometric pole, and it is uniform.*

As, by the last article, the polar distance PS, *Fig.* 1, of a fixed star S, does not sensibly change during the interval of a day, the apparent diurnal motion must be performed in a circle MSLU about the pole P. To prove that its motion is uniform, let the zenith distances of the star be observed, when it is on the meridian at M, and when in other positions S, S′, &c., and let the times of observation, as shown by a good sidereal clock, be also noted. The observed zenith distances, when corrected for refraction, give the true zenith distances ZM, ZS, ZS′, &c.; from the first of which the polar distance of the star becomes known. Then, since ZP is the complement of the latitude of the place, and PS, PS′, &c., the polar distance of the star, we know all the sides in each of the triangles PZS, PZS′, &c., and may compute the hour angles ZPS, ZPS′, &c. The hour angles thus computed are found to be proportional to the intervals between the times the star has the positions M and S, M and S′, &c. The hour angle ZPS, therefore, increases uniformly with the time, and consequently the apparent diurnal motion of the star is uniform.

If one of the positions of the star be near the horizon as at S″, a little discrepancy in the result is sometimes found. This depends on the uncertainty of the correction for refraction at these low altitudes; and observations of this kind serve to determine the proper amount of the correction. For, the rate of the clock having been ascertained, the true interval between the times at which the star has the positions M and S″ becomes known, and thence the hour angle ZPS″. Then, in the triangle PZS″, we have the sides ZP and PS″ and the included angle ZPS″ to find ZS″. The difference between the observed and computed values of ZS″ must be the refraction.

104. *The apparent diurnal revolutions of the heavenly bodies from east to west, is produced by a real, uniform revolution of the earth on its axis from west to east, during a sidereal day.*

The apparent diurnal motion of a heavenly body, is not one that becomes at once perceptible to the view, like that of a meteor through the air. But, in repeated observations at intervals of sufficient length, we see it at each succeeding observation, when it is to the east of the meridian, become more and more elevated and nearer the meridian, and when to the west, less and less elevated and farther from the meridian; and not feeling conscious of any motion ourselves, we impute this continued change of position to a westerly motion in the body. The change of position with regard to the horizon and meridian, and consequently the apparent motion of the body, must, however, be precisely the same, if, instead of the body revolving round the earth from east to west, the earth itself revolves round its axis from west to east, making a complete revolution in a sidereal day. Thus the hour angle MPS, *Fig.* 1, and therefore the apparent motion of a star S, will be exactly the same to an observer at A, whether we suppose the star to move *westwardly* from M to S in any observed time, or suppose that, in consequence of a rotation of the earth on its axis, the meridian PMP′, of the place A, moves, in the same time, *eastwardly* from the position PSP′ to the position PMP′. As the appearance is therefore the same on either supposition, it is more reasonable to assume this rotation of the earth on its axis than to suppose that all the heavenly bodies, situated at immense and various distances, should have motions so adjusted as to revolve round it in the same or nearly the same time. This assumption of the earth's rotation on its axis is confirmed by many astronomical facts.

An experimental confirmation of the earth's diurnal motion may be mentioned here. Assuming this motion, the top of an elevated tower must, in consequence of its greater distance from the earth's axis, move eastwardly faster than the bottom. Hence a stone, or other heavy body, let fall from the top of the tower, and retaining, by virtue of its inertia, the excess of the forward or eastwardly motion which it had at the top, must fall a little to the *east* of the vertical line through the point from which its fall commenced. Now, several experiments of this kind have been made, and the fall of the body has always been found to be in accordance with the assumed rotation of the earth.

CHAPTER VIII.

EARTH'S ANNUAL MOTION — THE SUN'S APPARENT PATH — OBLIQUITY OF THE ECLIPTIC—POSITIONS OF THE FIXED STARS —CONSTELLATIONS—CATALOGUES OF THE FIXED STARS.

105. *Sun's apparent motion, meridian altitude, and declination.* When the sun is observed on any day to be on the meridian at the same instant with some fixed star, he is found, on the next day, to be a little distance to the east when the star returns to the meridian, and comes to it about four minutes later than the star. On the third and succeeding days, he is still farther and farther to the east when the star returns to the meridian. He has therefore an apparent motion among the stars from west to east, as has been already stated (7).

The meridian altitude of the sun changes from day to day. On the 20th of March and 22d of September, it is the same or nearly the same, as that of the point of the equator which is on the meridian. He is therefore, at these times, in or nearly in the equator. From the 20th of March to the 22d of September, his meridian altitude is *greater* than that of the meridian point of the equator; and consequently, during this period, he is to the north of the equator and his declination is *north.* From the 22d of September to the 20th of March, his meridian altitude is *less* than that of the meridian point of the equator; he is therefore to the south of the equator and his declination is *south.* The greatest meridian altitude and north declination occur on the 21st of June; and the least meridian altitude and greatest south declination on the 21st of December. The greatest north and south declinations are found to be equal, each being about 23° 28′.

It follows, from the preceding observations, that about one half of the sun's apparent path among the fixed stars lies to the north of the equator, and the other to the south.

106. *The sun's apparent annual path is a great circle of the celestial sphere.*

On several consecutive days about the 20th of March, let the sun's polar distance, when on the meridian, be obtained (102), and also the interval of time, as shown by a well regulated sidereal clock, between the time of the passage of the sun's centre over the meridian and that of some fixed star. It will commonly be found that, on the first of some two consecutive days, the sun's polar distance is greater than 90°, and on the second, less than 90°. The sun must, therefore, in the intermediate time, have passed from the south to the north side of the equator.

Let EQFB, *Fig.* 18, be the equator, P and P′ its poles, and ECFD the sun's apparent path. Let *a* and *b* be the places of the sun in his apparent path, when on the meridian at the two noons preceding, and following his passage from the south to the north side of the equator, S, the star whose passages over the meridian were observed, and P*a′a*, P*bb′*, and PSG, arcs of declination circles. The intervals of time between the passages of the sun over the meridian and those of the star give, when converted into degrees (63), the angles GP*a′* and GP*b′*, or the arcs G*a′* and G*b′*, which are their measures. The difference between G*a′* and G*b′* gives *a′b′*. Then, the changes in the sun's polar distances and in the intervals of time being very nearly uniform, as will appear from examination of their values on several preceding and following days, we have, P*a* — P*b* : *aa′* : : *a′b′* : *a′*E. The arc *a′*E taken from G*a′* leaves GE, the distance of the point E from the declination circle through the star.*

Let *c* and *d* be the places of the sun in his apparent path when on the meridian at any subsequent times, and let the declinations *cc′* and *dd′* and the arcs G*c′* and G*d′* be obtained from observations as above. From the values of the latter and of GE, we know E*c′* = GE — G*c′*, and E*d′* = GE — G*d′*. Then, whatever be the sun's places *c* and *d*, it is found that the values of the quantities E*c′*, E*d′*, *cc′*, and *dd′* are such that the proportion, sin E*c′* : sin E*d′* : : tang *cc′* : tang *dd′*, is always true. But assuming ECFD to be a great circle, we have, from the right angled spherical triangles E*c′c* and E*d′d* (App. 48), tang *cc′* = tang E sin E*c′*, and tang *dd′* = tang E sin E*d′*; which gives the same proportion. Hence the

* From the observations of several consecutive days, the arc GE may be found with great precision, by a method of computation called *interpolation*. Some cases of this method will be found in the appendix.

sun's apparent path ECFD is a great circle, cutting the equator in two opposite points E and F.

107. *The apparent motion of the sun around the earth, is produced by a real annual motion of the earth round the sun.*

Let S and E, *Fig.* 19, be the situations of the sun and earth respectively, at any instant of time, fg, a part of the sun's apparent path in the celestial sphere, a, the apparent place of the sun, and s, a fixed star, supposed to be situated in the apparent path. Then will sEa be the angular distance of the sun from the star. If we suppose the sun to move from S to S′ in any given interval of time, his angular distance from the star will become sEb. But if, instead of supposing the sun to move, we suppose the earth to move, in the same interval of time, through the same angular distance from E to E′, the sun's angular distance from the star will then become sE′c. As the angles E′SE and SES′ are equal, E′S and ES′ are parallel, and the angle $s\text{E}b = s\text{F}c = s\text{E}'c + \text{E}s\text{E}'$. Hence the angular distance of the sun from the star, at the end of the interval, differs, on the two suppositions, by the angle EsE′; and consequently the sun's apparent motion, during the interval, differs by the same quantity.

If we assume the distance Es of the star to be so great that the distance from E to E′, whatever be their situations, is extremely small in comparison with it, the angle EsE′ will also be extremely small. Consequently, on this assumption, the sun's apparent motion will be sensibly the same, whether we suppose the sun to revolve round the earth, or the earth to revolve round the sun. But, as the bulk of the sun is more than a million times that of the earth (100), it seems highly improbable that the former revolves round the latter as the central body. The reasonable conclusion therefore is, that the earth revolves round the sun in the course of a year, in the same plane in which the sun appears to move, and thus produces the sun's apparent motion. This conclusion is confirmed by various astronomical facts; some of which will be noticed in their proper places. But although the earth's annual motion is fully established, astronomers frequently find it convenient to speak of the sun's motion; always, however, meaning the apparent motion.

As the earth revolves round the sun in about 365¼ days, at a distance of 95,000,000 miles (96), it may be found, by a simple computation, that its annual motion must be at the rate of about 19 miles in a second.

DEFINITIONS.

108. The *Ecliptic* is that great circle of the celestial sphere, which the sun appears to describe in his apparent annual motion.

The plane of this circle, which necessarily passes through the centres of the earth and sun, is called the *plane* of the ecliptic, or sometimes simply the *ecliptic.*

109. The *Zodiac* is a zone of the celestial sphere, extending 8° or 9° on each side of the ecliptic.

The sun and moon, and all the principal planets, have their motions within the limits of the zodiac.

110. The *Obliquity of the Ecliptic* is the angle which the ecliptic makes with the equator. Thus, the angle QEC, *Fig.* 18, is the obliquity of the ecliptic. It is evidently measured by CQ or BD, the greatest north or south declination of the sun. The obliquity of the ecliptic is about 23° 28′.

111. The *Equinoctial Points* or *Equinoxes* are the two points in which the ecliptic and equator intersect each other. The point at which the sun passes from the south to the north side of the equator is called the *vernal* equinox, and the other, the *autumnal.* Thus E is the vernal equinox, and F the autumnal equinox. The straight line, joining the equinoctial points E and F, is called the *line of the equinoxes.*

The *Solstitial Points* or *Solstices* are the two points in the ecliptic at which the sun's declination is the greatest, north and south. They are therefore 90° from the equinoxes. That to the north of the equator is called the *summer* solstice, and the other, the *winter.* Thus C is the summer and D the winter solstice.

The terms equinox and solstice are also used to denote the *times* at which the sun is at those points.*

* Formerly they were used only to refer to the times. At present they are applied to both the times and points: the context always indicating the sense in which they are used.

112. The *Signs of the Ecliptic* are twelve equal parts, into which the ecliptic is conceived to be divided, beginning at the vernal equinox and proceeding *eastward*. Each sign therefore contains 30°. They are designated by names or characters as in the following table.

1. Aries	♈	7. Libra	♎
2. Taurus	♉	8. Scorpio	♏
3. Gemini	♊	9. Sagittarius	♐
4. Cancer	♋	10. Capricornus	♑
5. Leo	♌	11. Aquarius	♒
6. Virgo	♍	12. Pisces	♓

The vernal equinox is sometimes termed *the First point of Aries.*

A body or a point is said to have a *direct* motion, when its motion is from *west* to *east*, according to the order of the signs of the ecliptic, and a *retrograde* motion when the motion is in a contrary direction, or from *east* to *west.*

113. The *Equinoctial* and *Solstitial Colures* are two declination circles passing through the equinoxes and solstices. Thus, EPFP′ is the equinoctial colure, and PCP′D is the solstitial colure.

It is evident that the solstitial colure passes through the poles p and p' of the ecliptic, as well as through those of the equator, and that the equinoctial points E and F are its poles.

114. The *Tropics* are two small circles parallel to the equator and passing through the solstices. That to the north of the equator is called the tropic of *Cancer*, and that to the south, the tropic of *Capricorn.* Thus CC′ is the tropic of Cancer, and DD′ the tropic of Capricorn. The distance of the tropics from the equator is evidently equal to the obliquity of the ecliptic.

The *Polar Circles* are two small circles parallel to the equator, and at a distance from its poles equal to the obliquity of the ecliptic. That about the north pole is called the *arctic* circle, and that about the south pole, the *antarctic.* Thus pq is the arctic, and $p'q'$ the antarctic, circle.

Circles corresponding to the tropics and polar circles, and bearing the same names, are conceived to be drawn on the earth's surface, dividing it into five portions called *zones.* The zone between the tropics is called the *torrid* zone; the two between the tropics and polar circles are called the *temperate* zones; and the two within the polar circles are called the *frigid* zones.

115. The *Right Ascension* of a body is the arc of the equator intercepted, to the *east*, between the vernal equinox and a declination circle passing through the body. Thus EG is the right ascension of the star S.

The right ascension and declination (27) of a body, designate its situation in reference to the equinoctial colure and the equator.

116. A *Circle of Latitude* is any great circle passing through the poles of the ecliptic. The arc pSH is part of a circle of latitude.

117. The *Longitude* of a body is the arc of the ecliptic intercepted, to the *east*, between the vernal equinox and a circle of latitude passing through the body.

The *Latitude* of a body is the arc of a circle of latitude intercepted between the body and the ecliptic. The latitude is *north* or *south*, according as the body is on the north or south side of the ecliptic. Thus, EH is the longitude, and HS the latitude, of the star S, *north*.

The longitude and latitude of a body designate its place in reference to the circle of latitude passing through the vernal equinox and the ecliptic.

PROBLEMS.

118. *To find the obliquity of the ecliptic.* The obliquity of the ecliptic may be found from the equation, $\text{tang } dd' = \text{tang E} \sin \text{E}d'$ (106), in which dd' and $\text{E}d'$ are known from observation, and the angle E is the obliquity of the ecliptic. This gives,

$$\text{tang E} = \frac{\text{tang } dd'}{\sin \text{E}d'}.$$

It may however be more accurately obtained from the sun's declination, found for several days at noon (102), about the time of either solstice. From these declinations, the value of CQ, the greatest declination, may be deduced by interpolation; and this expresses the obliquity of the ecliptic (110).

The obliquity is subject to some slight changes that will be noticed in the next chapter.

119. *To change the right ascension and declination of a body into longitude and latitude, or the contrary.*

Let S be the body, and let E and S be joined by an arc of a great circle. Put ε = angle QEC = obliquity of the ecliptic, which is supposed to be known, A = EG, the right ascension, D = GS, the declination, L = EH, the longitude, λ = HS, the latitude, N = the angle GES, and N′ = the angle HES.

Then, for the change from right ascension and declination to longitude and latitude, we have, from the triangles EGS and EHS (App. 48 and 49),

$$\text{tang N} = \frac{\text{tang D}}{\sin \text{A}}; \text{ and tang ES} = \frac{\text{tang A}}{\cos \text{N}}. \quad \text{Hence,}$$

$$\text{tang L} = \cos \text{HES tang ES} = \frac{\cos(\text{N} - \varepsilon) \text{ tang A}}{\cos \text{N}};$$

$$\text{tang } \lambda = \text{tang } (\text{N} - \varepsilon) \sin \text{L}.$$

For the change from longitude and latitude to right ascension and declination, we have, in like manner,

$$\text{tang N}' = \frac{\text{tang } \lambda}{\sin \text{L}}; \text{ and tang ES} = \frac{\text{tang L}}{\cos \text{N}'}. \quad \text{Hence,}$$

$$\text{tang A} = \cos \text{GES tang ES} = \frac{\cos(\text{N}' + \varepsilon) \text{ tang L}}{\cos \text{N}'};$$

$$\text{tang D} = \text{tang } (\text{N}' + \varepsilon) \sin \text{A}.$$

When the declination or latitude is *south* it must be taken *negative*. The subsidiary arc N or N′ may always be taken *affirmative* and less than 180°. When one of the quantities L and A is in either the fourth or first quadrant, it is evident the other must be in one of these two; and when one of them is in either the second or third quadrant, the other must be in one of these. With attention to these remarks and to the trigonometrical rules for the signs of the quantities, the preceding formulæ are applicable, whatever be the situation of the body.

For the sun, or any point in the ecliptic, as S′, we evidently have

$$\text{tang L} = \frac{\text{tang A}}{\cos \varepsilon}; \quad \text{tang A} = \cos \varepsilon \text{ tang L};$$

$$\text{tang D} = \text{tang } \varepsilon \sin \text{A}, \text{ or } \sin \text{D} = \sin \varepsilon \sin \text{L}.$$

FIXED STARS.

120. *Positions of the fixed stars.* When EG, the right ascension of one star S, has been obtained (106), the right ascension of any other body may be found from the observed interval in sidereal time, between its passage over the meridian and that of the star. This interval added to the right ascension of the star, expressed in time, or subtracted from it, according as the passage of the body is later or earlier than that of the star, will evidently give its right ascension in time. The method of finding the polar distance or declination has been already given (102).

When the right ascensions and declinations of the stars have been found from observations, their longitudes and latitudes may, if required, be computed by the last article.

121. *Constellations.* The ancients, in order to distinguish the various groups of stars, imagined figures of men, animals, and other objects, to be drawn around them in the concave surface of the celestial sphere. The group of stars contained within the contour of any one of these imaginary figures is called a *Constellation.* Each constellation bears the name of the figure which limits it.

The number of constellations formed by the ancients is 48. To these about 40 have since been added; some of them being small constellations, formed of stars not included in the ancient constellations, but most of them are in that part of the southern hemisphere not visible to the ancient observers. *Twelve* of the constellations follow one another along the ecliptic, and bear the same names as its signs. These are called *zodiacal* constellations.

122. *Stars of a constellation.* The stars of a constellation are distinguished from one another by the letters of the Greek alphabet, which are applied to them according to their apparent relative size or brightness. The principal star in the constellation is usually named α, the second β, the third γ, and thus on. When the number of stars in a constellation exceeds the number of letters in the Greek alphabet, as it generally does, the remainder are designated by the letters of the Roman alphabet or by numbers. The expres-

sion α Lyræ, denotes the star α in the constellation *Lyra*, a harp; and so of others.

Some of the stars have particular names, as *Sirius*, *Aldebaran*, *Arcturus*, &c.

123. *Definition. A Catalogue of fixed stars* is a table containing a list of stars with their right ascensions and declinations, or their longitudes and latitudes.

The first catalogue was formed by *Hipparchus*, about 130 years prior to the Christian era; and contained the positions of nearly 1000 stars. Various catalogues have since been formed; some of them containing the situations of many thousands of stars, most of which are only visible by the aid of a telescope.

CHAPTER IX.

PRECESSION OF THE EQUINOXES—ABERRATION—NUTATION.

124. *Position of the ecliptic and motion of the equinoxes.* From comparisons of catalogues of the stars, formed at different times, it is found that the latitudes of the stars continue always nearly the same. Hence the position of the ecliptic among the stars must be *fixed*, or nearly so.

But it is found, from these comparisons, that the longitudes of the stars are continually increasing at the rate of about 50″ in a year. This increase of longitude is common to all the stars, and, except for a few, is the same for each star. It cannot therefore be reasonably imputed to motions in the stars themselves. Hence it follows that the vernal equinox, the point from which longitude is reckoned, must have a backward or *retrograde* motion along the ecliptic, equal to the increase in the longitudes of the stars. Let ECFD, *Fig.* 20, be the ecliptic, p and p', its poles, E, the place of the vernal equinox at any time, and E′, its place at some subsequent time, it having, during the intermediate time, retrograded along the ecliptic through the arc EE′. Then must the longitude of any star S, be changed during this interval of time from EH to E′H; being increased by the quantity EE′.

As the autumnal equinox is always directly opposite to the vernal equinox, it must have the same motion.

125. *Definition.* The *Precession of the Equinoxes* is the *retrograde* motion which they have along the ecliptic. It is $50''.2$ in a year.

126. *The poles of the equator revolve with retrograde motions in small circles around the poles of the ecliptic, at distances equal to the obliquity of the ecliptic.*

As the ecliptic remains in a fixed position or nearly so (124), it is evident the equator must change its position, otherwise there could be no motion in the equinoctial points; and a motion of the equator must necessarily produce motions of its poles. Let ECFD, *Fig.* 20, be the ecliptic, p and p', its poles, and Pab, a small circle about the pole p, at a distance equal to the obliquity of the ecliptic. Then, since the distance between the poles of two great circles is equal to the angle they make with each other, if we suppose the obliquity of the ecliptic to continue the same, as it does nearly, the north pole of the equator must always be in the circle Pab.

Let EQFB be the position of the equator at any time. Then will the great circle pCp'D, having for its poles the equinoctial points E and F, be the position of the solstitial colure at that time (113), and P must therefore be the place of the north pole of the equator. Let E′Q′F′B′ be the position of the equator at some subsequent time. Then will the great circle pC′p'D′, having for its poles the equinoctial points E′ and F′, be the position of the solstitial colure, and P′, the position of the north pole of the equator. Hence, while the vernal equinox has retrograded from E to E′, the pole has retrograded from P to P′ in the small circle Pab. The south pole of the equator must evidently have a corresponding motion.

Cor. Since E and E′ are the poles of pPp' and pP′p', the arc EE′ is the measure of the angle PpP′. Hence the angular motion of the pole of the equator round the pole of the ecliptic is equal to the precession of the equinoxes: that is, it is $50''.2$ a year. It must therefore require nearly 26,000 years to make a complete revolution.

127. *Precession in Right Ascension.* If Em be perpendicular to E′G′, then will E′m be the retrograde motion of the equinox in right ascension, sometimes called the *precession in right ascension.* Taking EE′ = 50″.2, we find E′m = 46″, the annual precession in right ascension.

128. *Annual Variations in right ascension and declination.* As the longitudes of the stars are continually changing, their right ascensions and declinations must also change. These changes are, however, very different for different stars, depending on their positions. The change in the right ascension or declination of a star during a year, is called its *annual variation* in right ascension or declination. If we suppose E and E′ to be two positions of the vernal equinox at an interval of a year, and PsG and P′sG′ to be arcs of declination circles through a star at s, the annual variation of the star in right ascension will be the difference between EG and E′G′; and its annual variation in declination, the difference between sG and sG′.

Formulæ are easily investigated for computing the annual variations in right ascension and declination.* In catalogues of the stars, the values of the annual variations for each star, computed for the time for which the catalogue is formed, are annexed to the right ascension and declination of the star at that time. From these, the right ascension and declination of any star, contained in the catalogue, may be found for any given time, provided it be not many years distant from the time for which the catalogue was formed. In consequence, however, of small changes which the annual variations themselves undergo, from the changes in the positions of the stars in reference to the equator, it is requisite that new catalogues should be occasionally formed.

128 *a.* *Constellations of the zodiac and signs of the ecliptic.* At the time of the first catalogue of the stars, 130 years prior to the Christian era, the signs of the ecliptic corresponded very nearly to the constellations of the zodiac bearing the same names. But, in the interval of nearly 2000 years since that period, the vernal equinox has retrograded about 28°; so that the sign Taurus now

---

* See Appendix, art. 54.

nearly corresponds with the constellation Aries, the sign Gemini with the constellation Taurus, and so for the others.

129. *Visible effect of the precession of the equinoxes.* The effect of the precession of the equinoxes becomes, in the course of ages, very conspicuous in the northern and southern parts of the heavens. The poles of the heavens, in their slow retrograde revolutions about the poles of the ecliptic (126), must approach near to different stars in succession. At the time the first catalogue of the stars was formed, the north pole was nearly 12° distant from the present pole star, and its distance from it is now only about $1\frac{1}{2}$°. The pole will continue to approach this star till the distance between them is about *half a degree*, and will then recede. In a period of 12,000 years from the present time, the pole will have arrived within about 5° of a very bright star, α Lyræ, from which it is now more than 50° distant, and consequently will then be more than 40° from the present pole star.

This continual change in the position of the pole, must also make changes in the class of stars that are circumpolar at any given place. For a star cannot be circumpolar at any place, if its distance from the pole is greater than the altitude of the pole at the place, or (57) than the latitude of the place. In process of time our present pole star will cease to be a circumpolar star in the latitude of Philadelphia. The student will, however, observe that these changes must be periodical. At the termination of a period of 26,000 years (126), the position of the pole, with reference to the stars, and consequently the class of circumpolar stars at a place, will again become nearly the same as at the commencement of that period.

130. *Cause of the precession of the equinoxes.* Investigations in physical astronomy prove that the precession of the equinoxes is produced by the attractions of the sun and moon on that portion of the earth that is on the outer side of an imaginary sphere, conceived to be described about the earth's axis. The effect of these actions is a slow change in the direction of the earth's axis, and consequently corresponding changes in the positions of the equator and its poles.

ABERRATION.

131. *Dr. Bradley's Observations.* In the early part of the last century, Dr. Bradley, a celebrated English astronomer, commenced a series of accurate observations on the positions of some of the fixed stars, which he continued for a number of years. In the course of these observations, he found the apparent places of the stars to be subject to periodical changes, amounting, in some, to about 40″, and that the period of these changes was a year. After several unsuccessful attempts to account for these changes, it at length occurred to him that the annual motion of the earth, combined with the motion of light, must generally cause a star or other heavenly body to appear to be in a position different from its true position; and, on investigation, he found that the changes in the apparent positions of the stars which must thus be produced, corresponded with those he had observed.

132. *Effect of the combined motions of the earth and of light on the apparent place of a body.* From phenomena that will be hereafter noticed, it had been ascertained, prior to the time of Bradley, that the transmission of light, though inconceivably rapid, is not instantaneous. It occupies 8 m. 13 sec. in passing the distance from the sun to the earth, and consequently moves with a velocity of about 192,000 miles per second. The velocity of the earth in its annual motion is 19 miles per second (107). Disregarding the motion of an observer at the earth's surface, that is produced by the rotation on the axis, which is small in comparison with the annual motion, let BD, *Fig.* 21, be the path in which he is carried at any time by the annual motion of the earth, during an interval so short that the path may be regarded as straight. Let E be any point in BD, s, the position of a star, having the direction Es from the point E, AE, the distance through which the observer is carried during some small interval of time, and Ea, the distance through which light moves in the same time. Let A′a′, A″a″, &c., and Es′, be drawn parallel to Aa; and the former will divide EA and Ea proportionally. Then, if a particle of light in the ray sE be at a when the observer is at A, it will be at a′ when he is at A′, at a″ when he is at A″, &c. The particle therefore continues in the

same direction from him while he is moving from A to E, and will meet his eye at E, coming to it in the direction s'E, and consequently making the impression of having come from a star at s'. The same applies to a series of particles, or, in the undulatory theory of light, to a series of undulations. Hence the star s will appear to him to be at s', deviating from its true position by the angle sEs'.

This phenomenon may be illustrated by supposing a drop of rain falling in the direction of the line SE, and a hollow tube, with its axis in the position Aa at the instant the drop reaches the point a, the tube moving from Aa to Es' while the drop falls from a to E. It is plain that the drop will, in its descent along the line aE, describe the axis of the tube; and that to a person looking through the tube, and carried along with it, unconscious of his own motion and of that of the tube, the drop would seem to fall in the oblique direction of the tube.

In the triangle AEa we have Ea : EA : : sin EAa or sin DEs' : sin AaE or sin sEs'. But Ea : EA : : veloc. of light : veloc. of earth : : 192000 : 19. Hence 192000 : 19 : : sin DEs' : sin sEs'

$$= \frac{19}{192000} \sin DEs'. \quad \text{Or (App. 51),}$$

$$\frac{sEs'}{\omega} = \frac{19}{192000} \sin DEs';$$

$$sEs' = \frac{19\omega}{192000} \sin DEs' = 20''.36 \sin DEs',$$

or, $sEs' = 20''.36 \sin DEs$, without sensible error.

When the angle DEs is a right angle, the deviation sEs' is 20″.36, which is its greatest value.

The deviation evidently takes place in the direction in which the earth is moving, and will, therefore, have opposite directions at opposite seasons of the year. Thus if EFE′F′, *Fig.* 22, be the direction in which the earth revolves round the sun, the direction of its motion when at E, will be BD, and the deviation of the star s will be to the right; but when the earth is at E′, the direction of its motion will be B′D′, and the deviation of the star will be to the left. Therefore the whole change thus produced in the apparent place of a star may amount to twice the greatest value of sEs', or about 41″.

The direction in which the earth is moving at any time is known from the sun's longitude. For let EV be the direction of the vernal equinox. Then the angle VES expresses the sun's longitude, and VED, the direction of the point of the ecliptic towards which the earth is moving when at E. Now, regarding the earth's orbit as a circle, the angle SED is a right angle. Hence VED = VES — 90°; or the longitude of the point of the ecliptic towards which the earth is moving at any time, is less by 90° than the sun's longitude.

For a planet, the amount of deviation, at any time, is increased or diminished by its own motion during the time the light is passing from it to the earth.

133. *Aberration.* The deviation produced in the apparent place of a heavenly body by the combined motions of the earth and of light, is called *Aberration.*

The uranographical effect of aberration is that each star appears to describe annually an ellipse, having, for its centre, the true place of the star; the major axis of the ellipse being 40.″7; and, the minor axis being equal to the maximum aberration of the star in latitude or, 40.″7 × sine of star's latitude. For a star in the ecliptic, the latitude being nothing, the minor axis of the ellipse will be zero, and the star will appear to move along the ecliptic. For a star at the pole of the ecliptic, the latitude being 90°, the minor axis of the ellipse will be 40.″7 or, equal to the major axis, and the star will therefore appear to describe a circle about the pole of the ecliptic.

By formulæ which have been investigated for the purpose, the correction for aberration in the right ascension, declination, longitude or latitude of a body, may be computed for any given time.*

The computed aberrations are found to correspond with the annual deviations in the observed places of the stars. Hence, as the aberration would not have place if the earth had not an annual motion, this correspondence is a *strong confirmation of the earth's annual motion.*

---

* Formulæ for the aberration of a fixed star in right ascension and declination, are investigated in the Appendix.

NUTATION.

134. *Bradley's Observations.* Dr. Bradley found that, after allowances were made for precession and aberration, his observed places of the same star at different times were nearly the same. There were, however, deviations still too great to be ascribed to errors of observation. These deviations, unlike those due to aberration, varied from year to year, and appeared to require a period of about 19 years to go through their course. This led to new investigations; and he at length ascertained that these deviations were occasioned by an inequality in the precession of the equinoxes.

135. *Inequality in the precession of the equinoxes, and in the obliquity of the ecliptic.* When treating of the moon it will be shown, that, in her revolution round the earth, she does not move in the plane of the ecliptic, but in a plane, making with the former an angle of about 5° ; also, that the line, in which the plane of her orbit intersects that of the ecliptic, is continually changing its direction by a retrograde motion, making a complete revolution in a little less than 19 years. Now, the precession of the equinoxes is produced by the actions of the sun and moon on the protuberant part of the earth (130), but principally by that of the moon, in consequence of her comparative vicinity to the earth. The effect of the action of either of these bodies depends on its position with regard to the equator. As the protuberant part of the earth is equally divided by the equator, when either body is in the plane of the equator, its action can have no tendency to change the position of the earth about its centre, so as to affect the position of this plane, and consequently none to change the positions of the equinoctial points. Its effect in producing these changes increases with an increase in the distance of the body from the equator, and is greatest when that distance is greatest. Hence, from the continually varying positions of the sun and moon in reference to the equator, a small oscillatory motion of the equator is produced. This, necessarily, produces an inequality in the precession of the equinoxes and also in the obliquity of the ecliptic.

Cor. In consequence of the oscillatory motion of the equator, its poles, in their retrograde revolutions about the poles of the ecliptic (126), do not move strictly in circles, but in waving curves that pass alternately within and without the circles, somewhat similar to that in *Fig.* 23.

136. *Nutation.* The inequality in the precession of the equinoxes or in the obliquity of the ecliptic, produced by the varying effect of the actions of the sun and moon on the excess of the earth above an inscribed sphere, is called *Nutation.* That produced by the action of the sun is called *Solar* nutation; and that by the action of the moon is called *Lunar* nutation.

Formulæ have been investigated, by means of which the values of the nutations in the precession of the equinoxes and obliquity of the ecliptic, and in the right ascension, declination, or longitude of a body, may be computed for any given time.* No one of the nutations amounts to more than a few seconds.

137. *Diminution of the obliquity of the ecliptic.* From a comparison of the latitudes of the fixed stars, as determined at different periods, it is found that the plane of the ecliptic is subject to a slow progressive change of position. The direction of this change is such as to diminish the obliquity of the ecliptic. The diminution thus produced is found to be at the rate of 45″.7 in a century.

The change in the position of the ecliptic is occasioned by the actions of the planets on the earth. The planets revolve round the sun in planes making small angles with the plane of the ecliptic, and are therefore in the latter, only when passing from one side to the other. As the plane of the ecliptic is the plane in which the earth is at any time moving, the attraction of a planet, when not in this plane, must tend to draw the earth from it; or, which is equivalent, must tend to change the position of the plane. Investigations in physical astronomy prove that the whole combined effect of the planetary attractions, must be a progressive change in the position of the ecliptic. They also show that the rate of

* For the investigation of formulæ for the nutations in right ascension and declination, see Appendix.

diminution in the obliquity of the ecliptic produced by this change, after slightly increasing for a few centuries, must decrease; and that the whole change in the obliquity from its present value, can never exceed $1\frac{1}{2}$ degrees.*

138. *Parallax of a body in right ascension, declination, &c.* The apparent places of all the heavenly bodies, except the fixed stars, are more or less affected by parallax (90). The difference thus produced in the right ascension, declination, longitude, or latitude of a body, is called *parallax in right ascension, parallax in declination, &c.* By means of formulæ which have been investigated for the purpose, the values of these parallaxes for any body, may be computed for any given time and place.†

139. *Secular and periodic inequalities.* A *secular* inequality is one that requires many centuries to pass through its different values. Thus the change in the obliquity of the ecliptic, produced by the actions of the planets, is a secular inequality. The expression, *secular equation* or *secular variation*, generally implies the amount of the inequality for a century. Thus we say the secular diminution of the obliquity is 45″.7.

A *periodic* inequality is one that passes through its various values in a few days, months, or years, or at most in a few centuries. Thus the nutation of the equinoxes, frequently called the *equation of the equinoxes*, is a periodic inequality.

140. *Mean and apparent places.* The *mean* place of a body or point, the *mean* position of a plane, or the *mean* value of an angle, is that which it would have at any time, if the *periodic* inequality or inequalities to which it is subject did not exist. Thus the mean place of the vernal equinox at any time, frequently called the

---

* The change in the position of the ecliptic from its present position may be considerably greater than this; its limit, according to *Pontécoulant*, being a little over 6°. But in consequence of the change in the position of the equator, by which the procession of the equinoxes is produced, the variation in the oliquity cannot exceed the quantity mentioned in the text.

† For these formulæ and their investigations, the student is referred to larger treatises, such as those by Vince, Delambre, Woodhouse, &c.

mean equinox, is the place at which the equinox would be if there was no nutation. The same applies to the mean position of the equator, called the *mean equator;* and to the *mean obliquity* of the ecliptic. The *mean right ascension* of a body is the right ascension of the mean place of the body, reckoned from the mean equinox along the mean equator; and similarly for *mean* declination, longitude, or latitude.

The actual place of the vernal equinox at any time is called the *apparent* or *true equinox;* the actual obliquity of the ecliptic is called the *apparent* obliquity; and the equator in its actual position is called the *apparent* or *true equator.* The *apparent* or *true* right ascension of a body, is the right ascension of the apparent or true place of a body, reckoned from the apparent equinox along the apparent equator; and similarly for *apparent* or *true* declination, longitude, or latitude.

The point in which the arc of a declination circle, passing through the mean equinox, meets the apparent equator, is the *reduced* place of the mean equinox. The small distance between this point and the apparent equinox, is evidently the nutation of the equinox in right ascension, or the equation of the equinoxes in right ascension.

141. *Tables of reduction.* The exact *apparent* positions of the fixed stars, are so continually wanted by the practical astronomer in adjusting and examining the adjustments of his instruments, and as points of reference, that much attention has been devoted, to obtain concise and accurate methods of deducing these from the *mean* places given in catalogues. In the catalogue of 8377 principal fixed stars, published a few years since under the direction of the British Association for the Advancement of Science, besides the mean places, annual precessions, and secular variations, there are given certain constant logarithms for each star, by means of which, with others given, in the Nautical Almanac, for each day in the year, the apparent places of these stars may be found for any given time, with great facility.

Professor Bessel, in his *Tabulæ Regiomontanæ*, has given general formulæ and tables for reducing the mean places of the stars to apparent places.

CHAPTER X.

SIDEREAL AND SOLAR TIME—TROPICAL YEAR—SUN'S APPARENT ORBIT—KEPLER'S LAWS—SOLAR TABLES—EQUATION OF TIME —SUN'S SPOTS, AND ROTATION ON HIS AXIS—ZODIACAL LIGHT.

142. *Sidereal Time.* The *sidereal day*, as now used by astronomers, commences at the instant the apparent vernal equinox is on the meridian, and is reckoned through 24 hours to the return of the equinox to the meridian.* Consequently the sidereal time at any instant expresses the apparent right ascension of the meridian at that instant, or of any body that is then on the meridian. Thus if ε, *Fig.* 1, be the position of the apparent vernal equinox at any instant, the arc εQ, which expresses the sidereal time at the place A at that instant, expresses also the apparent right ascension of any body that is then on the meridian PZP′.

The sidereal clock is adjusted, or its error and rate determined, by observations of the passages over the meridian of certain fixed stars, whose apparent right ascensions are known, or may be computed with great precision. The apparent right ascension of any other body, when on the meridian, may then be found by observing the time of its passage as shown by the clock, and correcting this time for the error of the clock.

143. *Solar Time.* The interval between two consecutive returns of the sun's centre to the meridian, is called a *solar day;* and time reckoned by solar days is called *solar time.* The length of the solar day is found to be somewhat different at different seasons of the year; its mean or average length is called a *mean* solar day.

If on any day the sun is on the meridian of a place, at the same

* The equinox, having a precession in right ascension or westwardly motion of 46″ in a year (127) or $\frac{1}{8}$ of a second in a day, must return to the meridian sooner than a fixed star by $\frac{1}{120}$ of a second in time. The sidereal day, as here defined, is therefore shorter than as defined in Art. 53, by $\frac{1}{120}$ of a second. In consequence of the nutation of the equinoxes, it is not strictly uniform; but the deviation is extremely small.

instant with some fixed star, he will, in consequence of his apparent eastwardly motion (7), be to the east when the star returns to the meridian next day, and will not arrive at it till some minutes later than the star. Consequently, the solar day is longer than the sidereal day. The mean solar day is found to be equal to 24h. 3m. 56.555sec. of sidereal time.

144. *Tropical Year.* The interval between two consecutive returns of the sun to the vernal equinox is called a *tropical year.*

145. *Length of the tropical year.* From the sun's declination, and the sidereal time when he is on the meridian, obtained for a number of consecutive days about the 21st of March, the time when the declination is nothing: that is, the time when he is at the equinox, may be accurately determined. If this be done in successive years, the length of the year, in sidereal time, becomes known.

The length of the year, determined at different periods, is found to be subject to a slight variation. Its mean length at the present period, expressed in *mean solar time*, is 365d. 5h. 48m. 48sec.

146. *Sidereal year.* The time during which the sun, by his apparent motion, makes an entire revolution in the ecliptic, is called a *sidereal year.*

147. *Length of the sidereal year.* In consequence of the retrograde motion of the equinoxes (124), the arc of the ecliptic which the sun passes through during a tropical year, is less than 360° by 50″.2. Hence, as 360° — 50″.2 : 360° : : length of the tropical year : length of the sidereal year. The length of the sidereal year is thus found to be 365d. 6h. 9m. 10sec. It is therefore 20m. 22sec. longer than the tropical year.

148. *Sun's apparent orbit.* The path described by the sun's centre in the plane of the ecliptic, during an apparent revolution round the earth, is called the *sun's apparent orbit* or the *solar orbit.*

149. *The Solar Orbit is an ellipse, having the earth in one focus.* Let PSAB, *Fig.* 24, represent the sun's apparent orbit, E, the place of the earth, P, the sun's place in his orbit when his apparent diameter is greatest, A, his place when it is least, and S, his place at some intermediate time.

The obliquity of the ecliptic being known, the sun's longitude, at any time, becomes known from his observed right ascension (119). Hence, from a series of observations of the sun's right ascension and corresponding apparent diameter, continued throughout the year, we may obtain a series of corresponding longitudes and apparent diameters. From these, the two longitudes, corresponding to the greatest and least apparent diameters at P and A, may be easily found. These longitudes, thus determined, are found to differ 180°. Hence EA and EP are in the same straight line. It is also found, from a general examination of the corresponding longitudes and apparent diameters, that the apparent diameter at any place S, is equal to half the sum of the greatest and least diameters, *less* the product of half their difference by the cosine of the angle AES, which is the difference of the sun's longitudes at A and S.

Let AP be bisected in C, and let

δ = sun's apparent diameter at A,

δ' = do. P,

δ'' = do. S,

$m = \frac{1}{2}(\delta' + \delta), \qquad n = \frac{1}{2}(\delta' - \delta).$

Then, from the known relation between the angle AES and the apparent diameters, we have

$$\delta'' = m - n \cos \text{AES} \quad \text{(A)}$$

But (97) $\text{AE} : \text{EP} :: \delta' : \delta$ ..... (B)

or,
$$\text{AE} : \text{AE} + \text{EP} :: \delta' : \delta' + \delta$$
$$\text{AE} : \tfrac{1}{2}(\text{AE} + \text{EP}) :: \delta' : \tfrac{1}{2}(\delta' + \delta)$$
$$\text{AE} : \text{AC} :: \delta' : m = \frac{\text{AC}}{\text{AE}} . \delta' \quad \text{(C)}$$

Again from (B)
$$\text{AE} : \text{AE} - \text{EP} :: \delta' : \delta' - \delta$$
$$\text{AE} : \tfrac{1}{2}(\text{AE} - \text{EP}) :: \delta' : \tfrac{1}{2}(\delta' - \delta)$$
$$\text{AE} : \text{EC} :: \delta' : n = \frac{\text{EC}}{\text{AE}} . \delta' \quad \text{(D)}$$

Also (97)
$$\text{ES} : \text{EP} :: \delta' : \delta'' = \frac{\text{EP}}{\text{ES}} . \delta' \quad \text{(E)}$$

Substituting the values of m, n, and δ'' in (A), and dividing by δ', we have

$$\frac{\text{AC}}{\text{AE}} - \frac{\text{EC}}{\text{AE}} \cos \text{AES} = \frac{\text{EP}}{\text{ES}}$$

or, $ES\,(AC - EC \cos AES) = AE.EP = (AC + EC)\,(AC - EC)$
$$= AC^2 - EC^2.$$

$$\text{or, } ES = \frac{AC^2 - EC^2}{AC - EC \cos AES}.$$

This is the polar equation of an ellipse, of which AP is the transverse axis, C the centre, and E a focus.

150. *Earth's orbit.* Let S, *Fig.* 25, be the sun, and PEAE′ the earth's orbit, or path described by its centre in the plane of the ecliptic, during a revolution round the sun. Then substituting E for S, and the contrary, the demonstration in the last article proves that the earth's orbit is an ellipse, having the sun in one focus.

The discovery that the sun's apparent orbit, or the earth's real orbit, is an ellipse, was made in the early part of the 17th century by *Kepler*, a celebrated German astronomer. He first ascertained that the orbit of the planet Mars was an ellipse, and, pursuing his investigations, he found that the orbits of the earth and other planets were also ellipses. This, being one of several important discoveries made by him relative to the planetary motions, is called *Kepler's first law.*

151. *Definitions.* A *Radius Vector* is a straight line joining the centres of the sun and a planet, or the centres of a planet and satellite.

Perihelion, &c. In the orbit of the earth, or a planet, the point nearest the sun is called the *perihelion*, and that which is most distant, the *aphelion.* In the moon's orbit, or sun's apparent orbit, the point nearest the earth is called the *perigee*, and the most distant point, the *apogee.* These points have also the general appellation of *apsides:* the nearest point being called the *lower apsis*, and the most distant, the *higher apsis.* The transverse axis of the ellipse, or the line joining the apsides, is called the *line of the apsides.*

The *Eccentricity* of an elliptical orbit, as the term is generally used in astronomy, is the distance between the centre and a focus, expressed in terms of the semi-transverse axis regarded as a unit; or, which amounts to the same, it is the quotient of the distance between the centre and focus, divided by the semi-transverse axis.

152. *Sun's apparent motion in his orbit.* The sun's apparent motion or angular velocity, deduced from his longitude as determined from day to day, is found to be different in different parts of the orbit. The motion in longitude is least when the apparent diameter is least, which is about the 1st of July; and is then 57′ 11″ in a mean solar day. Its daily motion is greatest when the apparent diameter is greatest, which is about the 1st of January, and is then 61′ 10″. From a comparison of the daily motions and corresponding apparent diameters, it is found that the daily motion is always proportional to the square of the apparent diameter. But the apparent diameter is inversely proportional to the distance or radius vector (97). Hence, the daily motion or angular velocity is inversely proportional to the square of the radius vector.

Consequently, if r and r' be two radii vectores, and v and v' the corresponding angular velocities, we have $v : v' :: r'^2 : r^2$, or $vr^2 = v'r'^2$. This is also found to be true in the motion of a planet round the sun, and of the moon round the earth.

153. *The Sun's radius vector describes equal areas in equal times; and consequently, in unequal times, it describes areas proportional to the times.*

Let a and b, *Fig.* 26, be the sun's places in his orbit at some short interval prior and subsequent to his being at S, as, for instance, half an hour. Let ce and mn be arcs described about E, the place of the earth, as a centre; the former with a radius equal to a unit, and the latter with a radius equal to the radius vector from E to S. Then will the elliptical sector Eab be the area described by the radius vector during an hour when the sun is in the part of his orbit contiguous to S, and the angle aEb, or its measure, the arc ce, will be the sun's hourly motion or angular velocity.

Put $ce = v$, and the distance from E to S $= r$. Then $1 : r :: v : mn = vr$. Hence the area of the circular sector Emn $= \frac{1}{2}$En $\times\ mn = \frac{1}{2}vr^2$. But the sector E$mn$ does not sensibly differ from the elliptical sector Eab. Consequently, Eab $= \frac{1}{2}vr^2$. In like manner for another position of the sun as S′, taking v' and r' for the angular velocity and radius vector, we have E$a'b'$ $= \frac{1}{2}v'\,r'^2$.

But by the last article $vr^2 = v'r'^2$. Hence, Eab = E$a'b'$.

Kepler, who discovered this fact, found also that the radius vector of each planet describes about the sun equal areas in equal times This is called *Kepler's second law.*

154. *Kepler's third law.* In comparing the periods in which the planets revolve round the sun and their mean distances from him, Kepler discovered that *the squares of the periodical times of the planets are proportional to the cubes of their mean distances from the sun.*

155. *To find the position of the line of the apsides of the solar orbit.* Let B and D, *Fig.* 24, on opposite sides of the transverse axis AP, be corresponding points of the orbit. Then it is evident that the sun's daily or hourly motion at D must be the same as at B; the time in which he moves from P to D must be the same as that in which he moved from B to P; and the longitude of the perigee P must be midway between the longitudes of B and D. Hence, when from a series of the sun's longitudes determined from observation, two times and the corresponding longitudes are found, at which the sun's hourly or daily motion in longitude is the same, the longitude of the perigee and the time that the sun is at that point become known.

Another method. As AP, the line of the apsides, divides the orbit into two equal parts, the sun must be as long in passing from A to P as from P to A. The time in either case is therefore half a year; and in this time the sun passes through 180° of longitude. No other straight line through the earth's centre divides the orbit into two equal parts. It is, therefore, only in passing from one apsis to the other, that the sun employs just half a year in changing his longitude 180°. Hence two longitudes of the sun being found which differ 180°, and are separated by an interval of half a year, will be the longitudes of the perigee and apogee; and the corresponding times will be the times the sun is at those points.

156. *Motion of the apsides.* From observations made at distant periods it is found that the apsides have a slow *direct* motion. According to Prof. Bessel, from an examination of many observations made at various times, the longitude of the perigee at the beginning of the year 1800 was 279° 30′ 8″; and its yearly increase of longitude is 61″.52.

If from 61″.5, the annual motion of the perigee from the vernal equinox, we subtract 50″.2, the annual retrograde motion of the equinox, we have 11″.3 for the annual motion of the perigee.

Taking 180° from the longitude of the perigee in the year 1800, we have 99° 30′ 8″, for the longitude of the apogee at that time. If this be reduced to seconds and divided by 61″.52, the yearly motion of the apogee in longitude, the quotient is 5823. Hence it appears that about 5823 years anterior to the year 1800, the longitude of the apogee was nothing; and consequently the line of the apsides then coincided with the line of the equinoxes. It may be remarked, that this is about the period on which chronologists have fixed, as the time of the creation of the world.

157. *True and Mean Anomalies and Equation of the Centre.* The angular distance of a body from the perihelion or perigee of its orbit, reckoned to the eastward through the whole circumference of the circle, is called the *true anomaly.** The angular distance from the perihelion or perigee, at which the body would at any time be, if it moved with its mean or average angular velocity, is called its *mean anomaly.* The difference between the true and mean anomalies at any time is called the *equation of the centre.* Thus if S, *Fig.* 24, be the sun's place at any time, and *s* the place at which he would have been at that time, if he had moved from P to *s* with his mean angular velocity, then will the angle PES be his true anomaly, PE*s*′ his mean anomaly, and SE*s* the equation of the centre.

The equation of the centre expresses the difference between the mean and true longitudes. For, let EQ be the direction of the vernal equinox. Then the angle QES is the sun's true longitude when he is at S, and QE*s* his mean longitude; and these differ by the angle SE*s*.

158. *Anomalistic year.* The interval between two consecutive returns of the sun to the perigee is called an *anomalistic year.*

Hence as 360° — 50″.2 : 360° + 11″.3 : : length of tropical year : length of anomalistic year; which is thus found to be 365d. 6h. 13m. 46sec.

* Formerly the anomaly was reckoned from the aphelion or apogee.

159. *Sun's mean daily motions in longitude and anomaly.* As the length of the tropical year (145) : 1 day : : 360° : sun's mean daily motion in longitude; which is thus found to be 59′ 8″.33. In a similar manner, taking the length of the anomalistic year (158), we find the sun's mean daily motion in anomaly to be 59′ 8″.16.

When the longitude of the perigee of the solar orbit, and the time he is at that point, have been obtained (156), his mean longitude and mean anomaly may be found for any given time, by means of the known mean motions in longitude and anomaly.

160. *Greatest equation of the sun's centre.* At the perigee and apogee, the equation of the centre is evidently nothing. When the sun moves from the perigee, his true motion is greater than his mean motion, and therefore his true anomaly will be greater than his mean anomaly. The difference between these or the equation of the centre must necessarily increase, till the true motion diminishes so much as to become equal to the mean motion. It will then have its greatest value. For, from that time, the mean motion being greater than the true, the mean anomaly will gain on the true anomaly, and the equation of the centre must decrease till it becomes nothing at the apogee. The same in a reverse order takes place from the apogee to the perigee. As the orbit is symmetrical on each side of the line of the apsides, the greatest equation on one side must be equal to the greatest on the other.

By examining a daily series of the sun's true longitudes when on the meridian, obtained from observations, it will be found that about the first of April, the sun's true daily motion in longitude, which is then decreasing, is for several days nearly equal to the mean daily motion. The same will be found to have place about the first of October, except that then the true daily motion is increasing. At each period, let that noon at which the true and mean motions appear to be most nearly equal, be selected. Let L and L′ be the true longitudes of the sun at these noons respectively, M and M′, the mean longitudes, and E, the greatest equation of the centre. Then, since the difference between the true longitude and mean longitude is the same as that between the true anomaly and mean anomaly (157), we have

$$E = L - M, \text{ and also } E = M' - L'.$$

Hence, $2E = (M' - M) - (L' - L)$,

or, $$E = \frac{(M' - M) - (L' - L)}{2}.$$

In this expression for the greatest equation, $(M' - M)$ is the sun's mean motion in longitude during the interval between the two noons selected, and becomes known from the mean daily motion in longitude; and $(L' - L)$ is known from the given true longitudes.

The greatest equation of the sun's centre is thus found to be $1° 55'.3$, nearly.

161. *Eccentricity of the solar orbit.* Let $r = EP$ (*Fig.* 24), = the radius vector for the perigee, $r' = EA$ = radius vector for the apogee, v = sun's true daily motion at the perigee, v' = the same at the apogee, and e = the eccentricity. Then v and v', being the greatest and least daily motions, are known (152). By the same article, $r' : r :: \sqrt{v} : \sqrt{v'}$.

Hence, $r' + r : r' - r :: \sqrt{v} + \sqrt{v'} : \sqrt{v} - \sqrt{v'}$,

or, $$\frac{r' - r}{r' + r} = \frac{\sqrt{v} - \sqrt{v'}}{\sqrt{v} + \sqrt{v'}}.$$

But, $\dfrac{r' - r}{r' + r} = \dfrac{2EC}{2AC} = \dfrac{EC}{AC} = e$, (151).

Therefore, $$e = \frac{\sqrt{v} - \sqrt{v'}}{\sqrt{v} + \sqrt{v'}} = \frac{\sqrt{v} - \sqrt{v'}}{\sqrt{v} + \sqrt{v'}} \times \frac{\sqrt{v} - \sqrt{v'}}{\sqrt{v} - \sqrt{v'}},$$

or, $$e = \frac{v + v' - 2\sqrt{vv'}}{v - v'}.$$

Taking $v = 3670''$, and $v' = 3431''$ (152), we find,

$$e = .0168.$$

The eccentricity and greatest equation in any elliptical orbit evidently depend on each other. If either is given, the other may be obtained by mathematical investigation. From such investigation, it has been ascertained, that when the orbit does not differ greatly from a circle, the eccentricity is nearly equal to the quotient of half the greatest equation, expressed in seconds, divided by 206264″.8, the seconds in the radius of a circle.* Thus, with the value of E = 1° 55′.3 (160), we find for e the same value as above.

162. *Secular variations of the equation of the sun's centre and of the eccentricity of his orbit.* The greatest equation of the sun's centre, and consequently the eccentricity of his orbit, as determined at periods distant from one another, are found to be subject to a slow, but continued diminution. The secular diminution of the greatest equation is 18″.

Investigations in physical astronomy prove that the variation of the eccentricity, from which that of the equation of the centre results, is produced by the actions of the planets on the earth.

163. *Kepler's Problem.* When the eccentricity of the orbit and the mean anomaly of a body are given, the equation of the centre and true anomaly may be found. This problem, which was proposed and solved by Kepler, is one of some difficulty. Various solutions of it have, however, been since obtained; one of which is given in the appendix.

By formulæ obtained from the solution of the problem, tables have been computed for the sun, moon, and planets, which give the value of the equation of the centre, corresponding to any given mean anomaly.

The equation of the sun's centre for any given time, obtained from its table, and applied to his mean longitude at that time, gives his true longitude from the mean equinox, with the exception of some small corrections to be noticed in the next article.

164. *Perturbations.* The actions of the moon and planets cause the earth to deviate slightly from an elliptical orbit, and produce small periodic inequalities in its motion. These inequalities are called *perturbations.* The bodies which produce sensible perturba-

* Putting k = the quotient of the value of E, in seconds, divided by 206264″.8, the following formulæ have been obtained:—

$$k = 2e + \frac{11}{48}e^3 + \frac{599}{5120}e^5 + \&c.$$

$$e = \tfrac{1}{2}k - \frac{11}{768}k^3 - \frac{587}{983040}k^5 - \&c.$$

tions in the motion of the earth, or apparent motion of the sun, are, the moon, Venus, Mars, Jupiter, and Saturn. The whole amount of these perturbations, when greatest, is about 37″.

165. *Sun's latitude.* As the moon and planets are continually varying their positions with regard to the plane of the ecliptic, the effect of their actions in drawing the earth from this plane or changing its position (137), must also be continually varying. If, therefore, the plane of the ecliptic was regarded as always passing exactly through the centre of the earth, its progressive change of position would be subject to small periodic inequalities. Astronomers, however, find it more convenient to assume the change of position to be regular; and, consequently, to regard the earth's centre as deviating slightly from the plane of the ecliptic, sometimes on one side and sometimes on the other. Hence, as this plane passes through the sun's centre, when the centre of the earth is on one side of it, the centre of the sun must appear to be at an equal distance on the other side, and must have a small latitude. The greatest value of this latitude is only about one second. It may, therefore, be neglected, except in very accurate investigations and computations.*

166. *Solar Tables.* These are tables for computing the sun's longitude, latitude, radius vector, apparent semidiameter, and the apparent obliquity of the ecliptic at any given time. The best solar tables are those by Carlini, an Italian astronomer, published in the Milan Ephemeris for the year 1833.†

---

* The attractions of the different planets, depending on their masses and distances from the earth, are very different, and some of them extremely small. But a very small effect, if produced for a long time in the same direction, so as to accumulate, may at length become sensible. Investigations in physical astronomy show, that the attractions of all planets, except the asteroids, are sensibly operative in producing the progressive change in the position of the ecliptic. But with regard to the periodic inequalities in the effect produced, the case is different. Those inequalities in the attractions, which cause the earth's centre to deviate from the plane of the ecliptic and thus produce the sun's latitude, are only sensible for the moon, Venus, and Jupiter.

† A new set of Solar Tables has been computed by MM. Hansen and Olufsen, and just published by the Royal Society of Copenhagen. These tables have the advantage over Carlini's, of an additional twenty years' observations for their basis. They also surpass them in fulness and in convenience of arrangement, and will, doubtless, soon become the standard solar tables.

When the sun's longitude, and the apparent obliquity of the ecliptic, have been computed for a given time, his right ascension and declination may be found by spherical trigonometry (119). The right ascension may also easily be found by a table for the purpose, called a table of *reduction of the ecliptic to the equator*. A convenient one is given in Carlini's solar tables.

167. *Astronomical day.* The astronomical day commences at noon of the common day, and the numbering of the hours is continued to 24. The first 12 hours are the same as the afternoon hours of the common day; but the hours from 12 to 24 of the astronomical day correspond to the 12 hours before noon of the next following common day.

Hence, when a time in common reckoning is before noon, we must, in order to convert it into astronomical time, subtract a unit from the number of the day and add 12 to the number of the hour. Thus, May 12th, at 5 h. A. M., common reckoning, is the same as May 11th, at 17 h., in astronomical reckoning.

168. *Apparent time.* Time reckoned by the position of the sun's centre with reference to the meridian of a place, is called *apparent time.* And the instant at which the centre is on the meridian is *apparent noon.*

The length of the apparent day is variable, and from two causes. In the first place, the sun's motion in the ecliptic is not uniform, being affected by the equation of the centre and the planetary perturbations. In the second, as equal arcs of the equator pass the meridian in equal times, the arcs of the ecliptic which pass it in equal times must, in consequence of its obliquity to the equator, be unequal; so that, if the sun's motion in the ecliptic were uniform, the intervals between his consecutive returns to the meridian would not be equal.

As the length of the apparent day is not uniform, clocks or chronometers could not be made to exhibit apparent time without continually repeated adjustments.

169. *Mean time.* If we assume an imaginary sun to be at the reduced place of the mean vernal equinox (140), at the same instant that the real sun is at the apparent equinox, and to move from thence in the equator with a uniform motion, equal to the real

sun's mean motion in longitude, the time measured by the position of this imaginary sun, with reference to the meridian, is called *mean time.* And the instant at which the centre is at the meridian is *mean noon.*

The mean day, according to this definition, is of uniform length, and is evidently the same as that defined in a preceding article (143). Mean time, therefore, flowing uniformly, is that to which clocks are adjusted for the common purposes of society, and also for many astronomical purposes. Observatories are usually furnished with at least two clocks; one of which is adjusted to sidereal time, and the other to mean solar time.

170. *Equation of time.* The difference, at any instant, between apparent and mean time, is called the *equation of time.* It depends on the unequal motion of the sun in the ecliptic and the obliquity of the ecliptic to the equator (168).

171. *The equation of time is equal to the difference between the sun's true right ascension and the sum of his mean longitude and the equation of the equinoxes in right ascension, converted into time.*

Let EQ, *Fig.* 27, be an arc of the equator, P its pole, PZM an arc of the meridian of a place, of which Z is the zenith, EC an arc of the ecliptic, E the apparent or true equinox, E′ the mean equinox, S the true place of the sun in the ecliptic, and PE′G and PSH arcs of declination circles. Then G is the reduced place of the mean equinox, EG the equation of the equinoxes in right ascension, EH the sun's true right ascension, and the angle HPM, of which HM is the measure, is the hour angle for apparent time.

Let GS′ be equal to the sun's mean longitude. Then S′ is the place of the imaginary sun, assumed to move uniformly in the equator (169), $ES' = GS' + EG$, is the sum of the sun's mean longitude and the equation of the equinoxes in right ascension, and the angle S′PM, of which S′M is the measure, is the hour angle for mean time.

Put $T =$ the apparent time and $T' =$ the mean time. Then since each 15° of the hour angle corresponds to an hour, we have in hours or parts of an hour,

$$T = \frac{HM}{15} = \frac{EM - EH}{15}$$

$$T' = \frac{S'M}{15} = \frac{EM - ES'}{15}$$

Consequently, by subtraction, we have for the equation of time,

$$T - T' = \frac{ES' - EH}{15} \qquad \text{(F)}$$

or,

$$T' - T = \frac{EH - ES'}{15} \qquad \text{(G)}$$

From these we have, also, for the expression of either time in terms of the other and the equation of time,

$$T = T' + \frac{ES' - EH}{15} \text{ and } T' = T + \frac{EH - ES'}{15}.$$

The equation of time is given in the Nautical Almanac for every day in the year, with instructions whether to add or subtract in changing one time to the other.

172. *Times at which the equation of time is nothing.* When the effects of the two causes on which the equation of time depends (170), are opposed to each other and are equal, the equation of time must be nothing; apparent and mean time must then be the same. This occurs four times in the year; about the 15th of April, 15th of June, 1st of September, and 24th of December.

173. *Sidereal time.* The arc EM of the equator converted into time, expresses the sidereal time (142). But EM = EH + HM. Hence, the sidereal time is obtained by adding the apparent time to the sun's true right ascension, expressed in time.

Again, EM = GS′ + EG + S′M. Consequently, the sidereal time is also obtained, by adding the mean time to the sum of the sun's mean longitude and the equation of the equinoxes in right ascension, expressed in time.

As EM is the right ascension of the zenith Z, it follows that the sidereal time expresses the right ascension of the zenith in time.

174. *Solar Spots.* When the sun is viewed with a telescope furnished with a coloured glass to protect the eye, a number of dark spots are usually seen on his surface. Each spot generally consists of a central part of irregular form, which is black, surrounded by a margin or border, called the *penumbra*, of much lighter colour, as represented in *Fig.* 28. The spots differ greatly

in size and form. Some are occasionally observed so large as to be visible to the naked eye, when, in consequence of intervening light clouds, or of the sun being near the horizon, he can be thus viewed. The breadth of some of these, as ascertained from measurements with a micrometer, is found to be nearly or quite 50,000 miles.

When the spots are examined from hour to hour or day to day, they are found to vary in size and form, and also to change their positions on the disc; all of them moving towards the western edge. Some increase in size and others diminish. Some are formed and come into view where none were before visible; others, after gradually diminishing, entirely disappear; the central part being that which first becomes invisible. Sometimes a spot is seen suddenly to separate, or, as it were, to burst asunder. The change of form is sometimes gradual and sometimes rapid. None of the spots are permanent in their continuance. Some remain only for a few hours or days; and there are seldom any that continue longer than six or seven weeks. There are generally some spots to be seen on the disc, and at times as many as forty or fifty have been observed at once; but there have been periods of months, or even years, during which none were visible. When a spot remains for several weeks, it is seen to traverse the disc from east to west in about 13½ days, and after being invisible during about the same time, to re-appear on the eastern side; it being thus about 27 days in returning to the same position on the disc.

175. *Sun's rotation on his axis.* By observations of the right ascensions and declinations of the centres of the sun and of a spot, or by observations with a position micrometer (50), either being repeated from day to day, the positions of the spot with reference to the centre of the disc and the plane of the ecliptic, and thence the form of its path, may be determined. At two nearly opposite times in the year, in June and December, the paths are found to be nearly straight lines inclined in an angle of 7⅓° to the plane of the ecliptic. Except at these times, they are nearly semi-ellipses of great eccentricity; the eccentricity, however, though always great, varying with the season of the year.

From the motions of the spots and the paths they describe, it is inferred that the sun revolves from west to east on a fixed axis, inclined at an angle of about $7\frac{1}{3}°$ to a perpendicular to the plane of the ecliptic. Let EE′E″, *Fig.* 29, represent the earth's orbit, *epqp′* the sun, *fg* a perpendicular to the plane of the ecliptic through S, the sun's centre, *pp′* his axis, and *enq* a circle, called the *sun's equator*, the plane of which passes through his centre at right angles to the axis. A spot on the sun's surface must, by his revolution on his axis, describe either the circle *enq* or some other as *amb* parallel to it. These circles, being perpendicular to the axis *pp′*, must be inclined to the ecliptic in the same small angle in which the axis is inclined to the perpendicular *fg*. They are, therefore, in very oblique positions with regard to an observer on the earth, and, consequently, the spots describing them must appear to move in ellipses of large eccentricity. This would be accurately the case if the earth were at rest; but in consequence of its motion in its orbit, the apparent paths must deviate somewhat from accurate elliptical arcs, as they are found to do. When the earth is at those opposite points of its orbit through which the plane of the sun's equator passes, the observer being then in that plane, the spots must appear to move in straight lines, excepting so far as deviations are produced by the earth's motion.

It has been stated that a spot returns to the same position on the disc in about 27 days. This is not, however, the time of a revolution on the axis. For while the sun makes a revolution on his axis, the earth advances nearly a sign in the ecliptic; and, consequently, the sun must make more than a complete revolution, before the spot will appear to an observer on the earth to occupy its former position on the disc. By investigations and computations which may be omitted here, the real time of revolution can be determined from the observed positions of the same spot at different times. According to Delambre, the period of the sun's revolution on his axis is 25d. 0h. 17m. Some astronomers make it rather greater.

It is also ascertained that the position of the sun's axis is such, that when his longitude is about 170° 21′, it coincides with the plane ES*f*, passing through the centres of the earth and sun at right angles to the ecliptic, the northern half S*p* of the axis being then inclined towards the earth.

176. *Hypotheses relative to the solar spots.* Various hypotheses to account for the solar spots have been proposed. None of them are, however, entirely satisfactory. One of the most probable is that by Sir W. Herschel. He supposed that the mass of the sun is an opaque globular body, surrounded by an atmosphere of luminous matter; that this luminous atmosphere is not in contact with the body of the sun, but is sustained far above it by a transparent, elastic medium, in which floats a stratum of cloudy matter; and that, from the operation of local causes, cavities or openings are formed, both in the cloudy stratum and luminous part of the atmosphere, the opening in the latter being larger than that in the former. According to these assumptions, the part of the solid body of the sun that is under an opening, being shaded by the cloudy stratum, and thus receiving little or no light from the luminous part of the atmosphere, must appear as a black spot. The part of the cloudy stratum contiguous to its aperture, and under the larger opening, reflecting light received from the latter, would form the border or penumbra of the spot.

177. *Zodiacal Light.* A faint light, somewhat resembling that of the milky way, or more nearly that of the tail of a comet, and being nearly in the form of a cone with its base towards the sun, and its axis nearly in the direction of the ecliptic, is frequently seen at certain seasons of the year in the west after the close of twilight in the evening, or in the east before its commencement in the morning. This is called the *zodiacal light.*

The state of the air and other circumstances being the same, the zodiacal light is most distinctly seen when its direction or the direction of the ecliptic is most nearly perpendicular to the horizon. This, for places whose latitudes are from 40° to 50° north, occurs about the 1st of March for the evening, and about the 10th of October for the morning. In some years, the zodiacal light is very perceptible in the evening for several weeks contiguous to the former time, and in the morning for a like period contiguous to the latter time.

The distance to which the zodiacal light extends varies from 20° or 30°, to 70° or 80°. No very satisfactory explanation of this phenomenon has as yet been given. Sir William Herschel was of opinion, that the sun, viewed from one of the other stars, would appear to be surrounded by a nebulosity, similar to that in which some of the fixed stars appear to be enveloped, as seen from the earth.

CHAPTER XI.

PARALLELISM OF THE EARTH'S AXIS — VARIATIONS IN THE LENGTHS OF DAY AND NIGHT — SEASONS — ASTRONOMICAL PROBLEMS.

178. *Parallelism of the earth's axis.* In the annual motion of the earth round the sun, its axis continues very nearly in the same direction, or, in other words, it continues *parallel to itself* very nearly, the small deviation from parallelism being that which corresponds to the slow motion of the poles of the equator about those of the ecliptic (126). The axis being perpendicular to the plane of the equator, it is evident that it must continually make, with the axis of the ecliptic, an angle equal to the obliquity of the ecliptic. On this inclination of the axis, and on its parallelism, depend the variations in the lengths of day and night, and the changes of the seasons.

179. *Circle of Illumination.* The great circle in which a plane through the earth's centre, and perpendicular to its radius vector, intersects the surface of the earth, is called the *circle of illumination.* This circle separates the enlightened half of the earth's surface from the other half which is in the dark.*

180. *Different lengths of day and night.* Let S, *Fig.* 30, be the sun, ABCD the orbit of the earth, which may here be regarded as a circle, c the earth's centre, pp' its axis, making an angle of 23° 28′ with ab, a perpendicular to the ecliptic, and let p and p' be re-

---

* In consequence of the sun being much larger than the earth, and also of the refraction of the earth's atmosphere, the sun illuminates rather more than half the surface at the same time. But, in general explanations, this small excess is not noticed.

spectively the north and south poles of the earth. As the axis pp' continues parallel to itself during the earth's revolution round the sun, its position with regard to the radius vector Sc, and, consequently, with regard to the circle of illumination, which is perpendicular to Sc, must continually vary.

At the vernal and autumnal equinoxes, the sun's polar distance being then 90°, the angle pcS, which expresses this distance (27), must be a right angle, and consequently the axis pp' then coincides with the plane of the circle of illumination. This is represented in the positions of the earth near A and C, where $pbp'a$ is the circle of illumination. As the circle of illumination then passes through the poles of the earth, it must bisect not only the terrestrial equator, but also all circles on the earth's surface parallel to the equator. Hence, since, by the diurnal revolution of the earth on its axis, places on its surface move uniformly in circles parallel to the equator, each place must be as long on one side of the circle of illumination as on the other, and the length of the days and nights must be everywhere equal.

From the vernal equinox to the autumnal, the angle pcS is less than a right angle. The north pole is therefore turned towards the sun, and the south pole from him, and the circle of illumination cuts the parallels of latitude unequally; the longer parts, on the north side of the equator, being towards the sun, and on the south side, from him; the inequality in each case being greater as the latitude is greater. It consequently follows, that during this period, there must be continued day at the north pole and parts adjacent, and continued night at the south pole and adjacent parts; and that, at other places, the days and nights must be unequal; the days being longer than the nights in the northern hemisphere, and shorter in the southern; the difference evidently being greater as the latitude is greater. The portions of the earth around the poles, at which there is continued day or night, and the difference between the lengths of day and night at other places, continually increase from the vernal equinox to the summer solstice, at which time the angle pcS is least, and the inclination of the axis to the circle of illumination is greatest, being then equal to the obliquity of the ecliptic. This is represented at the place of the earth near B, where the circle of illumination, being seen edgewise, appears as the straight line *ba*. In this position of the earth, the whole of the northern frigid zone is illuminated, and the whole of the southern one is in the dark, and the parallels of latitude are most unequally divided; the days in northern latitudes having then their greatest lengths, and, consequently, the nights their least lengths. From the summer solstice to the autumnal equinox, the parts about the poles at which there is continued day or night, and the inequality in the lengths of day and night at other places, continually decrease.

From the autumnal equinox to the vernal, the angle *pc*S is greater than a right angle, the north pole is turned from the sun, and the south pole towards him, and the lengths of the days and nights are reversed; the nights being longer than the days in the northern hemisphere, and shorter in the southern; the greatest difference being at the winter solstice, when the angle *pc*S is greatest, and the inclination of the axis pp' to the circle of illumination is again equal to the obliquity of the ecliptic. This position of the axis is represented at the place of the earth near D.

As two great circles mutually bisect each other, the circle of illumination must always bisect the equator, and consequently, at places on the equator, the days and nights must be equal throughout the year, each being 12 hours long. And it follows from what has been said above, that from the equator to the polar circles the length of the longest day must vary from 12 hours to 24 hours; and from the polar circles to the poles, it must vary from 24 hours to six months.

181. *Seasons*. The amount of heat derived from the sun at any place, at different times in the year, depends, principally, on the length of the day, and the height to which he ascends above the horizon. The longer the day and the greater the altitude he acquires, the greater must be the amount of heat received. The change in the earth's distance from the sun must produce some effect; but on account of the small eccentricity of the earth's orbit, this change in distance is only a small part of the whole distance, and consequently the difference in the heat received, depending on this cause, cannot be great. Indeed, the sun is in perigee, or nearest the earth, about the 1st of January, which, in northern latitudes, is near the time at which we usually have our coldest weather.

The zenith of a place, at the terrestrial equator, is in the plane of the equator. It is, therefore, obvious, from a reference to *Fig.* 30, that as the earth revolves on its axis, the zenith of a place at the equator must, during one half the year, pass on one side of the sun's centre, and during the other half on the other side; or, in other words, the sun during one half the year must pass the meridian to the south of the zenith, and during the other half to the north of it; its meridian zenith distance not, however, at any time exceeding 23° 28′. Consequently, at places at the equator or near it, the intensity of the sun's heat during the middle part of the day must always be great; and as the days are at no time less or much less than 12 hours in length, the temperature is great throughout the year. The greatest intensity of the sun's rays at those places is at or near the equinoxes, when the sun passes the meridian at the zenith or very near it.

It is further obvious, that at all places within the torrid zone, the sun must, in the course of the year, pass the meridian on opposite sides of the zenith; and, at two periods in the year, it must pass at or very near the zenith. Consequently, not only at the equator, but throughout the torrid zone, there are two seasons in the year at which the sun, when on the meridian, is nearly or quite vertical, and the intensity of his heat is very great.

In either of the temperate zones, the sun passes the meridian at all times in the year, on the same side of the zenith. The difference, therefore, between the least meridian zenith distance, which, in the northern zone, occurs at the summer solstice, and its greatest, which occurs at the winter solstice, must be twice 23° 28′, or 47° nearly. In consequence of this large difference in the sun's meridian zenith distances, and of the increased length of the days at the period he approaches nearest the zenith, and diminished length at the time his meridian altitude is least, the difference in temperature at these opposite seasons is necessarily great.

In the frigid zones, the sun can never ascend far above the horizon, and consequently the temperature is always low.

It follows from the preceding paragraphs, that places at and near the equator may be regarded as having two summers and two winters in each year, without much difference in the temperature, which is always high, and that this is the case throughout the torrid zone; the difference, however, between the summers and one of the winters increasing as the distance from the equator increases.

It is usual in the north temperate zone to regard Spring as commencing at the vernal equinox, or 20th of March; Summer, at the summer solstice, or 21st of June; Autumn, at the autumnal equinox, or 22d of September; and Winter, at the winter solstice, or 21st of December.

182. *Duration of Twilight.* The time of twilight at any place is the time during which the sun, in his diurnal motion, is between the horizon of the place and a parallel to the horizon at the distance of about 18° below it (88). Let EPR, *Fig.* 31, be the meridian of a place, Z its zenith, HR the western half of its horizon, FG a parallel to HR at the distance of 18° below, EQ the equator, P its pole, and MN the western half of the sun's diurnal path. Then the time during which the sun is descending from A to B, or while the hour angle increases from ZPA to ZPB, is the time of the evening twilight. As the sun's declination changes but little during a day, the morning and evening twilights of the same day must evidently be nearly of the same length.

The angle APB, when converted into time at the rate of 15° to the hour, expresses the duration of twilight. The magnitude of this angle, and consequently the duration of twilight, evidently depend on the latitude of the place, and the declination of the sun. For the same declination, the twilight is longer as the latitude is greater. At places in northern latitudes, the twilight is longest at the time of the summer solstice; and shortest, when the sun has a few degrees of south declination.* At Philadelphia, and other places whose latitudes are about 40° N., the shortest twilight occurs about the 6th of March and 8th of October.

When the sun's diurnal path is M″N″, meeting the meridian above F, it is evident the twilight must continue all night, as the sun does not then descend so low as FG. This must take place when PN″, his distance from the elevated pole, is less than PF, or PH + HF, or the latitude of the place + 18°; or, which amounts to the same, when the latitude is greater than PN″ − 18°. Now, when the sun has his greatest declination, and of the same name

* This is shown in the Appendix, art. 53.

with the elevated pole, PN$''$ is about $66\frac{1}{2}°$, and, consequently, PN$'' - 18° = 48\frac{1}{2}°$. Hence, at a place whose latitude is more than $48\frac{1}{2}°$, the twilight continues all night at the time the sun's declination is greatest and of the same name with the latitude.

PROBLEMS.

183. *To find the latitude of a place.*

1st Method. Let M, M$'$, m or m', *Fig.* 31, be the point in which the sun, or a fixed star, passes the meridian PRN. Then we have the latitude ZQ $=$ ZM $+$ QM $=$ ZM$'$ $-$ QM $=$ Qm $-$ Zm $=$ Qm' $-$ Zm'. Hence, calling the zenith distance of the body north or south, according as the zenith is north or south of the body, when the declination of the body and its correct meridian zenith distance are of the same name, their sum will be the latitude, which will be of that name; and when they are of different names, their difference will be the latitude of the same name with the greater quantity; observing, however, that when the body passes the meridian below the pole, the supplement of the declination must be used instead of the declination itself. Consequently, when, from the observed meridian altitude of the sun or a star, the correct altitude has been found, by applying the proper corrections, the latitude is thus very easily obtained.

If two stars be selected, one of which passes the meridian to the south of the zenith, and the other to the north, at about the same altitude, and the latitude be obtained by each, the mean of the two results will be nearly free from any small errors depending on want of accuracy in the centering or adjustment of the instrument used in observing the altitudes, or in the table of refractions. For, as such errors would affect the observed altitudes equally, or nearly so, making them both too great or too small by the same quantity, it is obvious, from the expressions for the latitude in the two cases, that the latitude obtained by one star must be as much too great as that obtained by the other is too small.

2d Method. Let S, *Fig.* 33, be the position of a star out of the meridian, and let SD be an arc of a great circle perpendicular to the meridian. If the altitude of the star be observed and corrected for refraction, and the time at which the altitude is taken be also observed, we shall have given in the triangle ZPS, the two sides PS and ZS and the angle ZPS, to find PZ, the complement of the latitude. For PS is known from the declination of the star, ZS is the complement of the correct altitude, and the angle ZPS, the star's distance from the meridian, is the difference between the star's right ascension and the sidereal time of observation, expressed in degrees. If the observed time is mean solar time, the corresponding sidereal time must be obtained* (142).

In the right angled triangle PDS we have (App. 49),

$$\text{tang PD} = \cos \text{SPD} \text{ tang PS}.$$

And, from the right angled triangles PDS and ZDS, we have (App. 45),

$$\frac{\cos \text{PS}}{\cos \text{PD}} = \cos \text{SD} = \frac{\cos \text{ZS}}{\cos \text{ZD}},$$

or, $\cos \text{PS} : \cos \text{PD} :: \cos \text{ZS} : \cos \text{ZD}$.

The difference between PD and ZD, or their sum, when D falls between P and Z, gives PZ, the complement of the latitude.

It is best to make the observations when the star is near the meridian, as a slight inaccuracy in the observed time does not then sensibly affect the computed latitude. This is not, however, material when it is the Pole star that is observed, as its motion in altitude is, at all times, slow. The star selected should not be one that passes the meridian so near the zenith as to leave a doubt with regard to the side of it on which the perpendicular SD would fall.

There are various other methods of finding the latitude of a place; one of which has been given in a previous article (58).

183.*a* *Given the latitude of a place and the sun's declination, to find the time of his rising or setting.*

Let HWR, *Fig.* 31, be the western half of the horizon, Z the zenith, EQ the equator, and P the elevated pole. Also let NM be parallel to EQ, at a distance equal to the given declination. Then will A, its intersection with HR, be the point of the horizon at which the sun sets, and the hour angle APM, converted into

* In the Nautical Almanac, the sidereal time at mean noon, at Greenwich, is given for each day in the year; and the method of finding it, for any time, at any meridian, is given also.

time, will be the interval between noon and sunset. Supposing the declination not to change during the day, it is evident the interval between sunrise and noon will be just the same. The angle APM or its measure, the arc DQ, is called the *semi-diurnal arc*, and the arc ED, the *semi-nocturnal arc.* The arc DW, contained between a declination circle, passing through the point of setting or rising, and the western or eastern point of the horizon, is called the *ascensional difference.*

In the right angled spherical triangle ADW, we have (App. 48),

tang AD = sin DW tang EWH = sin DW cot PH,

or, sin DW = tang PH tang AD,

or, sin *ascen. diff.* = tang *lat.* × tang *declin.*

Now, as the angle WPQ, or its measure WQ, is 90° or 6 hours, it is evident that when the declination is of the same name with the latitude, the sun will set at some point A, between H and W, and the ascensional difference DW, converted into time and added to 6 hours, will give the semi-diurnal arc in time. When the declination is of a different name from the latitude, the sun will set at some point A′, between W and R, and the ascensional difference subtracted from 6 hours will give the semi-diurnal arc.

The semi-diurnal arc in time, evidently, expresses the time of sunset, and, subtracted from 12 hours, it gives the time of sunrise. These are in apparent time, which may be converted into mean, by applying the equation of time.

We may find an expression for the semi-diurnal arc without first obtaining the ascensional difference. For cos DQ = cos (WQ + DW) = cos (90° + DW) = — sin DW.

Hence, cos DQ = — tang PH tang AD,

or, cos *semi-diur. arc* = — tang *lat.* × tang *declin.*

184. *To find the time of the sun's apparent rising or setting.* When the sun's centre appears to be in the horizon at a point *a*, *Fig.* 31, we have (80 and 91), his zenith distance Z*b* = 90° + *refraction* — *parallax.*

Put, Z = Z*b* = sun's zenith distance,

L = PZ = complement of the latitude,

D = P*b* = sun's polar distance, and

P = angle ZP*b* = *d*Q = semi-diurnal arc.

Then in the spherical triangle ZP*b*, we have (App. 38),

$$\sin \tfrac{1}{2} P = \sqrt{\frac{\sin \tfrac{1}{2}(Z + D - L) \sin \tfrac{1}{2}(Z + L - D)}{\sin L \sin D}}.$$

If we take Z = 90° + *refraction* — *parallax* — *semi-diameter*, the above formula gives the semi-diurnal arc for the apparent rising or setting of the sun's upper limb.

185. *To find the time of beginning or end of twilight.*

At the beginning or end of twilight the sun is 18° below the horizon (88). Let B be the position of the sun when at this distance below the horizon. Then, in the triangle ZBP, we have L = PZ and D = PB as in the last article, and Z = ZB = 90° + 18° = 108°. Hence,

$$\sin \tfrac{1}{2} P = \sqrt{\frac{\sin \tfrac{1}{2}(108° + D - L) \sin \tfrac{1}{2}(108° + L - D)}{\sin L \sin D}}.$$

186. *Given the latitude of a place and the sun's declination and altitude, to find the time of day.*

Let S be the position of the sun. Taking D and L as above, and Z = ZS = 90° − SK, and P = the hour angle ZPS; the value of P may be found by the formula in article 184.

CHAPTER XII.

DEFINITIONS.—OF THE MOON.

DEFINITIONS.

187. *Conjunction, &c.* A body is said to be in *conjunction with the sun*, or simply in *conjunction*, when its position is such, that its longitude and that of the sun are the same;* to be in *opposition*, when their longitudes differ 180°; and to be in *quadrature*, when their longitudes differ 90° or 270°. The term *syzygy* is used to denote either conjunction or opposition.

---

* When any two of the heavenly bodies have the same longitude, they are said to be in conjunction.

When the body is in any of the four positions, midway between the *syzygies* and quadratures, it is said to be in *octant*.

As each of the planets Mercury and Venus revolves at a less distance from the sun than that of the earth (18), either of them may be in conjunction both on the same side of the sun with the earth and on the opposite side. The former is called *inferior*, and the latter, *superior*, conjunction.

Some of the preceding terms are frequently designated by characters, as follow :—

Conjunction ☌
Opposition ☍
Quadrature □.

188. *Nodes.* The two points in which the orbit of the moon or a planet is cut by the plane of the ecliptic, are called *nodes*. That node in which the body is, when passing from the *south* to the *north* side of the ecliptic, is called the *ascending* node, and the other the *descending* node. The nodes are frequently designated by the following characters :—

Ascending node ☊
Descending node ☋.

189. *Different revolutions of a body.* The *sidereal* or *periodic* revolution of a body is the time during which it makes a real revolution round the central body ; or, it is the period that elapses from the time the body and a fixed star have equal longitudes, supposing them to be observed at the central body, till their longitudes are again equal.

The *tropical* revolution of a body is the period that elapses from the time it is at the vernal equinox, or any given longitude, as seen from the central body, till its return to the same.

The *synodic* revolution of a body is the interval between two consecutive conjunctions or oppositions of the body. In the case of Mercury or Venus, it is the interval between two consecutive conjunctions of the same kind.

The *anomalistic* revolution of a body is the interval between two consecutive returns of the body to the perigee or perihelion of its orbit, or to the same anomaly.

The *nodical* revolution of a body is the interval between two consecutive returns of the body to the same node.

OF THE MOON.

189.*a Moon revolves round the earth.* The moon appears to make a complete circuit of the heavens in rather less than a month (6). Hence, either the moon really revolves round the earth, or the latter revolves round the former. That the moon's apparent motion is not, like that of the sun, produced by a motion of the earth, but that it is a real motion, follows from the relative sizes of the bodies: the bulk of the earth being nearly fifty times that of the moon (100).

Strictly speaking, the earth and moon both revolve about the *centre of gravity* of the two, which is a point in the line joining the centres, situated a small distance within the earth's surface. This follows from the principles of mechanics, and is in accordance with their motions as deduced from observations.

The distance of the moon from the earth is only about the 400th part of that of the sun (95, 96). While, therefore, the moon revolves round the earth, she at the same time revolves with the earth round the sun.

190. *Moon's tropical revolution.* From observations made when the moon is on the meridian, her right ascension and declination are easily obtained (142, 102). With these, and the known obliquity of the ecliptic, her longitude may be computed (119). By daily, or at least frequent, observations and computations of this kind, the interval, from the time at which the moon has any given longitude, till her return to the same, may be determined. This interval, which is the tropical revolution of the moon, is found to be subject to considerable variation. Its mean length is about 27.32 mean solar days.*

191. *Moon's path.* The moon's observed right ascension and declination serve to determine her latitude as well as longitude (119). Frequent determinations of both, during her revolution round the earth, show that her path does not coincide with the

* For more exact expressions of this and other periods, see tables at end of Part I.

ecliptic, but is inclined to it in a small angle, intersecting it in two opposite points or *nodes*.

192. *Moon's mean daily motion in longitude.* As the moon moves through 360° of longitude during a tropical revolution, we have the following proportion:—As moon's mean tropical revolution : 1 day : : 360° : moon's mean daily motion in longitude. The mean daily motion is thus found to be 13° 10′ 35″.

193. *Acceleration of moon's mean motion.* A comparison of modern observations with those of former periods proves that the moon's mean motion is subject to a small, but, thus far, a continual *acceleration.* The mean daily motion given in the preceding article is that which had place at the beginning of the present century. Investigations in physical astronomy, by *Laplace*, have made known the cause of the acceleration, and have shown that it is a periodical inequality in the moon's mean motion, requiring a great length of time to go through its different values. It is occasioned by the change in the eccentricity of the earth's orbit (162).

As the acceleration is a periodical inequality, the mean length of the moon's revolution which is *now diminishing* must, in process of time, cease to do so, and afterwards increase.

194. *Moon's sidereal revolution.* In consequence of the westerly or retrograde motion of the equinoxes (124), the moon returns to the vernal equinox, or to the same longitude, before she has quite completed a real or sidereal revolution. From the known precession of the equinoxes, which is 50″.2 in a year (125), their motion during a tropical revolution of the moon is found to be 3″.75 nearly. This, subtracted from 360°, leaves the arc of the ecliptic moved through by the moon during a tropical revolution. Hence, as this arc : 360° : : the tropical revolution of the moon : the sidereal revolution. It is thus found that the sidereal revolution exceeds the tropical by about 7 seconds.

195. *Moon's synodic revolution.* It is evident, that during the interval between two consecutive conjunctions of the moon with the sun, that is, during a synodic revolution of the moon, the excess of the moon's motion in longitude above that of the sun must be 360°. From the mean daily motions of the sun and moon in

longitude (159, 192), the excess of that of the moon above that of the sun, becomes known. Hence, as this excess : 360° : : 1 day : the moon's mean synodic revolution. Its length is thus found to be about $29\frac{1}{2}$ mean solar days.

196. *Positions of the moon's nodes.* Let E, *Fig.* 34, be the earth, VNL an arc of the ecliptic, *pmag* the moon's orbit, and BNH an arc of the great circle in which the plane of the moon's orbit meets the celestial sphere. Then since EN is in the plane of the ecliptic, the point *n* in which it meets the orbit is the moon's ascending node, and N is the place of the node referred to the celestial sphere.

From a series of the moon's longitudes and latitudes, computed from the observed right ascensions and declinations, we may find the longitude when the latitude is zero. This will evidently be the longitude of one of the nodes. If the latitude is then changing from south to north, it will be the longitude of the ascending node *n* or N. The longitude of the ascending node increased by 180° must give the longitude of the descending node.

197. *Retrograde movement of moon's nodes.* From the longitudes of the moon's nodes, repeatedly determined, it is found that they have a *retrograde* motion along the ecliptic, amounting to about 19° in a year. By this motion, which is not quite uniform, the nodes make a mean tropical revolution in 18 years and 224 days, nearly.

198. *Inclination of moon's orbit.* From a series of the moon's latitudes, the greatest latitude FF′ may be found.

This greatest latitude has place when NF′, the excess of the moon's longitude above that of the node, is 90°. It is, therefore, the measure of the angle LNH, which is the inclination of the moon's orbit to the ecliptic. The inclination of the orbit, thus obtained at different times, is found to be subject to some variation. Its least and greatest values are about 5° and 5° $17\frac{1}{2}'$.

199. *Orbit longitude.* When a body moves in an orbit inclined to the ecliptic, the sum of the longitude of the ascending node and the eastwardly angular distance of the body from the node, is called its *orbit longitude.* Thus, if V, *Fig.* 34, be the vernal equi-

nox, and M the moon's place referred to the celestial sphere, the sum of the angles VEN and NEM is the orbit longitude of the moon when at *m*. If V′ be a point in the orbit when referred to the celestial sphere, corresponding to the vernal equinox, that is, such that the angle V′EN is equal to VEN, then the angle V′EM or arc V′M, is the moon's orbit longitude when she is at *m*.

When the inclination of the orbit, the longitude of the node, and the longitude of the body are given, the orbit longitude is easily found. Let MD be an arc of a circle of latitude. Then VD is the longitude of the body. Subtracting VN, the longitude of the node, from VD, we have ND. Then in the right angled spherical triangle NDM, we have the base ND and the angle MND, to find NM, the measure of the angle NEM. The angle NEM added to V′EN or VEN, the longitude of the node, gives V′EM, the orbit longitude.

Conversely, the longitude may be found when the orbit longitude is given.

200. *Apsides of the moon's orbit.* Using the orbit longitudes of the moon, the orbit longitudes of the apsides of her orbit may be found by proceeding in the same manner as for the positions of those of the sun's apparent orbit (155). The orbit longitudes being found, the longitudes may be obtained by the preceding article.

201. *Motion of the apsides of the moon's orbit.* The longitudes of the apsides, obtained at different periods, are found to increase at the rate of about 41° in a year. They have, therefore, a *direct* motion, and make a mean tropical revolution in a little less than 9 years.

The motion of the moon's nodes, the variation in the inclination of the orbit, and the motion of the apsides are all effects of the sun's attraction on the moon.*

202. *Moon's orbit.* From the greatest and least parallaxes of the moon (95), and from her parallax when at any position *m*, *Fig.* 34, in her orbit, the least and greatest radius vectors E*p* and E*a*, and the radius vector E*m* become known (93). From the

---

* See Chapter XXII.

orbit longitudes of the perigee and of the moon, when at m, the value of the angle $p\text{E}m$ at that time is known, being equal to their difference.

Assuming the orbit to be an ellipse, of which ap is the transverse axis, E a focus, and C, the middle point of ap, the centre, we have $a\text{C} = \frac{1}{2}(a\text{E} + \text{E}p)$, $\text{EC} = \frac{1}{2}(a\text{E} - \text{E}p)$, and, by the property of the ellipse,

$$\text{E}m = \frac{a\text{C}^2 - \text{EC}^2}{a\text{C} + \text{EC} \cos p\text{E}m}.$$

Now, whatever be the position of m, the value of Em, obtained from this expression, is always found to be nearly equal to its value obtained from the parallax. Hence, the moon's orbit is nearly an ellipse, having the earth in one focus.

It may also be found in the same manner as for the sun (153), that the moon's radius vector describes round the earth nearly equal areas in equal times.

203. *Greatest equation of the centre and eccentricity of moon's orbit.* These may be obtained in the same manner as for the sun (160, 161). Or, the eccentricity may be found from the values of aC and EC, deduced from the greatest and least parallaxes, as in the preceding article. The eccentricity being known, the greatest equation may be computed by the formulæ in the note to article (161).

The greatest equation is found to exceed 6°, being more than three times that of the sun. Consequently, the eccentricity of the lunar orbit must also be more than three times that of the apparent orbit of the sun or orbit of the earth.

204. *Other equations of the moon's motion.* The moon's motion is subject to numerous inequalities besides the equation of the centre. The three principal ones are called, respectively, *Evection*, *Variation*, and *Annual Equation*. The Evection was discovered by Ptolemy. It depends on the angular distances of the moon from the sun and the perigee. When greatest it amounts to about $1\frac{1}{3}$°. The Variation was discovered by Tycho Brahe. It disappears when the moon is in the *syzygies* and quadratures, and is greatest when she is in octants. It then amounts to 35′.7. The

Annual Equation depends on the sun's mean anomaly, and, when greatest, amounts to 11′.2.

Investigations, in physical astronomy, by Laplace and others, have made known the causes of these inequalities, and have discovered various smaller ones with which the moon's motion is affected. By means of these investigations, and long continued accurate observations, the moon's motion is now known, and her place at any given time may be computed, with a very near approach to precision.

205. *Lunar Tables.* There are two sets of lunar tables, of nearly equal accuracy: one by *Burkhardt,* and the other by *Damoiseau.* The former, in which 36 equations are employed in finding the longitude, is used in computing the Nautical Almanac, Connaissance des Tems, and Berlin Jahrbuch.

206. *Moon's phases.* The different forms which the moon's visible disc presents, during a synodic revolution, are called *phases.*

The moon's phases are completely accounted for by assuming her to be an opaque globular body, rendered visible by reflecting light, received from the sun. Let E, *Fig.* 35, be the earth, and ABCD the orbit of the moon; the sun being supposed to be at a great distance in the direction ES. When the moon is in conjunction at A, the enlightened half* is turned directly from the earth, and she must then be invisible. It is then said to be *new moon.*

About 7½ days after new moon, when she is in quadrature at B, one half of her illuminated surface is turned towards the earth, and her enlightened disc then appears as a semi-circle. She is then said to be at her *first quarter.*

About 15 days after new moon, when she is in opposition at C, the whole of her illuminated surface is turned towards the earth, and she appears as a full circle. It is then said to be *full moon.*

About 7½ days after this, when she is again in quadrature, at D, one half of her illuminated surface being towards the earth, she again appears as a semi-circle. She is then said to be at her *last quarter.*

---

* As the sun is far greater than the moon, he enlightens rather more than half her surface. But this slight excess need not be here considered.

From new moon to first quarter, and from last quarter to new moon, her enlightened disc is called a *crescent.* This phase is represented near *a* and *d.* The two extremities of the crescent are called *cusps* or *horns.* From first quarter to full moon, and from full moon to last quarter, the form of her enlightened disc is said to be *gibbous.* This phase is represented near *b* and *c.*

207. *Lunation or Lunar Month.* The interval from new moon to new moon again, is called a *lunation* or *lunar month.* It is evidently the same as a synodic revolution of the moon.

208. *Mean New or Full Moon.* The time at which it would be new moon or full moon, according to the mean motions of the sun and moon, is called *mean* new moon or *mean* full moon.

209. *Obscure part of the moon's disc.* When the moon is first visible after new moon, the whole of her disc is quite perceptible, the part not fully illuminated appearing with a faint light. As the moon's *age*, that is, the time from new moon, increases, the obscure part becomes more and more faint; and it entirely disappears before full moon. This phenomenon depends on light reflected from the earth to the moon, and from the moon back to the earth.

When the moon is near to *a*, she evidently receives light from nearly the whole of the earth's illuminated surface; and this light being in part reflected back, renders visible that portion of the disc that is not directly illuminated by the sun. As the moon advances towards opposition at C, the quantity of light she receives from the illuminated surface of the earth must evidently decrease; and its effect in rendering the obscure part visible, is still further diminished by the increased size and, consequently, increased light of the directly illuminated part, which finally prevents the faint light of the former from making any impression.

210. *The earth as seen from the moon.* It is obvious, from the explanation in the preceding article, that to an observer at the moon, the earth must appear as a splendid moon, assuming all the phases of the latter body as seen from the earth, and having more than three times the apparent diameter (100).

211. *Moon's surface.* When the moon is viewed with a telescope, the line separating the enlightened part of the disc from the ob-

scure or dark part, is seen to be very irregular and serrated; and its form is often found to vary even during the time of observation. Bright spots are frequently seen on the dark part, near the line of separation. The light of these is observed to spread till they become united to the enlightened part. The whole enlightened surface, also, appears diversified with numerous spots of various shapes and different degrees of brightness.

These appearances clearly prove that the moon's surface must be very irregular, abounding in mountains and valleys; the bright spots seen on the dark part being evidently the elevated tops of mountains, which, being illuminated by the sun, are visible, while the lower part and intervening valleys remain in the dark, till by the moon's farther advance in her orbit they are reached by the sun's rays.

Luminous spots, entirely unconnected with the phases, or, in other words, spots that are not the reflection of the sun's light, have sometimes been observed on the moon's disc. These are supposed to be *volcanoes.*

As the line separating the enlightened from the dark part of the disc, is always extremely irregular, it follows that the moon cannot have any large tracts of water; for over a surface of water, the line would evidently be uniform.

212. *Height of the lunar mountains.* If the distance of the illuminated summit of a mountain from the enlightened part of the disc be observed with a micrometer, and the positions of the sun and moon at the time be obtained by observation or computation, its height may be computed.* Its height may also be found from the measured length of the shadow it casts when in a different position with respect to the sun. Dr. Herschel computed the heights of a number of the lunar mountains. The height of the highest of those he found to be about 1¾ miles. But, according to Schrœter, Mädler, and other distinguished selenographers, some of the lunar mountains are as high as any of those on our globe. The Appenine range rises to a height of 3½ miles. The abysses, however, are more remarkable, some of them being 20 or 30 miles in diameter, and 3 or 4 miles deep.

---

* See Appendix to Part I, art. 58.

213. *Same surface of the moon always towards the earth.* The various spots on the moon always occupy nearly the same positions on the disc. Hence it follows that nearly the same surface is always turned towards the earth. Hence, also, if we suppose the moon to be inhabited, the inhabitants on about one half the surface can never see the earth while they remain on that half.

214. *Of the moon's atmosphere.* The moon sometimes passes between the earth and sun, and, sometimes, between the earth and a star or planet, causing what in the former case is called an eclipse of the sun, and in the latter, an occultation of the star or planet. Assuming the moon to have no atmosphere, the durations of these phenomena may be very accurately computed by means of the known motions and apparent magnitudes of the bodies.

Now, if the moon was surrounded by an atmosphere, such as appertains to the earth, it would, by its action on the rays of light passing through it, produce a sensible effect on the duration of an eclipse of the sun or of an occultation. But no such effect has been observed. It is, therefore, inferred that the *moon has no atmosphere ; or that, if she has, it must be of very little density.*

From a full investigation of the subject, Professor Bessel draws the conclusion, that, if we assume the moon to have an atmosphere constituted like that of the earth, its density at the moon's surface cannot be more than about the 1000th part of that of the earth's, at the earth's surface.*

215. *Moon's rotation on her axis.* The moon revolves with a uniform motion, from west to east, about an axis nearly perpendicular to the plane of the ecliptic, in the *same time* that she makes a revolution in her orbit.

Let E, *Fig.* 36, be the centre of the earth, aa' a part of the moon's orbit, a and a' two successive positions of the moon's centre, and a'D a line parallel to aE. Then, since nearly the same surface of the moon is always turned towards the earth (213), that point in the surface which is at e when the moon's centre is at a, will be at e' or nearly so, when the centre is at a'. Assuming the point

* Astronomische Nachrichten, No. 263. This is an excellent astronomical periodical, originally edited by the late Professor Schumacher, at Altona, and at present conducted by Professor Hansen.

to be exactly at e', it must, during the interval, have moved about an axis perpendicular to the plane of the orbit, through the angle Ea'D, which is equal to aEa' the angular motion in the orbit. Hence, the angular motions about the axis and in the orbit being equal, the moon must revolve on her axis in the same time that she makes a revolution in her orbit.

The small changes observed in the position of the spots on the disc, are caused by the inequalities of her motion in her orbit, and an inclination of her axis; and not by any inequality in her rotation. For, assuming the rotation to be uniform, if the moon's motion from a to a' is greater than the mean motion, the angle aEa' must be greater than that through which a spot at e is carried in the same time by the rotation on the axis. Consequently, when the moon's centre is at a', the spot must be to the east of Ea'. If, on the contrary, the motion from a to a' is less than the mean motion, the spot must be to the west of Ea'. An inclination of the axis will, evidently, cause the spots to have an alternate north and south motion.

A minute investigation of the subject, founded on successive accurate observations of the positions of the spots, proves that the rotation on her axis is uniform.

216. *Inclination of moon's axis.* From the investigation mentioned in the preceding article, it is found that the axis is nearly perpendicular to the plane of the ecliptic. The plane of the moon's equator, that is, the plane through her centre, perpendicular to the axis, makes with the plane of the ecliptic, an angle of $1\frac{1}{2}°$; and it intersects the latter in a line parallel to the line of the nodes.

217. *Moon's librations.* The alternate east and west motion of the moon's spots, produced by the inequalities of her motion in her orbit (215), is called the moon's *libration in longitude;* and the alternate north and south motion, depending on the inclination of the axis, is called the *libration in latitude.*

The diurnal motion of the observer, by which his position with reference to the radius vector Ea is changed, as from c to d, produces a slight apparent motion in the spots. This is called the *diurnal* or *parallactic libration.*

In consequence of these librations of the moon, small portions of the surface to the east and west, and also to the north and south, alternately come into view and disappear.

218. *Moon's passage over the meridian.* The moon's mean daily motion in right ascension, which is the same as in longitude, is greater than that of the sun by more than 12°. Hence, if on any given day, we suppose the moon to be on the meridian at the same instant with the sun, she will, at the end of 24 hours, when the sun has again returned to the meridian, be more than 12° to the east, and will not, therefore, arrive at the meridian till nearly an hour later. On the next day she would arrive at the meridian nearly two hours later than the sun. Thus, her passage of the meridian is retarded from day to day. The mean retardation is about 52 minutes.

In consequence of the inequalities in the moon's motion in right ascension, depending in part on the inequalities of her motion in her orbit, but more on the inclination of the orbit to the equator, the daily retardation in her passage over the meridian is subject to considerable variation. It varies from about 38 to 66 minutes.

219. *To find the time of the moon's passage over the meridian on a given day.*

Let A and A′ be the right ascension of the moon and sun respectively, at noon of the given day, expressed in time, and reduced to seconds, m and m' their hourly variations in right ascension, also, in seconds of time, and let t be the required time of her passage over the meridian, in hours. Then, at the time t, we have the moon's right ascension $= A + tm$, and the sun's $= A' + tm'$. Hence, as the moon is on the meridian at the time t, if the latter right ascension be subtracted from the former, the remainder will be the time t in seconds; or, being divided by 3600, it will be the time t in hours. Consequently,

$$t = \frac{A + tm - (A' + tm')}{3600},$$

or,

$$3600t = A - A' + (m - m')\, t.$$

Hence,

$$t = \frac{A - A'}{3600 - (m - m')}.$$

The time of the moon's passage over the meridian of Greenwich, is given in the Nautical Almanac for every day in the year. It may easily be found for any other meridian by proportion.

Cor. If A and m be taken to represent the right ascension and hourly variation in the right ascension of a planet, the above formula will give the time of its passage over the meridian. For a fixed star, $m = o$, and the formula becomes,

$$t = \frac{A - A'}{3600 + m'}.$$

220. *To find the time of the moon's rising or setting.*

To find the true semi-diurnal arc of a body (183), the declination or polar distance when the body is in the horizon is required. For the sun, this is usually assumed to be the same as on the meridian, as it changes but little during the interval. But for the moon, the change is too great to be neglected. We may, however, by using the moon's meridian declination, or the declination* at an estimated time of rising or setting, obtain an approximate value of the semi-diurnal arc. This, subtracted from the time of passage over the meridian, or added to it, gives an approximate time of the rising or setting.

Then, to find the true time, let the declination be found for the approximate time, and again compute the semi-diurnal arc. This requires to be still further corrected, on account of the moon's increase in right ascension during the interval between her being on the meridian and in the horizon. To obtain this second correction, let D = the difference between the times of two consecutive passages over the meridian, one of which precedes, and the other follows, the required time of rising or setting. Then, as 24h. : semi-diurnal arc last computed : : D : the correction; which being added to this semi-diurnal arc, gives its true value very nearly. This applied to the time of the passage over the meridian, gives the time of rising or setting.

221. *Daily retardation of the moon's rising or setting.* The average daily retardation of the moon's rising or setting is the

---

* The moon's right ascension and declination are given in the Nautical Almanac for every hour.

same as that of her passage over the meridian. But the actual retardation, being affected by the moon's change in declination, as well as by the inequalities of her motion in right ascension, is subject to greater variation. In the latitude of Philadelphia, the least daily retardation is only about 20 minutes, and the greatest is about 1h. 20m.

The less or greater retardation of the moon's *rising*, attracts most attention when it occurs at the time of full moon, as it affects the succeeding period of moonlight evenings. When the retardation has its least or nearly its least value, the moon is up or rises early in the evening for five or six days following the full moon: whereas, when the retardation is greatest the moon ceases, in the course of two or three days, to be seen in the early part of the evening.

Supposing the moon's orbit to coincide with the ecliptic, from which it does not greatly deviate, the daily retardation in the moon's rising at full moon, has nearly its least value at the full moon, which occurs near the time of the autumnal equinox. As this is about the period of the English harvest, this rising of the moon is called the *Harvest Moon*.

CHAPTER XIII.

ECLIPSES OF THE SUN AND MOON.—OCCULTATIONS.

222. *Eclipses occur only at New and Full Moon.* As an eclipse of the sun is caused by the moon passing between the sun and earth (214), it can only occur when the moon is in conjunction with the sun, that is, at the time of new moon. An eclipse of the moon is caused by the interposition of the earth, between the sun and moon, which prevents, in whole or in part, the illumination of the latter by the former. It must therefore occur when the moon is in opposition, that is, at the time of full moon.

If the moon's orbit coincided with the plane of the ecliptic, there would necessarily be an eclipse of the sun at every new moon, since the moon would in that case pass directly between the sun and earth; and an eclipse of the moon at every full moon, as the earth would then be directly between the sun and moon. But as the orbit is inclined to the ecliptic, an eclipse can only occur when the moon, at the time of new and full moon, is at, or near one of its nodes. In other cases the moon is too far north or south of the ecliptic, to cause an eclipse of the sun or to be itself eclipsed.

ECLIPSES OF THE MOON.

223. *Earth's shadow and penumbra.* The magnitude of the sun being far greater than that of the earth, and both being globular bodies, the shadow of the earth must evidently be of a conical form. Let AB and *hg*, *Fig.* 37, be sections of the sun and earth by a plane passing through their centres S and E; and let AC and BC, and also, AH and BK, be tangents common to the two sections. Then will *g*C*h* be a section of the earth's conical shadow or *umbra*, as it is frequently called, and EC will be the *axis* of the shadow. If the plane CE*h*K, be supposed to revolve round the axis EC, the tangent *h*K will describe the convex surface of the frustum of a cone, within the whole of which, the light of the sun must be more or less obstructed by the earth. That part of the frustum, which is included between the umbra and convex surface, that is, the part of which H*g*C*h*K is a section, is called the earth's *penumbra.*

224. *Beginning or end of an eclipse of the moon.* An eclipse of the moon is regarded as beginning or ending at the instant her edge touches the earth's shadow. Thus, if *mn* be a part of the moon's orbit, the eclipse begins when the moon is at *a*, and ends when she is at *e*. Prior, however, to the beginning of an eclipse, while the moon is passing from the edge of the penumbra to the edge of the shadow, she must evidently suffer a gradual but increasing diminution of her light. This circumstance renders it difficult, if not impracticable, to observe with accuracy the instant at which the eclipse begins. On account of the gradual increase of the moon's light in passing from the shadow, the same difficulty occurs at the end.

Sometimes the moon, at full moon, though too far north or south of the ecliptic to come in contact with the shadow, may still be sufficiently near to pass through the penumbra. In this case the moon suffers a diminution of light without being eclipsed.

225. *Length of the earth's shadow.* The length of the earth's shadow exceeds three times the distance of the moon from the earth.

Let us assume the moon to be at one of her nodes at the instant she is in opposition at c. Then will the centres of the sun, earth, and moon, be in the same straight line SC.

Put π = moon's horizontal parallax,

π' = sun's " "

δ = moon's apparent semi-diameter,

δ' = sun's " "

Then the angle SEB $= \delta'$ and E$Bg = \pi'$. Hence, ECg = SEB − EBg $= \delta' - \pi'$. But, ECg is the parallax of the point C. Hence, (93 G),

$$ECg : \pi : : Ec : EC,$$

or, $$\delta' - \pi' : \pi : : Ec : EC = Ec \frac{\pi}{\delta' - \pi'}.$$

Hence, since (95, 96 and 100), π always exceeds $53'$ and $\delta' - \pi'$ is always less than $17'$, it follows that EC must be more than three times Ec, the moon's distance from the earth.

226. *Semi-diameter of the earth's shadow.* The apparent semi-diameter of the earth's shadow, at the distance of the moon, is called the semi-diameter of the shadow. Thus, the angle bEc is the semi-diameter of the shadow. The point c is called the *centre of the shadow.*

The semi-diameter of the earth's shadow is equal to the sum of the moon's and sun's horizontal parallaxes, *less* the sun's apparent semi-diameter.

Put S = bEc = semi-diameter of the earth's shadow. Then, since bEc = Ebg − ECg and (225), ECg $= \delta - \pi'$, we have S $= \pi - (\delta' - \pi')$, or S $= \pi + \pi' - \delta'$.

Taking $\pi = 57'\ 1''$, $\pi' = 9''$, and $\delta' = 16'\ 1''$, which are their mean values, very nearly, we have the mean value of S $= 41'\ 9''$.

Cor. Since S $= 41'\ 9''$, we have 2S $= 82'\ 18''$. The diameter of the shadow is, therefore, more than twice the moon's apparent diameter, and consequently, the moon may be entirely enveloped in the shadow.

Scholium. In obtaining the above expression for the semi-diameter of the shadow, the shadow is assumed to be limited by those rays of the sun which are tangents to the sun and earth. It is, however, found that the observed duration of an eclipse always exceeds the duration computed on this supposition. This is accounted for, by assuming that most of those rays which pass near the surface of the earth, are absorbed by the lower strata of the atmosphere. The extent of the obstruction to the passage of the light being thus increased, the diameter of the shadow, and consequently the duration of the eclipse, must also be increased.

In consequence of the difficulty in ascertaining the exact time of beginning and end of the eclipse (224), astronomers have differed as to the amount of the correction that should be made. According to the observations and computations of Dr. Mædler, a German astronomer, who has recently given particular attention to the subject, the computed semi-diameter should be increased by about a 50th part.* Hence,

$$S = \pi + \pi' - \delta' + \tfrac{1}{50}(\pi + \pi' - \delta').$$

227. *Moon visible when entirely enveloped in earth's shadow.* Another effect of the action of the earth's atmosphere is perceptible in eclipses of the moon. Those rays from the sun which enter the atmosphere and are so far from the surface as not to be absorbed, have their directions changed, and leave the atmosphere with a greater inclination to the axis of the shadow. In this way a sufficient quantity of light is generally thrown on the moon to render her visible, even when in the middle of the shadow. She appears, while in the shadow, with a dull red or copper-coloured light.

228. *Moon's angular distance, from the centre of the earth's shadow, at the beginning or end of a lunar eclipse.* This distance is equal to the sum of the semi-diameters of the earth's shadow and moon.

For, as the eclipse begins when the moon's centre is at a, the angle $a\mathrm{E}c$ expresses her distance from the centre of the shadow at that time. But $a\mathrm{E}c$ is equal to the sum of $b\mathrm{E}c$ and $a\mathrm{E}b$; or,

$$\textit{angular distance}, \ a\mathrm{E}c = S + \delta.$$

---

* Astr. Nach. Nos. 256, 286 and 338.

When the moon is first entirely in the shadow, or when she begins to emerge from it, her angular distance from the centre of the shadow will evidently be, $S - \delta$.

Cor. When at, or near, the time of full moon, the moon's angular distance from the centre of the shadow does not become less than $S + \delta$, there evidently cannot be an eclipse; and when it does become less, there must be an eclipse.

229. *Lunar ecliptic limits.* Referring the points and orbit to the celestial sphere, let c' *Fig.* 34, be the place of the centre of the earth's shadow in the ecliptic, and M′ the place of the moon's centre in her orbit NF, when the angular distance c'M′ is perpendicular to the orbit and is equal to $S + \delta$. Then it is evident, that, according as the distance of the centre of the shadow from the node N, or of the sun from the opposite node, is greater or less than Nc', the least distance of the centres of the moon and shadow must be greater or less than $S + \delta$. Hence, it follows (228 *Cor.*), that there can never be an eclipse of the moon when the distance of the sun from the nearest node is greater than the greatest value of Nc', and that there must always be one when this distance is less than the least value of Nc'. The greatest and least values of Nc' are, therefore, called the *lunar ecliptic limits.* Similar quantities for eclipses of the sun are called *solar ecliptic limits.*

Now, it is known, both from observations and from investigations in physical astronomy, that at the time of the syzygies the inclination of the moon's orbit has always nearly its greatest value of 5° 17′. Taking, therefore, this value of c'NM′ and the greatest and least value of $S + \delta$, which, including the correction of S, are about 63′ 17″ and 53′ 8″, the right angled spherical triangle c'M′N gives for the greatest value of Nc' or the greater limit 11° 32′, and for the less limit 9° 40′.

Taking into view the inequalities in the motions of the sun, moon and nodes, other limits corresponding to the mean motions, have been obtained. These are very convenient in determining when eclipses of the moon may or must occur. According to Delambre, if at the time of mean full moon, the mean longitude of the sun differs more than 12° 36′ from that of the nearest node, there cannot be an eclipse; but if it differs less than 9°, there must be an eclipse. These are, therefore, the lunar ecliptic limits for mean motions.

Now, by means of tables of the mean places and motions of the sun, moon and nodes, it is easy to find the time of mean full moon for any given month, and the mean longitude of the sun and node at that time. If the difference of these longitudes is greater than the greater ecliptic limit for mean motions, there cannot be an eclipse at that full moon; if it is less than the less limit, there must be one. When the difference falls between the limits, further computation is necessary to determine whether there will or will not be an eclipse. In this research it will not be necessary to make computations for all the full moons in the year, for it will at once be seen by the tables at what periods in the year the sun is near to either of the nodes, and it is only at these periods that eclipses can occur.

230. *Different kinds of lunar eclipses.* When the moon just touches the earth's shadow or passes through the penumbra without entering the shadow, the circumstance is called an *appulse.* When a part, but not the whole, of the moon enters the shadow, the eclipse is called a *partial* eclipse; when the moon enters entirely into the shadow, it is called a *total* eclipse; and when the moon's centre passes through the centre of the shadow, it is called a *central* eclipse. A central eclipse of the moon seldom, however, if ever, occurs.

It follows from a preceding article (227), that the moon does not in general entirely disappear even in total eclipses.

231. *Visibility of a lunar eclipse.* As in an eclipse of the moon there is a real loss of light at the moon, the eclipse must be visible, and present the same appearance at all places that have the moon above the horizon, during its continuance.

ECLIPSES OF THE SUN.

232. *Length of the moon's shadow.* The length of the moon's shadow is about equal to the distance of the moon from the earth; being, alternately, a little greater and a little less.

Suppose the moon, at new moon, to be at one of her nodes.

Her centre will then be in the plane of the ecliptic, and in the straight line passing through the centres of the sun and earth. Let AB, hg and DG, *Fig.* 38, be sections of the sun, moon and earth, by a plane passing through their centres S, M, and E. Also, let AC and BC, and AH and BK be tangents common to the sections of the earth and moon, and, therefore, limiting the sections of the shadow and penumbra.

Put d = ang. hCM = moon's app. semi-diam. as seen from C,
d' = " hSM = " " " " S,
d'' = " AhS = sun's " " " M,
R = Ec = earth's radius; and let π, π', &c. be as in Art. 225.

Then, since the parallaxes of bodies, and the apparent semi-diameters of the same body, seen at different distances, are inversely as the distances (93 *Cor.* and 97), we have,

$$\pi : \pi' : : \text{SE} : \text{ME},$$

or, $$\pi - \pi' : \pi : : \text{SM} : \text{SE} : : \delta' : d'' = \frac{\pi . \delta'}{\pi - \pi'} \text{.................... (A)}$$

and, $$\pi - \pi' : \pi' : : \text{SM} : \text{ME} : : \delta : d' - \frac{\pi' . \delta}{\pi - \pi'} \text{.................. (B)}$$

Also, $$d : \delta : : \text{ME} : \text{MC} = \text{ME}\, \frac{\delta}{d} \text{.................................... (C)}$$

Now, since hCM = AhS − hSM, or $d = d'' - d'$, we have,

$$d = \frac{\pi . \delta' - \pi' . \delta}{\pi - \pi'} = \delta' + \pi' . \frac{\delta' - \delta}{\pi - \pi'} \text{................ (D)}$$

By taking the greatest value of δ' and least of δ, and the corresponding values of π and π' (100, 95 and 96), the greatest numerical value of $\pi' . \frac{\delta' - \delta}{\pi - \pi'}$ will be obtained, and will be found to be less than three tenths of a second. Consequently d is always very nearly equal to δ'. We have therefore (C),

$$\text{MC} = \text{ME} . \frac{\delta}{\delta'}, \text{ very nearly.}$$

Hence the length of the shadow is greater than the moon's distance from the earth, equal to it, or less, according as δ is greater than δ', equal to it or less.

Cor. Since (C), MC $= \text{ME} . \frac{\delta}{d}$, we have,

$$EC = MC - ME = ME.\frac{\delta - d}{d}.$$

But, $$\delta - d = \delta - \frac{\pi.\ \delta - \pi'.\ \delta}{\pi - \pi'} = \pi.\ \frac{\delta - \delta'}{\pi - \pi'},$$

and (93.E), $ME = R.\frac{\omega}{\pi}$.

Hence, $$EC = R.\ \frac{\omega}{d}.\ \frac{\delta - \delta'}{\pi - \pi'} \qquad \text{(E)}$$

Taking the mean values of δ, δ', π, and π', and regarding d as equal to δ', we find EC $= -0.86$R. Hence, when the sun and moon are at their mean distances from the earth, the moon's shadow extends a little farther than the nearest part of the earth's surface. By taking the proper values of the quantities, it will be found that the shadow, when longest, extends beyond the earth's centre about three and a half times the earth's radius, and when shortest, does not reach the centre, by about six times the earth's radius.

233. *Breadth of the moon's shadow at the earth.* The greatest breadth of the moon's shadow at the earth, when it falls perpendicularly on the surface, is about 166 miles.

In the triangle ECa, we have,

$$Ea : EC : : \sin.\ d : \sin.\ EaC,$$

or, (232 E), $R : R.\frac{\omega}{d}.\ \frac{\delta - \delta'}{\pi - \pi'} : : \frac{d}{\omega} : \sin.\ EaC = \frac{\delta - \delta'}{\pi - \pi'}.$

Now the breadth of the shadow will evidently be greatest when the moon's distance from the earth is least, and the sun's distance, is greatest. Hence, taking the greatest value of δ and least of δ', and the corresponding values of π and π' we find,

$$\sin.\ EaC = \frac{60.45}{3680.5}$$

This gives the angle EaC $= 56'\ 28''$. Adding d or $\delta' = 15'\ 45''$, to EaC, we have aEc or the arc $ac = 1°\ 12'\ 13''$. Hence, $ab = 2°\ 24'\ 26'' = 2°.407$; and as each degree is $69\frac{1}{6}$ miles (71), the breadth of the shadow is 166 miles.

When the moon is at some distance from the node, the shadow falls obliquely on the earth, and its greatest breadth will evidently be increased.

234. *Breadth of the moon's penumbra at the earth.* The greatest breadth of the moon's penumbra at the earth's surface, when it falls perpendicularly on the surface, exceeds 4800 miles.

Put, δ'' = ang. hLM = moon's app. semi-diam. as seen from L.

Then, as hLM = ShB + hSM = $d'' + d'$, we have (232. A and B),

$$\delta'' = \frac{\pi.\delta' + \pi'.\delta}{\pi - \pi'} = \delta' + \pi'.\frac{\delta + \delta'}{\pi - \pi'} \dots\dots\dots\dots\dots\dots \text{(F)}$$

and $$\delta + \delta'' = \pi.\frac{\delta + \delta'}{\pi - \pi'}.$$

Now, $\delta'' : \delta : : \text{ME} : \text{ML} = \text{ME}.\frac{\delta}{\delta''}.$

Hence, $\text{EL} = \text{ME} + \text{ML} = \text{ME}.\frac{\delta + \delta''}{\delta''}.$

Or, by substituting the value of ME (93.E), and of $(\delta + \delta'')$,

$$\text{EL} = \text{R}.\frac{\omega}{\delta''}.\ \frac{\delta + \delta'}{\pi - \pi'}.$$

From the triangle LEK, we have,

$$\text{EK} : \text{EL} : : \sin\delta'' : \sin \text{EK}d,$$

or, $$\text{R} : \text{R}.\ \frac{\omega}{\delta''}.\ \frac{\delta + \delta'}{\pi - \pi'} : : \frac{\delta''}{\omega} : \sin \text{EK}d = \frac{\delta + \delta'}{\pi - \pi'}.$$

The breadth of the penumbra will evidently be greatest when the moon's distance from the earth is greatest and the sun's distance, least. The angle EKd being found with the values of δ, &c., corresponding to these distances, and δ'' being also found (F), the angle KEL or the arc Kc = EKd − δ'', becomes known. Twice Kc gives KH, which is thus found to be 69°.87, or 4833 miles.

235. *Different kinds of eclipses of the sun.* A *partial* eclipse is one in which a part, but not the whole of the sun, is obscured. A *total* eclipse is one in which the sun is entirely obscured. It must occur at all those places on which the moon's shadow falls. A *central* eclipse is one in which the axis of the moon's shadow or the axis produced, passes through the place at which the eclipse occurs.

An *annular* eclipse is one in which a part of the sun's disc is seen as a ring, surrounding the moon. To explain this, let AB, hg and DG, *Fig.* 39, be sections of the sun, moon and earth, when the moon's shadow does not extend to the earth. In this case, the tangents AC and BC, which limit the shadow, being produced, cross each other at C and meet the section of the earth at *a* and *b*. From *c*, or any other point between *a* and *b*, let tangents to the moon be drawn, as *ce* and *cf*. Then it is obvious, that the part of the sun's disc that is without the circle *ef*, described about the diameter *ef*, will be visible to an observer at *c*.

The greatest breadth of the part of the surface in which the eclipse is annular may be found in a similar manner to that of the shadow (233). It is about 200 miles.

236. *Visibility of an eclipse of the sun.* As the moon moves in her orbit from *m* to *n*, *Fig.* 38, her penumbra and its axis move over the earth's surface from west to east, passing in succession over different parts. At all places along the line in which the axis meets the surface, there must be a central eclipse. At all places contiguous to this line, on each side, there must be a total or an annular eclipse. And at places more remote from the central line, but within the limits of the penumbra, there will be a partial eclipse.

As the greatest breadth of the penumbra is less than half the semi-circumference of the earth, it is evident there must be a large part of the earth's enlightened hemisphere in which the eclipse is not visible, even when the extent of the penumbra is greatest.

When the moon is so far from the node at the time of new moon that the axis of the penumbra does not meet the earth, the eclipse cannot be central at any place; and the partial eclipse is only visible in a portion of the northern or southern hemisphere, according as the moon's latitude is north or south.

It follows from the above and a preceding article (231), that the visibility of an eclipse of the sun is of much less extent than that of an eclipse of the moon.

237. *General eclipse of the sun.* An eclipse of the sun, considered with reference to the whole earth and not to any particular place, is called the *general eclipse*.

The general eclipse commences at the first contact of the moon's penumbra with the earth, and ends at the last contact. Thus, *Fig.* 40, the general eclipse begins when the moon is at *u*, and ends when she is at *v*.

238. *Apparent distance of the centres of the sun and moon at the beginning or end of the general eclipse.* The angular distance of the centres of the sun and moon at the beginning or end of a general eclipse is equal to the sum obtained by adding the sum of the apparent semi-diameters of the sun and moon to the difference of their horizontal parallaxes.

For, *Fig.* 40, the apparent distance SEu = AEg + AES + gEu. But AEg = EgD − EAD = $\pi - \pi'$, AES = δ', and gEu = δ. Hence,

$$\textit{apparent distance}, \text{SE}u = \pi - \pi' + \delta' + \delta.$$

The apparent distance of the centres of the sun and moon, when the eclipse begins to be central for the earth in general, is equal to the difference of the horizontal parallaxes of the moon and sun.

For, the central eclipse must begin when the moon's centre is at x, in the line SD drawn from the sun's centre and tangent to the earth. Hence,

$$\textit{apparent distance}, \text{SE}x = \text{E}x\text{D} - \text{ESD} = \pi - \pi'.$$

239. *Solar ecliptic limits.* Taking the apparent distance of the moon from the sun at the beginning or end of a solar eclipse for the earth in general, found in the last article, the solar ecliptic limits may be found in the same manner as the lunar (229). They are 17° 21′ and 15° 25′.

According to Delambre, the solar ecliptic limits for mean motions are 19° 2′ and 13° 14′.

240. *Visible eclipses of sun and moon.* As the solar ecliptic limits exceed the lunar, eclipses of the sun occur more frequently than those of the moon. But, as the portion of the earth in which an eclipse of the sun is visible is much less than that in which an eclipse of the moon is visible, there are, for any given place, more *visible* eclipses of the moon than of the sun.

241. *Number of eclipses in a year.* There may be seven eclipses in a year, and cannot be less than two. When there are seven, five of them are of the sun and two of the moon; when there are but two, they are both of the sun.

Illustration. During a synodic revolution of the moon, the sun's mean motion in longitude is 29° 6′, and in this time the moon's nodes move backwards 1° 31′. Hence the moon's motion with reference to either of the nodes in one lunation is 30° 37′, and in half a lunation, 15° 18′. Calling one of the nodes A and the other B, let us suppose that the first full moon in the year occurs in the middle of January or a few days later, and that the sun's longitude is at that time about 2° less than that of the node A. The moon will then be near the node B, and must be largely eclipsed; and since the sun's motion in reference to the node during half a lunation is only 15° 18′, it is evident that at the previous and following new moons, the sun may be within the largest ecliptic limits from the node A, and may therefore be eclipsed at each of these new moons. At the full moon, which occurs in a little less than six months after the former, the sun having during six lunations moved nearly 184° with reference to the nodes, must be about 2° in advance of the node B. Consequently there must be a large eclipse of the moon, and, may be, two eclipses of the sun at this period. At the new moon which occurs five and a half lunations after this latter full moon, and, therefore, a little before the close of the year, the sun during the time having advanced more than 168° towards the node A, will be within about 10° of it, and must, therefore, be eclipsed. Thus, in the case assumed, there must be two eclipses of the moon in the year, and, may be, five of the sun.

If the sun passes one of the nodes two or three months after the commencement of the year, it cannot be at, or near, the same node again during the year. In this case, if it be new moon when the sun is near the node, but has one or two degrees less longitude than the node, there will be an eclipse of the sun at that new moon; but at the preceding and following full moons, the sun's distance from the node will be too great for an eclipse of the moon to occur. At the new moon, six lunations afterwards, the sun will be near the other node, and there will be an eclipse of the sun, but none of the moon, at either of the contiguous full moons. Hence, in this year there will be two eclipses of the sun and none of the moon.

The arc formed by taking the least true solar ecliptic limit on each side of the node is greater than the sun's motion in reference to the node during one lunation.* Hence, if there is not an eclipse

---

* It is a little greater than the sun's mean motion during this time, and considerably greater than the true motion. For when the ecliptic limit is least, the sun's apparent semi-diameter is least, and, consequently, his distance is greatest. His true motion is, therefore, then, less than his mean motion.

of the sun at the new moon prior to the sun's passage of either node, there must be one at the subsequent new moon. As, therefore, there must be an eclipse of the sun near the time of his passage of each node, there must be at least two in the year.

242. *Period of eclipses.* At the expiration of a period of 223 lunations, or a few days more than 18 years, eclipses both of the sun and moon return again in nearly the same order as during that period.

Illustration. A mean lunation or synodic revolution of the moon is 29.5306 days, and consequently 223 lunations in 6585.32 days. Now, the mean period in which the sun moves from one of the moon's nodes, to the same, is 346.62 days, very nearly, and consequently 19 of these periods is 6585.78 days. Hence at whatever distance the sun is from either of the nodes at any given new or full moon, he must at the end of 223 lunations be very nearly at the same distance from the same node. It, therefore, follows that after a period of 6585.32 days,* eclipses must occur again in the same order, or nearly so, as during that period.

This period was known to the Chaldean astronomers. It was by them called the *Saros*, and was used in predicting eclipses.

243. *Total and annular eclipses of the sun.* Although a large proportion of the eclipses of the sun are total, or annular, somewhere, on the earth, yet, for any given place, a total or annular eclipse is a phenomenon of rare occurrence.

That this must be the case may be inferred from the consideration that the tract across the earth's enlightened surface, in which an eclipse can be total or annular, is when widest, of but little breadth (233 and 235), and that a different latitude of the moon, at the time of the eclipse, must give a different tract.

The longest possible time that an eclipse of the sun can continue total, at any place, does not exceed 8 minutes; and the longest time that an eclipse can continue annular, does not exceed 12 minutes. In general, the times are much less than these.

---

* More accurately 6585.32128 days; which is 18 years, 11 d. 7 h. 42 m. 39 sec. when there are four Bissextile years in the period, or 18 y. 10 d. 7 h. 42 m. 39 sec. when there are five.

A total eclipse of the sun, especially when it occurs in a clear day and has several minutes duration, is a very interesting and impressive phenomenon. It is accompanied by a considerable reduction of the temperature of the air; and the obscurity is such that the principal stars and more conspicuous planets, above the horizon, become distinctly visible.

COMPUTATION OF LUNAR ECLIPSES.

244. *General Remarks.* The apparent distance of the centre of the earth's shadow and moon, and the arcs of the ecliptic and moon's orbit passed through by these during an eclipse, being necessarily small, they may without material error be regarded as straight lines. We may, also, regard the motion of the centre of the shadow in longitude, and the motions of the moon in longitude and latitude, as being uniform during the continuance of the eclipse.

Some of the quantities used in several of the subsequent articles will be designated as follows:

T = the time of the moon's opposition or of full moon, expressed in mean time,

t = any small interval of time, not exceeding two or three hours,

q = moon's latitude at the time of full moon,

p' = moon's hourly motion in longitude *less* sun's do.,

q' = moon's hourly motion in latitude,

$h = S + \delta$ = moon's distance from the centre of the earth's shadow at the beginning or end of an eclipse,

$h' = S - \delta$ = the distance at the time the eclipse begins or ceases to be total.

245. *Time of full moon, &c.* By means of a small set of tables, an approximate time of any full moon, that will not differ more than a few minutes from the true time, may very easily be found.* For this time, by means of solar and lunar tables, let the sun's longitude, hourly motion, apparent semi-diameter and horizontal parallax, and the moon's longitude, latitude, hourly motions in

---

* A set of tables of this kind, for finding the approximate time of new or full moon, adapted to the meridian at Greenwich, is included in the tables at the end of this work.

longitude and latitude, horizontal parallax and apparent semi-diameter, be computed. All these, except the sun's longitude, and the moon's longitude and latitude, may, without material error, be regarded as constant during the eclipse.

Let the sun's longitude at the approximate time of full moon be subtracted from the moon's, the latter being increased by 360°, if necessary. Then, if the remainder, which we will call R, is exactly 180°, the approximate time of full moon must be the true time. But if R differs from 180°, as will generally be the case, the difference, which may be called D, will be the moon's distance from opposition. Now, p', the difference of the hourly motions of the moon and sun, in longitude, expresses the moon's motion in longitude relative to the sun. Hence, if t be the interval between the approximate and true time of opposition or full moon, we have, $p' : D :: 1hr : t$. The interval t, found from this proportion, being added to the approximate time of full moon when R is less than 180°, but subtracted from it when R is greater than 180°, will give T, the true time of full moon.

The moon's motion in latitude, during the interval t, may be obtained from her hourly motion in latitude. This, added to the latitude at the approximate time of full moon, when the true time is later than the approximate time and the latitude is increasing, or when the true time is earlier than the approximate time and the latitude is decreasing, but subtracted in other cases, will give q, the latitude of the true time of full moon.

The value of S is found by a preceding article (226 *Schol.*). The values of h and h' then become known.

246. *Moon's relative orbit.* Let AB, *Fig.* 41, be an arc of the ecliptic, and FG an arc of the moon's orbit; and CM being drawn perpendicular to AB, let C and M be the places of the centres of the earth's shadow and moon respectively, at the time T, the true time of full moon. Also, let CC′ be the distance moved through by the centre of the shadow, and MM′ the distance moved through by the moon, during an interval of time t. Then will C′ and M′ be the places of the centres of the shadow and moon, at the time $T + t$.

But as the circumstances of an eclipse of the moon depend on the relative positions of the centres of the shadow and moon, it is more convenient to regard the centre of the shadow as fixed at C, and to use the moon's relative motion in reference to this centre. Draw, therefore, M′m parallel and equal to C′C. Then, will mC be parallel and equal to M′C′. Hence, m is the moon's relative place at the time T + t, in reference to C, the fixed position of the centre.

As CC′ is the motion of the centre of the shadow during the interval t, it must, evidently, be equal to the apparent motion of the sun during the same time. Let M′D and md be drawn parallel to MC, and ME parallel to AB. Then Ee = M′m = CC′. Consequently, Ee is equal to the sun's motion during the interval t; also ME is the moon's motion in longitude, and em = EM′, is her motion in latitude during the same time. Hence, Me, the moon's relative motion in longitude during the interval t, is equal to the difference between her motion in longitude and that of the sun, and em, her relative motion in latitude, is equal to her real motion in latitude.

Let t' be a different interval of time, and let m' be the moon's relative place at the end of this interval. Then, since the motions are regarded as uniform (244), Me' : Me : : t' : t, and $e'm'$: em : : t' : t, or, Me' : Me : : $e'm'$: em. Hence, m' must be in the straight line PQ, drawn through M and m. As, therefore, the moon's relative place moves along the line PQ, this line is called the moon's *relative orbit.*

247. *Inclination of moon's relative orbit.* Put I = the angle eMm = the inclination of the moon's relative orbit. Then, expressing the interval t in the last article in hours and decimal parts, we have,

$$\text{M}e = tp', \text{ and } em = tq'.$$

Hence,
$$\text{tang I} = \frac{em}{\text{M}e} = \frac{tq'}{tp'},$$

or,
$$\text{tang I} = \frac{q'}{p'}.$$

248. *Moon's hourly motion in her relative orbit.* Let n = moon's hourly motion in her relative orbit. Then, we have,

$$\text{M}m = tn, \text{ and } \text{M}e = \text{M}m \cos \text{I} = tn \cos \text{I}.$$

But (247), $Me = tp'$. Hence, $tn \cos I = tp'$, or,

$$n = \frac{p'}{\cos I}.$$

249. *Time of the middle of the eclipse.* Let AB, *Fig.* 42, be the ecliptic, PQ the moon's relative orbit, and C and M the places of the centres of the earth's shadow and moon at the time of opposition or full moon. Also, let the circle $KcLa$, described about the centre C with a radius equal to S (226 *Schol.*), represent the section of the earth's shadow, at the moon. With the same centre and a radius equal to $(S + \delta)$ or h, let arcs be described, cutting the relative orbit in D and E; and let CN be perpendicular to the orbit PQ, cutting it in H. Then, will D and E be the moon's places at the beginning and end of the eclipse. Hence, as CH evidently bisects DE, and as the moon's motion is regarded as uniform during the eclipse, the point H must be the moon's place at the middle of the eclipse.

Let T' = the time of the middle of the eclipse, and t = the interval between T' and T, the time of full moon.

Then, $MH = tn$. But, since CM is perpendicular to AB, and CH is perpendicular to PQ, the angle MCH is equal to the inclination of the relative orbit. Hence, from the right angled triangle CHM, we have $MH = CM \sin MCH = q \sin I$. Consequently,

$$tn = q \sin I, \text{ or } t = \frac{q \sin I}{n}.$$

Hence, as $T' = T \mp t$, we have,

$$T' = T \mp \frac{q \sin I}{n}.$$

The *upper* sign must be used when the latitude is *increasing*, and the *lower* when it is *decreasing*.

250. *Beginning and end of the eclipse.* From the triangle CHM, we have, $CH = CM \cos MCH = q \cos I$. Put,

B = the time of the beginning of the eclipse,
E = the time of the end,
t = the interval between the middle and beginning or end.

Then we have, $HD = tn$; and from the right angled triangle CHD, we have $HD = \sqrt{CD^2 - CH^2} = \sqrt{(CD + CH).(CD - CH)} = \sqrt{(h + q \cos I).(h - q \cos I)}$. Hence,

$$tn = \sqrt{(h + q \cos I).(h - q \cos I)}$$

or,
$$t = \frac{\sqrt{(h + q \cos I).(h - q \cos I)}}{n}.$$

Therefore, $B = T' - t$, and $E = T' + t$, become known.

251. *Times at which the eclipse begins and ceases to be total.* With the centre C and a radius equal to $S - \delta$, or h', let arcs be described cutting the relative orbit in F and G. Then, will F and G be the moon's places at the beginning and end of the total eclipse. Put,

B' = the time the total eclipse begins,

E' = the time it ends,

t = the interval between either of these and the time of the middle.

Then, we evidently have,

$$t = \frac{\sqrt{(h' + q \cos I).(h' - q \cos I)}}{n},$$

$$B' = T' - t, \text{ and } E' = T' + t.$$

When h' is less than CH or $q \cos I$, the eclipse cannot be total.

252. *Quantity of the eclipse.* The quantity of an eclipse, either of the sun or moon, is usually expressed in twelfths of the diameter, which are called *Digits.* In a total eclipse of the moon, the quantity of the eclipse is denoted by the number of digits contained in the distance between the inner edge of the moon and the nearest opposite edge of the shadow. Thus, in the eclipse represented in the figure, the number of digits contained in LN expresses the quantity of the eclipse. Let Q = the quantity of the eclipse. Then,

$$LR : LN :: 12 : Q,$$

or,
$$2\delta : LN :: 12 : Q = \frac{6LN}{\delta}.$$

But $LN = HN + HL = CN - CH + HL = CN + HL - CH = S + \delta - q \cos I = h - q \cos I$. Hence,

$$Q = \frac{6.(h - q \cos I)}{\delta}.$$

COMPUTATION OF SOLAR ECLIPSES.

253. *Time of New Moon.* The approximate time of new moon being found, and also the sun's and moon's longitudes, &c., at that time (245), the difference between the longitudes will be the moon's distance from conjunction. Whence, by means of the hourly motions in longitude, the true time of new moon may be obtained.

254. *General Eclipse.* Taking $h = \pi - \pi' + \delta' + \delta$, (238), the times of the middle, beginning, and end of an eclipse of the sun, for the earth in general, may be found in the same manner as those of a lunar eclipse.

255. *Eclipse for a given place.* Although the calculation of an eclipse of the sun for the earth in general, is equally simple with that of a lunar eclipse, it is quite different when the computation is to be made for a given place. This is much more difficult and tedious. For, the circumstances of the eclipse at a given place depend on the *apparent* relative positions of the sun and moon, that is, on their relative positions as seen at the given place. It therefore becomes necessary to take notice of the effect of parallax in changing the apparent relative positions of the bodies. Referring to the appendix for a more full and minute investigation of the subject, we shall here only give a general view of a method by which the computation for a given place may be made.

As the sun's parallax is very small, and it is only the apparent relative places of the sun and moon that are required, we may, without material error, refer the whole effect of parallax to the moon; that is, we may regard the sun's true place as being his apparent place, and then, in computing the moon's apparent place, use $\pi - \pi'$, the difference of the moon's and sun's horizontal parallaxes, instead of π, the moon's parallax.

Let T be time of the whole hour next preceding the approximate time of new moon, and for the time T, let the quantities mentioned in Article 245, be computed; and by means of the hourly motions, let the sun's longitude and the moon's longitude and latitude be found for the time T + 1hr. Then, using $\pi - \pi'$ instead of π, let the moon's parallaxes in longitude and latitude be computed for the times, T and T + 1hr, and, thence, let her apparent longitudes and latitudes for these times be found. The difference between the moon's apparent longitude, at the times, T and T + 1hr., will be her apparent hourly motion in longitude, and the difference between this and the sun's hourly motion will be the moon's apparent hourly motion relative to the sun, which may be called p'. The difference between the moon's apparent latitude, at the times, T and T + 1hr., will be q', the moon's apparent hourly motion in latitude. The difference between the sun's longitude at the time T, and the moon's apparent longitude at the same time, will be the moon's distance from apparent conjunction, at that time. Whence, from the value of p', the time of apparent conjunction may be found. The time, thus obtained, will however only be an approximate time, for the moon's apparent motions are not uniform. Let, now, the moon's apparent longitude and latitude be computed for the approximate time of the apparent conjunction, and the moon's distance from apparent conjunction at this time be found. This distance will, now, be very small. Hence, we may, by means of p' and q', find the true time of apparent conjunction, and the moon's apparent latitude q, at that time, very nearly. Also, with p' and q', we may, by Art. 247, find I, the inclination of the moon's apparent relative orbit.

Let AB, *Fig.* 43, be the ecliptic, and PQ the moon's apparent relative orbit. Then, SM being drawn perpendicular to AB and equal to q, the moon's apparent latitude at the time of apparent conjunction, S and M will be the place of the sun's and moon's centres, at that time. With the centre S and a radius equal to δ', the sun's semi-diameter, let the circle ab be described to represent the sun's disc. Let SD and SE be each equal to $\delta' + \delta$, the sum of the semi-diameters, of the sun and moon. Then will D and E be the moon's place, at the beginning and end of the eclipse. Hence, taking $h = \delta' + \delta$, the times of middle, beginning, and end, and the quantity of the eclipse, may be found, in exactly the same manner, as for an eclipse of the moon, except that, in finding the quantity, δ' must be used in the denominator of the fraction instead of δ.

But, as the moon's apparent motions in longitude and latitude are not uniform, the times of beginning and end, thus found, are only approximate times, and may differ some minutes from the true times. Let, therefore, the moon's apparent longitude and latitude be computed for the approximate time of beginning. We then, easily, obtain the average values of the moon's apparent hourly motions in longitude and latitude, between this time and the time of the middle of the eclipse. The latter hourly motion will be the average value of q', and the difference between the former and the sun's hourly motion, will be the average value of p'. Also, let the moon's augmented semi-diameter (97) be found for the approximate time of beginning,* and take $h =$ the sum of δ' and this augmented semi-diameter. Then, using the values of p', q' and h, let the time of beginning be again computed. This time will be very nearly the true time. If still greater accuracy is desired, it may be obtained by another repetition of the computation. In a similar manner, the true time of the end of the eclipse may be found.

OCCULTATIONS.

256. *Definition.* When the moon passes between the earth and a star or planet, she must, during the passage, render the body invisible to some parts of the earth. This phenomenon is called an *occultation* of the star or planet.

257. *Extent of visibility of an occultation of a fixed star.* The breadth of the portion of the earth's surface, in which an occultation of a star is visible, is much less than that for an eclipse of the sun. It is about 2150 miles.

Let E, *Fig.* 44, be the earth's centre, and E*s* the direction of the star; and, supposing the moon to pass directly between the star and centre of the earth, let M be the place of her centre when in that position. As the star has no sensible parallax, lines, as A*a* and B*b*, drawn from it and tangents to the moon, will be parallel to *s*E. Hence, in the portion of the surface whose breadth *ab* is limited by these parallel lines, the occultation must be visible.

* A formula for this purpose is always given with the formulæ for computing the parallaxes.

Now, as ha is parallel to ME, the angle aME = Mah = δ, the moon's semi-diameter, nearly. We, also, have (93 E), ME $= \frac{\text{E}a}{\sin \pi}$. Hence, from the triangle MEa, we have,

$$\text{E}a : \frac{\text{E}a}{\sin \pi} :: \sin \delta : \sin \text{M}a\text{E} = \frac{\sin \delta}{\sin \pi}. \text{ But (99), } \frac{\sin \delta}{\sin \pi} = .2730.$$

Therefore, the angle MaE = 164° 9′. Whence, taking aME $= \delta = 16'$, we obtain MEa = ac = 15° 35′; and for the length of ab, 2150 miles, nearly.

As the moon moves in her orbit from u towards n, the occultation will be visible in succession to different portions of the earth, lying in a direction, nearly, from west to east; the common breadth of the whole not differing greatly from that obtained above, except when the moon passes considerably north or south of the star.

The parallaxes and apparent diameters of all the planets are small. The extent of visibility of an occultation of any one of them will not, therefore, differ much from that of an occultation of a fixed star.

258. *Distance of moon's centre from the star, at the beginning or end of an occultation for the earth in general.* This distance is equal to the sum of the moon's horizontal parallax and semi-diameter.

Let CD, a tangent to the earth, be parallel to sE. Then the occultation must commence for the earth, in general, when the moon's edge comes to this line. Hence, the distance sEu = sEg + gEu = EgD + gEu = $\pi + \delta$.

The greatest value of $\pi + \delta$ is 78′ 20″.5. Hence, when the moon's least distance from the star exceeds this quantity, there cannot be an occultation at any place on the earth.

From the greatest and least values of $\pi + \delta$, and by taking into view the inequalities in the moon's motion, it has been found, that, when at the time of the moon's mean conjunction with a star, the difference between the mean latitude of the moon and that of the star is 1° 37′, there cannot be an occultation; but when the difference is less than 51′, there must be one. Between these limits there is a doubt, which can only be removed by computing the true place of the moon.

259. *Stars that may suffer an occultation.* As the sum of the moon's greatest latitude (198), and the greatest distance of the moon from a star, when an occultation can take place, is about 6° 36′, it follows that no star whose latitude is greater than this, can suffer an occultation, and that all those whose latitudes are a little less may be occulted.

260. *Computation of an occultation for a given place.* The computation of an occultation for a given place, either of a star or planet, differs but little from that of a solar eclipse. The star or planet takes the place of the sun. In the case of a star, it is to be observed that the star has no sensible parallax, apparent semi-diameter, or hourly motion. In the case of a planet, the moon's apparent relative hourly motion in latitude depends on the hourly motions in latitude of both the moon and planet. In making the computation, the difference between the latitude of the moon and star or planet, at the time of apparent conjunction, is used instead of the moon's latitude. Consequently, the arc AB, *Fig.* 43, which, in an eclipse of the sun, represents an arc of the ecliptic, in the case of an occultation, represents an arc passing through the star or planet, and parallel to the ecliptic. As the distance on this arc between two circles of latitude is less than on the ecliptic, the apparent distance of the moon in longitude from the star or planet, and the moon's apparent relative motion in longitude require small reductions. These are made by multiplying each, by the cosine of the latitude of the star or planet.

Instead of Longitudes and Latitudes, Right Ascensions and Declinations may be used in the calculation both of eclipses and occultations.*

261. *Irradiation and Inflexion.* Some astronomers have thought that the apparent diameter of the sun as obtained from observation, and given in the tables, is too great. This has been inferred from a comparison of the observed time of the beginning or end of a solar eclipse at a known meridian, with the time obtained by computation, after making allowances for the errors of the tables

---

* For the investigation of formulæ for computing eclipses and occultations, see the Appendix.

in other respects. To account for it, they have supposed that the apparent diameter of the sun is amplified by the very lively impressions so luminous an object makes on the organ of sight. This amplification is called *irradiation.* They have also supposed that the moon has an atmosphere which, by its action on the rays of light passing through it, inflects them, and produces an effect such as would be produced by a small diminution in the moon's semi-diameter. This is called the *inflexion* of the moon. Du Séjour, an astronomer of note of the last century, concluded, that in calculating solar eclipses, the sun's semi-diameter, as given in the tables, should be diminished $3\frac{1}{2}''$ for irradiation, and the moon's $2''$, for inflexion.

The subject of irradiation and inflexion is, however, involved in considerable uncertainty, and several eminent astronomers have doubted the existence of either.

262. *Scholium.* Eclipses of the sun and occultations are not only interesting phenomena, but when carefully observed, they are, also, practically useful. When observed at places whose latitudes and longitudes are truly known, they furnish means for detecting errors in the tables used in computing the places, parallaxes, and apparent diameters of the bodies. For, the difference between the observed and computed times must depend on these errors. When observed at places whose longitudes are not well known, they furnish the means of determining them more accurately. Their application for this latter purpose will be noticed in a subsequent chapter.

CHAPTER XIV.

GENERAL REMARKS ON THE PLANETS.—DEFINITIONS.—ELEMENTS OF THE ORBITS OF THE PLANETS.—CONVERSION OF THE HELIOCENTRIC PLACE OF A PLANET INTO ITS GEOCENTRIC PLACE.—RETROGRADE MOTIONS OF THE PLANETS.—REAL DISTANCE, ETC. OF THE PLANETS.

263. *General remarks.* Each of the planets, like the moon, is found to be, during about half its period, on one side of the ecliptic, and, during the other half, on the other side. Hence, we may infer that their orbits are all divided by the plane of the ecliptic in nearly equal parts. But the apparent motions of the planets differ essentially in one respect from that of the moon. The apparent motion of the latter is *always direct*, or from west to east; but the apparent motion of each planet, during a *part of its period*, is *retrograde*, or from east to west. When the motion is changing from direct to retrograde, or the contrary, the planet remains *stationary*, or nearly so, for some days. This difference between the motions of the moon and planets, is a consequence of their different centres of motion. As the latter revolve round the sun (17), their apparent motions must depend both on their own motions and on that of the earth.

DEFINITIONS.

264. *Geocentric and Heliocentric Places.* The *geocentric* place of a body is its place as seen from the centre of the earth; and the *heliocentric* place, is its place as seen from the centre of the sun.

265. *Curtate distance.* If a straight line be conceived to be drawn from the centre of a planet, perpendicular to the plane of the ecliptic, the distance from the point in which it meets this plane to the centre of the sun, is called the *curtate distance* of the planet. The point itself is called the *reduced place* of the planet. Thus, if P′SN, *Fig.* 46, be the plane of the ecliptic, S the sun's centre, NP a part of the orbit of a planet, P the place of the planet at

any time, and PP′ perpendicular to P′SN, then SP′ is the curtate distance of the planet at that time, and P′ is its reduced place.

266. *Elongation, &c.* If a plane triangle be formed by joining the reduced place of a planet, the centre of the sun, and centre of the earth, the angle at the earth is called the *elongation* of the planet, the angle at the sun is called the *commutation*, and the angle at the reduced place of the planet is called the *annual parallax.* Thus, SEP′ is the elongation, ESP′ the computation, and EP′S the annual parallax.

267. *Elements of the orbit of a planet.* There are seven different quantities necessary to be known in order to compute the place of a planet for a given time. These are called the *elements of the orbit.* They are, the longitude of the ascending node; the inclination of the plane of the orbit to that of the ecliptic; the periodic time, or the planet's mean motion; the mean distance of the planet from the sun, or, which is the same, the semi-transverse axis of its orbit; the eccentricity of the orbit; the longitude of the perihelion; and the time the planet is at the perihelion, or its mean longitude at a given time or *epoch.*

ELEMENTS OF THE ORBIT.

268. *Longitude of the ascending node.* When a planet is at either of its nodes, it is in the plane of the ecliptic, and, consequently, its latitude is then nothing. Let the right ascension and declination of the planet be observed on several consecutive days, at the period it is passing from the south to the north side of the ecliptic, and let its corresponding longitudes and latitudes be computed (119). From these, the time at which the planet's latitude is nothing, and its longitude at that time, may be obtained by proportion or interpolation. This longitude of the planet will evidently be the geocentric longitude of the node. Also, by means of the solar tables, let the longitude of the sun and the radius vector of the earth be found for the time the planet is at the node. By similar observations and computations when the planet returns to the node, let the values of the same quantities be again obtained. From these data, if we assume the node to remain in the same position, its heliocentric longitude may be found.

Let S, *Fig.* 45, be the sun, PQ a part of the orbit of the planet, N the node, E the place of the earth when the planet was found to be at the node N, from the first set of observations, and E′ its place at the time of the planet's return to the node. Also, let EV, E′V, and SV, all parallel to one another, represent the direction of the vernal equinox. Then, assuming the mean radius vector of the earth to be a unit, put r = SE = earth's radius vector, S = VES = sun's longitude, and G = VEN = geocentric longitude of the node, when the earth is at E; and let r', S′ and G′ represent the same quantities when the earth is at E′. Also, put R = SN = radius vector of the planet when at the node N, and N = VSN = the heliocentric longitude of the node. Then, we have, SEN = VES − VEN = S − G, and SNE = VAN − VSN = VEN − VSN = G − N. From the triangle SEN, we have,

$$\sin \mathrm{SNE} : \sin \mathrm{SEN} :: r : \mathrm{R},$$

or, $$\sin (\mathrm{G} - \mathrm{N}) : \sin (\mathrm{S} - \mathrm{G}) :: r : \mathrm{R},$$

or, $$r.\sin (\mathrm{S} - \mathrm{G}) = \mathrm{R} \sin (\mathrm{G} - \mathrm{N}) \ldots\ldots\ldots\ldots (\mathrm{A})$$

In like manner we have,

$$r'.\sin (\mathrm{S}' - \mathrm{G}') = \mathrm{R}.\sin (\mathrm{G}' - \mathrm{N}) \ldots\ldots\ldots\ldots (\mathrm{B})$$

Therefore, dividing (A) by (B), we have,

$$\frac{r.\sin (\mathrm{S} - \mathrm{G})}{r'.\sin (\mathrm{S}' - \mathrm{G}')} = \frac{\sin (\mathrm{G} - \mathrm{N})}{\sin (\mathrm{G}' - \mathrm{N})} = \frac{\sin \mathrm{G} \cos \mathrm{N} - \cos \mathrm{G} \sin \mathrm{N}}{\sin \mathrm{G}' \cos \mathrm{N} - \cos \mathrm{G}' \sin \mathrm{N}} = \frac{\sin \mathrm{G} - \cos \mathrm{G} \operatorname{tang} \mathrm{N}}{\sin \mathrm{G}' - \cos \mathrm{G}' \operatorname{tang} \mathrm{N}}.$$

Hence, we easily find,

$$\operatorname{tang} \mathrm{N} = \frac{r.\sin (\mathrm{S} - \mathrm{G}) \sin \mathrm{G}' - r'.\sin (\mathrm{S}' - \mathrm{G}') \sin \mathrm{G}}{r.\sin (\mathrm{S} - \mathrm{G}) \cos \mathrm{G}' - r'.\sin (\mathrm{S}' - \mathrm{G}') \cos \mathrm{G}}.$$

We, also, have (A), $\mathrm{R} = \dfrac{r.\sin (\mathrm{S} - \mathrm{G})}{\sin (\mathrm{G} - \mathrm{N})}$.

The heliocentric longitude of the descending node may be found in a similar manner.

269. *Retrograde motions of the nodes.* From observations made at distant periods, it is found that the heliocentric longitudes of the nodes of all the planets are slowly increasing. The greatest increase is about 70′ in a century. But, in consequence of the retrograde motion of the vernal equinox, the longitude of a fixed star increases, during a century, nearly 84′. Hence, as the increase in the longitude of each node is less than that of a fixed star, it follows that the nodes of all the planets have slow retrograde motions.

When the motion of the nodes of a planet has been found from observations at distant periods, the slight correction necessary in the longitude of the node, as determined by the last article on the assumption that the node did not move, may be easily made. It it is also obvious, that with the longitude of the node found for any known time and the motion of the node, the longitude may be easily obtained for any other given time.

270. *The plane of a planet's orbit.* When the heliocentric longitudes of the two nodes of the same orbit are obtained for the same instant of time, they are found to differ 180°. Hence, it follows that the line of the nodes, and, consequently, the plane of the orbit, pass through the centre of the sun.

271. *Inclination of the orbit.* To determine the inclination of the orbit, let the time at which the sun's longitude is the same as the heliocentric longitude of the node be found, by means of the solar tables; and let the longitudes and latitudes of the planet be found from its observed right ascension and declination, for several consecutive days, contiguous to this time. From these, its geocentric longitude and latitude at that time become known.

Let E, *Fig.* 46, be the earth, S the sun, and N the node, when the longitude of the sun and node are the same, and let P be the place of the planet in its orbit at that time. Let PP′ be perpendicular to the plane of the ecliptic, meeting it in P′ and P′D perpendicular to SN. Then will the angle PDP′ be the inclination of the plane of the orbit. Put $E = SEP' = VEP' - VES =$ geocentric longitude of the planet — sun's longitude, $\lambda = PEP' =$ geocentric latitude of the planet, and $I = PDP' =$ inclination of the orbit. Then,

$$DP' \text{ tang } I = PP' = EP' \text{ tang } \lambda.$$

But, $$DP' = EP' \sin E.$$

Hence, $$EP' \sin E \text{ tang } I = EP' \text{ tang } \lambda.$$

or, $$\text{tang } I = \frac{\text{tang } \lambda}{\sin E}.$$

The orbits of the planets, with one exception, have small inclinations. Those of Venus, Mars, Jupiter, Saturn, Uranus, and Neptune, are all less than 4°; that of Mercury is about 7°; that of Pallas is nearly 35°; and those of all the other asteroids are between 3° and 15°.

The inclinations are found to have small secular variations. Some of them are increasing and others decreasing.

272. *Periodic Time and Mean Motion.* The period which elapses from the time that the planet is at one of its nodes till its return to the same, allowance being made for the motion of the node, is the sidereal revolution of the planet.

Another method of finding the sidereal revolution, is to deduce it from the synodic revolution. The latter is obtained by observing the times of two consecutive conjunctions of the same kind, or two consecutive oppositions. The intervening period is the synodic revolution, from which the sidereal is easily obtained. Let s = the synodic revolution, p = the periodic time or sidereal revolution, and p' = the sidereal revolution of the earth, all expressed in mean solar days. Also, let m = mean daily motion of the planet, and m' = mean daily motion of the earth. Then we evidently have,

$$m' \backsim m : 360° :: 1 \text{ day} : s.$$

But, $m' = \dfrac{360°}{p'}$, and $m = \dfrac{360°}{p}$..................................(C)

Substituting these, and dividing the first and second terms of the proportion by 360, we have,

$$\frac{1}{p'} \backsim \frac{1}{p} : 1 :: 1 : s,$$

or, $p \backsim p' : pp' :: 1 : s,$ ....(D)

or, $p - p' : \pm pp' :: 1 : s.$

Hence, $sp - sp' = \pm pp'$, or, $p = \dfrac{sp'}{s \mp p}$

In which, the *upper* sign is to be used for a *superior* planet, and the *lower* for an *inferior*.

When p is found, m, the mean daily motion, becomes known (C).

Cor. From the proportion (D), we have, for the synodic revolution in terms of the periodic revolutions of the planet and earth,

$$s = \frac{pp'}{p \backsim p'}.$$

273. *Heliocentric Longitude and Latitude.* When the place

of the ascending node, and the inclination of the orbit of a planet are known, and the geocentric longitude and latitude of the planet at any time have been found, from its observed right ascension and declination, its heliocentric longitude and latitude, and also, its radius vector, at that time, may be obtained by computation.

Let the sun's longitude and radius vector, at the time, be calculated; and referring to *Fig.* 47, put,

$G = VEp =$ geocentric longitude of planet,
$\lambda = PEp =$ " latitude "
$L = VSp =$ heliocentric longitude "
$l = PSp =$ " latitude "
$L' = VES =$ sun's longitude,
$N = VSN =$ heliocentric longitude of ascending node,
$I = PDp =$ inclination of the orbit,
$E = SEp =$ elong., $S = ESp =$ commut., $p = SpE =$ an. parallax.
$D = NSE$, $x = NSp$,
$R = SE =$ earth's rad. vec., $r = SP =$ planet's radius vector.

Then, $N + D = VSN + NSE =$ earth's longitude $= L' + 180°$, or $D = L' + 180° - N$, and $E = VEp - VES = G - L'$, are known. We have also, $p = SpE = VSp - VAp = VSp - VEp = L - G$, $S = NSE - NSp = D - x$, and $L = VSN + NSp = N + x$. Now, by trigonometry, we have, $Dp = Sp \sin x$, and $Sp \tan l = Pp = Dp \tan I = Sp \sin x \tan I$,

or, $\tan l = \sin x \tan I$ ....(E)

Also, $Ep \tan \lambda = Pp = Sp \tan l$,

or, $\tan \lambda : \tan l :: Sp : Ep :: \sin E : \sin S$.............(F)

But, since $S = D - x$, we have, $\sin S = \sin (D - x) = \sin D \cos x - \cos D \sin x$. Substituting in the proportion (F), the values of $\sin S$ and $\tan l$, it becomes,

$\tan \lambda : \sin x \tan I :: \sin E : \sin D \cos x - \cos D \sin x$,
or, $\sin x \tan I \sin E = \tan \lambda \sin D \cos x - \tan \lambda \cos D \sin x$,
or, $\tan x \tan I \sin E = \tan \lambda \sin D - \tan \lambda \cos D \tan x$.

Hence, $$\tan x = \frac{\tan \lambda \sin D}{\tan I \sin E + \tan \lambda \cos D}.$$

Consequently, $L = N + x$, becomes known.

As $S = D - x$, is also known when x is known, we may obtain l from either (E) or (F), the latter of which gives,

$$\text{tang } l = \frac{\sin \text{S tang } \lambda}{\sin \text{E}}.$$

The triangles PpS and EpS, give Sp = SP cos PSp = r cos l, and, sin p : sin E : : R : Sp,
or, sin (L − G) : sin E : : R : r cos l.

Hence, $$r = \frac{\text{R} \sin \text{E}}{\cos l \sin (\text{L} - \text{G})}.$$

Let x' = NSP, and L″ = N + x' = VSN + NSP = heliocentric orbit longitude of the planet (199). Then DP = SD tang x'. Hence,

$$\text{SD tang } x = \text{D}p = \text{DP cos I} = \text{SD tang } x' \cos \text{I}.$$

Therefore,

$$\text{tang } x' = \frac{\text{tang } x}{\cos \text{I}}, \text{ or, tang } (\text{L}'' - \text{N}) = \frac{\text{tang } (\text{L} - \text{N})}{\cos \text{I}} \ldots\ldots \text{ (G)}$$

Consequently, L″ = N + x', becomes known.

274. *Longitude of the perihelion, &c.* Assuming the orbit of the planet to be an ellipse, if its heliocentric orbit longitude, and its radius vector be found for three different times (273), we may thence determine the longitude of the perihelion, the eccentricity, and the semi-transverse axis, of the orbit.

Let PDG, *Fig.* 48, be the orbit, P the perihelion, and D, E and F, the three positions of the planet in its orbit. Then, SD, SE and SF are known, and from the longitudes, the angles DSE and DSF are also known. Put r = SD, r' = SE, r'' = SF, θ = angle DSE, ϕ = DSF, x = PSD, a = PC = semi-transverse axis, and e = the eccentricity. Then, ae = SC (151). Hence (Conic Sections),

$$r = \frac{\text{PC}^2 - \text{SC}^2}{\text{PC} + \text{SC cos PSD}} = \frac{a^2 - a^2e^2}{a + ae \cos x},$$

or, $$r = \frac{a. (1 - e^2)}{1 + e \cos x} \ldots\ldots\ldots\ldots \text{ (H)}$$

$$r' = \frac{a. (1 - e^2)}{1 + e \cos (x + \theta)} \ldots\ldots\ldots\ldots \text{ (I)}$$

$$r'' = \frac{a. (1 - e^2)}{1 + e \cos (x + \phi)} \ldots\ldots\ldots\ldots \text{ (K)}$$

From (H) and (I), we have,

$$r + re \cos x = r' + r'e \cos (x + \theta)$$

or, $$e = \frac{r' - r}{r \cos x - r' \cos (x + \theta)} \quad \text{(L)}$$

In like manner, from (H) and (K), we have,

$$e = \frac{r'' - r}{r \cos x - r'' \cos (x + \phi)} \quad \text{(M)}$$

Put $r' - r = m$, and $r'' - r = n$; then from (L) and (M), we have,

$$\frac{m}{n} = \frac{r \cos x - r' \cos (x + \theta)}{r \cos x - r'' \cos (x + \phi)}$$

$$= \frac{r \cos x - r' \cos \theta \cos x + r' \sin \theta \sin x}{r \cos x - r'' \cos \phi \cos x + r'' \sin \phi \sin x}$$

$$\frac{m}{n} = \frac{r - r' \cos \theta + r' \sin \theta \operatorname{tang} x}{r - r'' \cos \phi + r'' \sin \phi \operatorname{tang} x}$$

Hence, we easily find,

$$\operatorname{tang} x = \frac{m (r - r'' \cos \phi) - n (r - r' \cos \theta)}{nr' \sin \theta - mr'' \sin \phi}.$$

The value of x being subtracted from the orbit longitude of the planet in the first position, the remainder must be the orbit longitude of the perihelion. Then, if L and L″ be the ecliptic and orbit longitudes of the perihelion, we have (273 G),

$$\operatorname{tang} (L - N) = \operatorname{tang} (L'' - N) \cos I.$$

Whence, L, the heliocentric ecliptic longitude of the perihelion, becomes known.

The value of e, the eccentricity, may be found from either of the expressions (L) and (M), and a, the semi-transverse axis, from (H), (I), or (K).

Scholium. In the above investigation it has been assumed, in accordance with Kepler's first law, that the orbits of the planets are ellipses. That they are so, or, at least, nearly so, is established by the fact, that different sets of observations, made on the planet in various parts of its orbit, give very nearly the same results for the longitude of the perihelion, the eccentricity, and the semi-transverse axis.

The semi-transverse axes of the orbits of the planets, or their mean distances determined as above, and the periodic times determined by a previous article (272), are found to accord with Kepler's third law (154), or nearly so. As the truth of this law has been confirmed by investigations in physical astronomy, and as the periodic times of the planets may be determined with great precision, we may, from these, obtain more accurate values of the semi-transverse axis. Thus, putting P and p for the periodic times of the earth and planet respectively, and A and a for their mean distances, we have $P^2 : p^2 : : A^3 : a^3$. Whence, a becomes known, when P, p, and A are known.

It is, however, usual to assume the earth's mean distance from the sun to be a unit, and to express the mean distances of the planets and the radii vectores of the earth and planets, in accordance with this assumption. We then have, $P^2 : p^2 : : 1 : a^3$; or,

$$a = \sqrt[3]{\frac{p^2}{P^2}}$$

275. *Motions of the perihelions and changes in the eccentricities.* From observations made on each of the different planets, at distant periods, it is found that the perihelions of all their orbits have slow motions. The motion of the perihelion of the orbit of Venus is retrograde. Those of the other planets are direct.

The eccentricities of the orbits are also subject to small secular variations. Some of them are at present increasing, and others decreasing.

276. *Semi-transverse axes of the orbits.* The semi-transverse axes of the orbits, or, which is the same, the mean distances of the planets from the sun, *do not change.* This fact was first discovered by Lagrange, from investigations in physical astronomy, and it accords with observation.

277. *Epoch of a planet's being at its perihelion.* From several observations of the planet about the time it has the same longitude as the perihelion, the exact time that it is at the perihelion may be obtained by proportion or interpolation.

278. *Scholium.* There are various other methods for determining most of the elements of the orbit of a planet, besides those given in the preceding articles. Those which are founded on ob-

servations of the planet when in conjunction or opposition, and at the nodes, are the most convenient and accurate. The elements of the orbit may, also, be determined with tolerable accuracy, by certain methods of estimation and computation, without extending the observations to the time of the planet's passage through the node. These methods were applied on the discovery of the new planets.

In determining the elements, many hundreds, or even thousands of observations have been employed; and, with the exception of those of the new planets, they are now known with a high degree of precision.

279. *Tables of the planets.* When the elements of a planet's elliptical orbit have been determined, its place in the orbit may be calculated by Kepler's Problem, or by a table of the equation of the centre, computed by means of that problem. But the motion of a planet is subject to small perturbations, produced by the actions of the other planets. Investigations in physical astronomy have furnished the means of computing these perturbations and forming tables by which their values for a given time may be easily obtained. A complete set of tables for any planet includes tables of the mean heliocentric places and motions of the planet, and of the perihelion and ascending node of the orbit, the equation of the centre, the values of the perturbations, the reduction of the planet's place in its orbit to the ecliptic, and the radius vector of the planet or its logarithm.*

280. *Geocentric longitude and latitude.* From the heliocentric longitude, latitude and radius vector of a planet as obtained from the tables, it is often required to find the geocentric longitude and latitude. Referring to *Fig.* 47, and designating the quantities as in article (273), we have, by trigonometry,

$$\mathrm{SE} + \mathrm{S}p : \mathrm{S} \backsim \mathrm{E}\,\mathrm{S}p :: \operatorname{tang} \tfrac{1}{2}(\mathrm{S}p\mathrm{E} + \mathrm{SE}p) : \operatorname{tang} \tfrac{1}{2}(\mathrm{S}p\mathrm{E} \backsim \mathrm{SE}p),$$

$$\text{or, } \mathrm{R} + r \cos l : \mathrm{R} \backsim r \cos l :: \operatorname{tang} \tfrac{1}{2}(p + \mathrm{E}) : \operatorname{tang} \tfrac{1}{2}(p \backsim \mathrm{E}),$$

$$\text{or, } 1 + \frac{r \cos l}{\mathrm{R}} :: 1 \backsim \frac{r \cos l}{\mathrm{R}} : \operatorname{tang} \tfrac{1}{2}(p + \mathrm{E}) : \operatorname{tang} \tfrac{1}{2}(p \backsim \mathrm{E}).$$

* The best tables of the planets are those by *Lindenau*, for Mercury, Venus and Mars, with the explanations in Latin; and those by *Bouvard*, for Jupiter, Saturn and Uranus, with the explanations in French.

Put, $\text{tang}\ \theta = \dfrac{r \cos l}{\text{R}}$.

Then, as (273), $p + \text{E} = \text{L} - \text{G} + \text{G} - \text{L}' = \text{L} - \text{L}'$, we have,
$1 + \text{tang}\ \theta : 1 \backsim \text{tang}\ \theta :: \text{tang}\ \frac{1}{2}\,(\text{L} - \text{L}') : \text{tang}\ \frac{1}{2}\,(p \backsim \text{E})$.

Hence, $\text{tang}\ \frac{1}{2}\,(p \backsim \text{E}) = \dfrac{1 \backsim \text{tang}\ \theta}{1 + \text{tang}\ \theta} \cdot \ \text{tang}\ \frac{1}{2}\,(\text{L} - \text{L}')$

or, (App. 15), $\text{tang}\ \frac{1}{2}\,(p \backsim \text{E}) = \text{tang}\ (45° \backsim \theta).\ \text{tang}\ \frac{1}{2}\,(\text{L} - \text{L}')$.

Then, $\frac{1}{2}\,(p \backsim \text{E})$, added to $\frac{1}{2}\,(p + \text{E})$, or its equal $\frac{1}{2}\,(\text{L} - \text{L}')$, for a superior planet, or subtracted from it for an inferior planet, gives E, the elongation. And E added to L′ gives G, the geocentric longitude.

As $(\text{L} - \text{L}')$, or its equal $(p + \text{E})$, is the supplement of S, and as the sine of an angle is the same as that of its supplement, we have for λ, the geocentric latitude (273, F),

$$\text{tang}\ \lambda = \frac{\sin \text{E}\ \text{tang}\ l}{\sin\,(\text{L} - \text{L}')} \qquad \text{(N)}$$

When the planet is in conjunction or opposition, this formula for the geocentric latitude is not applicable; for, then, E and $(\text{L} - \text{L}')$ are either 0° or 180°, and, consequently, their sines are, each, zero. But, as E, S and p are then in a straight line, we have $\text{E}p = \text{SE} \overset{+}{\backsim} \text{S}p = \text{R} \overset{+}{\backsim} r \cos l$. Hence,

$$r \sin l = \text{P}p = \text{E}p\ \text{tang}\ \lambda = (\text{R} \overset{+}{\backsim} r \cos l)\ \text{tang}\ \lambda,$$

or,
$$\text{tang}\ \lambda = \frac{r \sin l}{\text{R} \overset{+}{\backsim} r \cos l} \qquad \text{(O)}$$

The *upper* sign appertains to the conjunction of a superior planet or superior conjunction of an inferior, and the *lower*, to opposition or inferior conjunction.

281. *Distance of a planet from the earth.* For the distance of a planet from the earth, we have,

$$\text{EP} \sin \lambda = \text{P}p = \text{S}p \sin l = r \sin l,$$

or,
$$\text{EP} = \frac{r \sin l}{\sin \lambda} \qquad \text{(P)}$$

Another expression, more accurate in practice, especially when the latitudes are small, may be easily obtained. For,

$$\sin \text{E} : \sin \text{S} :: \text{S}p : \text{E}p,$$

or,
$$\sin \text{E} : \sin\,(\text{L} - \text{L}') :: r \cos l : \text{EP} \cos \lambda.$$

$$\text{Whence, EP} = \frac{r \cos l \sin (L - L')}{\cos \lambda \sin E} \text{..............................} (Q)$$

282. *Horizontal parallax of a planet.* Let r = radius vector of the planet, expressed in terms of the earth's mean distance from the sun regarded as a unit, π = the planet's horizontal parallax, and π' = the sun's mean horizontal parallax, that is, the parallax at the distance 1. Then (93),

$$EP : 1 :: \pi' : \pi = \frac{\pi'}{EP}.$$

Hence (281, P and Q),

$$\pi = \frac{\pi' \sin \lambda}{r \sin l}, \text{ or, } \pi = \frac{\pi' \cos \lambda \sin E}{r \cos l \sin (L - L')}.$$

283. *Apparent semidiameter of a planet.* The semidiameter of a planet, as obtained from observation with a micrometer or heliometer, when the planet is at a known distance, may be reduced to what it would be, if seen at the mean distance of the earth from the sun, that is, at the distance 1. Let δ' = this value of the semidiameter and δ = the value at any time. Then (97),

$$EP : 1 :: \delta' : \delta = \frac{\delta'}{EP}.$$

$$\text{Hence, } \delta = \frac{\delta' \sin \lambda}{r \sin l}, \text{ or, } \delta = \frac{\delta' \cos \lambda \sin E}{r \cos l \sin (L - L')}.$$

RETROGRADATIONS AND STATIONS OF A PLANET.

284. *Retrograde motion of a planet.* As the orbits of the planets are nearly circular, and do not deviate much from the plane of the ecliptic, we shall here regard them as circular and coinciding with that plane.

Let S, *Fig.* 49, be the sun, ACE the orbit of the earth, *ace* that of an inferior planet, A and *a* the places of the earth and planet when the latter is in inferior conjunction, B and *b* their places at some short time after conjunction, as, for instance, an hour, AV and BV the direction of the vernal equinox, and B*s* and *bt* perpendicular to SA. Also, let P and p be the periodic times of the earth and planet respectively, expressed in hours, M =

ASB = earth's hourly motion, and $m = aSb$ = planet's hourly motion. Then, by Kepler's third law,

$$SA^3 : Sa^3 :: P^2 : p^2$$

or,
$$SA\sqrt{SA} : Sa\sqrt{Sa} :: P : p.$$

$$\text{But, } P : p :: \frac{360°}{M} : \frac{360°}{m} :: m : M :: \sin m : \sin M.$$

$$\text{Hence, } SA\sqrt{SA} : Sa\sqrt{Sa} :: \sin m : \sin M \ldots\ldots\ldots\ldots\ldots\ldots\ldots\ldots (R)$$

$$\text{But, } Sa : SA :: Sa : SA.$$

$$\text{Therefore, } \sqrt{SA} : \sqrt{Sa} :: Sa \sin m : SA \sin M :: bt : Bs.$$

Consequently, as $\sqrt{SA}$ is greater than $\sqrt{Sa}$, it follows that bt is greater than Bs. The line Bb, therefore, inclines from AS. Hence, as BV is parallel to AV, the angle VBb, which is the planet's geocentric longitude when at b, is less than VAa, its geocentric longitude when at a. The apparent motion of the planet is, therefore, retrograde at the period of inferior conjunction.

Let E and e be the places of the earth and planet at the time of superior conjunction, and F and f their places an hour afterwards. Then, it is evident that VFf, the geocentric longitude of the planet at f, is greater than VEe, its geocentric longitude at e. The apparent motion is, therefore, direct at the period of superior conjunction.

As the direction of a from A, or b from B, is directly opposite to that of A from a, or B from b, it follows that, when the motion of the planet appears to be retrograde as seen from the earth, the motion of the earth must appear retrograde as seen from the planet; and the same must apply to the direct motions. Hence, regarding ace as the orbit of the earth, and ACE as that of a superior planet, it is obvious that the apparent motion of the superior planet must be retrograde at the period of its opposition, and direct at the period of conjunction.

It will appear, from the next article, that the period during which the apparent motion of a planet is retrograde, is much shorter than that during which it is direct.

285. *A planet sometimes appears to be stationary.* Let C and c be two corresponding places of the earth and planet, and D and d their places an hour afterwards. Then, if the places C and c be

such that Dd is parallel to Cc, the geocentric longitudes VDd and VCc will be equal, and, therefore, the planet must, at that time, appear to be stationary.

To find SCc, the angle of elongation when the planet appears stationary, put x = SCc and y = CcG; and regarding the earth's distance SC = 1, let a = Sc = the planet's distance. Then, SDd = Snc = SCc + CSD = x + M, and DdK = CkK = Sck + GSK = CcG + GSK = y + m. Hence, from the triangles SCc and SDd, we have,

$$\sin y : \sin x :: 1 : a :: \sin (y + m) : \sin (x + \text{M}), \ldots\ldots\ldots \text{(S)}$$

or, $\sin y : \sin y \cos m + \cos y \sin m :: \sin x : \sin x \cos \text{M} + \cos x \sin \text{M}$.

But, as M and m are both very small, we may regard $\cos \text{M} = 1$ and $\cos m = 1$. Hence,

$$\sin y : \sin y + \cos y \sin m :: \sin x : \sin x + \cos x \sin \text{M},$$

or, $\sin y : \sin x :: \cos y \sin m : \cos x \sin \text{M} :: \cos y : \dfrac{\sin \text{M}}{\sin m} \cos x$.

$$\text{But (284 R)}, \frac{\sin \text{M}}{\sin m} = \frac{\text{S}a\sqrt{\text{S}a}}{\text{SA}\sqrt{\text{SA}}} = \frac{\text{S}c\sqrt{\text{S}c}}{\text{SC}\sqrt{\text{SC}}} = a\sqrt{a}.$$

Hence, $\cos y : a\sqrt{a} . \cos x :: \sin y : \sin x :: 1 : a, \ldots\ldots\ldots\ldots$(S)

or, $\cos^2 y : a^3 \cos^2 x :: \sin^2 y : \sin^2 x :: 1 : a^2$.

Consequently, $a^3 \cos^2 x = a^2 \cos^2 y$ (T), and $\sin^2 x = a^2 \sin^2 y \ldots\ldots$(U)

Therefore (T and U), $a^3 - a^3 \sin^2 x = a^2 - a^2 \sin^2 y = a^2 - \sin^2 x^2$

or, $$(1 - a^3) \sin^2 x = a^2 - a^3 \ldots\ldots\ldots\ldots\ldots\ldots\ldots \text{(V)}$$

Again (U and T), $1 - \cos^2 x = a^2 - a^2 \cos^2 y = a^2 - a^3 \cos^2 x$,

or, $$(1 - a^3) \cos^2 x = 1 - a^2 \ldots\ldots\ldots\ldots\ldots\ldots\ldots \text{(W)}$$

Dividing (V) by (W), we have,

$$\text{tang}^2\, x = \frac{a^2 - a^3}{1 - a^2} = \frac{a^2}{1 + a},$$

or, $$\text{tang}\, x = \mp \frac{a}{\sqrt{(1 + a)}} \ldots\ldots\ldots\ldots\ldots\ldots\ldots\ldots\ldots\ldots \text{(X)}$$

The upper sign appertains to an inferior, and the lower to a superior, planet.

From (S), by dividing the first and third terms by $\cos y$, and the second and fourth by $\cos x$, we have,

$$1 : a\sqrt{a} :: \text{tang}\, y : \text{tang}\, x.$$

Hence, $$\text{tang}\, y = \frac{\text{tang}\, x}{a\sqrt{a}}.$$

The angles x and y being found, the angle CSc, which is equal to their difference, is known. Now, CSc is the difference between the angular motion of the earth and that of the planet during the interval from inferior conjunction, or from opposition, to the time the planet is stationary. Hence, if we put t = this interval, and d = the difference of the daily motions of the earth and planet, we have,

$$d : \mathrm{CS}c :: 1 \text{ day} : t = \frac{\mathrm{CS}c}{d}.$$

Now, it is evident, that the planet must appear stationary at a like interval prior to the conjunction or opposition, and that during the period from the prior to the subsequent stationary positions its apparent motion must be retrograde. Hence, $2t$ expresses the period during which the motion is retrograde. The value of $2t$, computed for each of the planets, is found to be much less than half the synodic revolution.

Scholium. The times of the stationary positions of the planets, and the periods of their retrogradations, computed as above, are found to agree, nearly, with those obtained from observation; and when more accurately computed by taking into view the inclinations and elliptical forms of their orbits, the agreement is complete. As these computations are founded on the arrangement of the sun, earth, and planets, according to the Copernican System, this agreement is a confirmation of the truth of that system.

286. *Real distances of the planets from the sun.* From the sun's horizontal parallax, the earth's distance from the sun becomes known (96). This distance, multiplied by the numbers respectively which denote the relative distances of the planets, obtained on the assumption that the earth's distance is a unit (274, *Schol.*), gives the real distances of the planets.

287. *Apparent and real diameters of the planets.* The apparent diameter of a planet is determined by measurements with a micrometer or heliometer. Then the planet's distance from the earth, at the time of observation, being computed (281), the real diameter becomes known (99 H).

An inferior planet is nearer the earth at inferior conjunction than at superior, by the whole diameter of its orbit; and a superior planet is nearer at opposition than at conjunction, by the diameter of the earth's orbit. Hence, as the apparent diameter is inversely as the distance, the apparent diameters of several of the planets are very variable. The greatest apparent diameter of Venus is about six times the least, and the greatest of Mars about five times the least.

288. *Rotations of the planets.* All the planets on which sufficiently accurate telescopic observations can be made to ascertain the fact, are found to revolve on their axes in the same direction as the earth's rotation; that is, from *west* to *east.*

A simple analogy in the times of rotation of the primary planets has been recently discovered by Mr. Kirkwood,* which, although not yet fully established as a physical law, has excited the interest of both European and American astronomers. It may be stated in the following manner. If P and P′ are the points of equal attraction between any planet and those next inferior and superior to it, respectively (the three being in conjunction), the distance between the points P and P′ will be the *diameter of the sphere of attraction* of the middle planet. Then, putting D and D′ = the diameters of the spheres of attraction of any two planets; and n and n' = the number of rotations in their sidereal revolutions, we have,

$$n^2 : n'^2 :: D^3 : D'^3,$$

Or, the square of the number of rotations made by a planet in its sidereal year is proportional to the cube of the diameter of its sphere of attraction. The accuracy of this proportion has been tested, as far as it can be in the present state of our knowledge of the masses and rotations of the planets, and found quite satisfactory.

289. *Bode's Law.* In the latter part of the last century, Professor Bode announced a remarkable relation among the distances of the planets from the sun, which is exhibited in the following table, the last column giving the true distances in whole numbers, on the supposition that the earth's distance is 10.

---

* Professor Daniel Kirkwood, of Delaware College.

				True rel. dist.
Mercury........	4	=	4...........	4
Venus...........	$4 + 3 \times 2^0$	=	7...........	7
Earth............	$4 + 3 \times 2^1$	=	10...........	10
Mars	$4 + 3 \times 2^2$	=	16...........	15
(Asteroids)	$4 + 3 \times 2^3$	=	28...........	(26)
Jupiter..........	$4 + 3 \times 2^4$	=	52...........	52
Saturn..........	$4 + 3 \times 2^5$	=	100...........	95
Uranus.........	$4 + 3 \times 2^6$	=	196...........	192
(Neptune)......	$4 + 3 \times 2^7$	=	388...........	300

By examining the table it will be seen that, prior to the discovery of the new planets, there was an abrupt increase from Mars to Jupiter. This break in the series of distances, led some astronomers to suppose there must be a planet occupying an intermediate position. Kepler had broached a similar opinion in one of his earliest publications, though it was not until after the discovery of Uranus, which seemed to strengthen the idea of a harmony in the planetary distances, that astronomers thought of searching for the body needed to complete the series. In the year 1800 an association of practical observers was established for the purpose of finding, if possible, this concealed planet, each member undertaking to examine minutely all the telescopic stars in a certain portion of the zodiac; and, on the first day of the present century, Piazzi, one of their number, discovered the planet Ceres, the first of thirteen small bodies discovered within the last half century, revolving around the sun at distances varying from 22 to 32. The average of the distances of the Asteroids is about 26 as inserted in the table. It was supposed that, if a planet existed exterior to Uranus, its distance would be about 388, but, unfortunately for Bode's *Law*, the distance of Neptune is only about 300.

CHAPTER XV.

INFERIOR PLANETS, MERCURY AND VENUS. — TRANSITS. — SUN'S PARALLAX.

291. *Greatest elongations of Mercury and Venus.* Mercury and Venus have their orbits so far within that of the earth, that their elongations are never great. They seem to accompany the sun, being seen in the west soon after sunset, or in the east a while before sunrise.

Let S, *Fig.* 50, be the place of the sun, ABC the orbit of Mercury, which we will here suppose to coincide with the plane of the ecliptic, FG a part of the earth's orbit, and A and *a* corresponding positions of the planet and earth, when the former is at its greatest elongation, at which time the angle *a*AS is a right angle. As the distances of the planet and earth from the sun both vary, the greatest elongation must also vary. Its value will evidently be greatest when SA is greatest, and at the same time S*a* least, that is, when, at the time of greatest elongation, Mercury is at the aphelion of his orbit and the earth in perihelion; and least, when the positions are reversed. With the least value of SA and greatest of S*a*, we find the least value of Mercury's greatest elongation to be about $17\frac{1}{2}°$, and with the greatest value of SA and least of S*a*, we find the greatest value to be about $28\frac{1}{3}°$. In a similar manner, we find the greatest elongation of Venus to vary from about 45° to nearly 48°.

292. *Synodic Revolutions of Mercury and Venus.* From the formula (272 Cor.), the synodic revolution of Mercury is found to be about 116 days, and that of Venus, 584 days.

293. *Phases of Mercury and Venus.* Regarding the planets as opaque globular bodies, which shine by reflecting the light of the sun, Mercury and Venus must assume the various phases of the moon. Referring to *Fig.* 50, let A and *a*, B and *b*, &c., be corresponding positions of one of these planets and the earth. Then, it is obvious that, while the planet is passing from its

greatest eastern elongation at A to its greatest western at C, the enlightened disc must have the crescent form, like the moon from third quarter to first; and while passing from C to the following greatest eastern elongation at E, it must have the gibbous form, like the moon from first quarter to third; excepting, however, the positions of inferior and superior conjunction at B and D, at the former of which, the enlightened surface is turned wholly from the earth, and at the latter, entirely towards it, like the moon at new and full moon.

When viewed with a telescope of sufficient power, both Mercury and Venus exhibit these phases. We have thus another confirmation of the truth of the Copernican System.*

MERCURY.

294. *General Remarks.* Mercury is the least of the planets, with the exception of the four lately discovered; and is much less than the earth. His apparent diameter varies from 5″ to 12″. He shines with a steady white light, appearing as a luminous point above the western or eastern part of the horizon. Being always near the sun and, therefore, but for a short interval above the the horizon when he is below it, and that principally during the twilight, telescopic observations of his appearance are difficult and considerably uncertain.

295. *Period, Distance from sun, &c.* Mercury revolves round the sun in a little less than three months, at a distance of 37 millions of miles. His diameter is about 3000 miles, and his volume or bulk about $\frac{1}{16}$ that of the earth. According to the observations of some astronomers, he revolves on his axis in 24h. 5m.; the axis making a large angle, with a perpendicular to the plane of the ecliptic.

296. *Visibility of Mercury.* In high latitudes, Mercury is seldom visible to the naked eye, in consequence of the increased

---

* It was objected to the system of Copernicus, that if it were true, Venus should sometimes appear horned like the moon. He admitted the conclusion, adding that, should we ever be able to see the actual shape, it would appear so.

duration of twilight (182). But, in latitudes not higher than those of the United States, he may, under favourable circumstances, be seen in the evening or morning, for a number of days about the period of his greatest east or west elongation. Supposing the atmosphere clear, the other circumstances that favour his visibility are, that the greatest elongation should occur during the period of shorter twilight, that he should then be near the aphelion of his orbit, or, at least, not very remote from it, and that his polar distance should be some degrees less than that of the sun.

VENUS.

297. *General Remarks.* Venus, the most brilliant of the planets, is frequently called the *morning* and *evening* star, as she is in general conspicuously visible at one or the other of these times. In remote periods, this planet was regarded as two different bodies; the morning star being called *Lucifer*, and the evening, *Hesperus*. The discovery that they were the same body is ascribed to Pythagoras.

The size of Venus is nearly the same as that of the earth, though a little less. Her apparent diameter varies from 10″ to 61″.

298. *Period, Distance, &c.* Venus revolves round the sun in about $7\frac{1}{2}$ months, at a distance of 69 millions of miles. Her diameter is about 7600 miles, and her volume $\frac{9}{10}$ that of the earth.

From observations of the motions of spots seen on the surface, it has been inferred that Venus revolves on her axis in 23h. 21m.; the axis making an angle of 75° with a perpendicular to the plane of the ecliptic, and 72° with a perpendicular to the plane of the orbit.

299. *Day and Night, and Seasons at Venus.* As the axis of Venus makes so large an angle with the axis of the orbit, it is evident that she must be subject to great and rapid changes in the lengths of her day and night, and correspondingly great vicissitudes in her seasons. The circles corresponding to our tropics must be within 18° of her poles, and those corresponding to our polar circles, within the same distance of her equator. It can, therefore, only be within a zone extending 18° on each side of her

equator, that each rotation on her axis will, throughout the year, bring a return of day and night. In other parts there may in the course of a year be, at the same place, alternate day and night for each rotation of the planet: a day lasting during many rotations; and a night of like duration. In the two zones of 54° breadth extending from the polar circles to the tropics, the sun will sometimes ascend to the zenith, and at others scarcely rise above the horizon, or not rise at all, for many consecutive days.

300. *Venus sometimes visible during the full light of day.* In consequence of the changes in the extent of the enlightened part of the disc and the varying distance of Venus from the earth, the intensity of her light is subject to considerable variation. By a simple investigation in Differential Calculus, in which the circumstances that influence the intensity of her light are noticed, it has been found that she gives the greatest light in about 36 days before and after inferior conjunction; her elongation being then about 40°, and the enlightened part of the disc not much over a third of the whole. At these periods the light is so great, that objects illuminated by it, cast perceptible shadows.* She may also, then, be very distinctly seen by the naked eye during the full light of day, even at mid-day, especially if at the time she has considerable north declination so as to rise far above the horizon.

301. *Mountains of Venus.* Some astronomers have thought that they had detected evidences of high mountains on Venus, and have computed the heights of some of them to be over 20 miles. But, as the intense light of this planet dazzles the sight and exaggerates the imperfections of the telescope, thereby, rendering observations difficult, and some of them quite uncertain, the existence of these mountains is not to be regarded as established.

* This is quite observable when the object is placed in an open window of a room, and the shadow is received on the opposite wall or on a white screen.

TRANSITS OF MERCURY AND VENUS.

302. *Definition.* When either Mercury or Venus, at the time of inferior conjunction, is near to either node of its orbit, or, which amounts to the same, when the longitudes of the sun and node are at that time nearly equal, the planet must pass between the sun and earth, appearing as a well-defined black spot traversing the disc. This phenomenon is called a *Transit* of the planet.

303. *Transits of Mercury.* The longitudes of Mercury's nodes are about 46° and 226°, and these longitudes vary but little more than a degree in a century. In the present age, therefore, transits of Mercury can only occur in the months of May and November, for it is only in these months that the sun can have the same longitude as the nodes.

When a transit has occurred at one node, there cannot be another at the same node, till the lapse of a period of time composed of whole synodic revolutions, and also, of whole years or nearly so. For they occur only at inferior conjunction, and those at the same node, nearly at the same time in the year. Hence, taking s to represent a synodic revolution of Mercury, p the periodic revolution of the earth or sidereal year, if m and n be two whole numbers such that $ns = mp$ nearly, or $\frac{n}{m} = \frac{p}{s}$ nearly, then will m be the number of the years in the period between two consecutive transits at the same node. Different values for m and n, less or more exact, may be obtained by the method of continued fractions.* It is thus found that transits at the same node occur at intervals of 6 or 7 years, 13 years, 33 years, &c.

A transit at one node is generally preceded or followed, at an interval of $3\frac{1}{2}$ years, by one at the other node.

The last two transits of Mercury, both of which were visible in this country, occurred in May, 1845, and November, 1848.†

---

* This method is frequently given in treatises on Algebra. A practical rule is given in Lewis's Arithmetic.

† Other transits that will occur during the present century, will happen November, 1861, November, 1868, May, 1878, November, 1881, May, 1891, and November, 1894. Of these the third and last will be visible in this country.

304. *Transits of Venus.* The longitudes of the nodes of Venus are about 75° and 255°, and the sun has these longitudes in June and December. Hence, it is only in June and December that transits of Venus occur.

A synodic revolution of Venus being about 584 days (292), a period of 5 synodic revolutions differs but little from 8 years. Hence, a transit at one node is generally preceded or followed, at an interval of 8 years, by another at the same node. A full investigation, with reference to both nodes, shows that, commencing with the last transit, which occurred in June, 1769, succeeding transits must occur at the terminations of the periods 105½ years, 8 years, 121½ years, and 8 years, taken in order and repeated in the same order. Thus, the last two transits were in June, 1761 and 1769, and the next two will occur in December, 1874 and 1882.

Transits of Venus occur, therefore, much less frequently than those of Mercury.

305. *Computation of a Transit.* The computation of a transit of Mercury or Venus, for any given place, is nearly like that of an eclipse of the sun; the data for the planet taking the place of those for the moon.

306. *Sun's Parallax.* A transit of Venus is a phenomenon of great interest and importance as affording the best means of determining with accuracy the sun's parallax, and thence, his distance from the earth. For a full investigation of the method by which the sun's parallax is deduced from observations of this phenomenon, the student must be referred to larger works. But the following illustration will enable him to understand the general principles on which the deduction depends.

Let the circle cDd, of which S is the centre, *Fig.* 51, represent the sun's disc, and let V be Venus, pq a part of her relative orbit, along which she appears to move in the direction from p to q, E the earth, and A and B the places of two observers, supposed to be situated at the opposite extremities of that diameter of the earth which is perpendicular to the ecliptic. Then, disregarding the earth's rotation, or, which is the same, supposing the positions A and B to remain fixed during the transit, the centre of the planet will, to the observer at A, appear to describe the chord *cd*, and to the observer at B, the parallel chord *ef*. Also, when, to the observer at A, the centre of the planet appears to be at *a*, it will, to the observer at B, appear to be at *b*. As AB is perpendicular to the plane of the ecliptic, and the plane of the sun's disc is for each observer very nearly so, the line *ba* may be regarded as being parallel to AB; and as the relative orbit, and, consequently, the chords *cd* and *ef* make but a small angle with the plane of the ecliptic, it may be regarded as perpendicular to these chords, and, therefore, as expressing the distance between them.

Now, the observers at A and B may determine the duration of the transit of the planet's centre as seen at these places; that is, the times of its appearing to describe the chords *cd* and *ef*. Then, as the relative hourly motion of Venus may be very accurately found from tables of the sun and planet, the values of the chords *cd* and *ef*, expressed in seconds, and, consequently, their halves *hd* and *kf*, may be obtained. Hence, *hd* and *kf*, and the sun's semi-diameter SD, being known, *h*D and *k*D, and, consequently, their difference *hk* or *ab*, are easily found.

As *ba* is parallel to AB, the triangles ABV and *ab*V are similar, and we have, $a\text{V} : \text{AV} :: ab : \text{AB}$. But, from the tables, we know the ratio of *a*V to *a*A, and, consequently, of *a*V to AV. This ratio is, at a mean, 72 to 28 very nearly, or 5 to 2 nearly. Hence, we have, approximately, $5 : 2 :: ab : \text{AB} = \frac{2}{5}\,ab$. But, AB, which is the measure of the angle A*a*B, is double the sun's horizontal parallax. Consequently, the sun's horizontal parallax $= \frac{1}{5}\,ab$, nearly. It follows that, whatever small error may be made in determining *ab*, the error in the parallax obtained will be only about *one-fifth* as great.

It is not necessary that the observers should be situated as supposed above; but it is important that they should be at places far distant from each other, in rather a north and south direction. The places being known, the complete investigation of the subject furnishes a method of deducing the parallax, taking into view the earth's rotation and every other circumstance that can influence the accuracy of the result.

307. *Determination of the sun's parallax.* Astronomers having made known the importance of having accurate observations of

the transits of Venus at different and distant places, expeditions on the most efficient scale were fitted out for the purpose previous to the last transit, in 1769, by the British, French, Russian, and other governments. From the observations then made, combined with some of those made in 1761, Professor Encke has found the sun's mean horizontal parallax to be 8″.5776.

CHAPTER XVI.

SUPERIOR PLANETS—SATELLITES OF JUPITER, SATURN AND URANUS.

308. *Superior Planets.* The superior planets, revolving in orbits without that of the earth, cannot exhibit to us phases similar to those of Mercury and Venus. The disc of Mars, however, about the period of his quadratures, appears decidedly gibbous. The other planets revolve so far without the earth's orbit that their enlightened surfaces are always turned almost entirely towards the earth, and the gibbous form is not perceptible.

MARS.

309. *General remarks.* Mars is easily distinguished from the other planets by the ruddy colour of his light. He is a small planet, next larger than Mercury. His apparent diameter varies from about 3″½ to 18″. In consequence of this great variation in apparent diameter, he appears at different times, except with regard to colour, as quite a different body.*

310. *Period, distance, &c.* Mars revolves round the sun in a little less than 23 months, at a distance of 144 millions of miles. His diameter is about 4000 miles, and his volume $\frac{1}{7}$ that of the earth. He revolves in 24 h. 39 m., about an axis that is inclined to the axis of the ecliptic, in an angle of 30° 18′.

---

* The change in the apparent diameter of Venus is still greater (297); but, in consequence of her phases, the change in the light received from her, while sufficiently remote from the sun to be visible, is much less.

311. *Spheroidal form.* According to the observations of some astronomers, Mars has perceptibly a spheroidal form. Arago makes his polar diameter to be less than the equatorial by $\frac{1}{39}$ of the latter.

JUPITER AND HIS SATELLITES.

312. *General remarks.* Jupiter is the largest of the planets, his volume exceeding the sum of all the others; and, with the exception of Venus, he is the most brilliant. His apparent diameter varies from 30″ to 45″.

313. *Period, distance, &c.* Jupiter revolves round the sun in rather less than 12 years, at a distance of 494 millions of miles. His diameter is 90,000 miles, which is more than 11 times the earth's diameter. Consequently, his volume is more than 1300 times that of the earth. He revolves in 9h. 56m. about an axis nearly perpendicular to the plane of the ecliptic.

314. *Spheroidal form of Jupiter.* The form of Jupiter is decidedly spheroidal. According to accurate micrometrical measurements, the polar diameter is less than the equatorial by nearly $\frac{1}{15}$ of the latter.

Investigations, founded on the principles of mechanics, prove, that if a globular body composed of yielding matter, such as may some time have been the state of the earth and planets, be made to rotate about an axis, it must assume a spheroidal form; differing less or more from a sphere, according to its magnitude and the rapidity of its rotation. These investigations, applied to the earth and Jupiter, assign to each, very nearly, the degree of oblateness it is found to have.

315. *Belts of Jupiter.* When Jupiter is examined with a telescope of considerable power, his disc is observed to be crossed in a direction parallel to his equator, by several dark bands which are called his *belts*. These belts do not always present exactly the same appearance. They vary slightly in breadth and position; but always remain parallel to one another and to the equator.

316. *Jupiter's Satellites.* It has already been mentioned (9), that Jupiter is attended by four moons or *satellites*. These revolve

round him in short periods, and at correspondingly small distances. The periods, approximately expressed, are 2, 3½, 7 and 17 days, and the distances 6, 10, 15 and 21 times the radius of Jupiter.

The orbits of the satellites are found to coincide nearly, but not exactly, with the plane of Jupiter's equator. Hence, as both the plane of his equator and that of his orbit have but small inclinations to the plane of the ecliptic, the satellites can never deviate far from either of these planes. They, therefore, always appear to be, and to move forward and backward, nearly in a straight line which crosses the centre of the disc in the direction of the belts.

The discovery of Jupiter's satellites by Galileo, was one of the first fruits of the invention of the telescope. They are visible with a telescope of small power.

317. *Disappearances of the satellites.* Frequently, not more than two or three of the satellites are visible; sometimes not more than one; and one instance is stated to have occurred, when they were all invisible at the same time. The causes of these disappearances will be easily perceived by reference to *Fig.* 52, in which S represents the sun, J Jupiter, E the earth, EK a part of the earth's orbit, and *abm* the orbit of one of the satellites, in which it moves in the direction *abm*. Jupiter is so remote from the sun, and so large a body, that his shadow extends to an immense distance. A portion of it, only, is represented in the figure. Now, supposing the earth to be at E, and the satellite to be beyond Jupiter, in the part of the orbit from *c* to *d*, it must evidently be invisible, being hid by the planet. When it is between the earth and Jupiter, in the part of the orbit from *e* to *f*, its light becomes blended with that of the planet, and it is generally invisible; if, however, it happens to be directly between the observer and one of the belts, it may sometimes be seen with a powerful telescope, appearing as a bright spot on the belt. When the satellite enters the shadow at *a*, it necessarily becomes invisible, as it then ceases to receive light from the sun. This last phenomenon is called an *eclipse* of the satellite.

While the satellite is passing from *m* to *n*, it casts a shadow on the planet. With a telescope of high power, the shadow may be distinctly seen as it traverses the disc.

318. *Eclipses of Jupiter's satellites.* The eclipses of Jupiter's satellites are phenomena of very frequent occurrence. For, in consequence of the great size of the planet, the small distances of the satellites, and the small inclinations of their orbits to that of their primary, the three interior satellites suffer an eclipse every synodic revolution; and the fourth very rarely passes opposition without being eclipsed.

Both the beginning and end of an eclipse of the third or fourth satellite, or the *immersion* and *emersion*, at a and b, may frequently be observed from the earth; both taking place on the same side of the planet. This is, also, sometimes the case with the second. But the orbit of the first is so near to Jupiter, that its immersion and emersion can never both be seen; one or the other taking place behind the planet. This will be perceived by supposing an orbit to be described, much smaller than that in the figure.

It is evident, from inspection of the figure, that the eclipses take place to the west of the planet, while the earth is to the west of SJ, that is, before the opposition of Jupiter; and to the east, while the earth is in the other half of its orbit, or after opposition.

319. *Revolutions and motions of the satellites.* From the observed times of immersion and emersion of a satellite, the time it is in opposition to the sun becomes known; for this time must evidently be the mean of the two former. It, therefore, follows that repeated observations of the eclipses of a satellite serve to determine its mean synodic revolution. From this, the periodic or sidereal revolution is easily found.

From the mean sidereal revolution, the mean motion or angular velocity becomes known.

The orbits of the satellites differ but very little from circles, and, consequently, their true elliptical motions differ but little from their mean motions. The mutual actions of the satellites produce, however, some perturbations in their motions. These have been carefully investigated by Laplace and others; so that their true motions are now quite accurately known.

320. *Curious relation in the mean motions of the first three satellites.* If the mean angular velocity of the first satellite be added to twice that of the third, the sum will be equal to three

times that of the second. From this, it follows that, if from the sum of the mean longitude of the first, and twice that of the third, three times that of the second be subtracted, the remainder will always be the same quantity; and, from observation, it is found that this quantity is 180°. Hence, it also follows that the first three satellites can never all be eclipsed at once.

321. *Use of the eclipses of Jupiter's satellites in determining the longitudes of places.* As a satellite, on entering the shadow, loses its light, and on leaving regains it, the same immersion or emersion must occur at the same instant for different places, however distant from one another. If, then, the times of immersion or emersion, as reckoned at two different places, be accurately observed, the difference of these times must be the difference of longitude of the two places. Consequently, if the longitude of one of them is known, that of the other becomes also known.

The times of the eclipses, computed from tables which have been formed for the purpose,* are given in the Nautical Almanac, for the meridian of Greenwich. These computed times differ but little from the times observed at that meridian. If, then, an eclipse of one of the satellites be observed at any place, the difference between the observed time, and the time given in the Nautical Almanac, expresses the longitude of the place from Greenwich.

This very simple method of finding the longitude is not so accurate as some others. For, as the light of the satellite gradually diminishes, while it is entering the shadow, and gradually increases as it is leaving it, like that of the moon when entering and leaving the earth's shadow, the *observed* time of disappearance or reappearance of the satellite, must depend on the power and perfection of the telescope used, and, in some measure, on the eye of the observer.

322. *Transmission of Light.* The grand discovery that the transmission of light is *not instantaneous*, but that it requires time proportionate to the distance, is due to *Roemer*, a Danish astronomer, who deduced it from observations of the eclipses of Jupiter's satellites. In 1675, Roemer examined and compared observations

* The best tables of Jupiter's satellites are those computed by Damoiseau.

of the eclipses of the satellites which had been made during a number of preceding years. He found that the eclipses which happened about the time of Jupiter's opposition, when he was nearest to the earth, all occurred some minutes *sooner* than they should do, according to the averages of the intervals between consecutive eclipses of each satellite; and that, when Jupiter was near conjunction, and, consequently, most remote from the earth, they all occurred as much *later* than they should do, according to these averages. The deviations appearing thus to be connected with the planet's distance from the earth, it occurred to him, while seeking for their cause, that they could be explained by assuming light to be uniformly transmitted in time: that is, by assuming that, when any very distant phenomenon happens, a measurable interval of time, proportionate to the distance, elapses between the actual occurrence of the phenomenon and the perception of it by the observer. Pursuing the inquiry, he found that the deviations he had noticed would be completely accounted for, by allowing 8m. 13sec. for the transmission of light through the distance between the sun and earth. This, since the sun's distance from the earth is 95,000,000 miles, gives to light the amazing velocity of more than 192,000 miles per second.

This conclusion, with regard to the transmission of light and its great velocity, subsequently received complete confirmation by Dr. Bradley's discovery of the abberration of light (131).

323. *Rotation of Jupiter's satellites.* From very frequently repeated observations of Jupiter's satellites, it has been ascertained that they are subject to marked periodical fluctuations with regard to brightness; and that the periods correspond respectively with the periodic revolutions of the satellites. Hence, it has been inferred that each satellite, like our moon, revolves on its axis in the same time that it revolves round the planet.

SATURN AND HIS SATELLITES AND RINGS.

324. *General remarks.* Saturn is a large planet, being next in size to Jupiter, and not greatly inferior. He is so remote from the sun, in comparison with the earth, that his apparent diameter is not subject to much variation. Its mean value is about 17″. Consequently, Saturn, though so remote, is, from his great size, a tolerably conspicuous object. He shines with rather a pale white light.

In addition to his eight satellites (9), Saturn is distinguished from all the other planets by being surrounded, at some distance, by two broad, flat, circular rings, situated in the same plane, and concentric with the planet and with each other.

325. *Saturn's period, distance, &c.* Saturn revolves round the sun in about 29½ years at the distance of 905 millions of miles. His diameter is about 79,000 miles, and his bulk nearly 1000 times that of the earth. He revolves in 10h. 29m., about an axis making an angle of 28° 40′ with the axis of the ecliptic.

326. *Saturn's Rings.* The rings of Saturn are opaque bodies, shining like the planet, by reflecting the light of the sun. This follows from the fact that they are observed to cast a shadow on the side of the planet next the sun, and to be shaded by it on the opposite side. *Fig.* 53, represents Saturn surrounded by these singular appendages; the body of the planet being striped by dark belts somewhat similar to those of Jupiter, but broader and less strongly marked.

From micrometrical measurements, it has been ascertained that the distance from the surface of Saturn to the inside of the nearest ring is a little over 19,000 miles; the breadth of this ring is about 17,000 miles; the interval between the two rings is 1,800 miles; and the breadth of the exterior ring is about 10,600 miles. The entire diameter of the exterior ring is 176,000 miles. The rings are extremely thin; their thickness, according to Sir J. Herschel, does not exceed 100 miles.

When the rings are examined with telescopes of moderate power, they appear as one, the interval between them not being perceptible; but this interval becomes distinctly seen when those of high power are used, appearing as a black line or narrow band as represented in the figure.*

327. *Inclination and Rotation of the Rings.* It has been ascertained that the rings coincide, or very nearly so, with the plane of Saturn's equator. They must, therefore, be inclined to the plane of the ecliptic in the same angle that the axis of the planet is inclined to the axis of the ecliptic; that is, in an angle of 28° 40′ (325). It is also found that the plane of the equator and rings, and, consequently, the line in which it intersects the plane of the ecliptic, remain parallel to themselves as Saturn makes his revolution in his orbit. From this it follows, that the axis of Saturn, like that of the earth, continues parallel to itself.

From observations of some parts of the rings less bright than others, it has been inferred that they revolve in their own plane, making a revolution in about 10h. 29m. It is worthy of remark, that this is nearly the time in which a satellite, at a distance from Saturn, corresponding to the middle of the rings, would revolve round the planet.

328. *Varying appearance of the Rings and their disappearances.* As the plane of the rings continues parallel to itself, and the angle of their inclination to the ecliptic is not large, the face of the rings can never be turned directly to the earth, or very nearly so; and they do not, therefore, ever present to us a circular appearance. Being seen obliquely, they must, like all circular rings when thus viewed, appear elliptical; the degree of ellipticity varying according to the greater or less obliqueness of their position, which, in consequence of the motions of Saturn and the earth, is continually changing.

Let S, *Fig.* 54, be the sun, *eae′* the orbit of the earth, and ABCD the orbit of Saturn, which we may here suppose to coincide with the plane of the ecliptic; and let the parallel lines in the figure

---

* Appearances of other lines of division have been seen occasionally by several observers, and sometimes under circumstances which seemed to leave no room to doubt the existence of one or more subdivisions in each ring.

The most interesting discovery recently made in reference to the rings of Saturn, is that of a new ring, of an obscure, dusky appearance, interior to the inner, principal one. This was observed at about the same time in November, 1850, by Mr. Bond of Cambridge, United States, and by Mr. Dawes of England. The breadth of the obscure ring is estimated to be 1″.7, or two-fifths of the interval between the surface of the planet and the inner, bright ring.

be the lines in which the plane of the rings intersects the plane of the ecliptic, in the positions of Saturn, to which they are drawn. Then, it is evident that when Saturn is in either of the positions A and C, the plane of the rings must pass through the sun, and only the edge of the exterior ring is illuminated. In these positions, the longitudes of which are 170° and 350°, the rings, in consequence of their being extremely thin, are *invisible*, except with a telescope of the very highest power. With such an instrument a fine line of light has been perceived, extending to some distance on each side of the planet.

It is not only at the positions A and C, that the rings are invisible. They usually disappear twice about each of these positions, remaining invisible some weeks at each disappearance. To understand this, suppose that, as Saturn approaches A, the earth is moving in the part $e''ea$ of its orbit. There must then be a time at which the line es, joining the earth and Saturn, will become parallel to CA. At this time, the plane of the rings must pass through the earth, and only the edge being towards it, they are invisible. After this, while the earth is moving from e to some position a, and Saturn from s to A, the plane of the rings passes between the sun and earth, and the enlightened face is turned from the earth. Hence, as, during this period, only the edge of the enlightened part of the rings is towards the earth, they remain invisible. When the planet has passed the position A, the sun and earth are both on the same side of the plane of the rings, the illuminated face is towards the earth, and the rings are again visible. This continues to be the case till the earth and planet attain the positions e' and s', when the plane of the rings again passes through the earth, and the rings become invisible. They continue so till the earth and planet arrive at the positions e'' and s'', when the plane of the rings a third time passes through the earth. After this, the illuminated face is turned towards the earth and the rings are visible till the planet approaches the opposite position C, when two other disappearances usually take place.*

The illuminated face of the rings must, obviously, be most

---

* It is obvious that the order and durations of the disappearances will be affected by the position of the earth when the plane of the rings first intersects the earth's orbit.

turned towards the earth when the planet is at or near the positions B and D, midway between A and C; and the rings must then appear most open. They have then nearly the appearance represented in *Fig.* 53.

While Saturn is in the part ABC of his orbit, that is, from 170° to 350° of longitude, the northern face of the rings is illuminated, and in the other part, the southern face.

329. *Period of the disappearances of the rings.* As the period of Saturn's revolution is about 29½ years, nearly 15 years must elapse from the time he is at A till he is at C, or from C to A, and this must be nearly the period from one set of disappearances to the next. The two to which the above illustration refers, took place about the position A, in the latter part of 1832 and towards the middle of 1833; the next occurred in 1847.

330. *Saturn's satellites.* The eight satellites of Saturn revolve round him in periods varying from 1 day to 79 days, and at distances varying from 3 to 64 radii of the planet. The eighth satellite is the most conspicuous; that, and the sixth, may be discerned with telescopes of moderate power. The third, fourth, and fifth can only be seen with a telescope of much higher power; and the first, second, and seventh only with a telescope of great power.

The eighth satellite, like those of Jupiter, exhibits periodic defalcations in its light, from observations of which it has been inferred, that it revolves on its axis in the same time that it makes a revolution round the planet (323).

The discovery of the sixth of these satellites was made by Huygens, in 1655; that of the third, fourth, fifth, and eighth by Cassini between the years 1670 and 1685; that of the first and second by Sir William Herschel, in 1789; and that of the seventh by Mr. Bond of Cambridge, Mass., and Mr. Lassell of Liverpool, in 1848. Mr. Bond having seen it about 48 hours earlier than Mr. Lassell.

Some authors have distinguished the satellites of Saturn by reversing the order of the numbers above and calling the exterior one the first, &c. Sir John Herschel, in a work published in 1847, in alluding to the inconvenience arising from this uncertainty, says: "Should an eighth satellite exist, the confusion of the old nomenclature will become quite intolerable," and proposes the following mythological names: 1st. Mimas; 2d. Enceladus; 3d. Tethys; 4th. Dione; 5th. Rhea; 6th. Titan; 8th. Iapetus. In accordance with this system, the seventh satellite recently discovered has been named Hyperion.

URANUS AND HIS SATELLITES.

331. *General remarks.* Uranus was discovered by Sir William Herschel in 1781, and was named by him the Georgium Sidus, in honour of his patron, King George III., which name, abbreviated to the Georgian, was retained in the Greenwich Nautical Almanac until the year 1851. By the French it was for a time called Herschel and by others Uranus. It is now universally recognised by its mythological name. The distance of Uranus is so great that, though a large planet, he is barely discernable by the sharpest sight without the aid of a telescope. His apparent diameter, which varies but little, is about 4″.

332. *Period, distance, &c., of Uranus.* Uranus revolves round the sun in about 84 years, at the distance of 1800 millions of miles. His diameter is about 35,000 miles, and his bulk about 80 times that of the earth.

333. *Satellites of Uranus.* According to the observations of Sir William Herschel with his great telescope, Uranus is attended by six satellites, revolving with *retrograde* motions in circular orbits nearly *perpendicular* to the plane of the ecliptic. These anomalies, in their motions and in the positions of their orbits, led some to doubt the correctness of the observations. But in 1833, Sir John Herschel confirmed his father's observations with regard to two of them; and in 1847, Struve at Pulkova, and Lassell at Liverpool, observed these two with a third several times. On one occasion, Mr. Lassell believed he saw still another.

NEPTUNE.

334. *General remarks.* Neptune is, so far as is known, the remotest planet in the solar system. The distance of this planet is so great that, though next to Saturn in size, it can never be seen

with the naked eye, and through ordinary telescopes it has the appearance of a small star. Its apparent diameter is about 3″ and it is only through telescopes of high power that it presents a measurable disc.

335. *Period, distance, &c., of Neptune.* Neptune revolves round the sun in about 164½ years, at the distance of 2850 millions of miles. Its diameter is not much less than 40,000 miles, and its volume about 100 times that of the earth.

336. *Satellite of Neptune.* Neptune is attended by at least one satellite, and analogy favours the presumption that there are several. This satellite was first seen on the 10th of October, 1846, by Mr. Lassell of Liverpool, and has since been observed many times by its discoverer and also by Mr. Bond of Cambridge. It revolves round its primary in 5 days and 21 hours at the distance of about 230,000 miles. From the motions of the satellite, the mass of the planet has been deduced with considerable exactness. It is $\frac{1}{19000}$ part of the sun's.

337. *History of the Discovery of Neptune.* Soon after the discovery of Uranus, by an examination of the catalogues of the fixed stars, it was found that the place of the planet had been recorded nineteen times, once as early as 1690, as that of a fixed star. In 1821, M. Bouvard of Paris published tables for computing the place of this planet (279), founded on the observations made since 1781, a period of about 40 years. In preparing these tables he discussed all the observations made during a period of 130 years; and, finding it impossible to represent the observed motion of Uranus during all this period by one set of elements and the perturbations (279) produced by the known planets, he rejected those made prior to 1781, attributing the discrepancies to imperfections in the ancient observations, or to "some extraneous and unknown influence which has acted on the planet." Soon after the construction of these tables, however, the planet was found to be departing from the path assigned by them to an extent that could not be ascribed to errors of observation. The difference between the observed and computed places amounted to nearly 1′.5 in 1840, by which time the belief in the existence of a trans-Uranian pla-

net had become general among astronomers, as there seemed to be no other way of accounting for the apparent anomalies in the motion of Uranus.

Mr. Adams of Cambridge, England, was the first to attempt the discovery of this unknown body. He communicated to Professor Airy in November, 1845, as the result of his investigation, the mass and the elements of the orbit of the disturbing planet. These results were not made public until some months afterwards. In the mean time, Mr. Le Verrier of Paris had undertaken a thorough investigation of this subject, and presented the results of his labours to the Academy of Sciences in three papers. The last of these, announcing the mass and orbit of the required body, was read on the 31st of August, 1846. Le Verrier soon after wrote to Dr. Galle of Berlin, stating that the longitude of his planet for the end of September was 325°, and requesting him to look for it. And on the evening of the 23d of September, 1846, the day on which the letter reached him, Dr. Galle found the planet in longitude 325° 53′, or within less than a degree of the place assigned by Le Verrier. At the suggestion of Professor Airy, Professor Challis of Cambridge had commenced, on the 29th of July, a systematic search for the planet, and had actually observed it twice prior to the 23d of September, as he ascertained by a subsequent reduction of his observations.

The orbits predicted by Adams and Le Verrier agree remarkably well with each other, but differ widely from the true orbit which has been deduced from three years observations made since the discovery of Neptune, and two observations made in 1795 by Lalande, who mistook it for a fixed star. These two were discovered among some 50,000 observations of Lalande, by Mr. Walker, of Washington, who computed the elements given in Article 411.

THE ASTEROIDS.

338. *General remarks.* On the first of January, 1801, the planet Ceres was discovered at Palermo, by Piazzi, one of an association of astronomers, engaged in searching for a planet between the orbits of Mars and Jupiter (289); and on the 28th of March, 1802, Dr. Olbers, of Bremen, discovered the planet Pallas, which

was found to have nearly the same mean distance from the sun as Ceres; their orbits approach each other very closely at the intersection of their planes, and both planets are extremely small. These facts led Dr. Olbers to conceive the idea, that they might be fragments of a large planet, which formerly revolved round the sun in nearly the same part of space, but which had been destroyed by some internal convulsion, and that more of these fragments might be found. Within five years, two more small planets were discovered: Juno, by M. Harding, of Lilienthal, in 1804, and Vesta, by Dr. Olbers, in 1807. The search was continued several years longer, but with no further success, and was abandoned in 1816. The discovery of another asteroid, Astræa, by M. Hencke, of Driesen, on the 8th of December, 1845, stimulated a number of observers to renew the search for other fragments, and their labours have been rewarded by the discovery of twenty-seven asteroids in a period of nine years; Astræa and Hebe, by Hencke; Iris, Flora, Clio, Irene, Melpomene, Fortuna, Calliope, Thalia, Euterpe, and one other, by Hind, of London; Metis, by Graham, of Ireland; Hygeia, Parthenope, Egeria, Eunomia, Psyche and Themis, by De Gasparis, of Naples; Thetis, Proserpine and Bellona, by Luther, of Bilk; Massalia and Phocea, by Chacornac, of Paris; Lutetia, by Goldschmidt, of Paris; Amphitrite, by Marth, of London; and one recently, by Ferguson, of Washington.

These discoveries have been greatly facilitated by the publication of the Berlin charts, containing all the stars to the 8th or 9th magnitude within 15° of the equator. When a star is noticed in the heavens, which is not on the chart, the observer, presuming it to be a planet, carefully notes its position relative to the surrounding stars; if, after the lapse of an hour or two, he finds it has moved, his suspicion is confirmed. In this way most of these small planets have been discovered. They closely resemble small stars, even when viewed with good telescopes; hence, they are called *Asteroids*. Owing to their extreme smallness, very little is known of their physical peculiarities. In Vesta and Pallas only have sensible discs been detected. The diameter of Vesta has been conjectured at about 270 miles.

339. *Periods and distances of the Asteroids.* The following table contains the times of revolution in days, and the mean distances in millions of miles, of the twenty-nine asteroids whose orbits are known.

	Period in Days.	Dist. in millions of miles.		Period in Days.	Dist. in millions of miles.		Period in Days.	Dist. in millions of miles.
Flora............	1193	209	Lutetia..........	1387	231	Proserpina.....	1578	252
Melpomene....	1270	218	Parthenope....	1401	233	Juno............	1593	254
Clio or Victoria	1303	222	Fortuna........	1408	234	Ceres............	1681	268
Euterpe........	1314	223	Thetis..........	1426	235	Pallas...........	1686	263
Vesta...........	1326	225	Amphitrite.....	1484	242	Bellona.........	1700	265
Iris..............	1345	227	Astræa..........	1511	245	Calliope.........	1815	277
Metis...........	1346	227	Egeria..........	1512	245	Psyche..........	1825	278
Phocea.........	1352	228	Irene............	1517	246	Hygeia.........	2043	299
Massalia.......	1365	328	Thalia..........	1554	249	Themis.........	2044	299
Hebe............	1379	230	Eunomia.......	1567	251			

CHAPTER XVII.

ON COMETS.

340. *General remarks.* Comets revolve about the sun, shine by reflecting his light, and are retained in their orbits by his attraction. But, in almost every thing else, they differ widely from the planets. They are not confined to the zodiac or adjacent regions, but traverse all parts of the heavens, and in all directions; the motions of some being direct, and others retrograde. They generally move in elliptical orbits of great eccentricity, alternately approaching comparatively near to the sun, and receding to immense distances; and some of them, moving off into the boundless regions of space, never return. They continue visible only for a few weeks or months, and some only for a few days; being only within the reach of observation while in those parts of their orbit that are adjacent to their perihelions. They are not solid bodies like the planets, but seem to consist mainly of masses of vaporous matter.

Some comets have presented very splendid appearances; and others, though less brilliant, have still been conspicuous objects; but, by far the most numerous class are barely discernible by the naked eye, or can only be seen by the aid of the telescope.

341. *Appearance of a comet.* A comet of the more conspicuous class, usually consists of a small, bright, central part or *nucleus*, enveloped to a considerable extent by an ill-defined nebulous mass of light, called the *coma;* the two together forming what is called the *head* of the comet. From the head, a stream or streams of light shoot out in a direction *opposite to the sun*, growing broader and more diffused as the distance from the head increases. This is called the *tail* of the comet.

The tail sometimes attains an immense length. That of the comet of 1680, one of the most celebrated of modern times, extended through an arc of more than 70°, or, according to some, more than 90°; which would make the real length to be more than 100 millions of miles. And there are records of others, in which the extent of this singular appendage was still greater.

A tail is not, however, by any means an invariable appendage of a comet. In some of the brightest, no tail has been perceptible; and in many it has been quite short. The smaller comets very frequently do not exhibit the least appearance of a tail. They appear only as round or somewhat oval masses of vaporous matter, increasing in density towards the centre; but without any distinct nucleus, or any thing that would indicate a central, solid mass. Stars, even of small magnitude, have been seen through what appeared to be the densest portion of their substance.

342. *Small quantity of matter in a comet.* Comets have been known to pass near to some of the planets and to have had their own motions much affected by the consequent attractions, without producing any sensible influence on the motions of these bodies. In one instance a comet passed among the satellites of Jupiter and was thrown by the attraction of the planet entirely out of the orbit it had been describing, and forced into another, quite different in extent; yet, not the least perceptible derangement of the motions of the satellites was produced. It is hence concluded that the quantity of matter in a comet must be very small.

343. *Orbit of a comet and its elements.* Investigations, founded on the law of gravitation, prove, that a body revolving about the sun and not influenced by the attraction of any other body, must move either in a circle or in some one of the three curves called Conic Sections. Comets are found generally to move in elliptical orbits of extremely great eccentricity; so great, that the part of the orbit described during the comet's visibility does not sensibly differ from a parabola. Some few have, however, been ascertained to have moved in hyperbolic orbits. These, after having passed their perihelions, must move off indefinitely, and cannot again return.

The elements of a comet's orbit are, *the perihelion distance, the* *longitude of the perihelion, the longitude of the node, the inclination of the orbit*, and *the time that the comet is at the perihelion.* The determination of these elements from observed geocentric places of a comet, is a problem of much difficulty, and the requisite computations are laborious. Various methods of making them have, however, been obtained, in some of which the labour is considerably lessened.* The computation is usually made, at least in the first place, on the assumption that the orbit is a parabola; which is equivalent to the assumption that it is an ellipse of great eccentricity. Three complete observed right ascensions and declinations of the comet, made at suitable intervals, with the times of observation, are sufficient; but a larger number is commonly employed in order that the results may be more independent of the unavoidable errors of observation.

When the elements of the orbits of a number of comets have been computed and arranged, and if, on comparing them, the same or nearly the same set of elements is met with at intervals of the same length, or nearly so, the presumption is, that they appertain to the same comet returning at these times. If the intervals are long, a difference in them of a year or more, may be the result of perturbations in the comet's motion, produced by the attractions of the planets.

344. *Halley's Comet.* In the early part of the last century, Halley, an eminent English astronomer, computed, from recorded observations, the elements of a number of comets. On comparing them, he found that the elements of a comet, which had appeared in 1680, and which he had himself observed, corresponded very nearly with those of two others, which had previously appeared at intervals, proceeding backwards, of about 75 and 76 years. This led him to suppose, that instead of three different comets, it might be the *same* comet, which had appeared at these times. Making

* Dr. Bowditch, in an appendix to the third volume of his translation of Laplace's Mécanique Céleste, has introduced several of the best methods in addition to that of the author, and has added tables which facilitate the computations. A more recent one by Airy, the present Astronomer Royal of England, is given in vol. XI. of the Memoirs of the Royal Astron. Society.

further researches, he became satisfied of the correctness of the supposition he had made, and concluded that the variation in period must have been produced by the attractions of the other heavenly bodies. Having, therefore, made a rough calculation of the effect which the attraction of Jupiter would produce on the revolution the comet was then performing, he ventured to predict its return in the latter part of 1758, or early part of 1759. Subsequently, Clairaut, an eminent French mathematician, calculated the effects of the attractions of both Jupiter and Saturn, and determined the time of the return to the perihelion, to be in the middle of April, 1759. It arrived there about a month prior to that time. In consequence of its return, nearly according to Halley's prediction, it has received his name.

With more ample means for correct computations, furnished by the observations during its appearance in 1759, and by the improvements in analysis, the recent return of Halley's comet in 1835, was much more accurately predicted. It arrived at the perihelion of its orbit, within less than two days of the time assigned for its return, by Pontécoulant, a distinguished French astronomer.

The least distance of Halley's Comet from the sun is 56 millions, and its greatest distance 3,350 millions of miles. The eccentricity of its orbit is 0.97 and its inclination to the ecliptic is 17° 44′. The motion of this Comet is retrograde.

345. *Encke's Comet.* The periodical character of this small comet, was discovered in 1819, by Professor Encke of Berlin, who identified the comet of that year with those that had been observed in 1786, 1795 and 1805, and which had been supposed to be different comets. He found its period to be only about 1207 days, or nearly 3½ years; and he predicted its return in 1822, which was verified by observation. Its subsequent returns have been predicted and observed.

This comet is sometimes called *the comet of short period.* Its perihelion distance is 31 millions, and its aphelion distance 390 millions of miles. The eccentricity of its orbit is 0.854, and the inclination to the plane of the ecliptic is 13° 22′. Its motion is direct.

346. *Resisting Medium.* The observations of the successive returns of Encke's comet, show that its period is subject to a small, but continued diminution. It also appears, that this diminution is not produced by the actions of the planets. Encke, therefore assumes, that instead of a perfect vacuum in space, there must exist an exceedingly rare medium, which, opposing no perceptible obstruction to the motions of dense bodies, sensibly resists the motion of a mere mass of vapour like that of the comet. The obvious effect of such a resistance would be a diminution of the comet's velocity, in consequence of which, it would move nearer the sun, and perform its revolution in less time.*

347. *Biela's Comet.* This is a very small comet without the least appearance of a nucleus. Captain Biela, of Josephstadt, discovered its periodic character on its appearance in 1825.

M. Gambart of Marseilles also made the discovery, and the comet is therefore frequently called *Gambart's.* It performs its revolution in about 6¾ years, in an ellipse, whose eccentricity is 0.75; the least and greatest distances of the comet from the sun being 86 millions and 586 millions of miles. The inclination of its orbit is about 13°, and its motion is direct.

At the return of this comet in 1832, which was predicted with considerable precision, some alarm was created by the announcement that it would pass very near the earth's orbit. It did, on the 29th of October, pass within a few thousand miles of a point at which the earth arrived about one month later. Owing to its proximity to a conjunction with the sun, during this perihelion passage, it was only visible through powerful telescopes. And, in

* Encke infers, from the observations made on the comet, that the resistance, is only sensible in a portion of space round the sun, not extending beyond the orbit of Venus. He thus accounts for the fact, that the motions of Halley's comet, and another periodical one, noticed in the next article, have not indicated any resistance; for, but a very small part of the orbit of the former, and none of that of the latter, are within that distance of the sun.

The positions of Encke's comet at its returns in 1842 and 1848, as observed on several evenings, by the editor, at the observatory of the Central High School in Philadelphia, were found to be within 30″ of space of its positions as given in Encke's Ephemeris, previously computed. This affords a striking evidence of the accuracy of the investigations and computations of its orbit and motion.

1839 it could not be observed at all. But, its last return was under more favourable circumstances, and it was observed from the 26th of November, 1845, until the 22d of April, 1846. On this occasion it presented the singular phenomenon of a double comet, or, two distinct comets moving through space, side by side. At first one was extremely small as compared with the other, but the smaller gradually increased, so that on the 13th of January the ratio of their magnitudes was as 1 to 8, and by the middle of February they were nearly equal in size; after which the *variable* comet began to diminish, and in about a month disappeared; while the other continued visible several weeks longer as a single comet.

The comet returned again in 1852, but under such unfavorable circumstances, as precluded the possibility of extended observations. Both nuclei were, however, observed by several persons in August and September. Fluctuations of relative brightness were noticed, similar to those of 1846, but much greater; so great, indeed, that for several days the two comets were *alternately visible*, —one nucleus being observed one day, and the other the next. Professor Hubbard, after a thorough discussion of all the observations made on this mysterious object in 1846 and 1852, found that the distance of the two nuclei apart, during their visibility in 1846, was about 200,000 miles, with but little variation from the 20th of January, to the 5th of March; after which, they sensibly approached each other until one disappeared, when their distance was 170,000 miles; whilst in 1852, they were nearly 1,800,000 miles apart. Professor Hubbard was unable to decide with certainty which of the nuclei of 1852 was identical with the principal one of 1846, but concluded, with a high degree of probability, that their relative apparent direction was reversed. By tracing the orbits back, he found that the separation probably occurred about 500 days before the perihelion passage of 1846.*

348. *Faye's Comet.* In 1843, M. Faye of the Paris Observatory discovered a comet and determined its orbit to be an ellipse with the surprisingly small eccentricity of 0.55. He found the period to be about 7 years. This comet is remarkable as having an orbit more closely resembling those of the planets in form than any other cometary orbit thus far known.

---

* The Imperial Academy of Sciences of St. Petersburg has offered a prize of 300 ducats for the best essay on the orbit of this remarkable comet, and the relation which the two parts bear to each other.

349. *De Vico's Comet.* This comet was discovered in 1844 by Sr. De Vico, Director of the Observatory at Rome. Its orbit was found to be an ellipse, with an eccentricity of 0.62, whose plane almost coincides with the ecliptic. Its period of revolution was computed to be about 5½ years. Le Verrier pronounced this comet probably identical with one which appeared in 1678.

350. *Lexell's Comet.* In the year 1770, a remarkable comet appeared, moving in an ellipse with the short period of 5½ years. By tracing back its motion, it was found that, early in 1767, it was very near to Jupiter, and that previous to that time it had been moving in an orbit requiring 50 years for a revolution. This change in its orbit was produced by the action of Jupiter. Again, in 1779 the comet passed so near to Jupiter that his attraction for it was 200 times greater than the Sun's, in consequence of which, its orbit was changed into one of long period. Some suppose that this comet and Faye's are identical.

351. *The Great Comet of* 1843. Of all the comets of recent years, no other has excited so much astonishment as did the one known as the Great Comet of 1843. It was first seen in many parts of the world on the 28th of February, in the day time, as a brilliant body quite near the Sun. Its distance from the nearest limb of the Sun, as measured with a sextant at 3 o'clock P. M. was 3° 36′. Soon after this it became visible after sun set as a very conspicuous object in the southwest. The apparent length of its tail varied from 50° to 70°, and the greatest real length was about 110 millions of miles. It continued visible to the naked eye but a short time, and the last telescopic observation of it was made on the 10th of April, at the Philadelphia High School Observatory. This comet passed its perihelion on the afternoon of the 27th of February, at which time it almost grazed the Sun's disc, being only 530,000 miles from his centre. According to the computations of Sir John Herschel, the heat it received when it was nearest the Sun must have been 47,000 times that received by the earth from a vertical sun. This will account for the intense brilliancy of this comet on the 28th of February.

The probable identity of this comet with that of 1668 is generally admitted by astronomers.

CHAPTER XVIII.

CLASSIFICATION OF THE FIXED STARS.—CLUSTERS AND NEBULÆ.—VARIABLE AND TEMPORARY STARS.—DOUBLE STARS.—BINARY SYSTEMS.—PROPER MOTIONS OF SOME STARS AND MOTION OF THE SOLAR SYSTEM.—ANNUAL PARALLAX AND DISTANCE OF THE STARS.—CATALOGUES OF THE STARS.

352. *Classification of the stars.* The stars are divided into classes, according to their apparent magnitudes or brightness. The most conspicuous stars form the first class, and are called stars of the *first magnitude;* those that are markedly less bright, form the second class, and are called stars of the *second magnitude;* and thus on, down to stars of about the *sixteenth magnitude*, which are the smallest that are distinctly visible with the most powerful telescopes. The stars that are visible to the naked eye, are included in the first six or seven magnitudes; principally, however, in the first six.

The magnitudes are denoted by the numbers 1, 2, 3, &c. A star that is regarded as intermediate in brightness between those of two consecutive classes, so as to render it doubtful in which it would be more appropriately placed, is frequently distinguished by two numbers with a point between them. Thus, 1.2 denotes a star intermediate between those of the first and second magnitudes.

353. *Number of stars in some of the classes.* The distribution of the stars into magnitudes, is arbitrary, and it has not been made on any definite principles. There is, therefore, some diversity in the distribution; different astronomers having differed in the magnitude they have attached to the same star.* On the whole,

---

*Also, from want of care in forming the constellations (121), some of them have been made more or less to overlap one another, so that the same star or stars are frequently included in two different constellations. The inconvenience resulting from these causes, has claimed the attention of the British Association, a scientific body, that meets annually in Great Britain; and it is probable that, ere long, we shall have a revision of the nomenclature of the stars, and of their division into constellation.

however, the first magnitude may be regarded as restricted to 18 or 20 principal stars; the second, to 50 or 60 next inferior; the third, to about 200 yet smaller; and thus on, the number in each class increasing rapidly as we descend in the scale of brightness. The number of stars in the first seven magnitudes, amounts, all together, to nearly 20,000. The whole number of stars visible with the best telescopes is not known; but it must amount to several millions.

The number of stars distinctly visible to the naked eye, is less than is generally supposed by those who only judge from the impression made, when viewing them on a fine evening. The number thus visible, at the same time above the horizon, does not greatly exceed a thousand. All the stars visible to the naked eye, with some others, are represented on celestial globes of 12 or 18 inches in diameter.*

---

* Students of astronomy who can have the use of a celestial globe, or celestial atlas, ought to make themselves familiar with the principal stars and constellations. To *rectify* the globe for this purpose, let the frame which supports it, be placed by estimation, or by the compass which is sometimes attached, so that the north and south points marked on its upper surface, called the *horizon* of the globe, may correspond to the north and south points of the horizon or nearly so. Then, let the brass ring in which the globe is suspended, called the *meridian* of the globe, be slid in its support, till the north pole of the globe, which is that situated in the constellation of *Ursa Minor*, is elevated above the northern point of the horizon by an arc equal to the latitude of the place.

Find the day of the month on the horizon, and the corresponding point in the contiguous graduated circle will be the sun's place in the ecliptic. Find this place in the ecliptic marked on the globe, and bring it to the graduated side of the meridian. Keeping the globe in this position, set the index placed at the north pole, to 12 on the hour circle around the pole; or if the globe has a moveable brass hour circle instead of an index, bring 12 on this hour circle to the graduated side of the meridian. Then turn the globe westwardly till the index points to the hour at which the globe is to be used; or when there is no index, till the hour on the brass hour circle comes to the graduated side of the meridian. The positions of the stars represented on the globe, will then correspond to their positions in the heavens; so that if a straight line be conceived to be drawn from the centre of the globe through the places of the stars marked on its surface, they will point to the stars themselves.

Kendall's Uranography and Atlas, Revised Edition of 1854, is a work adapted to give the astronomical student a satisfactory knowledge of the sidereal heavens.

354. *Relative light of stars of the different magnitudes.* According to the present classification of the stars, the light of an average star of the second magnitude, is about *one fourth* that of an average star of the first magnitude. For the other magnitudes, the light of a star of one magnitude is regarded as about *half* that of a star of the next higher magnitude. There is, however, considerable variety in the brightness of stars, that are classed as of the same magnitude ; especially those of the first magnitude. The light of Sirius, the brightest star in the heavens, is regarded as being from 15 to 20 times as great as some of the stars of the first magnitude; and more than 300 times as great as an average star of the sixth magnitude.

355. *Distribution of the stars.* The stars appear to be very unequally distributed over the heavens. This is observable by the naked eye, and becomes still more apparent by means of the telescope. There are various spaces which are faintly luminous, shining with a pale white light. Many of these, on applying telescopes of sufficient power, are found to consist of multitudes of small stars, distinctly separate, but very near to one another. These are called *Nebulæ.* The well known space called the *milky-way*, is of this kind ; and there are some others visible to the naked eye. In some of the nebulæ or clusters, the number of stars crowded into a small space, is immensely great. According to the estimation of Sir J. Herschel, there are some which contain more than ten thousand stars in a space that would be covered by a tenth part of the moon's disc. Again, there are many spaces, some of considerable extent, in which but few stars are seen, even with the best telescopes.

356. *Clusters of stars and nebulæ.* The beautiful cluster of stars called the *Pleiades*, in which six or seven are readily discernible by the naked eye, exhibits within the small space they occupy, fifty or sixty conspicuous stars, when viewed with a telescope of moderate power. The constellation called *Coma Berenices*, is another group more diffused, and composed of larger stars.

In the constellation *Cancer*, there is a luminous spot or nebula called *Præsepe*, or the bee-hive, which a telescope of moderate power resolves entirely into stars. In *Perseus*, is another spot crowded with stars, which become separately visible with a good telescope.

Most of the nebulæ, however, require a very powerful telescope to resolve them into stars; and there are many which have never been thus resolved, they being, it is probable, differently constituted. A prominent one of this class is situated near the star ν in *Andromedæ*. It is visible to the naked eye, and has, from its appearance, often been mistaken for a comet. It should be remarked that many of the most prominent objects hitherto regarded as belonging to the class of irresolvable nebulæ, have recently, by the aid of the gigantic telescope of Lord Rosse, been resolved into stars.

357. *Variable stars.* Some stars undergo periodical changes in their brightness, and are, therefore, called *variable* stars. One of the most remarkable of this class of stars, is *Mira*, or ο *Ceti*, which was discovered to be variable in the latter part of the 16th century. When brightest, it is of the second magnitude, and continues to exhibit nearly the same appearance for about three weeks. It then decreases, and in about two months ceases to be visible to the naked eye. After remaining thus invisible for six or seven months, it again appears, and in the course of six or seven weeks, is restored to its former brightness or nearly so. These periods, and also the greatest brightness of the star, are, however, subject to some variations. The average period of all the changes is about 11 months or, more exactly, 332 days. At the times of the least light of the star, it is frequently invisible, even with good telescopes.

Another very remarkable variable star is *Algol*, or β *Persei*, which was discovered to be such, in the latter part of the last century. It is usually of the second magnitude; but, after having continued so, during a period of about 60 or 61 hours, it suddenly decreases, and is reduced in about 4 hours to the fourth magnitude. Continuing thus, about a quarter of an hour, it then increases, and in about 4 hours more, it regains its usual magnitude. The period of these changes is 2 d. 20 h. 48 m. 58.5 sec.*

* According to the observation of Professor Argelander, a German Astronomer, given in the Astr. Nach., Nos. 416 and 417, the star ο Ceti had its greatest brightness in the year 1840, about the 3d of October; and the star β Persei, on the 22d of December in that year, had its least brightness at 9 h. 50½ min., mean time at Greenwich. With these epochs and the periods given above, the times of the

There are more than twenty other stars known to be variable to a greater or less extent; some of which have but recently been discovered to be so. The periods of the changes vary from a few days to more than a year.

358. *Temporary Stars.* Several instances are recorded of stars suddenly appearing, some of them of great splendour, where none had before been observed; and there are several stars noted in some of the ancient catalogues, that cannot now be found. One of the most noted of these *temporary* stars broke forth with great brilliancy on the 11th of November, 1572, in the constellation *Cassiopeia*, and was attentively observed by Tycho Brahe, the celebrated Danish astronomer. It was then as bright as Sirius, and increased in splendour so as to become distinctly visible at mid-day. It began to diminish in December of the same year, and in March 1574, it entirely disappeared.

In the years 945 and 1261, a brilliant star appeared in the same region of the heavens with that of 1572. Some have thought it must have been the same star that appeared in each of these years, and that it was, therefore, a variable star with a period a little over 300 years.

On the 27th of April, 1848, Mr. Hind of London, discovered a new star of the sixth magnitude, in the Serpent Bearer, which increased in brightness for a few days, then began to wane, and disappeared in less than two years. On the 5th of April, Mr. Hind had examined that part of the heavens with care, and was certain that at that time no star as bright as the *ninth* magnitude existed, where this one of the *sixth* was found three weeks later.

359. *Double Stars.* Many stars which when viewed with the naked eye or with telescopes of small power appear single, are by means of those of larger power resolved into two, three, or more stars distinctly separate but very near to one another. These are called *double* or *multiple* stars. Some of these are resolvable into separate stars by a telescope of moderate power, as *Castor* in the twins, which consists of two stars nearly equal, both being between

---

greatest light of the former star and least light of the latter, may be approximately determined for a few subsequent years.

the third and fourth magnitudes, at the distance of 5″ from each other. Many of them, however, require for their separation, a telescope of the superior class, and serve as good objects to *test* its perfection.

The individual stars forming a double star, are mostly very unequal in magnitude; and many of them exhibit the curious phenomenon of contrasted or *complementary* colours, that is colours which if combined would form white light. In such instances, the larger star is usually of a ruddy or orange hue, and the smaller is blue or green.* In the beautiful double stars α Herculis, and γ Andromedæ, which may be separated by a telescope of moderate power, this contrast is finely exhibited.

360. *Binary stars or systems.* Sir W. Herschel was the first that gave much attention to the subject of double or multiple stars. He observed a large number, and noted the distances by which the individual stars were separated, and their relative positions. Continuing and repeating his observations, he found that the distance and relative positions of these component stars were subject to slow but progressive changes. After having had his attention, frequently, thus directed for more than twenty years, he at length ascertained and announced the striking and interesting fact, that several, at least, of the double stars formed systems, in which one of the individuals revolved round the other, or rather, both round their common centre of gravity. These have received the appellation of *binary stars* or *binary systems*, to distinguish them from the other double stars whose apparent proximity probably proceeds from one being situated nearly behind the other, without their having any physical connection.

There are fifty or more of the double stars which are now known to form binary systems; a few of the more prominent of these are, *Castor*, or α Geminorum, γ Virginis, ξ Ursæ, α and ζ Herculis, σ and η Coronæ, ζ Cancri, and 61 Cygni.

* This probably depends on the well known optical fact, that when the retina of the eye is excited by any bright colour, a feeble light, which if seen by itself, might appear white, is affected with a tint complementary to that of the stronger light.

361. *Periods of some of the binary stars.* During the last twenty years, the interesting subject of double stars has claimed great attention, and they have been extensively observed, particularly by Sir J. South and Sir J. Herschel in England, and by Professors Struve and Mädler in Russia.* The observations on many of them completely establish a physical connection between the individuals of which they are composed. These are found to revolve about each other in elliptical orbits and to conform in their motions to the same dynamical laws that govern the motions of the planets round the sun.

The periods of several of the double stars have been computed, and appear to vary from about 40 to more than 1000 years. These computations are, however, to be regarded only as approximations; the more accurate observations not yet having been continued sufficiently long to give, with much accuracy, the lengths of the periods. According to Prof. Maedler, who has made some of the most recent computations, the period of Castor is about 230 years; γ Virginis 158; ζ Cancri 60 years; and ζ Herculis 36 years.†

362. *Proper motions of some of the stars.* A number of the stars, both single and binary, are found slowly to change their local situations in the heavens, having progressive apparent motions called their *proper* motions, varying in different stars from a small fraction of a second to 4″ or 5″ a year. From the directions and amounts of these motions, Sir W. Herschel drew the conclusion that they were only apparent motions, and were produced by a real motion of the Solar System towards a point of the heavens situated in the constellation Hercules.

This conclusion has recently been confirmed by Prof. Argelander.‡ He finds the point towards which the motion of our system is directed, to be that which, in 1800, had 260° right ascension and 32½° north declination.

---

* Prof. Struve made most of his observations at Dorpat; he is now Director of the Imperial Observatory at Pulkowa, near St. Petersburg. Prof. Maedler is his successor at Dorpat.

† Astron. Nachrichten. Nos. 317, 363, and 427.

‡ Astr. Nach. Nos. 363 and 364.

Independent determinations of this point have also been made by Struve, Luhndahl and Galloway, all obtaining nearly the same result. According to the calculations of Struve, the velocity with which the Solar System is moving, is about half as great as that of the earth in its orbit.

363. *Annual parallax of the stars.* The *annual parallax* of a star is the angle contained between two straight lines, conceived to be drawn from the star, one to the sun, and the other to the earth, when the earth is in such a part of its orbit that its radius vector is perpendicular to the latter line; or, in other words, it is the greatest angle at the star, that can be subtended by the semidiameter of the earth's orbit.

For each star, however situated, there must, it is evident, be some two opposite points of the earth's orbit, for each of which, the radius vector will be perpendicular to the right line joining the star and sun. The positions of the stars as seen from the earth, when at these points, must, therefore, differ by twice the annual parallax of the star. Hence, as the parallax must affect the right ascension and declination of the star, if these be observed when the earth is near these points, or in other favourable situations in its orbit, the parallax of the star may be determined, unless it is so small as to be within the limits of the probable errors of observation and the necessary corrections. Numerous observations, by several eminent astronomers, have been made for this purpose on some stars, which, from their apparent size and brightness, were supposed to be at a less distance than the generality of the stars. The results of these observations have been, that the parallax in each case was too small to be obtained with certainty by this method.

The apparent largeness or brightness of a star, is not, however, necessarily the most certain indication of its comparative proximity to the earth. A considerable proper motion produced by the motion of the Solar System (362), and, in case of a binary star, large apparent orbits, are probably stronger indications. Professor Bessel of Königsberg, therefore, made two entirely distinct series of observations on the binary star 61 Cygni, which has a large proper motion, amounting to $5''$ a year, and the components of which are about 16″ distant from each other. Instead of observing right ascensions and declinations, he measured with an excellent heliometer the distances from two contiguous, small, stationary stars, and thereby avoided the small errors to which the corrections for refraction, aberration and nutation are liable. The first series of observations gave, for the annual parallax of 61 Cygni, 0″.3136; and the second gave 0″.3483, differing only about $\frac{1}{30}$ of a second from the former.

The parallaxes of several other stars have since been determined, but that of 61 Cygni is considered by far the most reliable. The results will be found in the following table, of which the third column contains the distance from the sun in millions of millions of miles; and the fourth, the time the light occupies in passing from the star to the earth.

Star.	Parallax.	Distance.	Passage of light.	Author.
α Centauri	0.″913	21	3.54 yrs.	Henderson.
61 Cygni	0.″348	56	9.28 "	Bessel.
α Lyræ	0.″261	75	12.38 "	Struve.
Sirius	0.″230	85	14.04 "	Henderson.
1830 Groombridge	0.″148	132	21.80 "	Peters.
ι Ursæ Maj.	0.″133	147	24.30 "	Do.
Arcturus	0.″127	154	25.43 "	Do.
Polaris	0.″067	292	48.20 "	Do.
Capella	0.″046	426	70.00 "	Do.

364. *Distances of the stars.* When the annual parallax of a star has been determined, its distance becomes at once known; it being, in terms of the earth's distance from the sun, equal to the quotient of 206264″.8 (App. 51), divided by the annual parallax. Thus, taking 0″.3483 for the annual parallax of 61 Cygni, its distance is found to be about 592,000 times the distance of the earth from the sun. This is a distance so immense, that light, which moves with the amazing velocity of 192,000 miles in a second, would require more than nine years to come from the star to the earth. Yet, inconceivably great as this distance is, there are observable stars, whose distances are probably more than a hundred times as great, and the light of which would require more than a thousand years to traverse the space which separates them from the earth.

365. *Catalogues of stars.* Some of the most noted catalogues are, Bode's Catalogue and Atlas, containing the positions of 17,000 stars; Professor Bessel's Catalogue of 3222 stars, deduced from

observations made by Bradley, at the Royal Observatory, Greenwich; Piazzi's Catalogue of 7646 stars; and the Catalogue published by the Astronomical Society in their Memoirs, containing 2861 stars. In this last Catalogue, the mean right ascensions and declinations are given for the 1st of January, 1830; and they contain for each star, certain constant logarithms, by means of which, with other logarithms depending on the positions of the sun, moon and moon's node, given in the Nautical Almanac for each day in the year, the true apparent place of any of these stars, may be found, for a given time, with great facility.

Under the direction of the British Association, the Catalogue of the Astronomical Society has been revised and extended, so as to include 8377 stars, with the mean places reduced to the year 1850.

CHAPTER XIX.

DIFFERENT METHODS OF FINDING THE LONGITUDE OF A PLACE.

366. *General remarks.* The determination of the difference of longitude between two places, consists in finding the difference between the times reckoned at these places at the same instant of absolute time (63). When this has been done, if the longitude of one of the places is known, that of the other becomes also known. The method of finding the longitude of a place by means of a chronometer, has already been given (65). It is very simple, and is extensively used at sea. But as a chronometer is liable to change its rate of going during the voyage, especially if it is a long one, it is not safe to depend on this method alone.

367. *Lunar method of finding the longitude.* The *lunar method* is that by which the longitude of a place is found, from the measured angular distance of the moon from the sun, a star, or planet, situated nearly to the east or west of her place, at the time of observation. As the moon's motion is about half a degree an hour, she must change her angular distance from a body thus situated, at that rate, or nearly so. Hence, if the moon's true angular distance from the body at any instant, and also the time,

be obtained from observations at any place, and the time at the first meridian, when the moon has this true angular distance from the body, be found, the longitude becomes known. The Sun, Venus, Mars, Jupiter, Saturn, and nine stars situated contiguous to the moon's path, have been selected for observations of this kind. In the Nautical Almanac, the moon's computed true angular distances from several of these bodies, are given for each three hours, Greenwich time, of every day in the year; and also proportional logarithms, by which the distance for any intermediate time, or the time corresponding to an intermediate distance, may easily be obtained.

To apply this method, the distance of the enlightened limb of the moon, from the nearest limb of the sun, or, from one of the other bodies given in the Nautical Almanac, for the day, is measured with a sextant, and the time of observation noted. The altitudes of the moon and other bodies, at that time, are also observed with a quadrant or sextant by two assistants.* From these observations, the true distance of the moon's centre from that of the body, corrected for refraction, parallax and semidiameter, may be deduced, by methods given in treatises on Navigation.† When the true distance has been obtained, and then the time at Greenwich, corresponding to this distance, the difference between this time, and the time of observation, will be the longitude of the place; which will be *east* or *west* according as the Greenwich time is *earlier* or *later* than the time of observation.

This method of finding the longitude, is of great importance to the mariner, as all others, with the exception of that by the chronometer, require observations that cannot be made at sea.

368. *Longitude by moon culminating stars.* Certain stars situated contiguous to the moon's path and passing the meridian at short intervals before or after the moon, are called *moon culminating stars.* The moon's right ascension increases on an average a

---

* The observations may be made by one person, by first taking the altitudes, then the distance, and afterwards the altitudes again. From the two sets of altitudes, their values at the time of taking the distance may be obtained with sufficient accuracy.

† Dr. Bowditch's Navigation contains several of the best methods.

little more than half a degree, or two minutes in time during a sidereal hour, that is, during the interval that elapses from the time a star is on the meridian of any place till it is on the meridian of a place whose longitude is 15° or one hour west of the former. Hence, the intervals between the passages of the moon and a star over the meridians of two places differing an hour in longitude must differ about two minutes; and for other differences of longitude there must be a proportional difference in the intervals. It follows that, if the intervals between the passages of the moon and a star over the meridians of two places be accurately obtained by observations, the differences of their longitudes may be easily found by means of the moon's hourly variation in right ascension at the period of observation.

The Nautical Almanac contains a table in which are given for each day in the year, except a few near the times of new moon, the apparent right ascensions of several of the moon culminating stars, the apparent right ascension of the moon's enlightened limb at the instant it is on the meridian of Greenwich, and the hourly variation in the right ascension of the limb at that time. The difference between the right ascension of the star and the enlightened limb of the moon, is the interval between the passages of these over the meridian of Greenwich. From the computed interval for Greenwich, and the observed interval at any other place, the longitude of the latter may be obtained, but not with as much precision as from two observed intervals.

369. *Determination of the longitude of a place from observations of an Eclipse of the Sun, or of an Occultation.* The times of the beginning and end of an eclipse of the sun, or of an occultation of a star or planet, at any place, depend on the position of the place. Assuming the computed places of the bodies to be accurate, we may, from the carefully observed time of beginning or end of an eclipse or occultation at any place whose latitude is known, determine the corresponding time at the first meridian, and, consequently, the longitude of the place. If the phenomenon is also visible and the times of beginning and end are observed at places whose positions are accurately known, the determination of longitude by this means may be rendered nearly free from any

errors in the tabular places of the bodies. The investigations of formulæ for making the requisite computations will be given in the appendix.

370. *Longitude by the Eclipses of Jupiter's Satellites.* This method of finding the longitude of places has been already noticed (321). Although it is not so accurate as several others, its great simplicity, and the frequency of the occurrence of these phenomena, render it very convenient for approximate determinations of the longitude.

371. *Determination of longitude by means of the Electric Telegraph.* The Electric Telegraph affords the most direct and by far the best means of determining the difference of longitude between two places connected by it. This method has been extensively employed by Professor Bache, Superintendent of the Coast Survey, and with great success. The differences of longitude between Boston, New York, Philadelphia, Washington, and several other important points, have been determined with an unprecedented degree of precision. In the progress of these experiments the process has attained a high degree of perfection, and now consists in having an Astronomical Clock so connected with the Telegraph apparatus that each vibration of its pendulum either closes or breaks the galvanic circuit, so that the beats of the clock are transmitted through the entire line of telegraph. The beats of the clock are, moreover, recorded by the *Register*, on the fillet of paper, by dots at equal intervals; the space between two consecutive dots corresponding with a second of time. The date of any event may then be recorded by simply touching a key, in obedience to which the register makes a dot upon the graduated fillet of paper. The position of this dot between two of the seconds dots determines the fraction of a second with much greater accuracy than can be obtained in any other way. Provided with such an apparatus, the observer at the most eastern station records the time of the culmination of a certain star, by striking his telegraph key as it passes successively over each wire of his transit instrument. When the same star arrives at the meridian of the western station, the observer there goes through the same operation. Thus, the times, by the same clock, of the transits of the star over

the two meridians, are recorded upon the same paper. The difference between these times, after allowing for the rate of the clock, will evidently be the difference of longitude. This result will be independent of every important source of error except that of the imperfect adjustment of the transit instruments, the effects of which may be completely eliminated by a combination of several observations made with the instruments in different positions.

CHAPTER XX.

OF THE TIDES.

372. *Definitions.* The alternate rise and fall which take place in the surface of the ocean, seas, bays and contiguous rivers, twice in the course of each lunar day, or of 24h. 51m. mean solar time, are called the *Tides.* When the water is rising it is said to be *flood* tide, and when it is falling, *ebb* tide. When the water is at its greatest height it is said to be *high* water, and when at its least height, *low* water.

The swell in the waters of the ocean is called the *tide wave*, or, sometimes, the *primitive* tide wave; and that in a contiguous bay or river, proceeding from the former, is called a *derivative* tide wave. A curve line along the summit of the tide wave, or through different points or places that have high water at the same instant of time, is called a *cotidal* line.

373. *Causes of the tides.* The earth in its revolution round the sun is continually drawn, by the attraction of the moon, slightly aside from the place at which it would be, if this attraction did not exist. If the earth was entirely solid, all parts of it would necessarily be drawn aside to the same extent. But as the moon's attraction decreases in the same ratio that the square of the distance increases, and as a large portion of the external part of the earth is composed of water, which can yield to forces unequally impressed on it, it is evident that all parts will not be drawn aside equally. The portion of water nearest the moon, being most attracted,

will be drawn farther than the central and solid parts of the earth, and the central part farther than the opposite watery surface. Hence, the distance of the surface of the water from the centre of the earth must be increased, that is, there must be high water, both on the side of the earth nearest the moon and also on the opposite side. But a swell in the waters of some portions of the earth cannot take place without a corresponding depression in other portions. This depression, it is obvious, must be greatest in the vicinity of the great circle midway between the portions of the earth nearest the moon and most remote from her, and it must there be low water.

The sun's action must also produce similar effects. But although his whole attraction on the earth is far greater than the moon's, yet, as his distance is nearly 400 times that of the moon, the *inequality* of his attraction at the surface and centre is less; and consequently, his influence in producing a tide is also less. The height of the solar tide is only about *one third* of that of the lunar tide.

In the open ocean, the average rise and fall of the tides, or height of high water above low water, is about $2\frac{1}{2}$ feet.

374. *Spring and Neap Tides.* At the time of new moon, the attractions of the sun and moon are nearly in the same direction, and their actions are, therefore, united in producing the tides. They are also united at the time of full moon, when the moon is in opposition; for each body produces a tide not only on the side of the earth nearest it, but also on the opposite side (364). Hence, at the times of the syzygies, the tides must rise above their average height. The tides occurring at or near these times are called *spring* tides.

At the times of the quadratures, the action of the sun tends to produce low water, where that of the moon produces high water, and the contrary. The tides occurring at these times will not, therefore, rise to their average height. These are called *neap* tides.

As the greatest effect of a varying action does not take place at the instant the action itself is greatest, but some time afterwards, so it is with the tides. The most marked spring and neap tides occur about a day and a half after the times of the syzygies and quadratures.

The effects of the separate actions of the sun and moon, being nearly as 1 to 3, their joint effect must be to their effect when acting in opposition to each other, nearly as 4 to 2. Hence, the height of the spring tides above the medium surface of the water must be about double that of the neap tides. This result is confirmed by observation.

375. *Perigean and Apogean Tides.* The moon's influence in producing the tides, must evidently be greatest when her distance from the earth is least, and least when the distance is greatest. Consequently, other circumstances being the same, the tides will be higher a short time after the moon is in perigee, and lower, a short time after she is in apogee, than at other times.

Unusually high tides occur, when the moon is in perigee, at or near the time of a new or full moon.

The variation in the earth's distance from the sun, has also a slight influence on the height of the tides.

376. *Effect of the moon's declination on the tides.* The height of the tide at a given place is influenced by the declination of the moon. When the moon has no declination, the highest tides must evidently occur along the equator; and the height must diminish from thence towards the north and south. When she has north declination, the highest tides on the side of the earth next the moon, will be at places having a corresponding north latitude, and on the opposite side, at those which have an equal south latitude. From these parallels of latitude, the height of the tide will gradually diminish to the north and south. It therefore follows, that, when the moon's declination is north, the height of the tide at a place in north latitude will be greater when the moon is above the horizon than when she is below it; and at a place in south latitude it will be just the reverse. This is illustrated by *Fig.* 55, in which the exterior curve is an exaggerated representation of the oval form of the curve through the summit of the tide wave, on the supposition that the whole earth is covered by water.

When the moon's declination is south, the whole is reversed. The tide at a place in north latitude is then higher when the moon is below the horizon than when she is above it; and at a place in south latitude it will be just the contrary.

377. *Position and motion of the tide wave.* The full effect of the moon's action at any place, occurring after her passage over the meridian, the tide wave or cotidal line in the open ocean is always to the east of the moon, and generally at the distance of about 30°. It must, therefore, have a westwardly motion, following the moon in her apparent diurnal motion round the earth; and it would thus, if the whole earth was covered with water, make a complete circuit in the course of a lunar day. This motion of the tide wave is not, however, a continued forward motion of the same portion of water, but merely an undulation of successive portions.

It follows from the preceding, that in the open ocean it must be high water about two hours after the moon's passage over the meridian. This is, however, subject to some variation, depending on the relative positions of the sun and moon.

378. *Tides not perceptible in lakes and inland seas.* As the tides result from the unequal actions of the sun and moon on different parts, it requires a great extent of surface to render them sensible. No perceptible tides are, therefore, observed even in the largest lakes of this continent or in the inland seas of the eastern continent.

379. *Tides in bays, rivers, &c.* The tides in bays, rivers, narrow seas, and generally on shores far from the main body of the ocean, are not produced by the direct actions of the sun and moon, but are derivative waves propagated from the great tide wave. These derivative waves are usually attended by a current, which in some situations is quite rapid. Its velocity is, however, far less than that of the tide wave.

The interval between the moon's passage over the meridian, and the time of high water at places situated on the shores of continents, or on bays and rivers, depends principally on the distances the derivative tide waves have to pass, and on the less or greater obstructions to their motions, resulting from shoals and indentations of the coast. It is, therefore, very different at different places. At the same place, however, this interval has a mean value, from which it seldom deviates more than an hour; the deviation depending mainly on the moon's position with reference to the sun.

380. *Establishment of a port.* The mean interval between the moon's passage over the meridian and high water at any port on the days of new and full moon, is called the *establishment* of the port. When, by careful observations at any port or other place on tide water, the establishment has been determined, the time of high water at that place, on any given day, may be easily computed. This is done by adding the value of the establishment to the time of the moon's passage over the meridian, obtained from the Nautical Almanac, and then applying the correction due to the moon's position with regard to the sun. The correction is obtained from a small table calculated by a formula deduced by Laplace.*

381. *Rise of the tides at different places.* The rise of the tide, or difference between the heights of high and low water, is very different at different places, being affected by various local causes. Thus, at New York, the mean rise of the spring tides is about 5ft.; at Boston, 11ft.; at Brest in France, 19ft.; at Bristol in England, 42ft.; and at Cumberland at the head of the Bay of Fundy, 71ft.

According to Professor Whewell, the great tide wave of the South Atlantic Ocean moves northwardly along the coast of North America to the mouth of the Bay of Fundy, where it is met by another tide wave, moving in the opposite direction; this accounts for the extraordinary high tides in that Bay.

The height of the tides in many situations is considerably influenced by the direction of the wind on the coast, especially when it is strong, and continues for a length of time in the same direction.

382. *Unit of altitude.* The *unit of altitude* of a place, is the mean rise of the spring tides at that place, that is, it is the rise of the tide about a day and a half after the syzygies, on the supposi-

* The theory of the tides, a subject of great difficulty, has been elaborately treated by Laplace in the 4th Book of the Mécanique Céleste. The formula referred to, is contained in the 42d sect. of the 3d chap. of the book.

The table of corrections and another table containing the establishment of the port for various places, are given in treatises on Navigation.

tion that the sun and moon are then both in the plane of the equator, and at their mean distances from the earth. When, by a series of observations, the unit of altitude of a place has been determined, the rise of the spring tides, as affected by the distances and declinations of the sun and moon, may be computed by a formula deduced for the purpose.*

In many places it is of great importance to be able thus to know, by previous computations, when unusually high spring tides are likely to occur, in order to guard against damages which might otherwise be the result.

CHAPTER XXI.

OF THE CALENDAR.

383. *Calendar*. The *Calendar* is a distribution of time into periods of different lengths, as years, months, weeks, and days.

384. *Julian Calendar*. It has been shown that the tropical year contains 365d. 5h. 48m. 48sec. (145). But, in reckoning time for the common purposes of life, it is most convenient to have the year contain a certain number of *whole* days. In the calendar established by Julius Cæsar, and thence called the *Julian Calendar*, three successive years are made to consist of 365 days each; and the fourth, of 366 days. The year which contains 366 days, is called a *Bissextile* year. It is also frequently called *Leap* year. The others are called *common* years. The added day in a bissextile year is called the *Intercalary* day.

According to the Julian calendar, and reckoning from the epoch of the Christian era, every year, the number of which is exactly divisible by 4, is a bissextile; and the others are common years.

---

* Méc. Cél. Bk. 4, Chap. 3, Sect. 41. The American Almanac contains tabular numbers for all the spring tides that occur in the year. The product of the unit of altitude of a place, by any of the numbers, is the height of the corresponding spring tide.

385. *Julian Year.* It is evident that the reckoning by the Julian calendar supposes the length of the year to be 365¼ days. A year of this length is called a *Julian Year.* A Julian year, therefore, exceeds the true astronomical year by 11m. 12sec. This difference amounts to about 3¼ days in the course of 400 years.

386. *Gregorian Calendar.* At the time of the *Council of Nice*, which was held in the year 325, the vernal equinox fell on the 21st of March, according to the Julian calendar. But by the latter part of the 16th century, in consequence of the excess of the Julian year above the true solar year, it came ten days earlier, that is, on the 11th of March. It was observed that, by continuing to reckon according to the Julian calendar, the seasons would fall back, so that in process of time they would correspond to quite different times of the year. This reckoning also led to irregularity in the times of holding certain festivals of the church. The subject, claiming the attention of Pope Gregory XIII., he, with the assistance of several astronomers, reformed the calendar. To allow for the 10 days, by which the vernal equinox had fallen back from the 21st of March, he ordered that the day following the 4th of October, 1582, should be reckoned the 15th instead of the 5th. And in order to keep the vernal equinox to the 21st of March, in future, it was concluded that three intercalary days should be omitted every four hundred years. It was also concluded that the omission of the intercalary days should take place in those centurial years, the numbers of which were not divisible by 400. Thus, the years 1700, 1800, and 1900, which, according to the Julian calendar, would be bissextile, would, according to the reformed calendar, be common years.

The calendar, thus reformed, is called the *Gregorian Calendar.* It is easy to perceive, by a short calculation, that time reckoned by the calendar, agrees so nearly with that reckoned by true solar years, that it will require 3600 years to produce a difference of one day.

387. *Adoption of the Gregorian calendar.* The Gregorian calendar was at once adopted in Catholic countries; but, in those where the Protestant religion prevailed, it did not obtain a place till some time after. In England and her colonies, it was not in-

troduced till the year 1752.* It is now used in all Christian countries except Russia.

388. *Old and New Styles.* The Julian and Gregorian calendars are also designated by the terms *Old Style* and *New Style.* In consequence of the intercalary days, omitted in the years 1700 and 1800, there is now 12 days difference between them.

389. *Months.* The year is divided into 12 portions, called calendar months. Each of these contains either 30 or 31 days, except the second month, February, which in a common year contains 28 days, and in a bissextile, 29 days; the intercalary day being added at the last of this month.

390. *Dominical Letter.* It was formerly customary to designate the days of the week in the calendar, by the first seven letters of the alphabet, always placing them so that A corresponded to the first day of the year, B to the second, C to the third, D to the fourth, E to the fifth, F to the sixth, G to the seventh, A to the eighth, B to the ninth, and so on. According to this arrangement, whatever letter designates any given day of the week in the first part of the year, continues to designate the same throughout the year. The letter designating the first day of the week, or Sunday, is called the *Dominical Letter.*

As a common year consists of 365 days, or 52 weeks and 1 day, the last day of each common year must fall on the same day of the week as the first, and the next year must commence one day later in the week. Consequently, the day of the week which was

---

* At this time there was a difference of 11 days between the Julian and Gregorian calendars, in consequence of the suppression, in the latter, of the intercalary day in 1700. It was, therefore, enacted by parliament, that 11 days should be left out of the month of September, of the current year, 1752, by calling the day following the 2d of the month, the 14th, instead of the 3d.

Previous to this, years commencing at two different times had been in use in England. The *historical* year commenced on the 1st of January, as at present. But the *civil* or *legal* year commenced on the 25th of March. Dates in the interval between these times, were frequently expressed by naming both years. Thus, in books printed prior to 1752, we often meet with dates expressed as follows: Feb. 2d, 1735–6, or 173$\frac{5}{6}$. The same act that introduced the Gregorian calendar, established the 1st of January, as the commencement of the civil, as well as of the historical year.

the first day of the former year, and was designated by A, is the seventh day of the second year, and is designated by G; that which was the second, and was designated by B, in the former year, is the first, and is designated by A in the second, and so on. It therefore follows, that whatever letter is the dominical letter in any common year, the letter next preceding it in the alphabet is the dominical letter in the following year; except the former was A, in which case the second is G.

In every common year, the first day of March is the 60th day of the year, and consequently corresponded to the letter D. In bissextile years, on account of the intercalation, the 1st of March is the 61st day of the year; but the letter D was still made to correspond to it, and the letters for the remaining part of the year were arranged accordingly. It therefore follows, that, after the 29th of February, any given day of the week was designated by the letter of the alphabet next preceding that by which it was designated in the first two months. Consequently, a bissextile had two dominical letters, one of which appertained to January and February, and the other, which was the next preceding letter in the alphabet, appertained to the other ten months.

From what has been said it follows, that the dominical letters succeed one another in a retrograde order, that is, in the order G, F, E, D, C, B, A, G, F, *&c.*; and that each bissextile has two in the same order.

It is now usual to retain only the dominical letter in the calendar, and to designate the other days of the week by numbers or by their names.

391. *Determination of the dominical letter.* The year 1800, which was a common year, commenced on the fourth day of the week, and, consequently, the dominical letter was the 5th of the alphabet, which is E. From thence, taking into consideration that every four years, in which a bissextile is included, requires five dominical letters in a retrograde order, it is easy to find the dominical letter for any year in the present century. To do this, multiply the number of years above 1800, by 5, and divide the product by 4, neglecting the remainder. Divide the quotient by 7, and subtract the remainder from 5; or from 12, when the remainder is equal to or greater than 5. The last remainder is the

number of the dominical letter. In bissextile years, the dominical letter thus obtained is that for the last ten months of the year. The dominical letter for the first two months is the next following letter in the alphabet.

Delambre, in the 38th chapter of his Astronomy, has given the investigation of a formula for finding the dominical letter in any century, according to the Gregorian calendar.

392. *Solar Cycle.* The *Solar Cycle* is a period of 28 years, in which, according to the Julian calendar, the days of the week return to the same days of the month, and in the same order. The first year of the Christian era was the 10th of this cycle. Consequently, if 9 be added to the number of any year, and the sum be divided by 28, the remainder will be the number of the year of the solar cycle. When there is no remainder, the year is the 28th of the cycle.

393. *Lunar Cycle.* The *Lunar Cycle*, or, as it is sometimes called, the *Metonic Cycle*, is a period of 19 years, in which the conjunctions, oppositions, and other aspects of the moon, return on the same days of the year. The synodic revolution of the moon being 29.5305885 days, 235 revolutions are 6939.688 days; which differs only an hour and a half from 19 Julian years. The number by which the year of the lunar cycle is designated, is frequently called the *Golden Number.*

The first year of the Christian era was the 2d of the lunar cycle. Hence, to find the year of the cycle, for any given year, add 1 to the number of the year, and divide by 19. The remainder expresses the year of the cycle. If nothing remains, the year is the 19th of the cycle.

394. *Cycle of the Indiction.* The *Cycle of the Indiction* is a period of 15 years. This period, which is not astronomical, was introduced at Rome, under the emperors, and had reference to certain judicial acts.

To find the cycle of the indiction for a given year, add 3, and divide by 15. The remainder expresses the year of the cycle.

395. *Julian Period.* The *Julian Period* is a period of 7980 years, obtained by taking the continued product of the numbers, 28, 19, and 15. After one Julian period, the different cycles of

the sun, moon, and indiction, return in the same order, so as to be just the same in a given year of the period, as in the same year of the preceding period. The first year of the Christian era was the 4714th of the Julian period. Hence, if 4713 be added to the number of a given year, the result will be the year of the Julian period.

396. *Epact.* The *Epact*, as an astronomical term, is the mean age of the moon at the commencement of a year, or, in other words, it is the interval between the commencement of the year and the time of the last mean new moon; and is expressed in days, hours, minutes and seconds.

The *Epact*, as given in the calendar, is nearly the age of the moon at the commencement of the year, expressed in whole days, and was introduced for the purpose of finding the days of mean new and full moon throughout the year, and thence the times of certain festivals. Without entering into any explanation of the reason of the rule, it must suffice here to observe, that the *Epact* for any year during the present century may be found by multiplying the golden number of the year by 11, adding 19 to the product and dividing the sum by 30. The remainder is the *Epact* for the year.

CHAPTER XXII.

UNIVERSAL GRAVITATION. — TABLES OF THE ELEMENTS OF THE ORBITS OF THE PLANETS AND OF THEIR MASSES AND DENSITIES.

397. Physical astronomy, in which the principle of universal gravitation is applied to the investigation of the motions of the heavenly bodies, and the various effects of their actions on one another, is a very extensive and, in many of its parts, very difficult department of science. A few propositions of an elementary character, and some general remarks and results, are all that will be here introduced.*

---

* The celebrated *Principia* of Newton was the first work on physical astronomy. At the present time, the prominent works on this subject generally, or on the moon

398. *The moon is retained in her orbit by the force of gravity diminished in proportion to the square of the distance from the earth's centre.*

Let E, *Fig.* 56, be the centre of the earth, A a point on its surface, and GH a part of the moon's orbit, assumed to be circular. When the moon is at any point M in her orbit, she would, by the first law of motion, move on in the direction of the line MF, a tangent to the orbit at M, if she was not acted on by some force to turn her aside. Let L be her place in her orbit one second of time after she has been at M, and let LC and LD be drawn parallel to EM and MF respectively; and joining LM, let EI be drawn perpendicular to it, and, therefore, bisecting it in I. The line CL, or its equal MD, is the distance the moon has been drawn, during one second, from the tangent towards the earth at E. Now, as the distance a body moves in a given time is proportional to the force by which it is moved, MD may be taken as a measure of the force by which the moon is drawn towards the earth.

Put g = MD, G = the force of gravity at the earth's surface, or the distance a heavy body falls there, by this force, in a second, r = EA, the earth's radius, d = EM, the moon's distance, p = moon's sidereal revolution in seconds, π = moon's horizontal parallax, and $\varpi = 3.14159$, &c. Then, assuming that the force MD, or g, is that of gravity diminished in proportion to the square of the distance from the earth's centre, we have,

$$r^2 : d^2 :: g : \mathrm{G} = \frac{d^2}{r^2} \cdot g \qquad \text{(A)}$$

Now, by similar triangles, we have EM : IM :: LM : MD, or 2EM : 2IM, or LM :: LM : MD, that is,

$$2d : \mathrm{LM} :: \mathrm{LM} : g \qquad \text{(B)}$$

But the chord LM does not sensibly differ from the arc LM, which is the distance described by the moon in one second. Hence, as $2d\varpi$ is the circumference of the moon's orbit, we have,

$$p : 1 :: 2d\varpi : \mathrm{LM} = \frac{2d\varpi}{p}.$$

Substituting the value of LM in (B), it becomes,

---

in particular, are Laplace's *Mécanique Céleste*, Pontécoulant's *Systēme du Monde*, and Planâ's *Théorie de la Lune.*

$$2d : \frac{2d\varpi}{p} :: \frac{2d\varpi}{p} : g = \frac{2d\varpi^2}{p^2} \qquad \text{(C)}$$

Hence, from (A), we have,

$$G = \frac{d^2}{r^2} \times \frac{2d\varpi^2}{p^2} = \frac{2d^3\varpi^2}{r^2p^2}$$

Or, by substituting for d, its value $\frac{r}{\sin \pi}$,

$$G = \frac{2r\varpi^2}{p^2 \sin^3\pi} \qquad \text{(D)}$$

Taking the mean values of r, p, and π,* we easily find the value of G to be 16.22 ft.; which is very nearly equal to its known value as determined by experiment. This conformity of the computed result with that obtained by experiment, may be regarded as establishing the truth of the proposition.

399. *The planets are retained in their orbits about the sun, and the satellites in theirs, about their respective primaries, by forces directed in each case to the central body and varying inversely as the square of the distance from that body.*

Assuming the planets and satellites to be retained in their orbits by forces directed and varying as stated in the proposition, it is proved, by a series of investigations that we shall omit, that their motions and periods must be in conformity with Kepler's Laws. Hence, as these laws were deduced from observations, and have been fully confirmed by subsequent observations, it follows that the proposition must be true.

400. *Determination of the relative masses or quantities of matter in the sun and planets.*

For a planet that has a satellite, let D be the mean distance of the planet from the sun, d the mean distance of the satellite from the planet, P and p the periodical revolutions of the planet and satellite respectively, and m the mass of the planet, that of the sun being regarded as a unit or 1.

Also, let f = force of gravity of a unit of mass at a unit of

---

* These are, $2r = 41776044$ ft., p, in seconds, $= 2360585$, and $\pi = 57' \; 1''$.

distance. Then, since the whole force of gravity at a given distance is proportional to the mass, we have mf = force of gravity of the mass m at a unit of distance. Hence, g being taken for its force at the distance d, we have,

$$d^2 : 1^2 :: mf : g = \frac{mf}{d^2}.$$

But (398 C), $g = \dfrac{2d\varpi^2}{p^2}$

$$\text{Hence, } \frac{mf}{d^2} = \frac{2d\varpi^2}{p^2}, \text{ or } m = \frac{2d^3\varpi^2}{fp^2} \quad \text{(E)}$$

In like manner for the sun and planet, the mass of the sun being 1, we have,

$$1 = \frac{2D^3\varpi^2}{fP^2} \quad \text{(F)}$$

Dividing (E) by (F), there results,

$$m = \frac{d^3}{D^3} \times \frac{P^2}{p^2} \quad \text{(G)}$$

In this investigation, the attraction of the planet on the sun, and that of the satellite on the planet, have both been omitted. But as the mass of the planet is very small in comparison with that of the sun, and the mass of the satellite in comparison with that of the planet, the result is but little affected by the omission. We have, thus, a very simple formula for computing, with considerable accuracy, the mass of a planet that is attended by a satellite.

Applying this formula to the planet Neptune, we have (335 and 336), $P = 164\frac{1}{2}$ years, $p = 5$ days, 21 hours, $D = 2{,}850{,}000{,}000$ miles, and $d = 230{,}000$ miles, which give, for the mass of Neptune, $m = \frac{1}{18000}$ nearly. This result accords very well with the value given in Article 336.

The masses of the planets which have no satellite, and also that of the moon, are deduced, by more difficult investigations, from the ascertained effects of their actions on other bodies.

401. *The Densities of the Sun and Planets.* The densities of bodies are proportional to their masses, divided by their volumes.

Hence, from the known masses and volumes of the sun and planets, their densities are easily obtained.*

402. *The density of the earth increases towards the centre.* Supposing the earth to have been once in a fluid state and homogeneous throughout, it is ascertained by investigation that, in consequence of its revolution on its axis, it would have taken the form of an oblate spheroid, having the polar radius to the equatorial in the ratio of 229 to 230, and, consequently, have an ellipticity of $\frac{1}{230}$. If, instead of being homogeneous, it is composed of strata increasing in density towards the centre, the form would still be that of an oblate spheroid, but of *less* ellipticity. Hence, as the actual ellipticity of the earth, which is only $\frac{1}{300}$, is considerably less than $\frac{1}{230}$, and as it is probable the earth was once in the state supposed, it is inferred that the density increases towards the centre.

This inference is confirmed by very accurate observations made at the sides of the mountain Schehallion in Scotland, by Dr. Maskelyne. From the effect of the mountain in changing the direction of the plumb line of a plummet suspended near it, and from the known figure and volume of the mountain determined by a survey, it was found that the mean density of the mountain was to that of the whole earth, nearly as 5 to 9.

403. *Kepler's Laws, though very nearly, are not rigorously true.* The deviation from entire accuracy is caused by the attractions of the planets on the sun and on one another, and also by the attractions of the satellites on their primaries. But, as the masses of all the planets taken together are very small in comparison with that of the sun, and those of the satellites in comparison with those of their primaries, the deviation with regard to either of the laws is also small.

404. *The sun's action increases the gravity of the moon to the earth at the quadratures, and diminishes it twice as much at the syzygies; the effect, on the whole, being a diminution of her gravity to the earth by about the 358th part.*

* The masses and densities of the sun, planets and moon, as deduced from the most accurate investigations, are given in the tables at the end of this part of the work.

Let ACBO, *Fig.* 57, represent the orbit of the moon, which may in this investigation be considered as coinciding with the plane of the ecliptic. Also let S be the sun, E the earth, M the place of the moon in her orbit, and AB perpendicular to SE, the line of the quadratures. Let the line SE represent the force which the sun exerts on the earth at E, or on the moon, when in quadratures, at A and B.* Then, $SM^2 : SE^2 :: SE : \frac{SE^3}{SM^2}$ = the force with which the sun acts on the moon at M. In the line MS, produced if necessary, take $MD = \frac{SE^3}{SM^2}$; then MD represents the force which the sun exerts on the moon at M. Let the force MD be resolved into the two, MH and MG, one of which, MH, is equal and parallel to ES. Then since the force MH is equal and parallel to ES, it will have no tendency to change the relative motions or positions of the earth and moon. The other force MG, will therefore represent, in quantity and direction, the whole effect of the sun's action in disturbing the moon's motion in her orbit. Let SM be produced to meet the diameter AB in N. Then, because the angle ESN is always very small, being when greatest only about 7′, the line SN may be considered equal to SE. Hence,

$$MD = \frac{SE^3}{SM^2} = \frac{SN^3}{SM^2} = \frac{(SM + MN)^3}{SM^2}$$

$$= \frac{SM^3 + 3SM^2 \times MN + 3SM \times MN^2 + MN^3}{SM^2}$$

$$= SM + 3MN + \frac{3MN}{SM} \times MN + \frac{MN^2}{SM^2} \times MN.$$

But, as MN is very small in comparison with SM, the last two terms may be omitted without material error.

Therefore, $MD = SM + 3MN$; or, $SD = 3MN$.

As the angle ESM is very small, and SD is also small, the line DG must very nearly coincide with SE, and, consequently, the

* Strictly speaking, as the quantity of matter in the earth is greater than that in the moon, the forces which the sun exerts on the earth and moon, when at equal distances, are not equal. But the effects of those forces, in moving the bodies, are equal, and it is these effects which is the subject under consideration.

point G with the point L. We may therefore consider ML as the force by which the sun disturbs the motion of the moon. Now, EL + LS=ES=HM=DG=SD + LS, very nearly, or, EL=SD, very nearly.

Hence, if MK be perpendicular to SE, we have,

$$\mathrm{EL} = 3\mathrm{MN} = 3\mathrm{EK}.$$

Let the force ML be resolved into two others, one MQ in the direction of the radius vector, and the other MP in the direction of a tangent to the orbit at M. Then the force MQ increases or diminishes the gravity of the moon to the earth, according as the point Q falls between E and M, or in EM produced. The other force MP increases or diminishes the moon's angular motion about the earth. Since the moon's orbit does not differ much from a circle, the angle QMP may be considered as a right angle. Put a = SE, r = EM, x = the angle AEM, m = mass of the earth, that of the sun being 1, and f = force of gravity of the sun, or of a unit of mass, at a unit of distance. Then,

$$\mathrm{EK} = \mathrm{EM} \cos \mathrm{MEK} = r \sin x,$$

$$\mathrm{EL} = 3\mathrm{EK} = 3r \sin x,$$

$$\mathrm{MP} = \mathrm{LQ} = \mathrm{EL} \cos x = 3r \sin x \cos x.$$

But (App. 7), $\sin x \cos x = \frac{1}{2}\sin 2x$. Hence,

$$\mathrm{MP} = \frac{3r}{2} \sin 2x.$$

Also, $\mathrm{EQ} = \mathrm{EL} \sin x = 3r \sin^2 x$, and $\mathrm{MQ} = \mathrm{EQ} - \mathrm{EM} = 3r \sin^2 x - r = -r\,(1 - 3 \sin^2 x)$.

Or, using the affirmative sign to denote an increase in the moon's gravity to the earth.

$$\mathrm{MQ} = + r\,(1 - 3 \sin^2 x).$$

Now the analytical expression of the force represented by ES, is evidently $\frac{f}{a^2}$. Hence, since ES : MQ : : the force ES : the force MQ, we have,

$a : r\,(1 - 3 \sin^2 x) : : \frac{f}{a^2}$: the force MQ. Consequently, the

force $\mathrm{MQ} = \frac{fr}{a^3}(1 - 3 \sin^2 x)$.. (H)

In like manner,

$$\text{the force MP} = \frac{3fr}{2a^3}\sin 2x \ldots\ldots\ldots\ldots \text{(I)}$$

When the moon is in quadratures, $x = 0$, or $180°$. Consequently, then,

$$\text{the force MQ} = +\frac{fr}{a^3} \ldots\ldots\ldots\ldots \text{(K)}$$

But when the moon is in syzygies, $x = 90°$ or $270°$. Hence, then,

$$\text{the force MQ} = -\frac{2fr}{a^3} \ldots\ldots\ldots\ldots \text{(L)}$$

The first part of the proposition is, therefore, proved.

Now, it is evident, from (H), that the force MQ $= 0$, when $3\sin^2 x = 1$, or $\sin x = \surd\frac{1}{3}$; that is, when $x = 35°\ 15'\ 52''$. The moon's gravity to the earth is, therefore, increased while she is within about 35° of her quadratures, on either side, and is diminished in all the remaining part of the orbit; and the greatest diminution is double the greatest increase. It follows, therefore, that in the whole the moon's gravity to the earth is diminished by the action of the sun. An easy investigation, with the aid of differential calculus, proves that the mean or average diminution is $\frac{fr}{2a^3}$; r representing in this case the mean distance of the moon from the earth.

Dividing $\frac{fr}{2a^3}$, the mean *diminution* of the moon's gravity to the earth by $\frac{mf}{r^2}$, which expresses the *whole* gravity, the quotient $\frac{r^3}{2ma^3}$, is the mean diminution of the moon's gravity expressed as a fraction of the whole. Substituting, in this, the value of m (391 G), and observing that a and r are used here in the place of D and d, we have,

$$\frac{r^3}{2ma^3} = \frac{p^2}{2\text{P}^2} = \frac{1}{358}\text{ nearly.}$$

405. *The inequality in the moon's motion, called the* Annual Equation (204), *proceeds from an inequality in the sun's disturbing force, depending on the variation in the earth's distance from the sun.*

The expression $\frac{fr}{2a^3}$ designates the mean diminution of the moon's gravity to the earth for a given distance of the earth from the sun. Hence, as the distance a varies with the time in the year, or with the sun's anomaly, the mean diminution of the moon's gravity must also vary. This variation causes a change in the moon's distance from the earth, and, consequently, in her velocity. The change in her place resulting from this change of velocity, is the annual equation.

406. *The* Evection *is produced by an inequality in the sun's disturbing force, depending on the position of the line of the apsides of the moon's orbit with regard to the line of the syzygies.*

Let R and r denote the distances of the moon from the earth, in apogee and perigee, when the line of the apsides coincides with the line of the syzygies, X and x, the distances at which the moon would be from the earth, in apogee and perigee, if she was not acted on by the sun, and G and g, the perigean and apogean gravities in that case. Also, put $n = \frac{f}{a^3}$, and supposing the earth's distance from the sun to remain constant, n will be constant. Then (404 L), $G - 2rn$ and $g - 2Rn$ will be the perigean and apogean gravities of the moon, when the line of the apsides coincides with the line of the syzygies. Hence,

$$X^2 : x^2 :: G : g,$$

$$\text{and } R^2 : r^2 :: G - 2rn : g - 2Rn.$$

Consequently,

$$\frac{X^2}{x^2} = \frac{G}{g},$$

$$\text{and } \frac{R^2}{r^2} = \frac{G - 2rn}{g - 2Rn}$$

Now, as G is greater than g, and $2rn$ less than $2Rn$, it is evident that

$$\frac{G - 2rn}{g - 2Rn} \text{ is greater than } \frac{G}{g}.$$

$$\text{Hence, } \frac{R^2}{r^2} \text{ is greater than } \frac{X^2}{x^2}.$$

It therefore follows, that, when the line of the apsides coincides with the line of the syzygies, the ratio of the apogean distance of the moon to the perigean distance, and, consequently, the eccentricity of the orbit, is increased by the action of the sun. In like manner it may be shown, that, when the line of the apsides coincides with the line of the quadratures, the sun's action diminishes the eccentricity of the orbit. The change in the eccentricity of the orbit, produces a change in the equation of the centre; and this change is the Evection.

407. *The* Variation *is produced by the resolved part of the sun's disturbing force that acts in the direction of a tangent to the moon's orbit.*

It has been shown (404 I), that MP, the part of the sun's disturbing force that acts in the direction of a tangent to the moon's orbit, and therefore changes her motion in her orbit, is equal to $\frac{3fr}{2a^3} \sin 2x$. Hence, supposing the earth's distance from the sun, and the moon's distance from the earth, to remain constant, this force is proportional to sin $2x$; that is, to the sine of twice the distance of the moon from the quadratures. It is, therefore, greatest in the octants, and is nothing in the syzygies and quadratures. The inequality in the moon's motion thus produced is the Variation.

408. *The motion of the Apsides of the moon's orbit is produced by the action of the sun in diminishing the moon's gravity to the earth.*

If the moon was only acted on by the earth's attraction, she would describe an ellipse, and her angular motion, would be just 180°, from one apsis to the other; or, which is the same, from one place where the orbit cuts the radius vector at right angles, to the other. But, in consequence of the change produced in the moon's gravity to the earth, by the action of the sun, the moon's path is not truly an ellipse. When the effect of the sun's action is a diminution of the moon's gravity, she will continually recede from the ellipse that would otherwise be described, her path will be less curved, and she must move through a greater distance before the radius vector intersects the path at right angles. She must, therefore, move through a greater angular distance than 180°, in going from one apsis to the other, and, consequently, the apsides will advance. On the contrary, when the gravity is increased by the sun's action, the moon's path will fall within the ellipse which she would otherwise describe, its curvature will be increased, and the distance through which she must move before the radius vector intersects her path at right angles, will be less. The apsides will, therefore, move backwards. Now, it has been shown (404), that the sun's action alternately diminishes and increases the moon's gravity to the earth. The motion of the apsides will, therefore, be alternately direct and retrograde. But, as the diminution has place during a much longer part of the moon's revolution, and is besides greater than the increase, the direct motion will exceed the retrograde. Consequently, in an entire revolution of the moon, the apsides have a progressive motion.

409. *Motion in the moon's nodes and change in the inclination of her orbit.* The direction in which the sun's disturbing force acts on the moon, does not, except in some particular cases, coincide with the plane of her orbit. This force, therefore, causes the moon to leave the plane of her orbit, or, which is equivalent, causes this plane itself to change its position, varying both the line in which it intersects the plane of the ecliptic and the angle it makes with that plane. By a simple but tedious investigation, it may be shown, that, in consequence of the sun's action, the nodes must, during each synodic revolution of the moon, move alternately backwards and forwards; the backward motion being, however, the greater, so that, on the whole, they must have a retrograde motion. It may also be shown, that the inclination of the orbit must alternately increase and diminish, vibrating thus, about its mean value, from which it never widely deviates.

410. *Stability of the solar system.* The mutual actions of the planets and satellites, and the inequality of the sun's action on a planet and its satellite in different positions, produce continual changes in the motions of the bodies, and in the eccentricities and inclinations of their orbits. Although some of these changes are

ascertained from observations to be periodical, and it is found that the quantities subject to them, alternately increase and decrease, so that their mean or average values remain the same, yet there are others which have always been accumulating from the period of the earliest observations to the present time. One of these, the acceleration of the mean motion of the moon (193), has long attracted attention. If this acceleration of her motion, and the consequent diminution of her distance, were perpetually to continue, it would follow that she would eventually be precipitated to the earth. Such a result, if it were a necessary consequence of the structure and working of the system, would seem to imply some imperfection in the works of the all-wise Creator of the universe. But the profound investigations of Lagrange and Laplace have shown, that, with the system constituted as it is, no such result can have place; that not only *some*, but *all*, the changes produced in the motions and orbits, by the mutual attractions of the bodies, must be periodical; and that, though some of the quantities in which these changes are produced must continually increase, or continually decrease, for many thousands of years, they cannot perpetually do so. Through the operation of the very same causes, the quantities that are now increasing, must in process of time decrease, and those that are decreasing, must increase. None of them can ever widely deviate from their average values. Thus, notwithstanding the many perturbations and seeming irregularities, the *stability of the system* is preserved.

411. *Tables relative to the planets and satellites.* The following tables contain the elements of the orbits of the planets, and their masses and densities as far as they are known. The longitudes are reckoned from the mean equinox of the epoch.

The fourth table, page 218, contains the elements of the first twenty-seven asteroids, which have been collected from the most reliable sources.

ELEMENTS OF THE PRINCIPAL PLANETS.

PLANET'S NAME.	Epoch. Greenwich, M. N.	Mean Longitude at the Epoch.	Longitude of the Perihelion.	Secular Variation.
Mercury	1801, Jan. 1.	166° 0′ 48.″6	74° 21′ 46.″9	+ 9′ 44″
Venus....................	"	11 33 3. 0	128 43 53. 1	— 4 28
The Earth..............	"	100 39 10. 2	99 30 5. 0	+ 19 41
Mars	"	64 22 55. 5	332 23 56. 6	+ 26 22
Jupiter..................	"	112 15 23. 0	11 8 34. 6	+ 11 5
Saturn	"	135 20 6. 5	89 9 29. 8	+ 32 17
Uranus	"	177 48 23. 0	167 31 16. 1	+ 4 0
Neptune	1847, Jan. 1.	328 32 44. 2	47 12 6. 5	......

PLANET'S NAME.	Longitude of the Ascending Node.	Secular Variation.	Inclination to the Ecliptic.	Secular Variation.
Mercury................	45° 57′ 30.″9	— 13′ 2″	7° 0′ 9.″1	+ 18.″1
Venus....................	74 54 12. 9	— 31 11	3 23 28. 5	— 4. 5
The Earth	......	......	......	......
Mars	48 0 3. 5	— 38 49	1 51 6. 2	— 0. 3
Jupiter..................	98 26 18. 9	— 26 21	1 18 51. 3	— 22. 6
Saturn	111 56 37. 4	— 32 22	2 29 35. 7	— 15. 5
Uranus..................	72 59 35. 3	— 59 59	0 46 28. 0	+ 3. 1
Neptune	130 4 20. 8	......	1 46 59. 0	......

PLANET'S NAME.	Sidereal Revolution in Mean Solar Days.	Mean Distance from the Sun, or Semi-major Axis.	Eccentricity.	Secular Variation.
Mercury................	87.969258	0.3870981	0.2055149	+ 0.00000387
Venus....................	224.700787	0.7233316	0.0068607	— 0.00006275
The Earth..............	365.256361	1.0000000	0.0167836	— 0.00004359
Mars	686.979646	1.5236923	0.0933070	+ 0.00009019
Jupiter..................	4332.584821	5.2027760	0.0481621	+ 0.00016036
Saturn	10759.219817	9.5387861	0.0561505	— 0.00031240
Uranus..................	30686.820830	19.1823900	0.0466794	— 0.00002521
Neptune	60125.483	30.0359000	0.0087195	......

ELEMENTS OF THE MINOR PLANETS, OR ASTEROIDS.

PLANET'S NAME.	Epoch, Mean Time. (G), Greenwich. (B), Berlin.	Mean Longitude at the Epoch.	Longitude of the Perihelion.	Longitude of the Ascending Node.	Inclination to the Ecliptic.	Sid. Revolution in M. S. Days.	Mean Distance from the Sun.	Eccentricity.
Flora	1848, Jan. 1, (B)	68° 48′ 47.″5	33° 0′ 23.″2	110° 18′ 3.″8	5° 53′ 6.″2	1193.222	2.201653	0.1565408
Melpomene	1852, July 10, (G)	302 19 23. 4	15 15 36. 5	150 0 47. 6	10 9 2. 2	1269.633	2.294670	0.2165090
Clio	1850, Sept. 19, (B)	339 11 55. 9	301 52 31. 4	235 26 54. 4	8 23 6. 8	1303.197	2.334935	0.2181896
Euterpe	1854, Jan. 1, (G)	74 46 25. 6	87 15 29. 0	93 30 18. 7	1 36 9. 0	1314.029	2.347853	0.1713700
Vesta	1850, Jan. 9, (B)	116 30 51. 0	250 46 32. 2	103 23 31. 6	7 8 29. 7	1325.147	2.361081	0.0895694
Iris	1853, March 23, (B)	162 49 27. 0	41 18 59. 9	259 14 48. 7	5 28 16. 0	1345.315	2.384977	0.2317452
Metis	1853, Oct. 9, (B)	26 48 39. 8	71 40 41. 6	68 29 59. 7	5 35 55. 3	1346.111	2.385917	0.1233821
Phocea	1854, Jan. 1, (G)	314 12 54. 9	302 41 49. 7	214 4 5. 7	21 34 54. 2	1352.166	2.393066	0.2507330
Massalia	1853, Jan. 1, (G)	44 54 11. 8	98 19 0. 5	206 53 29. 3	0 41 4. 5	1365.150	2.408360	0.1457463
Hebe	1853, Jan. 20, (B)	97 16 27. 6	15 13 58. 8	138 32 8. 7	14 46 35. 1	1379.457	2.425160	0.2020361
Lutetia	1853, Jan. 9, (B)	43 29 52. 0	326 33 2. 1	80 26 50. 2	3 5 36. 9	1387.100	2.434106	0.1624455
Parthenope	1853, Jan. 1, (G)	129 54 44. 9	316 14 19. 3	124 57 44. 7	4 36 1. 2	1400.803	2.450113	0.0998142
Fortuna	1852, Sep. 27.354, (B)	357 27 6. 8	32 21 19. 1	211 17 34. 8	1 32 35. 0	1408.038	2.458544	0.1702998
Thetis	1852, May 31, (B)	214 32 51. 0	259 13 18. 0	125 26 25. 2	5 35 39. 3	1425.857	2.479240	0.1308650
Astræa	1851, April 29.5, (B)	197 37 6. 8	135 42 31. 7	141 27 47. 5	5 19 23. 0	1511.370	2.577402	0.1887517
Egeria	1852, Dec. 21, (B)	229 43 5. 4	119 36 46. 4	43 18 46. 6	16 32 59. 5	1512.106	2.578238	0.0853696
Irene	1852, Aug. 3, (G)	328 49 52. 4	178 46 2. 5	86 49 55. 3	9 6 42. 2	1517.455	2.584312	0.1683269
Thalia	1853, Jan, 0, (B)	89 5 3. 8	123 11 57. 0	67 55 4. 3	10 13 59. 1	1554.212	2.625877	0.2353947
Eunomia	1852, Dec. 21, (B)	64 5 33. 0	28 9 5. 7	293 55 9. 3	11 43 55. 0	1566.768	2.640004	0.1874427
Proserpine	1854, Jan. 1, (G)	273 55 31. 7	235 38 29. 3	45 55 54. 6	3 35 39. 6	1578.167	2.652790	0.0861356
Juno	1850, April 8, (B)	178 55 23. 6	54 24 12. 8	170 54 45. 6	13 3 22. 1	1594.296	2.670837	0.2548847
Ceres	1850, Sept. 25, (B)	6 52 41. 9	147 46 12. 4	80 48 46. 6	10 37 4. 4	1682.125	2.768051	0.0766523
Pallas	1850, Aug. 23, (B)	338 52 59. 1	121 21 48. 5	172 43 59. 7	34 37 33. 1	1686.510	2.772858	0.2398150
Calliope	1853, Jan. 0, (B)	77 6 24. 4	58 49 24. 2	66 36 50. 5	13 44 48. 7	1814.765	2.911710	0.1036109
Psyche	1853, Jan. 1, (G)	202 58 0. 6	12 57 14. 4	150 39 47. 6	3 3 52. 0	1825.315	2.922984	0.1376471
Hygeia	1851, Sept. 28.5, (B)	356 45 11. 9	228 2 28. 7	287 38 26. 6	3 47 10. 8	2043.390	3.151392	0.1009159
Themis	1854, Jan. 1, (G)	215 34 49. 0	139 30 0. 1	35 35 1. 1	0 49 13. 7	2043.992	3.152007	0.1176300

Different Revolutions of the Moon.

		Days.
Tropical revolution		27.3215255
Sidereal	"	27.3215830
Synodic	"	29.5305885
Anomalistic	"	27.5545704
Nodal	"	27.2122222

Sidereal revolutions of the satellites, and their mean distances from the planets about which they revolve. The distances are expressed in terms of the equatorial radius of the planet.

JUPITER.

		Mean Distance.	Sider. Revolution. Days.
1st Satellite		6.04853	1.7691378
2d	"	9.62347	3.5511810
3d	"	15.35024	7.1545528
4th	"	26.99835	16.6887697

SATURN.

			Mean Distance.	Sider. Revolution. Days.
1st	Satellite,	Mimas	3.351	0.94271
2d	"	Enceladus	4.300	1.37024
3d	"	Tethys	5.284	1.88780
4th	"	Dione	6.819	2.73948
5th	"	Rhea	9.524	4.51749
6th	"	Titan	22.081	15.94530
7th	"	Hyperion	26.5	21.18
8th	"	Iapetus	64.359	79.32960

URANUS.

	Mean Distance.	Sider. Revolution. Days.
1st Satellite	13.120	5.8926
2d "	17.022	8.7068
3d "	19.845	10.9611
4th "	22.752	13.4559
5th "	45.507	38.0750
6th "	91.008	107.6944

Masses and densities of the sun and planets, the mass of the sun and density of the earth being each assumed = 1.

	Masses.	Densities.
Sun	1	0.252
Neptune	$\frac{1}{19000}$	unknown.
Uranus	$\frac{1}{24905}$	0.242
Saturn	$\frac{1}{3502}$	0.138
Jupiter	$\frac{1}{1048}$	0.238
Mars	$\frac{1}{2680337}$	0.948
Earth	$\frac{1}{359551}$	1.000
Venus	$\frac{1}{401839}$	0.923
Mercury	$\frac{1}{4865751}$	1.12

Denoting the earth's mass by a unit, the moon's mass is about $\frac{1}{75}$, and her density about 0.615.

Remark. The masses of the planets given above, except that of Neptune, are taken from a table in the *Astr. Nach. No.* 443. That of Mercury has been very recently obtained by Prof. Encke, from the effects of this planet in disturbing the motion of the comet which bears his name.

APPENDIX TO PART I.

TRIGONOMETRICAL FORMULÆ.

A NUMBER of the formulæ included in the following collection are used in the present work. The demonstrations may be found in any good work on Trigonometry.* They are introduced here, and numbered in order to facilitate the references.

From a single arc or angle a, *the radius being* = 1.

1. $\sin^2 a + \cos^2 a = 1$
2. $\sin a = \text{tang } a \cos a$
3. $\sin a = \dfrac{\text{tang } a}{\sqrt{(1 + \text{tang}^2 a)}}$
4. $\cos a = \dfrac{1}{\sqrt{(1 + \text{tang}^2 a)}}$
5. $\text{tang } a = \dfrac{\sin a}{\cos a}$
6. $\text{cot. } a = \dfrac{1}{\text{tang } a} = \dfrac{\cos a}{\sin a}$
7. $\sin a = 2 \sin \tfrac{1}{2} a \cos \tfrac{1}{2} a$
   or, $\sin 2a = 2 \sin a \cos a$
8. $\cos a = 1 - 2 \sin^2 \tfrac{1}{2} a$
9. $\cos a = 2 \cos^2 \tfrac{1}{2} a - 1$
10. $\text{tang } \tfrac{1}{2} a = \dfrac{\sin a}{1 + \cos a}$
11. $\text{tang } \tfrac{1}{2} a = \dfrac{1 - \cos a}{\sin a}$
12. $\text{tang}^2 \tfrac{1}{2} a = \dfrac{1 - \cos a}{1 + \cos a}$

For two arcs a *and* b *of which* a *is supposed to be the greater.*

13. $\sin (a \pm b) = \sin a \cos b \pm \cos a \sin b$
14. $\cos (a \pm b) = \cos a \cos b \mp \sin a \sin b$
15. $\text{tang } (a \pm b) = \dfrac{\text{tang } a \pm \text{tang } b}{1 \mp \text{tang } a \text{ tang } b}$
16. $\sin a \cos b = \tfrac{1}{2} \sin (a + b) + \tfrac{1}{2} \sin (a - b)$
17. $\cos a \sin b = \tfrac{1}{2} \sin (a + b) - \tfrac{1}{2} \sin (a - b)$
18. $\sin a \sin b = \tfrac{1}{2} \cos (a - b) - \tfrac{1}{2} \cos (a + b)$
19. $\cos a \cos b = \tfrac{1}{2} \cos (a - b) + \tfrac{1}{2} \cos (a + b)$

---

* The best treatise upon that subject, in our language, is that by Chauvenet, recently published.

20. $\sin a + \sin b = 2 \sin \frac{1}{2} (a + b) \cos \frac{1}{2} (a - b)$

21. $\sin a - \sin b = 2 \cos \frac{1}{2} (a + b) \sin \frac{1}{2} (a - b)$

22. $\cos b + \cos a = 2 \cos \frac{1}{2} (a + b) \cos \frac{1}{2} (a - b)$

23. $\cos b - \cos a = 2 \sin \frac{1}{2} (a + b) \sin \frac{1}{2} (a - b)$

24. $\text{tang}\, a + \text{tang}\, b = \dfrac{\sin (a + b)}{\cos a \cos b}$

25. $\text{tang}\, a - \text{tang}\, b = \dfrac{\sin (a - b)}{\cos a \cos b}$

26. $\cot b + \cot a = \dfrac{\sin (a + b)}{\sin a \sin b}$

27. $\cot b - \cot a = \dfrac{\sin (a - b)}{\sin a \sin b}$

28. $\text{tang} \frac{1}{2} (a + b) = \dfrac{\sin a + \sin b}{\cos a + \cos b}$

29. $\text{tang} \frac{1}{2} (a - b) = \dfrac{\sin a - \sin b}{\cos a + \cos b}$

30. $\cot \frac{1}{2} (a + b) = \dfrac{\sin a - \sin b}{\cos b - \cos a}$

31. $\cot \frac{1}{2} (a - b) = \dfrac{\sin a + \sin b}{\cos b - \cos a}$

32. $\dfrac{\text{tang} \frac{1}{2} (a + b)}{\text{tang} \frac{1}{2} (a - b)} = \dfrac{\sin a + \sin b}{\sin a - \sin b}$

For a Spherical Triangle, in which A, B, C, are the angles, and a, b, *and* c, *the opposite sides, as in* Fig. 58.

33. $\sin A \sin b = \sin B \sin a$

34. $\cos a = \cos A \sin b \sin c + \cos b \cos c$

35. $\cos A = \cos a \sin B \sin C - \cos B \cos C$

36. $\cot a = \dfrac{\cot A \sin B + \cos B \cos c}{\sin c}$

37. $\cot A = \dfrac{\cot a \sin b - \cos C \cos b}{\sin C}$

38. $\sin \frac{1}{2} A = \sqrt{\dfrac{\sin \frac{1}{2} (a + b - c) \sin \frac{1}{2} (a + c - b)}{\sin b \sin c}}$

39. $\text{tang} \frac{1}{2} (b + a) = \text{tang} \frac{1}{2} c \dfrac{\cos \frac{1}{2} (B - A)}{\cos \frac{1}{2} (B + A)}$

40. $\text{tang} \frac{1}{2} (b - a) = \text{tang} \frac{1}{2} c \dfrac{\sin \frac{1}{2} (B - A)}{\sin \frac{1}{2} (B + A)}$

41. $\text{tang} \frac{1}{2} (B + A) = \cot \frac{1}{2} C \dfrac{\cos \frac{1}{2} (b - a)}{\cos \frac{1}{2} (b + a)}$

42. $\text{tang } \frac{1}{2} (B - A) = \cot \frac{1}{2} C \frac{\sin \frac{1}{2} (b - a)}{\sin \frac{1}{2} (b + a)}$

43. $\begin{cases} \cot \frac{1}{2} C = \text{tang } \frac{1}{2} (B - A) \frac{\sin \frac{1}{2} (b + a)}{\sin \frac{1}{2} (b - a)} \\ \cot \frac{1}{2} C = \text{tang } \frac{1}{2} (B + A) \frac{\cos \frac{1}{2} (b + a)}{\cos \frac{1}{2} (b - a)} \end{cases}$

44. $\begin{cases} \text{tang } \frac{1}{2} c = \text{tang } \frac{1}{2} (b - a) \frac{\sin \frac{1}{2} (B + A)}{\sin \frac{1}{2} (B - A)} \\ \text{tang } \frac{1}{2} c = \text{tang } \frac{1}{2} (b + a) \frac{\cos \frac{1}{2} (B + A)}{\cos \frac{1}{2} (B - A)} \end{cases}$

For a right angled spherical triangle in which C is the right angle, and the opposite side c, *the hypotenuse, as in* Fig. 59.

45. $\cos c = \cos a \cos b$ 48. $\text{tang } a = \sin b \text{ tang } A$
46. $\cos c = \cot A \cot B$ 49. $\text{tang } a = \cos B \text{ tang } c$
47. $\sin a = \sin c \sin A$ 50. $\cos A = \sin B \cos a$

51. *If any small arc or angle* a, *not exceeding, or not much exceeding a degree, be expressed in seconds, and if* $\omega = 206264''.8$ *we have,*

$$\sin a = \frac{a}{\omega}, \text{ very nearly.}$$

For the sine of a small arc is very nearly equal to the length of the arc itself; and to obtain the length of an arc, expressed in seconds, we have this proportion. As the number of seconds in the whole circumference is to the seconds in the arc, so is the length of the circumference to the length of the arc. Hence,

$$1296000'' : a : : 6.2831853 : \text{length of } a,$$

or, $$\text{length of } a = \frac{6.2831853\ a}{1296000''} = \frac{a}{206264''.8} = \frac{a}{\omega}.$$

Consequently, $$\sin a = \text{length of } a = \frac{a}{\omega}.$$

As the circumference of a circle, divided by 6.283, &c., gives the radius, it is evident that 206264″.8 are the seconds in the radius.

Cor. The number of seconds in an arc is equal to the product of ω by the length of the arc; the radius being unity.

INVESTIGATIONS OF VARIOUS FORMULÆ.

52. *To find formulæ for the values of the rectangular co-ordinates CD and AD, of a place A, on the earth's surface, Fig. 4, and its distance AC from the centre, in terms of the latitude and eccentricity.*

Put $\rho =$ AC, the equatorial radius Cq being assumed $= 1$.
$\phi =$ angle ZGq, the latitude,
$\phi' =$ " zCq, the geocentric latitude,
$e =$ earth's eccentricity.

Then, we have, CD $= \rho \cos \phi'$, and AD $= \rho \sin \phi'$.

Let the semicircle eBq be described on eq, and let DA be produced to meet it in B. Then, by Conic Sections,

$$Cp : Cq :: DA : DB,$$

or, (74), $\sqrt{(1 - e^2)} : 1 :: \rho \sin \phi' : DB = \rho \sin \phi' . \dfrac{1}{\sqrt{(1 - e^2)}}.$

But, $DB^2 + DC^2 = CB^2 = Cq^2 = 1.$

$$\text{Hence, } \rho^2 \sin^2 \phi' . \frac{1}{1 - e^2} + \rho^2 \cos^2 \phi' = 1,$$

$$\text{or } \frac{1}{1 - e^2} \tan^2 \phi' + 1 = \frac{1}{\rho^2 \cos^2 \phi'}.$$

But (76 A), $\tan^2 \phi' = (1 - e^2)^2 \tan^2 \phi$.

$$\text{Consequently, } (1 - e^2) \tan^2 \phi + 1 = \frac{1}{\rho^2 \cos^2 \phi'}.$$

$$\text{or, } (1 - e^2) \frac{\sin^2 \phi}{\cos^2 \phi} + 1 = \frac{1}{\rho^2 \cos^2 \phi'}$$

$$\sin^2 \phi - e^2 \sin^2 \phi + \cos^2 \phi = \frac{\cos^2 \phi}{\rho^2 \cos^2 \phi'}$$

$$1 - e^2 \sin^2 \phi = \frac{\cos^2 \phi}{\rho^2 \cos^2 \phi'}.$$

$$\text{Whence, } \rho \cos \phi' = \frac{\cos \phi}{\sqrt{(1 - e^2 \sin^2 \phi)}} \quad . \; . \; . \; . \quad \text{(A)}$$

Again (App. 2), $\rho \sin \phi' = \rho \cos \phi' \tan \phi' = \rho \cos \phi' \tan \phi \, (1 - e^2)$

$$= \frac{(1 - e^2) \tan \phi \cos \phi}{(1 - e^2 \sin^2 \phi)}$$

$$\text{or, } \rho \sin \phi' = \frac{(1 - e^2) \sin \phi}{\sqrt{(1 - e^2 \sin^2 \phi)}} \quad . \; . \; . \; . \; . \; . \; . \quad \text{(B)}$$

Adding together the squares of the equations (A) and (B) we have,

$$\rho^2 = \frac{(1 - e^2)^2 \sin^2 \phi + \cos^2 \phi}{1 - e^2 \sin^2 \phi} = \frac{\sin^2 \phi - 2\, e^2 \sin^2 \phi + e^4 \sin^2 \phi + \cos^2 \phi}{1 - e^2 \sin^2 \phi}$$

$$= \frac{1 - 2\, e^2 \sin^2 \phi + e^4 \sin^2 \phi}{1 - e^2 \sin^2 \phi}$$

$$\text{or, } \rho = \sqrt{\frac{1 - (2 - e^2)\, e^2 \sin^2 \phi}{1 - e^2 \sin^2 \phi}} \quad . \quad . \quad . \quad . \quad . \quad . \quad (C)$$

The value of ρ, multiplied by the number of miles in the equatorial radius, gives the value of AC in miles.

53. *To find the times of longest and shortest twilight at a given place.* Let HZG, *Fig.* 32, be the meridian of the place, Z its zenith, HR its horizon, FG, parallel to HR, 18° below, P the elevated pole, AB the part of sun's diurnal path included between HR and FG, PA and PB arcs of declination circles, and ZA and ZB arcs of vertical circles. Put $L = PH$ = latitude of the place, D = sun's declination, and $2a = 18°$. Then, (App. 34), we have,

$$\cos ZPA = \frac{\cos AZ - \cos PZ \cos AP}{\sin PZ \sin AP} = \frac{\cos 90° - \sin L \cos (90° \pm D)}{\cos L \sin (90° \pm D)}$$

$$= \frac{\pm \sin L \sin D}{\cos L \cos D} = \pm \text{tang } L \text{ tang } D \quad . \quad . \quad . \quad . \quad . \quad . \quad . \quad . \quad . \quad . \quad (A)$$

$$\cos (ZPA + APB) = \cos ZPB = \frac{\cos BZ - PZ \cos BP}{\sin PZ \sin BP} =$$

$$\frac{\cos (90° + 2a) - \sin L \cos (90° \pm D)}{\cos L \sin (90° \pm D)} = -\frac{\sin 2a}{\cos L \cos D} \pm \text{tang } L \text{ tang } D.$$

Hence, $\cos ZPA - \cos (ZPA + APB) = \dfrac{\sin 2a}{\cos L \cos D}$,

or, (App. 23), $2 \sin \frac{1}{2} APB \sin (ZPA + \frac{1}{2} APB) = \dfrac{\sin 2a}{\cos L \cos D}$.

Let now, $H = ZPA$, and $x = APB$, when AP is greater than 90°, that is, when the declination is of a different name from the latitude, and $H' = ZPA$, and $x' = APB$, for an equal declination, when of the same name with the latitude. Then, it is evident from the expression for cos ZPA, that H will be less than 90°, and that H′ will be the supplement of H. Hence, we have,

$$2 \sin \tfrac{1}{2} x \sin (H + \tfrac{1}{2} x) = \frac{\sin 2a}{\cos L \cos D}$$

and $$2 \sin \tfrac{1}{2} x' \sin (180° - H + \tfrac{1}{2} x') = \frac{\sin 2a}{\cos L \cos D}$$

or, $$2 \sin \tfrac{1}{2} x' \sin (H - \tfrac{1}{2} x') = \frac{\sin 2a}{\cos L \cos D} \quad . \quad . \quad . \quad . \quad . \quad (B)$$

Consequently, $\sin \frac{1}{2} x' \sin (H - \frac{1}{2} x') = \sin \frac{1}{2} x \sin (H + \frac{1}{2} x)$, or (App. 18), $\frac{1}{2} \cos (H - x') - \frac{1}{2} \cos H = \frac{1}{2} \cos H - \frac{1}{2} \cos (H + x)$, or, $\cos (H - x') + \cos (H + x) = 2 \cos H$.

But, (App. 22), $\cos(H - x) + \cos(H + x) = 2 \cos H \cos x$.

Hence, $\cos(H - x') - \cos(H - x) = 2 \cos H (1 - \cos x) = 4 \cos H \sin^2 \frac{1}{2} x$.

Now, since H is less than 90°, the second member is affirmative. Consequently, $\cos(H - x')$ is greater than $\cos(H - x)$, and therefore, x' is greater than x; that is, *the twilight is longer when the latitude and sun's declination are of the same name, than when they are respectively of the same values, but of different names.* And it is easy to perceive from equations (A) and (B), that *the longest twilight at a place, occurs when the declination is greatest and of the same name with the latitude.*

For the shortest twilight. Let the triangle BPC, having the side BP = AP, have also, PC = PZ, and BC = AZ = 90°; then its three sides being respectively equal to those of the triangle APZ, we have the angle CPB = ZPA. Taking CPA from each, we have APB = ZPC. Hence, the twilight will be shortest when the angle ZPC is least.

Let ZD be a vertical circle through C, and PE an arc of a great circle bisecting the angle ZPC. Since PZ and PC are constant, the angle ZPC will be least when ZC is least; and since BZ and BC are constant, ZC will be least when the angle ZBC is least, that is, when it becomes Zero, or when BZ and DZ coincide. But, when BZ and DZ coincide, we must have DC = BC = 90°. Hence, as DZ = 90° + $2a$, it follows, that, when the twilight is shortest, ZC = $2a$.

Now, as the triangle ZPC is isosceles, and PE bisects the vertical angle, it must also bisect the base ZC and be perpendicular to it. Hence, in the right-angled triangle ZEP, we have (App. 47),

$$\sin ZPE = \frac{\sin ZE}{\sin PZ} = \frac{\sin a}{\cos L}.$$

Twice the angle ZPE converted into time, gives the duration of shortest twilight. From the right-angled triangles ZEP and DEP, we have (App. 45),

$$\frac{\cos PZ}{\cos ZE} = \cos PE = \frac{\cos PD}{\cos DE}, \text{ or, } \frac{\sin L}{\cos a} = \frac{\cos PD}{\cos (90° + a)} = -\frac{\cos PD}{\sin a}.$$

Hence, $\qquad \cos PD = -\sin L \tang a.$

As cos PD is negative, PD, the sun's distance from the elevated pole, must be more than 90°, and, consequently, the declination must be of a name *contrary* to that of the latitude.

When PD has been found from the above expression, the sun's declination is known; and by a Nautical Almanac, the times in the year when the sun has this declination are easily found.

54. *To find the annual variations of a star in right ascension and declination.*

Referring to *Fig.* 20, described in previous Articles (126 and 128), as it is evident that mg does not sensibly differ from EG, and as the difference between the complements of two arcs is the same as the difference between the arcs themselves, we have,

$$E'G' - EG = E'G' - mg = E'm + gG',$$

and, $G's - Gs = Ps - P's,$

or, ann. var. in R. Ascen. $= E'm + gG'$. . . . . (A)

ann. var. in Declin. $= Ps - P's$. . . . . (B)

Draw $P'r$ perpendicular to the declination circle PsG. Then, as PP′ and $P'r$ are very small, we may, without material error, regard PpP' as a spherical triangle, right angled at P, and P′, PrP' and $E'mE$ as right angled plane triangles, and $sP' = sr$. Put

$$A = EG = \text{right ascen. of the star}$$
$$D = Gs = \text{declination do.}$$
$$\varepsilon = pP = QEC = \text{obliq. of the ecliptic.}$$

Then, taking small arcs or angles instead of their sines, and observing that $PpP' = CC' = EE' = 50''.2$ (126, *Cor.*), and $P'Ps = 90° - GPQ = EQ - GQ = EG = A$, we have, from the triangles PpP' and PrP',

$$PP' = PpP' \sin pP = 50''.2 \sin \varepsilon,$$
$$P'r = PP' \sin P'Ps = PP' \sin A = 50''.2 \sin \varepsilon \sin A,$$

and, $Pr = PP' \cos P'Ps = PP' \cos A = 50''.2 \sin \varepsilon \cos A.$

Consequently, $Ps - P's = Ps - rs = Pr = 50''.2 \sin \varepsilon \cos A$. (C)

We have, also, from the right angled spherical triangles $P'rs$ and $gG's$, the latter of which may be regarded as right angled at g as well as at G′, we have,

$$P'r = P'sr \sin P's = gsG' \cos G's, \text{ and } gG' = gsG' \sin G's.$$

From these, we have,

$$gG' = \frac{P'r \sin G's}{\cos G's} = P'r \text{ tang } G's = 50''.2 \sin \varepsilon \sin A \text{ tang } G's.$$

But, since the quantity which is multipled by tang $G's$ is small, we may, without sensible error, put tang Gs or tang D, instead of tang $G's$. We then have,

$gG' = 50''.2 \sin \varepsilon \sin A \text{ tang } D$ . . . . (D)

From the triangle $E'mE$, we have,

$E'm = E'E \cos mE'E = 50''.2 \cos \varepsilon$ . . . . . . (E)

Substituting, in formulæ (A and B), the values of $E'm$, gG' and $Ps - P's$ (E, D, and C), we have,

Ann. var. in R. Ascen. $= 50''.2 \cos \varepsilon + 50''.2 \sin \varepsilon \sin A \text{ tang } D$

Ann. var. in Declin. $= 50''.2 \sin \varepsilon \cos A.$

In applying these formulæ, the declination is to be regarded as *affirmative* or *negative*, according as it is *north* or *south*.

When the mean right ascension and declination of a star is known for a given time, and the annual variations have been computed by the above formulæ, its mean right ascension and declination at a subsequent time not distant from the former more than 20 or 30 years, or, in case of the pole star or other star near the pole, 10 or 15 years, are obtained by adding to the given right ascension and declination, the product of the corresponding annual variation by the number of years, and parts of a year, in the interval. In like manner, the right ascension and declination may be obtained for a prior time, only subtracting the products instead of adding them.

55. *To find the aberration of a fixed star in right ascension and declination for a given time.*

Let s, *Fig.* 60, be the star, E the earth, BQ the equator, P its pole, LC the ecliptic, V the vernal equinox, N the pole of the declination circle PsG, and sNM an arc of a great circle, which, passing through the pole N, must be perpendicular to PsG. Also, let D be the point of the ecliptic towards which the earth is moving at the given time, and sD an arc of the great circle, in which the plane sED cuts the celestial sphere. Then (132), the direct effect of aberration is to make the star appear to be at a point s' between s and D, such that $ss' = 20''.36 \sin Ds$.

Let $s'd$ be perpendicular to PG. Then, as ss' is very small, we may regard the triangle $s'ds$ as rectilineal. We have, also, without sensible error, $G's' = Gd$, $Pe = Ps$, and $s'd = es$. Now, since $G's' = Gd = Gs - sd$, it is evident the effect of aberration on the star s, which is in the first quadrant, is to diminish the right ascension by the quantity G′G, and the declination by sd. Consequently,

$$\text{Aber. in right ascen.} = - G'G \quad \ldots \ldots \quad (A)$$

$$\text{Aber. in declin.} = - sd \quad \ldots \ldots \quad (B)$$

Put $a = 20''.36$, $A = VG =$ star's right ascension, $D = Gs =$ star's declination, $S =$ sun's longitude, and $\varepsilon = GVF =$ obliquity of the ecliptic. Then, using small arcs and angles instead of their sines, we have,

$$es = ePs \,.\, \sin Pe = ePs \,.\, \sin Ps = G'G \cos D;$$

also, $es = s'd = ss' \sin DsF = 20''.36 \sin Ds \sin DsF = a \sin Ds \sin DsF$.

Hence, $G'G \cos D = a \sin Ds \sin DsF$.

But, $\sin DsF : \sin F :: \sin DF : \sin Ds$, or, $\sin Ds \sin DsF = \sin F \sin DF$; and in the right angled spherical triangle VGF,

$$\sin VG = \sin F \sin VF, \text{ or, } \sin F = \frac{\sin VG}{\sin VF} = \frac{\sin A}{\sin VF}.$$

Therefore, $G'G \cos D = a \sin F \sin DF = \dfrac{a \sin A}{\sin VF} \sin DF$

or, $$-GG' = -\frac{a \sin A}{\cos D \sin VF} \sin DF$$

Put $$m = -\frac{a \sin A}{\cos D \sin VF} \quad . \; . \; . \; . \; . \; . \; . \; . \; . \; . \; . \quad (C)$$

Then (A), Aber. in right ascen. $= - G'G = m \sin DF$ . . (D)

Again, in the right angled triangle VGF, we have (App. 49),

$\tan VG = \cos \varepsilon \tan VF$, or, $\cot VF = \cos \varepsilon \cot VG = \cos \varepsilon \cot A$ (E)

Put $VF = 90° - \phi$. Then, since $VD = S - 90°$ (132), we have, $DF = VF - VD = 180° - (S + \phi)$, $\sin DF = \sin (S + \phi)$, $\cot VF = \tan \phi$, and $\sin VF = \cos \phi$. Hence (E, C, and D),

$$\left.\begin{aligned} \tan \phi &= \cos \varepsilon \cot A \\ m &= -\frac{a \sin A}{\cos D \cos \phi} \\ \textit{Aber. in right ascen.} &= m \sin (S + \phi) \end{aligned}\right\} \quad . \; . \; . \; . \; . \; . \quad (F)$$

Now, in the small triangle $s'ds$, we have,

$$sd = ss' \cos s'sd = ss' \sin MsD = a \sin Ds \sin MsD.$$

But, $\sin MsD : \sin M :: \sin MD : \sin Ds$, or, $\sin Ds \sin MsD = \sin M \sin MD$; and, $\sin MV : \sin NV :: \sin MNV$, or, $\sin GNs : \sin M = \dfrac{\sin NV \sin GNs}{\sin MV} = \dfrac{\cos A \sin D}{\sin MV}$.

Hence, $-sd = -a \sin Ds \sin MsD = -a \sin M \sin MD = -\dfrac{a \cos A \sin D}{\sin MV} \sin MD$

Put $$n = -\frac{a \cos A \sin D}{\sin MV} \quad . \; . \; . \; . \; . \; . \; . \; . \; . \; . \; . \quad (G)$$

Then (B), Aber. in declin. $= - sd = n \sin MD$ . . (H)

Now, in the triangle MNV, we have (App. 36),

$$\cot MV = \frac{\sin \varepsilon \cot MNV + \cos \varepsilon \cos NV}{\sin NV} = \frac{-\sin \varepsilon \cot D + \cos \varepsilon \sin A}{\cos A};$$

or, $$-\cot MV = \frac{\sin \varepsilon \cot D}{\cos A} - \cos \varepsilon \tan A \quad . \; . \; . \; . \; . \; . \quad (I)$$

Put

$MV = 90 + \theta$.* Then, $\sin MD = \sin (MV + VD) = \sin (S + \theta)$, $\sin MV = \cos \theta$, and $\cot MV = -\tan \theta$, or, $-\cot MV = \tan \theta$.

---

*When MV is less that 90°, as in the figure, it is $90° + \theta - 360°$ that is to be regarded as equal to MV.

Hence (I, G, and H),

$$\left.\begin{aligned}\tan\theta &= \frac{\sin\varepsilon\cot D}{\cos A} - \cos\varepsilon\tan A \\ &= \frac{a\cos A\sin D}{\cos\theta} \\ \textit{Aber. in declin.} &= n\sin(S+\theta)\end{aligned}\right\} \quad . \quad . \quad . \quad (K)$$

The increase or diminution of an arc by 180°, changes the sign of its sine or cosine, but does not affect its numerical value. It is, therefore, evident that if the value of the arc ϕ be increased or diminished by 180°, the expression for the aber. in right ascen. (F) will still be true; for the signs of both factors m and $\sin(S+\phi)$ being thus changed, there will be no change in the sign of the product. Hence, we may always take ϕ affirmative, and not exceeding 180°. Similar observations apply to the arc θ.

The quantities ϕ, θ, m and n change but little for a number of years, and therefore, when once computed for any star, they serve for a long time in computing the aberration of that star. Table IX contains the values of these quantities for 30 principal fixed stars. The values of m and n are all made affirmative by increasing the values of ϕ and θ by 180°, when requisite.

56. *To find formulæ for the lunar nutation in right ascension and declination.*

To obtain these formulæ we have recourse to certain results established by Physical Astronomy. It has been proved that the phenomena of nutation may be explained on the supposition that the pole of the equator, instead of moving strictly in a circle about the pole of the ecliptic (126), moves in a small ellipse about the mean place of the pole, that is, around that point in the circle at which the pole would be if the nutation did not exist, and in a period equal to that of the moon's nodes. The major axis of this ellipse is situated in the solstitial colure, and is to the minor axis in the ratio of the cosine of the obliquity of the ecliptic to the cosine of twice the obliquity. The major axis has been found to be equal to 18″.44;* and hence, the minor axis is 13″.73.

Let ELF, *Fig.* 61, be the ecliptic, N its pole, NLM the solstitial colure, EMF the mean equator, P the mean place of the pole, A*b*C*d* the ellipse in which the pole is assumed to move, and ABCD a circle about the centre P. Then, according to the investigation in Physical Astronomy, if the arc ABO be made equal to the longitude of the moon's node, and O*c* be

* This is the value given by Struve, in No. 426 of the *Astr. Nach.*, as recently obtained from a series of observations made by him at Dorpat.

drawn perpendicular to NL, the point p, in which it meets the ellipse, will be the true place of the pole, and the great circle E′G′F′, described about the pole p, will be the true equator. Let s be the situation of a star, and let pr be drawn perpendicular to the declination circle PG, and Em perpendicular to E′G′F′. Then, observing that p stands here in place of P′, *Fig.* 20, we have, as for the annual variations (App. 54),

$$\text{Nut. in right ascen.} = \mathrm{E}'m + g\mathrm{G}' \quad \text{(A)}$$
$$\text{Nut. in declin.} = \mathrm{P}s - ps = \mathrm{P}r \quad \text{(B)}$$
$$pr = p\mathrm{P} \sin p\mathrm{P}s$$
$$\mathrm{E}'m = \mathrm{E}'\mathrm{E} \cos \varepsilon \quad \text{(C)}$$
$$g\mathrm{G}' = pr \text{ tang } \mathrm{G}'s = pr \text{ tang D} = p\mathrm{P} \sin p\mathrm{P}s \text{ tang D} \quad \text{(D)}$$
$$\mathrm{P}r = p\mathrm{P} \cos p\mathrm{P}s \quad \text{(E)}$$

Now in the triangle pPN, we have,

$$\sin \mathrm{PN} : p\mathrm{P} :: \sin \mathrm{P}p\mathrm{N} : p\mathrm{NP}, \text{ or LL}', \text{ or E}'\mathrm{E}.$$

But, as Npc may be regarded as a right angle, or Np parallel to Nc, we have, PpN = APp. Hence, as PN = ε,

$$\sin \varepsilon : p\mathrm{P} :: \sin \mathrm{AP}p : \mathrm{E}'\mathrm{E} = \frac{p\mathrm{P} \sin \mathrm{AP}p}{\sin \varepsilon}.$$

Therefore, $\mathrm{E}'m = \mathrm{E}'\mathrm{E} \cos \varepsilon = p\mathrm{P} \sin \mathrm{AP}p \cot \varepsilon$ . . . . (F)

As $p\mathrm{P}s = \mathrm{AP}p + \mathrm{AP}s = \mathrm{AP}p + \mathrm{MG} = \mathrm{AP}p + \mathrm{EG} - \mathrm{EM} = \mathrm{AP}p + \mathrm{A} - 90°$, and as $\sin(\mathrm{A} - 90) = -\cos \mathrm{A}$, and $\cos(\mathrm{A} - 90) = \sin \mathrm{A}$, we have (App. 13 and 14),

$$\sin p\mathrm{P}s = \sin \mathrm{AP}p \sin \mathrm{A} - \cos \mathrm{AP}p \cos \mathrm{A}$$
$$\cos p\mathrm{P}s = \cos \mathrm{AP}p \sin \mathrm{A} + \sin \mathrm{AP}p \cos \mathrm{A}.$$

Hence (D and E),

$$g\mathrm{G}' = p\mathrm{P} \sin \mathrm{AP}p \sin \mathrm{A} \text{ tang D} - p\mathrm{P} \cos \mathrm{AP}p \cos \mathrm{A} \text{ tang D} \quad \text{(G)}$$
$$\mathrm{P}r = p\mathrm{P} \cos \mathrm{AP}p \sin \mathrm{A} + p\mathrm{P} \sin \mathrm{AP}p \cos \mathrm{A}. \quad \text{(H)}$$

Put $a = \mathrm{PC} = 9.22$

$b = d\mathrm{P} = 6.865$

N = arc ABO = longitude of moon's node;

then, 360° — N = APO.

Now, in the small right angled triangles OcP and pcP, we have,

$$\mathrm{O}c = \mathrm{OP} \sin \mathrm{APO} = a \sin(360° - \mathrm{N}) = -a \sin \mathrm{N},$$
$$pc = p\mathrm{P} \sin \mathrm{AP}p.$$

But, by a property of the ellipse,

$$\mathrm{DP} : d\mathrm{P} :: \mathrm{O}c : pc;$$

or, $a : b :: -a \sin \mathrm{N} : pc = -b \sin \mathrm{N}.$

Hence, $p\mathrm{P} \sin \mathrm{AP}p = -b \sin \mathrm{N}.$

We have, also, $p\mathrm{P} \cos \mathrm{AP}p = \mathrm{P}c = \mathrm{OP} \cos \mathrm{APO} = a \cos \mathrm{N}.$

Substituting the values of pP sin APp and pP cos APp in (F, G and H), we have,

$$E'm + gG' = -b\,(\cot \varepsilon + \sin A \operatorname{tang} D) \sin N - a \cos A \operatorname{tang} D \cos N \quad \text{(I)}$$

$$Pr = -b \cos A \sin N + a \sin A \cos N. \quad \text{(K)}$$

Put
$$-b\,(\cot \varepsilon + \sin A \operatorname{tang} D) = m' \cos \phi'$$
$$-a \cos A \operatorname{tang} D = m' \sin \phi'.$$

Then, $\operatorname{tang} \phi' = \dfrac{a \cos A \operatorname{tang} D}{b\,(\cot \varepsilon + \sin A \operatorname{tang} D)},\ m' = -\dfrac{a \cos A \operatorname{tang} D}{\sin \phi'}$ (L, M)

and, $E'm + gG' = m'\,(\sin N \cos \phi' + \cos N \sin \phi') = m' \sin (N + \phi')$

or (A), *nut. in right ascen.* $= m' \sin (N + \phi')$. . . . . . (N)

Put
$$-b \cos A = n' \cos \theta'$$
$$a \sin A = n' \sin \theta'.$$

Then, $\operatorname{tang} \theta' = -\dfrac{a}{b} \operatorname{tang} A,\ n' = \dfrac{a \sin A}{\sin \theta'},$ . . . (P, Q)

and, $Pr = n'\,(\sin N \cos \theta' + \cos N \sin \theta') = n' \sin (N + \theta'),$

or (B), *nut. in declin.* $= n' \sin (N + \theta')$ . . . . (R)

The observations in the last Article relative to ϕ, θ, m and n, apply equally here to ϕ', θ', m' and n'.

To find the solar nutation in right ascension and declination. It has been found that the solar nutation may be explained by assuming the pole of the equator to describe a small ellipse about its mean place, in like manner as for the lunar nutation. For the solar nutation, we have, *Fig.* 61, PC $= 0''.555$, P$d = 0''.500$, and the arc ABO = twice the sun's longitude. Hence, if S be the sun's longitude, the formulæ for the solar nutation in right ascension and declination will be the same as for the lunar, except that, instead of the values of a and b in the last Article, we shall have, $a = 0''.545$ and, $b = 0''.5$, and, instead of N, we shall have, 2S.

Note. As these values of a and b are about $\frac{1}{16}$ of their values for the lunar nutations, we may obtain approximate values of the solar nutations, by computing as for the lunar, only using, in (N and R), 2S instead of N, and taking $\frac{1}{16}$ of the results.

57. *Given the eccentricity of the orbit of a planet and the mean anomaly, to find the true anomaly.*

Let the semi-ellipse PDA, *Fig.* 62, represent one-half the orbit, C the centre, S the place of the sun in one focus, D the place of the planet in its orbit at any time, and F the place at which it would have been at that time if its angular motion had been uniform. On the diameter AP let the semi-circle AGP be described, and let CL be drawn parallel to SF, and GDH perpendicular to AP. Then, the angle PCL = PSF is the mean anomaly, and PSD is the true anomaly. The angle PCG is called the *eccentric* anomaly.

Assuming AC, the mean distance of the planet, to be a unit, put

e = SC = eccentricity
m = PCL = PSF = mean anomaly
u = PSD = true anomaly
x = PCG = eccentric anomaly
T = time of describing semi-ellipse PDA
t = time of describing arc PD.

By the property of the ellipse,

area PGA : area PRA : : AC : CR : : area PGS : area PDS,

or, area PGA : area PGS : : area PRA : area PDS.

But, by Kepler's second law (153),

area PRA : area PDS : : T : t. Hence,

area PGA : area PGS : : T : t : : 180° : PCL : : area PGA : sect. PCL.

Hence, sect. PCL = area PGS

sect. PCG — sect. PCL = sect. PCG — area PGS,

or, sect. GCL = triang. GCS . . . . (A)

But, sect. GCL = ½ AC × arc GL, and triang. GCS = ½ AC × CS × sin PCG.

Hence, arc GL = CS × sin PCG = $e \sin x$.

Or, for the arc GL, or angle GCL, in seconds, we have (App. 51, *Cor.*),

angle GCL = $e\omega \sin x$ . . . . (B)

Also, since PCG = PCL + GCL, we have,

$x = m + e\omega \sin x$ . . . . (C)

For most of the planets the value of e is less than 0.1, and it is not for any of them more than about 0.25. Hence, it is evident (B), that the angle GCL is generally quite small, and that it is never large. The sector GCL differs, therefore, but very little from the triangle GCL, and we have (A), triang. GCL = triang. GCS, nearly. Consequently, SL is nearly parallel to CG, and the angle PSL is nearly equal to PCG, the eccentric anomaly.

Put $2d$ = 180° — diff. of angles CSL and CLS
p = angle PSL = eccentric anomaly, nearly,
$x = p + z$.

By trig., we have,

CL — CS : CL + CS : : tang ½ (CSL — CLS) : tang ½ (CSL + CLS),

or, 1 — e : 1 + e : : tang ½ (180° — $2d$) : tang ½ (180° — m), or, $1 - e : 1 + e :: \cot(90° - \frac{1}{2}m) : \cot(90° - d) :: \text{tang}\,\frac{1}{2}m : \text{tang}\,d$.

Hence, $$\text{tang}\,d = \frac{1+e}{1-e}\,\text{tang}\,\tfrac{1}{2}m \quad . \quad . \quad . \quad . \quad . \quad (D)$$

But, $\frac{1}{2}m + d = 90° - \frac{1}{2}(CSL + CLS) + 90° - \frac{1}{2}(CSL - CLS)$ $= 180° - CSL = PSL = p$;

or, $$p = \tfrac{1}{2}m + d \quad . \quad . \quad . \quad . \quad . \quad . \quad . \quad . \quad . \quad . \quad (E)$$

Now, substituting $p + z$ instead of x in formula (C) and observing that, as z must be very small, we may regard $\cos z = 1$, and $\omega \sin z = z$, in seconds, we have,

$p + z = m + e\omega \cdot \sin(p + z) = m + e\omega \sin p + ez \cos p$, very nearly

Hence, $$z = \frac{m + e\omega \sin p - p}{1 - e \cos p}, \text{ very nearly.} \quad . \quad . \quad . \quad (F)$$

And, since $x = p + z$, we have,

$$x = p + \frac{m + e\omega \sin p - p}{+1 - e \cos p}, \text{ very nearly,} \quad . \quad . \quad . \quad . \quad . \quad (G)$$

If x is desired with still greater accuracy, it may be obtained by taking p equal to the value of x found from formula (G), and recomputing with this value. This repetition is, however, seldom if ever necessary.

Now, as $AC = 1$, $SC = e$, and $PSD = u$, we have, the property of the ellipse,

$$SD = \frac{1 - e^2}{1 + e \cos u}, \text{ and } SH = SD \cos u = \frac{(1 - e^2)\cos u}{1 + e \cos u}.$$

But, $SH = CH - CS = CG \cos PCG - CS = \cos x - e$.

Hence, $$\frac{(1 - e^2)\cos u}{1 + e \cos u} = \cos x - e, \text{ or, } \cos u = \frac{\cos x - e}{1 - e \cos x}.$$

But (App. 12),

$$\text{tang}^2\,\tfrac{1}{2}u = \frac{1 - \cos u}{1 + \cos u} = \frac{1 + e - (1 + e)\cos x}{1 - e + (1 - e)\cos x} = \frac{1 + e}{1 - e} \cdot \frac{1 - \cos x}{1 + \cos x}$$

$$= \frac{1 + e}{1 - e}\,\text{tang}^2\,\tfrac{1}{2}x.$$

Hence, $$\text{tang}\,\tfrac{1}{2}u = \text{tang}\,\tfrac{1}{2}x\sqrt{\frac{1 + e}{1 - e}} \quad . \quad . \quad . \quad . \quad . \quad (H)$$

Having found the value of p from the expressions (D and E), we find x from (G), and then the true anomaly u, from (H).

58. *To determine the height of a lunar mountain.*

Let ABO, *Fig.* 68, be the enlightened hemisphere of the moon, E the situation of the earth, ES′ the direction of the sun from the earth, and SM a solar ray, touching the moon in O, which will be one of the points in the curve separating the enlightened from the dark part of the moon. Also, let M be the summit of a mountain, situated near to O, and just sufficiently elevated to receive the sun's light on its top. To an observer at E, the summit M, of the mountain, will appear as a bright spot on the dark part of the moon. Let the angle MEO be measured with a micrometer, and let the angle of elongation CES′, be found from the positions of the sun and moon at the time.

We may, without material error, regard ES′ as parallel to MS, EC as equal to EO, and the angle MES′ as equal to CES′. Consequently, sin OME = sin MES′ = sin CES′. Hence, from the triangle EMO, we have,

$$\sin \mathrm{OME} : \sin \mathrm{MEO} :: \mathrm{EO} : \mathrm{MO},$$

or,

$$\sin \mathrm{CES}' : \sin \mathrm{MEO} :: \mathrm{EC} : \mathrm{MO} \quad . \quad . \quad . \quad . \quad . \quad (\mathrm{A})$$

Put δ = app. diameter of the moon,
d = real diameter "
C = angle MCO.

Then, since δ and C are both small, we have (97), $\mathrm{OC} = \mathrm{EC} \sin \frac{1}{2} \delta = \frac{1}{2} \mathrm{EC} \sin \delta$, and $\mathrm{MO} = \mathrm{OC} \operatorname{tang} \mathrm{C} = \frac{1}{2} \mathrm{EC} \sin \delta \operatorname{tang} \mathrm{C} = \mathrm{EC} \sin \delta \operatorname{tang} \frac{1}{2} \mathrm{C}$. Hence (A),

$$\sin \mathrm{CES}' : \sin \mathrm{MEO} :: \mathrm{EC} : \mathrm{EC} \sin \delta \operatorname{tang} \tfrac{1}{2} \mathrm{C},$$

or,

$$\operatorname{tang} \tfrac{1}{2} \mathrm{C} = \frac{\sin \mathrm{MEO}}{\sin \delta \sin \mathrm{CES}'} = \frac{\mathrm{ang\ MEO}}{\delta \sin \mathrm{CES}'} \quad . \quad . \quad . \quad . \quad . \quad (\mathrm{B})$$

$$\text{Now, } \mathrm{M}a = \mathrm{MC} - \mathrm{OC} = \frac{\mathrm{OC}}{\cos \mathrm{C}} - \mathrm{OC} = \mathrm{OC}\, \frac{1 - \cos \mathrm{C}}{\cos \mathrm{C}} = (\text{App. } 11)$$

$$\mathrm{OC} \,.\, \frac{\sin \mathrm{C} \operatorname{tang} \frac{1}{2} \mathrm{C}}{\cos \mathrm{C}} = \mathrm{OC} \operatorname{tang} \mathrm{C} \operatorname{tang} \tfrac{1}{2} \mathrm{C} = d \operatorname{tang} \tfrac{1}{2} \mathrm{C} \operatorname{tang} \tfrac{1}{2} \mathrm{C} = d \operatorname{tang}^2 \tfrac{1}{2} \mathrm{C}.$$

Observing that Ma is the height of the mountain, we have, by substituting the value of tang $\frac{1}{2}$ C (B),

$$\textit{Height of the mountain} = d \left(\frac{\mathrm{ang\ MEO}}{\delta \sin \mathrm{CES}'} \right)^2.$$

When the moon's apparent diameter was 32′ 5″.2 and her elongation CES′ = 125° 8′, the angle MEO for one of the mountains was found by Dr. Herschel to be 40″.625. Hence, taking the moon's diameter = 2160 miles, we easily obtain, from the preceding formula, the height of the mountain = 1.45 miles.

INTERPOLATION.

59. *Interpolation* is a method by which, from several given consecutive values of a quantity, corresponding to given values of another quantity, on which it depends or with which it is connected, an intermediate value of

the first may be found, corresponding to a given intermediate value of the second. Thus, if the values of an astronomical quantity x, whose variation is not very irregular, be given for the times T, T + 1hr., T + 2hrs., &c., its value for an intermediate time, as for instance, for the time T + $1\frac{1}{3}$hr., may be found by interpolation.

60. Let the values of the quantity x, at the times T, T + 1, T + 2, &c., be a, a', a'', &c. If the first of these values be subtracted from the second, the second from the third, and so on, the remainders are called *first differences.* If the first of these be subtracted from the second, the second from the third, and so on, the remainders are called *second differences.* In like manner we obtain *third differences, fourth differences, &c.* Let the values of x, at the times T, T + 1, &c., and the successive orders of differences be arranged as in the following table; in which the first differences are denoted, by Δa, $\Delta a'$, &c., the second by $\Delta^2 a$, $\Delta^2 a'$, &c., and the others in a similar manner.

	vals. of x	1st. diffs.	2nd. diffs.	3rd. diffs.	4th. diffs.	5th. diffs.
T	a	Δa	$\Delta^2 a$	$\Delta^3 a$	$\Delta^4 a$	$\Delta^5 a$
T + 1	a'	$\Delta a'$	$\Delta^2 a'$	$\Delta^3 a'$	$\Delta^4 a'$	
T + 2	a''	$\Delta a''$	$\Delta^2 a''$	$\Delta^3 a''$		
T + 3	a'''	$\Delta a'''$	$\Delta^2 a''''$			
T + 4	a''''	$\Delta a''''$				
T + 5	a'''''					

Then, we evidently have,

$\Delta^4 a' = \Delta^4 a + \Delta^5 a$

$\Delta^3 a' = \Delta^3 a + \Delta^4 a$

$\Delta^3 a'' = \Delta^3 a' + \Delta^4 a' = \Delta^3 a + 2\Delta^4 a + \Delta^5 a$

$\Delta^2 a' = \Delta^2 a + \Delta^3 a$

$\Delta^2 a'' = \Delta^2 a' + \Delta^3 a' = \Delta^2 a + 2\Delta^3 a + \Delta^4 a$

$\Delta^2 a''' = \Delta^2 a'' + \Delta^3 a'' = \Delta^2 a + 3\Delta^3 a + 3\Delta^4 a + \Delta^5 a$

$$\Delta a' = \Delta a + \Delta^2 a$$
$$\Delta a'' = \Delta a' + \Delta^2 a' = \Delta a + 2\Delta^2 a + \Delta^3 a$$
$$\Delta a''' = \Delta a'' + \Delta^2 a'' = \Delta a + 3\Delta^2 a + 3\Delta^3 a + \Delta^4 a$$
$$\Delta a'''' = \Delta a''' + \Delta^2 a''' = \Delta a + 4\Delta^2 a + 6\Delta^3 a + 4\Delta^4 a + \Delta^5 a.$$

Hence, since $a' = a + \Delta a$, $a'' = a' + \Delta a'$, &c., we have,

$$a' = a + \Delta a$$
$$a'' = a + 2\Delta a + \Delta^2 a$$
$$a''' = a + 3\Delta a + 3\Delta^2 a + \Delta^3 a$$
$$a'''' = a + 4\Delta a + 6\Delta^2 a + 4\Delta^3 a + \Delta^4 a$$
$$a''''' = a + 5\Delta a + 10\Delta^2 a + 10\Delta^3 a + 5\Delta^4 a + \Delta^5 a'.$$

&c. &c.

The co-efficients of the terms in the values of a, a', a'', &c., which are the values of x at the times T, T + 1, T + 2, &c., are, therefore, the same as the co-efficients of the terms in the first, second, third, &c., powers of a binomial. Hence, if we take t for an interval of time, expressed in hours, or hours and parts of an hour, we shall have, for the value of x at the time $\mathrm{T} + t$,

$$x = a + t.\Delta a + \frac{t\,(t-1)}{2}\,\Delta^2 a + \frac{t\,(t-1)\,(t-2)}{2.3}\,\Delta^3 a + \frac{t\,(t-1)\,(t-2)\,(t-3)}{2.3.4.}\,\Delta^4 a + \&c. \quad \text{(A)}$$

61. The process of interpolation can only be employed with advantage when the differences of some order, not very high, become equal, or very nearly equal. If we suppose the fourth differences to be equal, then the the fifth and higher differences must $= 0$. Hence, we would have, in this case, for the value of x at the time $\mathrm{T} + t$, the rigorous formula,

$$x = a + t\,\Delta a + \frac{t\,(t-1)}{2}\,\Delta^2 a + \frac{t\,(t-1)\,(t-2)}{2.3}\,\Delta^3 a + \frac{t\,(t-1)\,(t-2)\,(t-3)}{2.3.4}\,\Delta^4 a \quad \text{(B)}$$

If the differences of the fifth and higher orders, though not absolutely $= 0$, are very small, the formula will still be very nearly accurate for values of t not exceeding, or not much exceeding, 4hrs. For, in this case, the co-efficients of the terms in which these higher differences enter, being each less than a unit, the terms omitted will be very small.

62. *Another Formula.* We may obtain a formula, that is frequently more convenient than those above, by taking the times and values of x, as in the following table.

	vals. of x	1st. diffs.	2nd diffs.	3rd diffs.	4th diffs.
T — 2	$a_{\prime\prime}$				
		$\Delta a_{\prime\prime}$			
T — 1	$a_{\prime}$		$\Delta^2 a_{\prime\prime}$		
		$\Delta a_{\prime}$		$\Delta^3 a_{\prime\prime}$	
T	a		$\Delta^2 a_{\prime}$		$\Delta^4 a_{\prime\prime}$
		Δa		$\Delta^3 a_{\prime}$	
T + 1	a'		$\Delta^2 a$		$\Delta^4 a_{\prime}$
		$\Delta a'$		$\Delta^3 a$	
T + 2	a''		$\Delta^2 a'$		$\Delta^4 a$
		$\Delta a''$		$\Delta^4 a'$	
T + 3	a'''		$\Delta a''$		
		$\Delta a'''$			
T + 4	a''''				

Put $\Delta a + \Delta a_{\prime} = 2b$, $\Delta^2 a_{\prime} = c$, $\Delta^3 a_{\prime} + \Delta^3 a_{\prime\prime} = 2d$, and $\Delta^4 a_{\prime\prime} = e$. Then, since half the difference of two quantities, added to half the sum, gives the greater quantity, we have, $\Delta a = b + \frac{1}{2}\,c$, and $\Delta^3 a_{\prime} = d + \frac{1}{2}\,e$. Hence, assuming the fourth differences to be equal, $\Delta^4 a = e$, $\Delta^3 a = \Delta^3 a_{\prime} + e = d + \frac{3}{2}\,e$, and $\Delta^2 a = \Delta^2 a_{\prime} + \Delta^3 a_{\prime} = c + d + \frac{1}{2}\,e$. Substituting these values of Δa, $\Delta^2 a$, $\Delta^3 a$, and $\Delta^4 a$, in the formula (B), it becomes,

$$x = a + t\,(b + \tfrac{1}{2}\,c) + \frac{t\,(t-1)}{2}\,(c + d + \tfrac{1}{2}\,e) + \frac{t\,(t-1)\,(t-2)}{2.3}$$

$$(d + \tfrac{3}{2}\,e) + \frac{t\,(t-1)\,(t-2)\,(t-3)}{2.3.4}e.$$

Performing the multiplications indicated, and reducing, we have for the value x, at the time $T + t$,

$$x = a + tb + \frac{t^2}{2}c + \frac{t\,(t^2-1)}{6}d + \frac{t^2\,(t^2-1)}{24}e, \quad . \quad . \quad . \quad (C)$$

$$\text{or, } x = a + t\,(b - \tfrac{1}{6}d) + \frac{t^2}{2}\,(c - \tfrac{1}{12}e) + \frac{t^3}{6}d + \frac{t^4}{24}e \quad . \quad . \quad . \quad (D)$$

The interval t, may be taken of any value between $-\,2$ and $+\,2$.

63. The *hourly variation* of a quantity at any time, is the variation or change in its value which would take place in an hour, if the rate of change at that time were to be the same, throughout the hour. The *average hourly variation* of a quantity between two times T and $T + t$, is that hourly variation which, if continued during the interval t, would produce the change in the value of the quantity, that actually takes place during this interval.

64. *Average hourly variation.* It follows, from the definition, that the difference between the values of a quantity at the times T and $T + t$, divided by the interval t, must give the average hourly variation during the interval. Hence, if we put $x' =$ this average hourly variation, we have, from the formula (D),

$$x' = b - \tfrac{1}{6}d + \frac{t}{2}\left(c - \frac{1}{12}e\right) + \frac{t^2}{6}d + \frac{t^3}{24}e. \quad . \quad . \quad . \quad . \quad (E)$$

The average hourly variation or the values of x' for the whole hours, are,

$$\left.\begin{array}{lll} \text{at,} & T-2 & b - c + \tfrac{1}{2}d - \tfrac{1}{4}e \\ \text{"} & T-1 & b - \tfrac{1}{2}c \\ \text{"} & T & b - \tfrac{1}{6}d \\ \text{"} & T+1 & b + \tfrac{1}{2}c \\ \text{"} & T+2 & b + c + \tfrac{1}{2}d + \tfrac{1}{4}e \end{array}\right\} \quad . \quad . \quad . \quad (F)$$

65. *Hourly variation.* If in the formula (D) we put $t + t'$ instead of t, we shall have, for the value of x at the time $T + t + t'$,

$$x = a + t\,(b - \tfrac{1}{6}\,d) + \frac{t^2}{2}\,(c - \tfrac{1}{12}e) + \frac{t^3}{6}d + \frac{t^4}{24}e + t'\,(b - \tfrac{1}{6}d)$$

$$+ \frac{2tt' + t'^2}{2}(c - \tfrac{1}{12}e) + \frac{3\,t^2t' + 3\,tt'^2 + t'^3}{6}d +$$

$$\frac{4\,t^3t' + 6\,t^2t'^2 + 4\,tt'^3 + t'^4}{24}e.$$

From this value of x, subtracting its value at the time $T + t$ (D), and dividing the remainder by t', we find the average hourly variation between the times $T + t$, and $T + t + t'$, to be,

$$b - \tfrac{1}{6}d + \frac{2\,t + t'}{2}(c - \tfrac{1}{12}e) + \frac{3\,t^2 + 3\,tt' + t'^2}{6}d +$$
$$\frac{4\,t^3 + 6\,t^2\,t' + 4\,tt'^2 + t'^3}{24}e.$$

Now, it is evident that the smaller the interval t' is, the nearer will this average hourly variation approach to the hourly variation at the time $T + t$. Hence, if we now take x' to stand for the hourly variation at the time $T + t$, we have, by taking $t' = 0$, in the above expression,

$$x' = b - \frac{1}{6}d + t\left(c - \frac{1}{12}e\right) + \frac{t^2}{2}d + \frac{t^3}{6}e. \quad . \quad . \quad . \quad . \quad . \quad \text{(G)}$$

The hourly variations, or the values of x' at the whole hours, are,

$$\left.\begin{array}{ll|l} \text{at,} & T - 2 & b - 2c + \frac{11}{6}d - \frac{7}{6}e \\ \text{''} & T - 1 & b - c + \frac{1}{3}d - \frac{1}{12}e \\ \text{''} & T & b - \frac{1}{6}d \\ \text{''} & T + 1 & b + c + \frac{1}{3}d + \frac{1}{12}e \\ \text{''} & T + 2 & b + 2c + \frac{11}{6}d + \frac{7}{6}e \end{array}\right\} \quad . \quad . \quad . \quad \text{(H)}$$

INVESTIGATION OF FORMULÆ FOR COMPUTING SOLAR ECLIPSES, OCCULTATIONS, AND TRANSITS.

66. Let O, *Fig.* 64, be the centre of the earth, A a place on its surface, S the centre of the sun, M that of the moon, and S′, M′, A′, s, and m, the points in which OS, OM, OA, AS, and AM, produced, meet the celestial sphere. Then will S′ and M′ be the *true* places of the sun and moon, s and m, their *apparent* places, and A′, the geocentric zenith of the place A.

Let a be the zenith of the place A, OZ a straight line parallel to MS, meeting the celestial sphere in Z, EQ the equator, E the vernal equinox, P the north pole of the equator, and PB, PC, PF, and PK, declination circles through Z, S′, M′, and a and A′. Also, let BX and ZY be each a quadrant. Then, since BX is a quadrant, X is the pole of the declination circle YB; and, consequently, OX is perpendicular to OY and OZ. Also, since ZY is a quadrant, OY is perpendicular to OZ. Hence OX, OY, and OZ form a system of rectangular axes, having their origin at O, the centre of the earth, and having the axis OZ parallel to MS, the line joining the centres of the moon and sun.

67. Taking the equatorial radius of the earth $= 1$, let

x, y, z, be the co-ordinates of M,
$x' y', z'$, " " " S,
x'', y'', z'', " " " A,

Also let ρ = OA = distance of place A from earth's centre,
R = OM = " the moon from " "
R′ = OS = " sun " "
R″ = sun's mean distance,
G = MS = distance between the centres of the moon and sun,
A = EF = right ascension of the moon,
A′ = EC = " " sun,
a = EB = " " point Z,
μ = EK = " " zenith, or the sidereal time,
D = FM′ = declination of the moon,
D′ = CS′ = " " sun,
d = BZ = " " point Z,
ϕ = Ka = " " point a, or geogr. Lat. of A,
ϕ' = KA′ = " " point A′, or geocen. lat. of A,
π = moon's equatorial horizontal parallax,
π' = sun's " "
π'' = sun's " " " at mean distance,
δ = moon's appar. semidiameter for earth's centre,
δ' = sun's " " " "
δ'' = sun's " " " at mean distance,

$$g = \frac{G}{R'}, r = \frac{R}{R'}, k = \frac{\sin \delta}{\sin \pi} = 0.2730 \ (99),$$

$$r' = \frac{R'}{R''} = \text{sun's radius vector to mean dist., a unit.}$$

Then we have (93. E.), $R = \frac{1}{\sin \pi}$, and $R'' = \frac{1}{\sin \pi''}$. Consequently,

$$\left.\begin{aligned} r &= \frac{R}{R'} = \frac{R}{r' R''} = \frac{\sin \pi''}{r' \sin \pi} \\ G &= R'g = R''r'g = \frac{r' g}{\sin \pi''} \end{aligned}\right\} \quad \ldots\ldots\ldots\ldots \text{(A)}$$

We have also (99. *Cor.*), $\left.\begin{aligned} \text{Sun's radius} &= \frac{\sin \delta''}{\sin \pi''} \\ \text{Moon's radius} &= \frac{\sin \delta}{\sin \pi} = k \end{aligned}\right\} \quad \ldots\ldots \text{(B)}$

68. *To find the values of a, d, and g.* Let EX′ be a quadrant. Then OX′, OP, and OE will evidently be another system of rectangular axes, having the same origin as the former system. On OZ, take OL = MS,

and let MG, SH, and LI be perpendicular to EOQ, the plane of the equator, meeting it in G, H, and I. Also let GU, HV, and IW be perpendicular to the axis OE, and let GN be parallel to it. As MS and OL are parallel and equal, their projections GH and OI are parallel and equal. Hence, as GN is parallel to OW, the right angle triangles GNH and OWI, are equal, and we have OW = GN = UV = OV — OU. Consequently, that ordinate of the point L, which is parallel to the axis OE, is equal to the difference between the ordinates of S and M, parallel to the same axis. The same relation must evidently have place for the ordinates parallel to the axes OX′ and OP. Hence, if we put α, α', and α'' for the ordinates of M, S, and L, parallel to OX′; β, β', and β'', for those parallel to OP; and γ, γ', and γ'', for those parallel to OE, we shall have, $\alpha'' = \alpha' - \alpha$; $\beta'' = \beta' - \beta$; and $\gamma'' = \gamma' - \gamma$.

Now, since OL = MS = G, we have OI = OL cos BOZ = G cos d; and, consequently, γ'' = OW = OI cos EOB = G cos d cos a. Also β'' = LI = OL sin BOZ = G sin d: and α'' = IW = OI sin EOB = G cos d sin a. The co-ordinates of S and M will evidently have similar expressions. Hence, we have,

$$\begin{array}{lll} \gamma'' = G \cos d \cos a & \beta'' = G \sin d & \alpha'' = G \cos d \sin a \\ \gamma' = R' \cos D' \cos A' & \beta' = R' \sin D' & \alpha' = R' \cos D' \sin A' \\ \gamma = R \cos D \cos A & \beta = R \sin D & \alpha = R \cos D \sin A. \end{array}$$

Consequently,

$$G \cos d \cos a = R' \cos D' \cos A' - R \cos D \cos A$$
$$G \cos d \sin a = R' \cos D' \sin A' - R \cos D \sin A$$
$$G \sin d = R' \sin D' - R \sin D.$$

Multiplying the first of these last three equations by cos A′, and the second by sin A′, and adding the products; and multiplying the first, by sin A′, and the second by cos A′, and subtracting the first product from the second, we obtain,

$$G \cos d \cos (a - A') = R' \cos D' - R \cos D \cos (A - A')$$
$$G \cos d \sin (a - A') = - R \cos D \sin (A - A')$$
$$G \sin d = R' \sin D' - R \sin D.$$

Dividing by R′, and putting g for its equal $\frac{G}{R'}$, and r for its equal $\frac{R}{R'}$, we have,

$$\left.\begin{array}{l} g \cos d \cos (a - A') = \cos D' - r \cos D \cos (A - A') \\ g \cos d \sin (a - A') = - r \cos D \sin (A - A') \\ g \sin d = \sin D' - r \sin D \end{array}\right\} \quad (C)$$

From these, we easily obtain,

$$\left.\begin{aligned} \text{tang}\,(a - A') &= -\frac{r \cos D \sin (A - A')}{\cos D' - r \cos D \cos (A - A')} \\ \text{tang}\,d &= \frac{(\sin D' - r \sin D) \cos (a - A')}{\cos D' - r \cos D \cos (A - A')} \\ g &= \frac{\cos D' - r \cos D \cos (A - A')}{\cos d \cos (a - A')} \end{aligned}\right\} \quad . \quad . \quad (D)$$

These formulæ for determining the values of a, d, and g, are perfectly rigorous. But since r is always a small quantity, and since, at the time of an eclipse, A′ is nearly equal to A, and D′ to D, it is evident that a must be very nearly equal to A′, and d to D′. Taking this into view, we may deduce from (C), the following more simple expressions, which will give the values of a and d, true to a small fraction of a second, and the value of g with corresponding precision.

$$\left.\begin{aligned} a &= A' - r (A - A') \\ d &= D' - r (D - D') \\ g &= 1 - r \end{aligned}\right\} \quad . \quad . \quad . \quad . \quad . \quad . \quad . \quad . \quad . \quad (E)$$

69. *To find the values of* x, y, *and* z; *and* x'', y'', *and* z''. Let Z and M′ be joined by the arc of a great circle. Then, in the spherical triangle ZPM′, we have (App. 34),

$$\cos ZM' = \cos M'P \cos ZP + \sin M'P \sin ZP \cos M'PZ,$$

$$\text{or,} \quad \cos ZOM' = \sin D \sin d + \cos D \cos d \cos (A - a).$$

The right ascensions of X and Y are, $a + 90°$, and $a + 180°$, and their declinations are, 0°, and $90° - d$. These right ascensions and declinations being substituted for a and d, in the expression for cos ZOM′, will evidently give the expression for cos XOM′ and cos YOM′. We shall thus have,

$$\begin{aligned} \cos XOM' &= \cos D \sin (A - a) \\ \cos YOM' &= \sin D \cos d - \cos D \sin d \cos (A - a) \\ \cos ZOM' &= \sin D \sin d + \cos D \cos d \cos (A - a). \end{aligned}$$

If perpendiculars be let fall from the point M, on the axes OX, OY, and OZ, three right angled plane triangles will be formed, the common hypotenuse of which will be $OM = R = \frac{1}{\sin \pi}$, and the bases will be x, y, and z. Hence, $x = \frac{1}{\sin \pi} \cos XOM'$, $y = \frac{1}{\sin \pi} \cos YOM'$, and $z = \frac{1}{\sin \pi} \cos ZOM'$; or,

$$\left.\begin{aligned} x &= \frac{\cos D \sin (A - a)}{\sin \pi} \\ y &= \frac{\sin D \cos d - \cos D \sin d \cos A - a)}{\sin \pi} \\ z &= \frac{\sin D \sin d + \cos D \cos d \cos (A - a)}{\sin \pi} \end{aligned}\right\} \quad . \quad . \quad . \quad . \quad (F)$$

In like manner we find

$$\left.\begin{aligned} x'' &= \rho \cos \phi' \sin (\mu - a) \\ y'' &= \rho \sin \phi' \cos d - \rho \cos \phi' \sin d \cos (\mu - a) \\ z'' &= \rho \sin \phi' \sin d + \rho \cos \phi' \cos d \cos (\mu - a). \end{aligned}\right\} \quad . \quad . \quad (G)$$

By substituting, $1 - 2 \sin^2 \frac{1}{2} (A - a)$, in the formulæ (F), instead of its equal $\cos (A - a)$, (App. 8), and reducing, we obtain the following more simple formulæ for computing, x, y, and z.

$$\left.\begin{aligned} x &= \frac{\cos D \sin (A - a)}{\sin \pi} \\ y &= \frac{\sin (D - d)}{\sin \pi} + x \operatorname{tang} \tfrac{1}{2} (A - a) \sin d \\ z &= \frac{\cos (D - d)}{\sin \pi} - x \operatorname{tang} \tfrac{1}{2} (A - a) \cos d \end{aligned}\right\} \quad . \quad . \quad . \quad (G')$$

70. *To find the equations of contact.* Let the points O, A, M, and S, and the axes OX, OY, and OZ in *Fig.* 63, be the same as in *Fig.* 64. Conceive a conical surface FEBCG, having its vertex at E, in the line SM, to circumscribe the sun and moon; and another F′E′BCG′ having its vertex E′ in SM produced, to do the same. Then, it is evident, that whenever the surface FEBCG meets the place A, there must be, at that instant, an *external contact* of the limbs of the sun and moon; that is, the eclipse at that place must be just *beginning* or *just ending.* And if during the eclipse, the surface F′E′BCG′ meets the place A, there must be at the instant at which this occurs, an *internal contact* of the limbs; that is, the eclipse at the place A must then be just commencing or just ceasing to be either *annular* or *total.*

Hence, it follows that, if a straight line AD, drawn from A perpendicular to SM produced, meets the surface FEBCG in a, and the surface F′E′BCG′ in a', there must be an *external* contact when AD $= a$D, and an *internal* contact when AD $= a'$D.

Now, since AD is perpendicular to DM, and DM is parallel to OZ, it follows that AD is parallel to the plane XOY; consequently, the co-ordinates of the point D, are x, y, and z''. Let AX″, AY″, and AZ″, be respectively parallel to OX, OY, and OZ. Then, since AX″, AY″, and AD, are each parallel to the plane XOY, they are all in one plane. Hence, if Dp be perpendicular to AX'', we shall have $Ap = x - x''$, and $Dp = y - y''$. Put

$U = AD$; $h = aD$; $h' = a'D$;
$f =$ ang. SEC; $f' =$ ang. $SE'C$;
$P =$ ang. $DAX'' =$ the angle which the plane $Z''AMS$ makes with the plane $Z''AX''$.

Then, $Ap = AD \cos DAX'' = U \cos P$, and $Dp = AD \sin DAX'' = U \sin P$. Hence,

$$\left.\begin{array}{l} U \cos P = x - x'' \\ U \sin P = y - y'' \end{array}\right\} \quad . \quad . \quad . \quad . \quad . \quad . \quad . \quad (H)$$

In the right angled triangles ECS and EdM, we have the angle $dEM = SEC = f$. Hence,

$$CS = ES \sin f, \qquad dM = EM \sin f,$$

and $\quad CS + dM = (ES + EM) \sin f = G \sin f.$

But (App. 67 B & A), $CS = \dfrac{\sin \delta''}{\sin \pi''}$, $dM = k$, and $G = \dfrac{r'g}{\sin \pi''}$.

Consequently, $\dfrac{\sin \delta''}{\sin \pi''} + k = \dfrac{r'g}{\sin \pi''} \sin f$,

or, $$\sin f = \frac{1}{r'g} (\sin \delta'' + k \sin \pi'') \quad . \quad . \quad . \quad . \quad . \quad . \quad (I)$$

Again, in the right angled triangles $E'CS$ and $E'd'M$, we have the angle $d'E'M = SE'C = f'$. Whence, as $G = MS = E'S - E'M$, we find,

$$\sin f' = \frac{1}{r'g} (\sin \delta'' - k \sin \pi'') \quad . \quad . \quad . \quad . \quad . \quad . \quad (K)$$

Since, $EM = \dfrac{dM}{\sin f} = \dfrac{k}{\sin f}$, $E'M = \dfrac{k}{\sin f'}$, and $DM = z - z''$, we have $ED = DM + EM = z - z'' + \dfrac{k}{\sin f}$, and $E'D = DM - E'M = z - z'' - \dfrac{k}{\sin f'}$. Consequently $aD = ED \text{ tang } f = z \text{ tang } f - z'' \text{ tang } f + k \sec f$, and $a'D = z \text{ tang } f' - z'' \text{ tang } f' - k \sec f'$. Or,

$$\left.\begin{array}{l} h = z \text{ tang } f + k \sec f - z'' \text{ tang } f \\ h' = z \text{ tang } f' - k \sec f' - z'' \text{ tang } f' \end{array}\right\} \quad . \quad . \quad . \quad . \quad . \quad (L)$$

As an external contact takes place when $AD = aD$, that is, when $U = h$, and an internal contact when $U = h'$, we have, from equations (H),

$$\text{for external contact} \left\{\begin{array}{l} h \cos P = x - x'' \\ h \sin P = y - y'' \end{array}\right\}$$

$$\text{for internal contact} \left\{\begin{array}{l} h' \cos P = x - x'' \\ h' \sin P = y - y'' \end{array}\right\}$$

The preceding formulæ may be more concisely expressed in the following manner. Put

$$\left.\begin{aligned} \sin f &= \frac{1}{r'g}(\sin \delta'' \pm k \sin \pi'') \\ l &= z \tan f \pm k \sec f \\ h &= l - z'' \tan f \end{aligned}\right\} \quad \ldots\ldots\ldots\ldots \quad (M)$$

In these, the *upper* sign is to be taken for an *external* contact, and the *lower* for an internal contact. Also, if we take $\delta'' = 15' \, 59''.788$, $k = 0.2725$, and $\pi'' = 8''.5776$, as determined by *Bessel*, *Burkhardt*, and *Encke*, respectively, we have,

$$\log. (\sin \delta'' + k \sin \pi'') = 7.6688050$$
$$\log. (\sin \delta'' - k \sin \pi'') = 7.6666896.$$

Then will the equations of contact be,

$$\left.\begin{aligned} h \cos P &= x - x'' \\ h \sin P &= y - y'' \end{aligned}\right\} \quad \ldots\ldots\ldots\ldots\ldots \quad (N)$$

71. *Solution of the equations of contact.* Let the values of x, y, x'', and y'', found for a time T, taken to a whole hour near the time of new moon, be p, q, u, and v, respectively, and let their average hourly variations between the time T and another time $T + t$, be p', q', u', and v'. Then, at the time $T + t$, we shall have, $x = p + p't$, $y = q + q't$, $x'' = u + u't$, and $y'' = v + v't$. Consequently, if $T + t$ be the time of contact, the equations of contact (N), will become,

$$\left.\begin{aligned} h \cos P &= p - u + (p' - u')\, t \\ h \sin P &= q - v + (q' - v')\, t \end{aligned}\right\} \quad \ldots\ldots \quad (O)$$

The values of p' and q' may be found by the formula (App. 64 E). To obtain expressions for u' and v', let c be the hourly variation of $(\mu - a.)$ Taking for $(\mu - a)$ and d, their values at the time T, we have, for that time (App. 69 G),

$$x'' = \rho \cos \phi' \sin (\mu - a)$$
$$y'' = \rho \sin \phi' \cos d - \rho \cos \phi' \sin d \cos (\mu - a).$$

And, disregarding the slight changes in the value of $\cos d$ and $\sin d$, we have, at the time $T + t$,

$$x'' = \rho \cos \phi' \sin (\mu - a + tc)$$
$$y'' = \delta \sin \phi' \cos d - \rho \cos \phi' \sin d \cos (\mu - a + tc).$$

Hence (App. 64),

$$u' = \rho \cos \phi' \; \frac{\sin (\mu - a + tc) - \sin (\mu - a)}{t}$$

$$v' = \rho \cos \phi' \sin d \; \frac{\cos \,(\mu - a) - \cos \,(\mu - a + tc)}{t}.$$

But (App. 21 & 23), $\sin (\mu - a + tc) - \sin (\mu - a) = 2 \sin \frac{1}{2} tc \cos (\mu - a + \frac{1}{2} tc)$, and $\cos (\mu - a) - \cos (\mu - a + tc) = 2 \sin \frac{1}{2} tc \sin (\mu - a + \frac{1}{2} tc)$. Hence,

$$\left.\begin{aligned} u' &= \rho \cos \phi' \frac{2 \sin \frac{1}{2} tc}{t} . \cos (\mu - a + \tfrac{1}{2} tc) \\ v' &= \rho \cos \phi' \sin d \frac{2 \sin \frac{1}{2} tc}{t} . \sin (\mu - a + \tfrac{1}{2} tc). \end{aligned}\right\} \quad . \quad (P)$$

When the arc $\frac{1}{2} tc$ is very small, we may take the arc itself instead of its sine. Hence, in this case, if the arc be expressed in seconds, we shall have (App. 51), $\frac{2 \sin \frac{1}{2} tc}{t} = \frac{tc}{206265'' t} = \frac{c}{206265''}$. Consequently, taking $t = o$, in $(\mu - a + \frac{1}{2} tc)$, we have, for the hourly variation at the time T,

$$\left.\begin{aligned} u' &= \rho \cos \phi' \cos (\mu - a) \cdot \frac{c}{206265''} \\ v' &= \rho \cos \phi' \sin d \sin (\mu - a) \cdot \frac{c}{206265''} \end{aligned}\right\} \quad . \quad . \quad . \quad . \quad . \quad . \quad (Q)$$

Assume $\left.\begin{aligned} p - u &= m \sin \text{M}, \quad p' - u' = n \sin \text{N} \\ q - v &= m \cos \text{M}, \quad q' - v' = n \cos \text{N} \end{aligned}\right\} \quad . \quad . \quad . \quad (R)$

Then the equations of contact (O), will become,

$$\left.\begin{aligned} h \cos \text{P} &= m \sin \text{M} + nt \sin \text{N} \\ h \sin \text{P} &= m \cos \text{M} + nt \cos \text{N} \end{aligned}\right\} \quad . \quad . \quad . \quad . \quad . \quad . \quad (S)$$

Multiplying the first of these by cos N, and the second by sin N, and subtracting the second product from the first; and multiplying the first by sin N, and the second by cos N, and adding the products, we have, by putting $\text{P} + \text{N} = \psi$,

$$\left.\begin{aligned} h \cos \psi &= m \sin (\text{M} - \text{N}) \\ h \sin \psi &= m \cos (\text{M} - \text{N}) + nt \end{aligned}\right\} \quad . \quad . \quad . \quad . \quad . \quad . \quad (T)$$

Hence,

$$\cos \psi = \frac{m \sin (\text{M} - \text{N})}{h}$$

$$t = - \frac{m \cos (\text{M} - \text{N})}{n} + \frac{h \sin \psi}{n}.$$

As the cosine of a negative arc is the same as that of an affirmative arc of the same numerical value, the arc ψ, and, consequently, the last term of the expression for t, may be either affirmative or negative. We may, however, always take the arc ψ affirmative and not exceeding 180°, provided we prefix the double sign to the last term of the expression for t. We shall thus have,

$$t = - \frac{m \cos (\text{M} - \text{N})}{n} \mp \frac{h \sin \psi}{n}.$$

If, instead of h, its value, found from the first of the equations (T), be substituted, the expression for t becomes,

$$t = - \frac{m \cos (\text{M} - \text{N} \mp \psi)}{n \cos \psi}.$$

Collecting the equations, we have,

$$\left.\begin{aligned} \cos \psi &= \frac{m \cos (\mathrm{M} - \mathrm{N})}{h} \\ &= -\frac{m \cos (\mathrm{M} - \mathrm{N})}{n} \mp \frac{h \sin \psi}{n} \\ \text{or,} \quad t &= -\frac{m \cos (\mathrm{M} - \mathrm{N} \mp \psi)}{n \cos \psi} \\ \text{and, } \mathrm{T} + t &= \text{time of contact} \end{aligned}\right\} \quad \ldots\ldots \quad (\mathrm{U})$$

The *upper* sign is to be taken for the *beginning* of the eclipse, and the *lower* for the *end*.

Taking for h, its value for internal contact, these formulæ also serve to find the times at which the eclipse, when it is annular or total, begins or ceases to be so.

Remarks. As the values of N and n depend on those of p', q', u', and v' (R), and these depend on the required time T + t, it is evident that this time cannot be truly obtained by a simple computation of the formulæ. But, by computing the values of N and n with the values of p', q', u', and v', found for the time T (App. 64 F & 71 Q), and taking the value of h for the same time, an approximate value of T + t will be obtained, which will never deviate more than a few minutes from the true value. Then, by repeating the computation with the values of p', q' u', and v', found for this approximate time T + t, (App. 64 E & 71 P), and with the value of h for the same time, a second approximate time T + t, will be obtained, which will be very nearly true. If greater accuracy is desired, the com putation may be again repeated.

When great precision is desired, it is best, after the second approximate value of T + t has been found as above, to take p and q to stand for the values of x and y, computed for this time (App. 61 D), and u and v for the values of x'' and y'', computed with the values of $(\mu - a)$ and d, at the same time. Then using these values of $(\mu - a)$ and d, find p', q', u', and v' by the formulæ (App. 65 G & 71 Q), and with the values thus obtained, compute M, m, N and n. Taking, now, T to represent the second approximate value of T + t, and completing the computation, using the value of h at this time, we obtain T + t, the time of contact, with great accuracy. The error arising from disregarding the changes in the values of cos d and sin d is thus avoided.

When only an approximate calculation of the eclipse is intended, it will be sufficient to calculate the values of x and y, only for the times T and T + 1hr. The value of x, at the time T will be p, and that of y at the same time, will be q. The first value of x, subtracted from the second, will give p', and the first value of y, subtracted from the second, will give q'. These values of p' and q' may be regarded as constant during the eclipse.

Second Solution. By substituting in equations (O), only the assumed values of $(p' - u')$ and $(q' - v')$, we have,

$$h \cos P = p - u + nt \sin N$$
$$h \sin P = q - v + nt \cos N.$$

Multiplying the first by cos N, and second, by sin N, and subtracting the second product from the first, we obtain,

$$h \cos (P + N) = (p - u) \cos N - (q - v) \sin N,$$

or, $$\cos \psi = \frac{(p - u) \cot N - (q - v)}{h} \sin N.$$

And from the first of equations (O), we have,

$$t = \frac{h \cos P - (p - u)}{p' - u'}.$$

But since $P + N = \mp \psi$, we have $P = -N \mp \psi = -(N \pm \psi)$, and $\cos P = \cos (N \pm \psi)$. Consequently, collecting the equations, we have,

$$\left.\begin{aligned} \cos \psi &= \frac{(p - u) \cot N - (q - v)}{h} \sin N \\ t &= \frac{h \cos (N \pm \psi)}{p' - u'} - \frac{p - u}{p' - u'} \\ T + t &= \text{time of contact} \end{aligned}\right\} \quad \ldots\ldots \quad (U)$$

72. *Nearest apparent approach of the centres.* Expressing the values of x, y, x'' and y'' at a time $T + t'$, as for the time $T + t$, in the last article, and substituting them in (App. 70 H), we have,

$$U \cos P = p - u + (p' - u').\, t'$$
$$U \sin P = q - v + (q' - v').\, t'$$

or, $$U \cos P = m \sin M + nt' \sin N$$
$$U \sin P = m \cos M + nt' \cos N.$$

From which we obtain,

$$U \cos \psi = m \sin (M - N)$$
$$U \sin \psi = m \cos (M - N) + nt'.$$

Adding together the squares of these equations, we have,

$$U^2 = m^2 \sin^2 (M - N) + \left(m \cos (M - N) + nt' \right)^2.$$

Now as the squares of real quantities are always affirmative, it is evident that U must have its *least* value when $m \cos (M - N) + nt' = 0$. And then $U = \pm m \sin (M - N) = \pm h \cos \psi$. But when U or AD

Fig. 63, is least, the apparent distance of the centres of the sun and moon must be least. Hence, for the time of the nearest approach of the centres, we have,

$$\left.\begin{array}{l} t' = -\dfrac{m \cos (\mathrm{M} - \mathrm{N})}{n}, \\ \text{or,} \quad t' = -\dfrac{(q - v) \cot \mathrm{N} + (p - u)}{p' - u'} \sin \mathrm{N} \\ \mathrm{T} + t' = \text{time of nearest approach} \\ \mathrm{U} = \pm\, h \cos \psi \end{array}\right\} \quad \ldots\ldots \quad (\mathrm{V})$$

As the least value of U is $\pm\, m \sin (\mathrm{M} - \mathrm{N})$, it evidently follows, that if, at the time of nearest approach of the centres, the value of $m \sin (\mathrm{M} - \mathrm{N})$ is greater than the value of h, for external contact, there cannot be an eclipse. If at that time it is less than the value of h for internal contact, the eclipse will be either annular or total.

73. *Quantity of the eclipse.* Let the straight line $a''c$ be drawn in the plane ADS, *Fig.* 63, touching the moon near d. Then it is manifest, that when the point a'' coincides with A, the circular segment mcnC of the sun's disc, must be eclipsed at that place. We may, without material error, assume that the three lines aC, $a''c$, and bB, all cut one another at the point d, and that the small angles BdC and Cdc, or their equals adb and ada'', are proportional to the sides subtending them. Hence, putting w for the digits eclipsed, we have, $ab : aa'' :: adb : ada'' :: \mathrm{B}d\mathrm{C} : \mathrm{C}dc ::$ 2SC : Cc : 12 : w. Or,

$$w = 12\,\frac{aa''}{ab} = 12.\frac{a\mathrm{D} - a''\mathrm{D}}{a\mathrm{D} + \mathrm{D}b} = 12.\frac{a\mathrm{D} - a''\mathrm{D}}{a\mathrm{D} + a'\mathrm{D}} = \frac{12(h - a''\mathrm{D})}{h + h'}.$$

But regarding f' as equal to f, from which it differs but little (App. 70 I & K), we have (App. 70 L), $h + h' = 2\,(z \text{ tang } f - z'' \text{ tang } f) = 2\ h - k \sec f)$. Also, when a'' coincides with A, we have $a''\mathrm{D} = \mathrm{U}$. Hence,

$$w = \frac{6\ (h - \mathrm{U})}{h - k \sec f}.$$

Substituting the value of U at the time of nearest approach of the centres, (71 V), we have,

$$w = \frac{6\ (h \mp h \cos \psi)}{h - k \sec f} = \frac{6\ h\ (1 \mp \cos \psi)}{h - k \sec f}.$$

Or (App. 8 & 9),

$$w = \frac{12\ h \sin^2 \tfrac{1}{2}\psi}{h - k \sec f}, \text{ or, } \frac{12\ h \cos^2 \tfrac{1}{2}\psi}{h - k \sec f} \quad \ldots\ldots\ldots \quad (\mathrm{W})$$

The *first* or *second* expression is to be used, according as ψ is *less* or *greater* than 90°.

74. *Position of the point of contact.* As the earth's radius is insensible in comparison with the distance of the celestial sphere, AX, AY, and AZ, *Fig.* 64, may be regarded as parallel to OX, OY, and OZ, and, consequently, as corresponding to AX″, AY″, and AZ″, *Fig.* 63. The line AZ being parallel to OZ, it must be parallel to MS, and must, therefore, be in the same plane with Am and As. Consequently, the plane ZAm, which passes through the moon's apparent centre m, passes also through the sun's apparent centre s, and, therefore, through the point of apparent contact.

Let Zsm and Za be arcs of great circles. Then the angle PZm will be the angular distance of the point of contact from the declination circle PB, passing through the point Z, and the angle aZm will be its angular distance from the vertical circle aZ, passing through the same point. But the distance of Z from s, at the time of an eclipse, never exceeds a few seconds.* We may, therefore, regard the angles PZm and aZm as expressing respectively the angular distances of the point of contact from a declination circle and a vertical circle, both passing through the apparent centre of the sun.

Now the angle PZm is the angle made by the plane ZAm with the plane ZAY, which is equal to the angle made by the plane Z″AMS, *Fig.* 63, with the plane Z″AY″. Therefore, the angle PZm is equal to the complement of the angle made by the plane Z″AMS with the plane Z″AX″, and, consequently, it is equal to $90° - \mathrm{P}$. But $\mathrm{P} = \psi - \mathrm{N}$; or, since ψ is negative for the beginning of the eclipse, and affirmative for the end, $\mathrm{P} = \mp \psi - \mathrm{N}$. Hence, $\mathrm{PZ}m = 90° + \mathrm{N} \pm \psi$.

Put $\mathrm{P}' = \mathrm{PZ}m$ = angular distance of point of contact from the north point of the sun's disc, to the *left*, $\mathrm{V} = a\mathrm{Z}m$ = angular distance of the point of contact from the sun's vertex, to the *left*, and $\mathrm{Q} = \mathrm{PZ}a$. Then, $\mathrm{V} = \mathrm{P}' - \mathrm{Q}$. Consequently, we have the following expressions, in which, the upper sign appertains to the beginning of the eclipse, and the lower, to the end.

---

* Since MS is parallel to AZ, the angle $s\mathrm{AZ} = \mathrm{ASM}$. But when a coincides with A, *Fig.* 63, we have $\mathrm{ASM} = a\mathrm{SM} = a\mathrm{EM} - \mathrm{S}a\mathrm{C} = f - \delta'$, very nearly. Hence, *Fig.* 64, $s\mathrm{Z} = \text{ang. } s\mathrm{AZ} = f - \delta'$. Now, $\sin f = \frac{1}{r'g}(\sin \delta'' + k \sin \pi'') = \frac{1}{g}(\sin \delta' + k \sin \pi') = \frac{\sin \delta' + k \sin \pi'}{1 - r} = \sin \delta' + \frac{r \sin \delta' + k \sin \pi'}{1 - r}$; or, taking arcs instead of sines, $f = \delta' + \frac{r\delta' + k\pi'}{1 - r}$, and $f - \delta' = \frac{r\delta' + k\pi'}{1 - r}$. Consequently, $s\mathrm{Z} = \frac{r\delta' + k\pi'}{1 - r}$; the greatest value of which is about 5″.

$$\left.\begin{array}{l} P' = 90° + N \pm \psi \\ V = 90° + N \pm \psi - Q \end{array}\right\} \quad . \quad . \quad . \quad . \quad . \quad . \quad . \quad . \quad . \quad . \quad (X)$$

To find Q, we have, from the spherical triangle ZPa (App. 37),

$$\cot PZa = \frac{\cot Pa \sin PZ - \cos ZPa \cos PZ}{\sin ZPa} = \frac{\text{tang}\, \phi \cos d - \sin d \cos(\mu - a)}{\sin(\mu - a)}$$

$$\text{or, tang}\, Q = \frac{\sin(\mu - a)}{\text{tang}\, \phi \cos d - \sin d \cos(\mu - a)} \quad . \quad . \quad . \quad . \quad . \quad . \quad . \quad (Y)$$

If we multiply both numerator and denominator of the value of tang Q, by $\rho \cos \phi$, the products will only differ from the values of x'' and y', by having ϕ instead of ϕ'. Hence,

$$\text{tang}\, Q = \frac{x''}{y'} \text{ nearly} \quad . \quad . \quad . \quad . \quad . \quad . \quad . \quad . \quad . \quad . \quad . \quad . \quad . \quad (Z)$$

Remark. The data for computing an eclipse by the preceding formulæ are easily obtained from the Nautical Almanac, as the moon's right ascension and declination are there given for every hour. In the Berlin Ephemeris, the values of x, y, l, log. z, and log. tang f, for all important eclipses, are given for several consecutive whole hours near to the time of new moon. This very much facilitates the computation.*

75. *Other formulæ for finding the values of the quantities contained in the equations of contact.* The equations of contact depend on the relative positions of the sun and moon. Hence, as the sun's parallax is very small, it is evident that the equations must still be very nearly true, if, in finding the values of the quantities contained in them, the moon's parallax be assumed to be equal to the difference of the parallaxes of the moon and sun, and then the sun be regarded as having no parallax. The error in the computed time of beginning or end, resulting from these assumptions, will not exceed a small fraction of a second.

Taking, therefore, $\pi - \pi'$ instead of π, and then assuming $\pi' = 0$, we have, $r = \frac{R}{R'} = \frac{\sin \pi'}{\sin \pi} = 0$, and (E), $a = A'$, $d = D'$, and $g = 1$.

---

* The preceding method of computing a solar eclipse, is derived from an excellent investigation of the subject by Professor Bessel, *Astr. Nach.* No. 321.

Prof. Hansen has shown, *Astr. Nach.* No. 347, that, in small eclipses near the horizon, refraction produces a sensible, though very slight influence on the times of beginning and end. In other cases the effect of refraction is quite insensible.

Therefore (G′ and F),

$$\left.\begin{aligned} x &= \frac{\cos D \sin (A - A')}{\sin (\pi - \pi')} \\ y &= \frac{\sin D \cos D' - \cos D \sin D' \cos (A - A')}{\sin (\pi - \pi')} \\ z &= \frac{\sin D \sin D' + \cos D \cos D' \cos (A - A')}{\sin (\pi - \pi')} \end{aligned}\right\} \quad . \quad . \quad . \quad (a)$$

or,

$$\left.\begin{aligned} x &= \frac{\cos D \sin (A - A')}{\sin (\pi - \pi')} \\ y &= \frac{\sin (D - D')}{\sin (\pi - \pi')} + x \operatorname{tang} \tfrac{1}{2} (A - A') \sin D' \\ z &= \frac{\cos (D - D')}{(\sin \pi - \pi')} - x \operatorname{tang} \tfrac{1}{2} (A - A') \cos D' \end{aligned}\right\} \quad . \quad . \quad . \quad (a')$$

Or, without sensible error,

$$\left.\begin{aligned} x &= \frac{\cos D \sin (A - A')}{\sin (\pi - \pi')} \\ y &= \frac{\sin (D - D')}{\sin (\pi - \pi')} + \tfrac{1}{2} x \sin (A - A') \sin D' \\ z &= \frac{\cos (D - D')}{\sin (\pi - \pi')} - \tfrac{1}{2} x \sin (A - A') \cos D' \end{aligned}\right\} \quad . \quad . \quad . \quad (a'')$$

$$\left.\begin{aligned} x'' &= \rho \cos \phi' \sin (\mu - A') \\ y'' &= \rho \sin \phi' \cos D' - \rho \cos \phi' \sin D' \cos (\mu - A') \\ z'' &= \rho \sin \phi' \sin D' + \rho \cos \phi' \cos D' \cos (\mu - A') \end{aligned}\right\} \quad . \quad . \quad (b)$$

We have also (M), $\sin f = \frac{\sin \delta''}{r'} = \sin \delta'$, or, $f = \delta'$. And instead of k, or its equal $\frac{\sin \delta}{\sin \pi}$, we have, $\frac{\sin \delta}{\sin (\pi - \pi')}$ or $\frac{k \sin \pi}{\sin (\pi - \pi')}$. Hence, rejecting, in the value of l, the factor sec f, which differs extremely little from a unit, we have (M),

$$\left.\begin{aligned} l &= z \operatorname{tang} \delta' \pm k \frac{\sin \pi}{\sin (\pi - \pi')} \\ h &= l - z'' \operatorname{tang} \delta' \end{aligned}\right\} \quad . \quad . \quad . \quad . \quad . \quad . \quad . \quad . \quad . \quad (c)$$

Taking, for c, the hourly variation of $(\mu - A')$, we have, for the average hourly variation of x'' and y'', between the times T and T + t (P),

$$\left.\begin{aligned} u' &= \rho \cos \phi' \frac{2 \sin \frac{1}{2} tc}{t} \cos (\mu - A' + \tfrac{1}{2} tc) \\ v' &= \rho \cos \phi' \sin D' \frac{2 \sin \frac{1}{2} tc}{t} \cos (\mu - A' + \tfrac{1}{2} tc) \end{aligned}\right\} \quad . \quad . \quad (d)$$

And for the hourly variations at the time T, we have (Q),

$$\left.\begin{aligned} u' &= \rho \cos \phi' \cos (\mu - A') \frac{c}{206265''} \\ v &= \rho \cos \phi' \sin D' \sin (\mu - A') \frac{c}{206265''} \end{aligned}\right\} \quad . \; . \; . \; . \; . \; . \; (e)$$

76. *The expressions for* x, y, *ana* z, *may be changed into others depending on the moon's longitude and latitude.* Let EQ, *Fig.* 65, be the equator, EC the ecliptic, P and p their poles, and M and S the places of the moon and sun respectively, on the supposition that the moon's latitude is equal to the difference of their latitudes, and that the sun's latitude is nothing. Also, let pPD, pMB, PMA, PSG, ME, MS, and MD, be arcs of great circles. Then, since the arc pD passes through p and P, the poles of EC and EQ, it is perpendicular to each of these circles, and the arcs EF and ED are quadrants. Put

L = EB = moon's longitude,
L′ = ES = sun's "
λ = BM = moon's latitude — sun's latitude,
ε = ang. DEF = apparent obliquity of the ecliptic.

Then in the triangle pPM, we have (App. 34),

$$\cos PM = \sin pP \sin pM \cos PpM + \cos pD \cos pM$$

or, $\sin D = \sin \varepsilon \cos \lambda \sin L + \cos \varepsilon \sin \lambda.$

In the triangle MpD, we have,

$$\begin{aligned} \cos DM &= \sin pD \sin pM \cos PpM + \cos pP \cos pM \\ &= \cos \varepsilon \cos \lambda \sin L - \sin \varepsilon \sin \lambda. \end{aligned}$$

But, $\cos DM = \cos AM \cos AD = \cos D \sin A.$

Hence, $\cos D \sin A = \cos \varepsilon \cos \lambda \sin L - \sin \varepsilon \sin \lambda.$

Again, $\cos AM \cos AE = \cos EM = \cos BM \cos BE,$

or, $\cos D \cos A = \cos \lambda \cos L.$

Hence, collecting the results, we have,

$$\begin{aligned} \sin D &= \sin \varepsilon \cos \lambda \sin L + \cos \varepsilon \sin \lambda \\ \cos D \sin A &= \cos \varepsilon \cos \lambda \sin L - \sin \varepsilon \sin \lambda \\ \cos D \cos A &= \cos \lambda \cos L. \end{aligned}$$

In like manner, the sun's latitude being assumed $= o$, we have,

$$\begin{aligned} \sin D' &= \sin \varepsilon \sin L' \\ \cos D' \sin A' &= \cos \varepsilon \sin L' \\ \cos D' \cos A' &= \cos L'. \end{aligned}$$

Now, in the triangle MPS, we have,

$$\begin{aligned} \cos MS &= \cos PM \cos PS + \sin PM \sin PS \cos MPS \\ &= \sin D \sin D' + \cos D \cos D' \cos (A - A). \end{aligned}$$

But, $\cos MS = \cos BM \cos BS = \cos \lambda \cos (L - L').$

Hence, $\sin D \sin D' + \cos D \cos D' \cos (A - A') = \cos \lambda \cos (L - L').$

W

And, therefore (a), $z = \dfrac{\cos \lambda \cos (L - L')}{\sin (\pi - \pi')}$.

Multiplying the second of the equations (a), by $\sin (\pi - \pi') \cos D'$, we have,

$$y \sin (\pi - \pi') \cos D' = \sin D \cos^2 D' - \cos D \cos D' \sin D' \cos (A - A')$$
$$= \sin D - \left(\cos D \cos D' \cos (A - A') + \sin D \sin D'\right) \sin D'$$
$$= \sin D - \cos \lambda \cos (L - L') \sin D'$$
$$= \sin \varepsilon \cos \lambda \left(\sin L - \cos (L - L') \sin L'\right) + \cos \varepsilon \sin \lambda$$
$$= \sin \varepsilon \cos \lambda \cos L' \sin (L - L') + \cos \varepsilon \sin \lambda.$$

Also, multiplying the first of equations (a), by $\sin (\pi - \pi') \cos D'$, we have, $x \sin (\pi - \pi') \cos D' = \cos D \cos D' \sin (A - A')$

$$= \cos D \sin A \cos D' \cos A' - \cos D \cos A \cos D' \sin A'$$
$$= \cos \varepsilon \cos \lambda (\sin L \cos L' - \cos L \sin L') - \sin \varepsilon \sin \lambda \cos L'$$
$$= \cos \varepsilon \cos \lambda \sin (L - L') - \sin \varepsilon \sin \lambda \cos L'.$$

Hence we have,

$$\left.\begin{aligned} x &= \frac{\cos \varepsilon \cos \lambda \sin (L - L') - \sin \varepsilon \sin \lambda \cos L'.}{\cos D' \sin (\pi - \pi')} \\ y &= \frac{\sin \varepsilon \cos \lambda \cos L' \sin (L - L') + \cos \varepsilon \sin \lambda}{\cos D' \sin (\pi - \pi')} \\ z &= \frac{\cos \lambda \cos (L - L')}{\sin (\pi - \pi')}. \end{aligned}\right\} \quad . \quad . \quad (f)$$

As at the time of an eclipse, λ and $(L - L')$ are always small arcs, we may, for an *approximate* calculation, regard the cosine of each as equal to a unit. We shall then have, by taking λ, $(L - L')$ and $(\pi - \pi')$ in seconds, instead of their sines, and putting $C = \dfrac{\cos \varepsilon}{(\pi - \pi') \cos D'}$,

$$\left.\begin{aligned} x &= C. (L - L') - C. \lambda. \text{tang}\, \varepsilon \cos L' \\ y &= C. (L - L') \text{tang}\, \varepsilon \cos L' + C. \lambda \\ z &= \frac{1}{\sin (\pi - \pi')} \end{aligned}\right\} \quad . \quad . \quad . \quad . \quad . \quad (f')$$

77. *Central Eclipse.* If, during an eclipse, the line SM, produced, *Fig.* 63, meets the earth, there must evidently be a central eclipse along the line in which it meets the illuminated surface, in its passage across this surface. Let SM, produced, meet the plane XOY in D′, and let A′ be the point in which OD′ intersects the earth's surface. Then, since D′S is parallel to OZ, it must be perpendicular to the plane XOY, and consequently OD′ is perpendicular to D′S. The central eclipse must, therefore, begin or end when D′ coincides with A′. Draw $A'p'$ parallel to YO. Then, when D′ coincides with A′, we have $x = Op' = x''$, and $y = A'p'$

$= y''$. Put P = angle D′OX. Then $Op' = OA' \cos P = \rho \cos P$, and $A'p' = OA' \sin P = \rho \sin P$. Hence, for the beginning or end of the central eclipse we have,

$$\left.\begin{array}{l}\rho \cos P = x = x'' \\ \rho \sin P = y = y''\end{array}\right\} \quad \ldots \ldots \ldots \ldots \quad (g)$$

Take T, p, and q, as in (App. 71), and let p' and q' be the hourly variations of x and y, at the time T. Then, as great accuracy is not important in investigations and computations relative to the general eclipse, we may regard p' and q' as constant. We shall, therefore, at a time $T + t$, have $x = p + p't$, and $y = q + q't$. Put

$$\left.\begin{array}{ll}p = m \sin M, & p' = n \sin N \\ q = m \cos M, & q' = n \cos N\end{array}\right\} \quad \ldots \ldots \ldots \quad (h)$$

Then $x = m \sin M + nt \sin N$, and $y = m \cos M + nt \cos N$. Hence, if $T + t$ be the time of beginning or end of the central eclipse, we have,

$$\left.\begin{array}{l}\rho \cos P = m \sin M + nt \sin N \\ \rho \sin P = m \cos M + nt \cos N\end{array}\right\} \quad \ldots \ldots \ldots \quad (i)$$

Consequently, as, in (App. 71), we obtain

$$\left.\begin{array}{l}\cos \psi = \dfrac{m \sin (M - N)}{\rho} \\ t = -\dfrac{m \cos (M - N)}{n} \mp \dfrac{\rho \sin \psi}{n} \\ P = \psi - N\end{array}\right\} \quad \ldots \quad (k)$$

Since, for the place A′, $z'' = 0$, and (g), $x'' = \rho \cos P$, and $y'' = \rho \sin P$, we have (b),

$$\left.\begin{array}{l}\cos P = \cos \phi' \sin (\mu - A') \\ \sin P = \sin \phi' \cos D' - \cos \phi' \sin D' \cos (\mu - A') \\ 0 = \sin \phi' \sin D' + \cos \phi' \cos D' \cos (\mu - A')\end{array}\right\} \quad \ldots \quad (l)$$

Multiplying the second equation by $\cos D'$, and the third by $\sin D'$, and adding the products, we have,

$$\sin \phi' = \sin P \cos D' \quad \ldots \ldots \ldots \ldots \quad (m)$$

From the third equation of (l), we have, $-\sin \phi' \operatorname{tang} D' = \cos \phi' \cos (\mu - A')$. But (m), $-\sin \phi' \operatorname{tang} D' = -\sin P \sin D'$. Hence, $-\sin P \sin D' = \cos \phi' \cos (\mu - A')$. Dividing this into the first of equations (l), we obtain,

$$\operatorname{tang.} (\mu - A' = -\frac{\cot P}{\sin D'} \quad \ldots \ldots \ldots \ldots \quad (n)$$

Dividing the third of the equations (l), by $\cos \phi' \sin D'$, and transposing, we have,

$$\operatorname{tang.} \phi' = -\cot D' \cos (\mu - A') \quad \ldots \ldots \ldots \quad (p)$$

Taking $\rho = 1$, we find, from the equations (k), the times very nearly, at which the central eclipse for the earth in general, begins and ends. The values of $(\mu - A')$, found from (n), subtracted from its values at the first meridian, at the times $T + t$, give the longitudes of the places at which the eclipse begins and ceases to be central; the longitude being *west* or *east*, according as the remainder is *affirmative* or *negative.* The geocentric latitudes of the places may be found either from equation (m), or (p).

If greater accuracy is desired, we may, after finding the value of ψ (k), compute the value of ψ' (m), and then, after having found the corresponding value of ρ from a table of its value,* make the computation with this value.

78. *To find a series of places, at which the eclipse will be central.* It is evident that, at the time an eclipse is central at any place A, we have $x = x''$, and $y = y''$. Hence, taking the values of x and y, as in the last article, we have,

$$m \sin M + nt \sin N = \rho \cos \phi' \sin (\mu - A')$$
$$m \cos M + nt \cos N = \rho \sin \phi' \cos D' - \rho \cos \phi' \sin D' \cos (\mu - A').$$

Multiplying the first by cos N, and second by sin N, and subtracting the first product from the second, and then multiplying the first by sin N, and second by cos N, and adding the products, we obtain,

$$-m \sin (M - N) = \rho \sin \phi' \cos D' \sin N - \rho \cos \phi' \left(\sin (\mu - A') \cos N + \sin D' \sin N \cos (\mu - A')\right)$$

and $$m \cos (M - N) + nt = \rho \sin \phi' \cos D' \cos N + \rho \cos \phi' \left(\sin (\mu - A') \sin N - \sin D' \cos N \cos (\mu - A')\right)$$

Put $\sin D' \sin N = m' \sin M'$, $\sin D' \cos N = n' \sin N'$,
$\cos N = m' \cos M'$, $\sin N = n' \cos N'$.

Then, $-m \sin (M - N) = \rho \sin \phi' \cos D' \sin N - \rho m' \cos \phi' \sin(\mu - A' + M')$,
$m \cos (M - N) + nt = \rho \sin \phi' \cos D' \cos N + \rho n' \cos \phi' \sin (\mu - A' - N')$.

Put $\left.\begin{array}{r} m' \sin (\mu - A' + M') = b \sin B \\ \cos D' \sin N = b \cos B \end{array}\right\}$ or $\tan B = \dfrac{m' \sin (\mu - A' + M')}{\cos D' \sin N}$

Then, $-m \sin (M - N) = \rho b \sin (\phi' - B) = \dfrac{\rho \sin (\phi' - B) \cos D' \sin N}{\cos B}$,

$m \cos (M - N) + nt = \rho \sin \phi' \cos D' \cos N + \rho n' \cos \phi' \sin (\mu - A' - N')$.

Hence, $\sin (\phi' - B) = -\dfrac{m \sin (M - N)}{\rho \cos D' \sin N} \cos B$

$$t = -\frac{m \cos (M - N)}{n} + \rho \sin \phi' \frac{\cos D' \cos N}{n} + \rho \cos \phi' \sin (\mu - A' - N') \frac{n'}{n}$$

---

* See table XVI.

Let H = the value of the hour angle $(\mu - A')$, at a required place, at the time the eclipse is central there, H' = its value at the first meridian at the time T, and λ = the longitude of the required place. Then, $H' + 15t$ $= H' - \frac{15\, m \cos(M - N)}{n} + \rho \sin \phi' \frac{15 \cos D' \cos N}{n} + \rho \cos \phi' \sin (H - N') \frac{15n'}{n}$ = the hour angle at the first meridian, at the time T. Consequently,

$$\lambda = H' - \frac{15m \cos(M - N)}{n} + \rho \sin \phi' \frac{15 \cos D' \cos N}{n} + \rho \cos \phi' \sin (H - N') \frac{15n'}{n} - H$$

$$= H' - \frac{15m \cos (M - N)}{n} + M' + \rho \sin \phi' . \frac{15 \cos D' \cos N}{n}$$

$$+ \rho \cos \phi' \sin (H - N') \frac{15n'}{n} - (H + M').$$

Put $H'' = H' - \frac{15\, m \cos(M - N)}{n} + M', L = H + M' = \mu - A' + M', S = M' + N', B' = \frac{m'}{\cos D' \sin N'}, C' = - \frac{m \sin (M - N)}{\cos D' \sin N}, E' = \frac{15 \cos D' \cos N}{n}$, and $F' = \frac{15n'}{n}$. Then, $L - S = (H + M') - (M' + N') = H - N'$. Hence, assuming $\rho = 1$, we have, by substitution,

$$\left.\begin{array}{l} \tan B = B' \sin L \\ \sin (\phi' - B) = C' \cos B \\ \lambda = H'' + E' \sin \phi' + F' \cos \phi' \sin (L - S) - L \end{array}\right\} \quad (q)$$

The quantities H'' and S and the logarithms of B', C', E', and F', being computed for the time T, may be regarded as constant throughout the eclipse. Let the values of L, at the times of beginning and end of the central eclipse be obtained, by adding M' to the value of $(\mu - A')$, as found by (n) of the last article, for each of these times. Then, assuming for L, any intermediate value, and finding B, from the first of equations (q), we obtain ϕ' from the second, and λ from the third; and these are the geocentric latitude and the longitude of the place at which the eclipse is central, when L has this assumed value. By assuming for L a series of values between the extremes mentioned above, a corresponding series of places at which the eclipse will be central may be found.

Taking $h = z \tan f - k \sec f$, which is its value for internal contact, except that the small term $z'' \tan f$ is omitted, and then putting $C' = - \frac{m \sin (M - N) \pm h}{\cos D' \sin N}$, the formulæ (q), serve to find a series of places in the northern or southern limit of visibility of the annular or total eclipse. For an annular eclipse, the upper sign corresponds to a place in the northern limit, and the lower, to one in the southern limit. The contrary has place for a total eclipse.

When a series of places at which the eclipse will be central has been found, if a curve line be drawn through their positions on a map, it will represent the line of the central eclipse.

If a series of places in the northern, and also in the southern limit of annular or total visibility, be found, and lines be drawn through their positions, they will bound the narrow portion of the earth's surface, within which, the eclipse is annular or total, as in *Fig.* 66, which applies to the eclipse in May, 1836.*

Note. The arcs B and (ϕ' — B) in formulæ (q), may each be taken less than 90°, being marked affirmative or negative according to the sign of the tangent or sine.

79. *Occultations.* If, instead of the quantities referring to the sun, those referring to a star or planet be taken, the formulæ obtained for computing an eclipse of the sun will also be applicable to the computation of an occultation of the star or planet.

For a *star*, as its diameter and parallax are insensible, we have, $r = o$, $a = A'$, $d = D'$, $f = o$, $l = k$, $h = l = k = 0.2725$. Also, as the star's right ascension A$'$, does not sensibly change during the continuance of an occultation, we have, the hourly variation of $(\mu - A') = 15° \, 2' \, 27''.84 = 54147''.84$. Hence,

$$x = \frac{\cos D \sin (A - A')}{\sin \pi}$$

$$y = \frac{\sin (D - D')}{\sin \pi} + x \operatorname{tang} \tfrac{1}{2} (A - A') \sin D'$$

$$x'' = \rho \cos \phi' \sin (\mu - A')$$

$$y'' = \rho \sin \phi' \cos D' - \rho \cos \phi' \sin D' \cos (\mu - A')$$

$$u' = \rho \cos \phi' \, . \, \frac{2 \sin (t . 27073''.92)}{t} \cos (\mu - A' + t . 27073''.92)$$

$$v' = \rho \cos \phi' \sin D' \frac{2 \sin (t . 27073''.92)}{t} \sin (\mu - A' + t . 27073''.92)$$

$$k \cos P = x - x''$$

$$k \sin P = y - y''.$$

---

* For the investigation of formulæ for determining the entire limits of visibility and other circumstances relative to the general eclipse, as represented in *Fig.* 66, the student may be referred to Woolhouse's Tract on Eclipses, which forms the Appendix to the Nautical Almanac for 1836. This subject has also been very fully investigated by Prof. Hansen in the *Astr. Nach.* Nos. 339 to 342.

The position of the point of contact, in an occultation, is usually denoted, by giving its distance from the north point, or from the vertex of the moon's disc. The expression for this, will evidently be obtained, very nearly, by subtracting 180° from the expression for the position of the point of contact in an eclipse of the sun. We shall thus have, if P′ and V now refer to the north point and vertex of the moon's disc,

$$P' = N \pm \psi - 90°$$
$$V = N \pm \psi - 90° - Q.$$

The first expresses the distance to the *left* of the north point of the moon's disc, and the latter, the distance to the *left* of the vertex.*

80. *Transits of Mercury and Venus.* Instead of the quantities which have referred to the moon, using those that refer to the planet, and taking $k = 0.3766$ for Mercury, and 0.9617 for Venus, the formulæ obtained for an eclipse of the sun, will serve to calculate a transit of either of these planets; observing, however, that the values of a, d, and g, must be obtained from the formulæ (D), and not from the approximate formulæ (E).

81. *Formulæ for computing an* observed *eclipse of the sun.* Let T′ be the observed mean time of beginning or end of the eclipse at a place whose latitude is known, T a mean time at the first meridian, taken to a whole hour near to the time of new moon, $T + t$ the mean time at the first meridian, corresponding to the time T′, and $d' = T' - (T + t)$. Then will d' be the longitude in time of the place at which the eclipse is observed; it being *east* if *affirmative*, but *west* if *negative*.

Let p and q be the values of x and y at the time T, and p' and q' their average hourly variations between the times T and $T + t$. Then, at the time $T + t$, we have, $x = p + p't$, and $y = q + q't$. Consequently, taking for x'', y'', and h, their values at this time, the equations of contact will be (App. 70 N),

$$\left.\begin{aligned} h \cos P &= p - x'' + p't \\ h \sin P &= q - y'' + q't \end{aligned}\right\} \quad \ldots\ldots\ldots\ldots \quad (r)$$

Put
$$\left.\begin{aligned} p - x'' &= m \sin M, \qquad & p' &= n \sin N \\ q - y'' &= m \cos M, \qquad & q' &= n \cos N \end{aligned}\right\} \quad \ldots\ldots\ldots \quad (s)$$

Then,
$$\left.\begin{aligned} h \cos P &= m \sin M + nt \sin N \\ h \sin P &= m \cos M + nt \cos N \end{aligned}\right\} \quad \ldots\ldots\ldots\ldots \quad (t)$$

---

* Data are given in the Berlin Ephemeris, by which the computations of the principal occultations that occur in the year, is greatly facilitated. These also include data, adapted to formulæ investigated by Prof. Hansen, in the *Astr. Nach.* No. 360, by means of which the position of the point of contact with reference to the contiguous spots on the moon's disc, is easily computed.

Hence we have, as in (App. 71),

$$\left.\begin{array}{l} \cos \psi = \dfrac{m \sin (\mathrm{M} - \mathrm{N})}{h} \\ t = -\dfrac{m \cos (\mathrm{M} - \mathrm{N} \mp \psi)}{n \cos \psi} \\ \text{And, consequently, } d' = \mathrm{T}' - \mathrm{T} + \dfrac{m \cos (\mathrm{M} - \mathrm{N} \mp \psi)}{n \cos \psi} \end{array}\right\} \quad . \quad . \quad (u)$$

To make the computation, the observed time of beginning or end may be reduced to the time $\mathrm{T} + t$, at the first meridian, by using an *assumed* longitude of the place. Then, having found the values p', q', x'', y'', z'', and h, for this time, and computed the values M, m, N, and n, from (s), we find d' from (u). If d', thus found, does not differ more than a few minutes from the assumed longitude, it may be regarded as the true longitude, as obtained from the observation. But if d' differs considerably from the assumed longitude, the computation should be repeated, taking d' as the assumed longitude. When the beginning and end have both been observed, the computation should be made for each, and the mean of the two results be taken as the longitude of the place.

82. As the solar and lunar tables cannot be regarded as perfectly accurate, the longitude obtained as above is liable to a small error depending on little errors in the elements used in the computation. But when the eclipse has also been observed at Observatories or other places whose positions are accurately known, the means are afforded of correcting the result for the principal errors in the elements. Those liable to the greatest errors, though these are but small, are the right ascension and declination of the moon.

Let $\Delta\mathrm{A}$ and $\Delta\mathrm{D}$ be the corrections which ought to be applied to A and D, the computed right ascension and declination of the moon, so that $\mathrm{A} + \Delta\mathrm{A}$, and $\mathrm{D} + \Delta\mathrm{D}$, may be the true values. The values of x and y, or their representatives p and q, will require corrections depending on the corrections $\Delta\mathrm{A}$ and $\Delta\mathrm{D}$. In obtaining them, we may, without material error, take, for p and q, the approximate expressions $p = \dfrac{\cos \mathrm{D}\,(\mathrm{A} - a)}{\pi}$ and $q = \dfrac{\mathrm{D} - d}{\pi}$, deduced from the equations (App. 69 F). Substituting, in the first of these, $\mathrm{A} + \Delta\mathrm{A}$ for A, and, in the second, $\mathrm{D} + \Delta\mathrm{D}$ for D, we have,

$$\frac{\cos \mathrm{D}\,(\mathrm{A} - a + \Delta\mathrm{A})}{\pi} = \frac{\cos \mathrm{D}\,(\mathrm{A} - a)}{\pi} + \frac{\cos \mathrm{D}\,\Delta\mathrm{A}}{\pi}$$

$$\frac{\mathrm{D} - d + \Delta\mathrm{D}}{\pi} = \frac{\mathrm{D} - d}{\pi} + \frac{\Delta\mathrm{D}}{\pi}.$$

Consequently, neglecting the extremely small correction that would be produced by the correction of D, in the factor cos D, the terms $\frac{\cos D \ \Delta A}{\pi}$ and $\frac{\Delta D}{\pi}$ are the corrections of p and q. Or, putting $\omega \sin \pi$ for π (App. 51), we have,

$$\text{correction of } p = \frac{\cos D}{\sin \pi} \cdot \frac{\Delta A}{\omega}.$$

$$\text{“} \qquad q = \frac{1}{\sin \pi} \cdot \frac{\Delta D}{\omega}$$

The correction of z may be omitted, as producing scarcely any influence on the value of h, (App. 70 M).

Let β and δ be two unknown quantities whose values depend on those of ΔA and ΔD, and assume,

$$\left.\begin{aligned} n \sin N. \beta - n \cos N. \delta &= \frac{\cos D}{\sin \pi} \cdot \frac{\Delta A}{\omega} \\ n \cos N. \beta + n \sin N. \delta &= \frac{1}{\sin \pi} \cdot \frac{\Delta D}{\omega}. \end{aligned}\right\} \quad . \ . \ . \ . \ . \ . \ . \ (v)$$

Applying the corrections of p and q, thus expressed, to the equations (t), we have,

$$\left.\begin{aligned} h \cos P &= m \sin M + nt \sin N + n \sin N. \beta - n \cos N. \delta \\ h \sin P &= m \cos M + nt \cos N + n \cos N. \beta + n \sin N. \delta \end{aligned}\right\} \quad (w)$$

Whence we obtain,

$$h \cos (P + N) = m \sin (M - N) - n\delta$$

$$h \sin (P + N) = m \cos (M - N) + nt + n\beta$$

The squares of these, added together, give,

$$h^2 = \Big(m \cos (M - N) + nt + n\beta\Big)^2 + \Big(m \sin (M - N) - n\delta\Big)^2$$

Hence, putting $h \cos \psi$ instead of its value, $m \sin (M - N)$, we have,

$$\Big(m \cos (M-N) + nt + n \beta\Big)^2 = h^2 - (h \cos \psi - n \delta)^2 = h^2 \sin^2 \psi + 2 \, hn \delta \cos \psi - n^2 \delta^2$$

Extracting the square root, and observing that, as δ is a very small quantity, the terms involving its square and higher powers may be omitted, we obtain,

$$m \cos (M-N) + nt + n \beta = h \sin \psi + \frac{n \delta \cos \psi}{\sin \psi} = h \sin \psi + \frac{n \delta}{\text{tang } \psi}.$$

Hence, regarding ψ as always affirmative and less than 180°, and placing the double sign before the terms involving its sine or tangent, as in (App. 70), we have,

$$t = - \frac{m \cos (M - N)}{n} \mp \frac{h \sin \psi}{n} - \beta \mp \frac{\delta}{\text{tang } \psi},$$

or, $$t = -\frac{m \cos (M - N \mp \psi)}{n \cos \psi} - \beta \mp \frac{\delta}{\operatorname{tang} \psi},$$

and, $$d' = T' - T + \frac{m \cos (M - N \mp \psi)}{n \cos \psi} + \beta \mp \frac{\delta}{\operatorname{tang} \psi}.$$

Let $s = 3600 =$ number of seconds in an hour, and let the terms connected with $T' - T$ be multiplied by s. Then will the terms be expressed in seconds, and the expression for d' will be,

$$d' = T' - T + \frac{ms}{n} \cdot \frac{\cos (M - N \mp \psi)}{\cos \psi} + \beta s \pm \frac{\delta s}{\operatorname{tang} \psi} \qquad (x)$$

Now, from equations (v), we obtain,

$$\beta = \frac{1}{\omega\, n \sin \pi} \cdot (\sin N \cos D . \Delta A + \cos N . \Delta D)$$

$$\delta = \frac{1}{\omega\, n \sin \pi} \cdot (\sin N . \Delta D - \cos N \cos D . \Delta A)$$

Put $$\left.\begin{array}{l} \varepsilon = \sin N \cos D . \Delta A + \cos N . \Delta D \\ \zeta = \sin N . \Delta D - \cos N \cos D . \Delta A \\ a = \dfrac{s}{\omega\, n \sin \pi}, \; b = \dfrac{a}{\operatorname{tang} \psi} \end{array}\right\} \; . \; . \; . \; . \; . \; . \; . \qquad (y)$$

Then, by substitution in (x), we have,

$$d' = T' - T + \frac{ms}{n} \cdot \frac{\cos (M - N \mp \psi)}{\cos \psi} + a\, \varepsilon \pm b \zeta \quad . \; . \; (z)$$

When the times of beginning and end of the eclipse have been observed at a place whose longitude d' is accurately known, we shall, by making the computations for these times and this place and taking for d' its known value, have two equations containing the unknown quantities ε and ζ, from which their values may be found. When the eclipse has been well observed at a number of places whose positions are accurately known, it is better to make the computations for several of these, and thus obtain a number of equations containing ε and ζ, from a proper combination of which, their values may be more accurately obtained.

When the values of ε and ζ have been found by computation for places whose positions are known, and substituted in (z), this formula will give the longitude for any other place at which the eclipse has been observed, with the corrections for the errors in the moon's computed right ascension and declination.

With the values of ε and ζ, the values of ΔA and ΔD may be found from the first two equations of (y).

83. *Observed occultation of a star.* The formulæ that have been obtained for computing an observed eclipse of the sun, serve also for the

computation of an observed occultation of a star, using the data for the star instead of those for the sun.*

---

* The formulæ for computing an observed eclipse or occultation, have been derived from an excellent Tract by Prof. Bessel, *Astr. Nach.* Nos. 151 and 152. In this tract, the investigations are extended so as to notice errors in some of the other elements obtained from the tables.

For all important eclipses of the sun, the values of N and n are given in the Berlin Ephemeris, for several consecutive whole hours, near to the time of new moon.

AN

ELEMENTARY TREATISE

ON

ASTRONOMY.

PART II.

Catalogue of the Tables, with occasional observations.

TABLES I. and II.

Logarithms and logarithmic Sines and Tangents, to four decimal figures. To avoid an extra line of figures, the 10 in the index of the tangents and cotangents has been rejected, when the index exceeded 10.

TABLES III., IV., and V.

Log. tangent of the Obliquity of the Ecliptic.—Log. A = log. cosine of obliquity of the ecliptic *less* log. of the difference of the moon's and sun's parallaxes, and log. B = arith. comp. log. sine of difference of the parallaxes.—Log. tangent of sun's semidiameter.

TABLE VI.

Latitudes of a number of places with their longitudes from the meridian of Greenwich.

The latitudes and longitudes of several of the places in the United States are given according to the determinations of R. T. Paine, the former editor of the astronomical part of the *American Almanac*, a valuable work, published annually in Boston.

TABLE VII.

Mean Refractions with the corrections due to given changes in the states of the barometer and thermometer.

TABLE VIII.

Sun's Parallax in Altitude.

TABLE IX.

Mean Right Ascensions and Declinations of 30 principal Fixed Stars for the beginning of the year 1850, with their Annual Variations; also, auxiliary quantities to facilitate the computations of their aberrations and nutations. North declination is indicated by the sign *plus*, and South declination by the sign *minus*.

TABLES X. and XI.

These serve to convert intervals of mean solar time into equivalent intervals of sidereal time, and the contrary.

TABLES XII. to XV., inclusive.

Auxiliary tables, for the computations of Solar Eclipses, and Occultations.

TABLE XVI.

Reductions of the Moon's Parallax and of the latitude of a place, and also the logarithms of the earth's radius, according to the compression $\frac{1}{300}$.

TABLE XVII.

Logarithms to be added to the logarithmic cosine and sine of the geographic latitude of a place, to obtain the logarithms of $\rho \cos \phi'$ and $\rho \sin \phi'$; in which ρ is the radius of the earth at the place, and ϕ' the geocentric latitude.

TABLES XVIII. to XXI., inclusive.

These serve to find the time of New or Full Moon in any month approximately, or within a few minutes of the true time.

The time of mean new moon in January of each year, as given in table XVIII., has been diminished by 15 hours. These 15 hours have been added to the equations in table XXI. Thus, 4h. 20m. has been added to the first equations; 10h. 10m. to the second; 10 minutes to the third; and 20 minutes to the fourth. By this means, the equations are all made additive.

TABLES XXII. to XXXI., inclusive.

These are approximate Solar Tables, by which the sun's true longitude, hourly motion, semidiameter and radius vector, and the apparent obliquity of the ecliptic, may be determined for a given time, very nearly.

The Sun's Mean Longitude, the longitude of the perigee, and Arguments for finding some of the small equations of the sun's place given in table XXII., are all computed for mean noon at the meridian of Greenwich, on the first of January for common years, and on the second of January for bissextiles. The sun's longitudes and the longitudes of his perigee have each been diminished by 2°. As each is diminished by the same quantity, the mean anomaly, which is obtained by subtracting the longitude of the perigee from the sun's longitude, and which is the argument for the equation of the centre, is not affected. The Argument I. is for the equation depending on the action of the moon; Argument II. is for that depending on the action of Jupiter; Argument III. is for that depending on the action of Venus; and Argument N, is for the Nutation, or equation of the equinoxes.

Of the 2° which has been subtracted from the sun's mean longitudes, 1° 59′ 30″ is added to the equation of the centre, and 10″ to each of the small equations due to the actions of the Moon, Jupiter, and Venus.

TABLES XXXII. to LXIV., inclusive.

Approximate Lunar Tables, by which the moon's true longitude, latitude, horizontal parallax, semidiameter and hourly motions in longitude and latitude for a given time, may be determined, very nearly.

The Epochs of the Moon's Mean Longitude, and of the Arguments for finding the Equations which are necessary in determining the True Longitude and Latitude of the Moon given in table XXXII., are all computed for mean noon at the meridian of Greenwich, on the first of January for common years, and on the second of January for bissextiles. The Argument for the Evection is diminished by 29′, the Anomaly by 1° 59′, the Argument for the Variation by 8° 59′, the Mean Longitude by 9° 44′, and the Supplement of the Node is increased by 7′. This is done to balance the quantities which are applied to different equations to render them affirmative.

TABLE LXV.

Five pages of the Nautical Almanac, for the month of May, 1836.

TABLES LXVI., LXVII., and LXVIII.

Tables of Second, Third, and Fourth Differences; useful in finding, from the Nautical Almanac, the moon's longitude or latitude for any intermediate time between noon and midnight.

TABLES LXIX., to LXXIX., inclusive.

Approximate tables for the planet Mercury; including also a small table containing the Heliocentric Longitude, Latitude, &c., of the planet Venus at the times of Transit over the sun's disc in 1874 and 1882.

TABLE LXXX.

Logistical Logarithms. This table is convenient in working proportions when the terms are minutes and seconds, or degrees and minutes, or hours and minutes.

TABLE LXXXI.

Reduction to the Meridian. (See Problem XXX.)

PRELIMINARY OBSERVATIONS.

It is frequently convenient to regard quantities as separated into two classes; those of one class being called *affirmative*, and those of the other *negative*. Thus, a right line or an arc of a circle, taken in one direction, being regarded as *affirmative*, a line or arc taken in the opposite direction, is regarded as *negative*. An affirmative quantity is denoted by having the sign +, called the *affirmative* or *plus* sign, prefixed to it, and a negative quantity by having the sign —, called the *negative* or *minus* sign, prefixed to it. Before an affirmative quantity the sign is frequently omitted, it being understood to be affirmative if neither sign is prefixed; but before a negative quantity the sign must always be expressed.

If an affirmative arc, and a negative arc, equal to the supplement of the former to 360°, both commence at the same point in the circumference of a circle, they must also both terminate at the same point. We may, therefore, denote the position of a point in the circumference with reference to a given or fixed point, either by an affirmative arc or by a negative one equal to its supplement to 360°. Thus, supposing the affirmative arc to be 294° 47′, we may substitute in place of it, — 65° 13′.

To add quantities, having regard to their signs. When all the quantities have the same sign, add them as in common arithmetic, and prefix that sign to the sum. When the quantities have different signs, add the affirmative quantities into one sum, and the negative into another. Then take the difference between these two sums and prefix the sign of the greater.

When several arcs are to be added together, if the sum exceeds 360°, we may reject 360°, or any multiple of it, and regard the result as the sum of the arcs.

EXAMPLES.

		′	″
Add,	—	2′	11″
	—	7	2
	—	12	57
Sum,	—	22	10

Add,	+ 3.72
	— 7.56
	+ 2.41
Sum,	— 1.43

Add,	— 28.4
	+ 75.2
	+ 33.9
Sum,	+ 80.7

		°	′	″
Add,	+	179°	4′	12″
	—	15	9	30
	+	236	27	10
	—	25	59	11
Sum,	+	14	22	41

		°	′	″
Add,	—	3°	7′	10″
	+	2	8	5
	+	317	29	47
	—	12	15	20
Sum,	+	304	15	12
or,	—	55	44	48.

To subtract quantities having regard to their signs. Suppose the sign of the quantity, which is to be subtracted, to be changed, that is, if it is affirmative, suppose it to be negative, or if it is negative, suppose it to be affirmative. Then proceed as in the above rule for adding quantities.

When one arc is to be subtracted from another, and the latter is the less of the two, we may increase it by 360°.

EXAMPLES.

	′	″
From	4′	11″
Subt.	7	27
Rem.	— 3	16

From	27.5
Subt.	— 12.3
Rem.	+ 39.8

From	— 8.273
Subt.	— 3.197
Rem.	— 5.076

		°	′	″
From		312°	17′	39″
Subt.		17	51	47
Rem.	+	294	25	52
or,	—	65	34	8

		°	′	″
From		21°	17′	25″
Subt.		156	54	13
Rem	—	135	36	48
or,	+	224	23	12

x 2

To find the Logarithmic Sine, Cosine, Tangent, or Cotangent of an arc, with its proper Sign, from Tables that extend only to each minute of the quadrant.

When the given arc does not exceed 180°. With the given arc, or when it exceeds 90°, with its supplement to 180°, take out from the table the required Sine or Tangent, &c. When there are *seconds*, take out the quantity corresponding to the given degrees and minutes; also take the difference between this quantity and the next following one, in the table. Then 60″ : the odd seconds of the given arc : : the difference : a fourth term. This fourth term, *added* to the quantity taken out, when it is *increasing*, but *subtracted* when it is *decreasing*, will give the required quantity.

When the given arc exceeds 180°. Subtract 180° from it, and proceed as before. When the arc exceeds 270°, it is more convenient, and amounts to the same, to subtract it from 360°.

To determine the Sine of the quantity. Call the arc from 0° to 90°, the *first* quadrant; from 90° to 180°, the *second* quadrant; from 180° to 270°, the *third* quadrant; and from 270° to 360°, the *fourth* quadrant. Then,

The *Sine* of an *affirmative* arc is *affirmative* for the first and second quadrants; and *negative* for the third and fourth. For a *negative* arc it is just the reverse; the sine being negative in the first and second quadrants and affirmative in the third and fourth.

The *Cosine* of an *affirmative* arc is *affirmative* for the first and fourth quadrants, and *negative* for the second and third. It is the same for a *negative* arc.

The *Tangent* or *Cotangent* of an *affirmative* arc is *affirmative* for the first and third quadrants, and *negative* for the second and fourth. For a *negative* arc it is just the reverse; the tangent and cotangent being *negative* in the first and third quadrants, and *affirmative* in the second and fourth.

Note. Negative logarithms or logarithmic sines, &c., are frequently designated by a small *n*, placed at the right hand, instead of the sign —, before them.

By attending to the preceding rules, the student will easily find the Sine, Cosine, &c., of an arc in either quadrant, with its appropriate sign—, as exemplified in the following table:

Arc.			Log. sine.	Log. cosine.	Log. tangent.	Log. cotang.
37°	18′	21″	9.78252	9.90060	9.88193	10.11807
— 37	18	21	9.78252*n*	9.90060	9.88193*n*	10.11807*n*
114	35	10	9.95872	9.61916*n*	10.33956*n*	9.66044*n*
— 114	35	10	9.95872*n*	9.61916*n*	10.33956	9.66044
247	12	36	9.96470*n*	9.58811*n*	10.37659	9.62341
314	17	50	9.85475*n*	9.84409	10.01065*n*	9.98935*n*

The logarithmic Sine, Cosine, Tangent, or Cotangent of an arc being given, to find the arc.

When the given quantity can be found in the table, under or over its name, take out the corresponding arc. When the given quantity is not found exactly in the table, and the arc is required to seconds, take out the degrees and minutes corresponding to the next less quantity, when that quantity is increasing; but to the next greater when it is decreasing. Take the difference between the quantity corresponding to the degrees taken out, and the next following one in the table; also, take the difference between the same quantity and the given one. Then, the first difference : the second : : 60″ : the number of seconds which is to be annexed to the degrees and minutes. Then,

For a Sine. When it is *affirmative*, the required affirmative arc will be, either the arc found in the table, or its supplement to 180°. When the sine is *negative*, the required arc will be, either the arc found in the table, increased by 180°, or its supplement to 360°.

For a Cosine. When it is *affirmative*, the required affirmative arc will be, either the arc found in the table, or its supplement to 360°. When the cosine is *negative*, the required arc will be, either the supplement of the arc found in the table, to 180°, or that arc increased by 180°

For a Tangent or Cotangent. When it is *affirmative*, the required affirmative arc will be, either the arc found in the table, or that arc increased by 180°. When the tangent or cotangent is *negative*, the required arc will be, either the supplement of the arc, found in the table, to 180°, or its supplement to 360°.

When the required arc comes out more than 180°, the equivalent negative arc is frequently taken.

These rules are exemplified by the quantities in the following table:—

Log. sine	9.78252	arc	37°	18′	21″	or	142°	41′	39″
Log. sine	9.85475*n*	arc	225	42	10	or	314	17	50
Log. cosine	9.90060	arc	37	18	18	or	322	41	42
Log. cosine	9.61916*n*	arc	114	35	11	or	245	24	49
Log. tangent	9.88193	arc	37	18	21	or	217	18	21
Log. tangent	10.33956*n*	arc	114	35	11	or	294	35	11
Log. cotangent	9.62341	arc	67	12	36	or	247	12	36
Log. cotangent	9.98935*n*	arc	134	17	51	or	314	17	51

Note. Tables which extend only to five decimals, will give the arc, for a tangent or cotangent, true to the nearest second, for a few degrees, near to 0°, 90°, 180°, or 270°; for a sine, near to 0° or 180°; and for a cosine near to 90° or 270°. In other cases they cannot be depended on to give the seconds accurately. They are, however, sufficient for many calculations; particularly, when the nature of the problem does not make it necessary that the required arc or angle should be determined with great accuracy.

As most mathematical students are furnished with a set of such tables, and as an example worked by them will serve as well to illustrate a rule as if worked by those which are more extensive, they will generally be used in working the examples and questions in the following problems.

Observations relative to the Signs and Indices of Logarithms. A logarithm is affirmative when the natural number is affirmative, and negative when it is negative.

When several logarithms, or logarithms and the arithmetical complements of logarithms, are added together, if they are all affirmative, or if there is an *even* number of *negative* ones, the resulting logarithm will be affirmative; but if there is an *odd* number of negative ones, the resulting logarithm will be negative.

Instead of the negative index of the logarithm of a decimal number, the index increased by 10, is frequently used. Thus, when there is no cipher between the decimal point and first significant figure, 9 is put for the index; when there is one cipher between them, 8; when there is two, 7; and so on. When this is done, and the resulting logarithm of a computation is the logarithm of a natural number, if the index is 9, the number will be a decimal without any cipher between the decimal point and first significant figure; if it is 8, there must be one cipher between them; if it is 7, there must be two; and so on. If the index is near to 0, the resulting number is generally integral.

Rejection of the tens *in the index of the sum of logarithms.* In working the following problems, when several logarithms or logarithms and the arithmetical complements of logarithms are added together, the *tens* in the index of the sum are to be rejected. When, however, the sum is the log. tangent or log. cotangent of an arc, and a table of log. tangents is used in which the 10 in the index has not been rejected, one 10 should be retained in the index of the sum, if its rejection would reduce this index below 5.

PROBLEMS FOR MAKING VARIOUS ASTRONOMICAL CALCULATIONS.

PROBLEM I.

To work, by logistical logarithms, a proportion, the terms of which are minutes and seconds of a degree, or of time, or hours and minutes.

With the minutes at the top and seconds at the side, or if a term consists of hours and minutes, with the hours at the top and minutes at the side, take from table LXXX., the logistical logarithms of the three given terms, and proceed in the usual manner of working a proportion by logarithms. The quantity, in the table, corresponding to the resulting logarithm, will be the fourth term.

Note 1. The logistical logarithm of 60′ is 0.

2. The student will easily perceive that proportions that are worked by logistical logarithms, may also be worked by the common rule in arithmetic.

EXAM. 1. When the moon's hourly motion is 31′ 57″, what is its motion in 39m. 22sec.? *Ans.* 20′ 58″.

As 60 m.	0
: 39 m. 22 sec.	1830
: : 31′ 57″	2737
: 20′ 58″	4567

2. If the moon's declination change 2° 29′ in 12 hours, what will be the change in 8h. 21m.? *Ans.* 1° 44′.

As 12h.	6990
: 8h. 21m.	8565
: : 2° 29′	13831
	22396
: 1° 44′	15406

3. When the sun's hourly motion is 2′ 31″, what is its motion in 17m. 18sec.? *Ans.* 0′ 44″.

4. When the sun's declination changes 22′ 14″ in 24 hours, what is its change in 19h. 25m.? *Ans.* 17′ 59″.

35

PROBLEM II.

From a table in which quantities are given, for each Sign and Degree of the circle, to find the quantity corresponding to Signs, Degrees, Minutes, and Seconds.

Take out, from the table, the quantity corresponding to the given signs and degrees; also take the difference between this quantity and the next following one. Then 60′ : odd minutes and seconds : : this difference : a fourth term. This fourth term added to the quantity taken out, when the quantities in the table are increasing; but subtracted, when they are decreasing, will give the required quantity.

Note 1. When the quantities change but little from degree to degree, the required quantity may frequently be estimated, without the trouble of making a proportion.

Note 2. The given quantity with which a quantity is taken from a table, is called the *Argument.*

Note 3. In many tables, the argument is given in parts of the circle, supposed to be divided into 100, 1000, or 10,000, &c., parts. The method of taking quantities from such tables is the same as is given in the above rule; except that, when the argument changes by 10, the first term of the proportion must be 10, and the second, the odd units; when the argument changes by 100, the first term must be 100, and the second, the odd parts between hundreds; and so on.

EXAM. 1. Given the argument $1^s\ 9°\ 31'\ 26''$, to find the corresponding quantity in table XLIV. *Ans.* 11° 13′ 57″.

$1^s\ 9°$ gives 11° 11′ 15″.

The difference between 11° 11′ 15″ and the next following quantity in the table is 5′ 9″.

As 60′ : 31′ 26″ : : 5′ 9″ : 2′ 42″.*

To	11° 11′ 15″
Add	2 42
	11 13 57

2. Given the argument $10^s\ 13°\ 16'\ 54''$, to find the corresponding quantity in table XLVII. *Ans.* 93° 32′ 37″.

$10^s\ 13°$ gives 93° 33′ 40″.

* The student can work the proportion, either by common arithmetic, or by logistical logarithms, as he may prefer.

The difference between 93° 33′ 40″ and the next following quantity in the table, is 3′ 43″.

As 60′ : 16′ 54″ : : 3′ 43″ : 1′ 3″

From 93° 33′ 40″
Take 1 3

93 32 37

3. Given the argument 4^s 11° 57′ 10″, to find the corresponding quantity in table XXVII. *Ans.* 3° 24′ 6″.

4. Given the argument 3721, to find the corresponding quantity in table XXXVII. *Ans.* 4′ 52″.

PROBLEM III.

To convert Degrees, Minutes, and Seconds of the Equator into Time.

Multiply the quantity by 4, and call the product of the seconds, thirds; of the minutes, seconds; and of the degrees, minutes.

EXAM. 1. Convert 72° 17′ 42″ into time.

72° 17′ 42″
4

4h. 49m. 10sec. 48‴. = 4h. 49m. 11sec. nearly.

2. Convert 117° 12′ 30″ into time. *Ans.* 7h. 48m. 50sec.

3. Convert 21° 52′ 27″ into time. *Ans.* 1h. 27m. 30sec.

PROBLEM IV.

To convert Time into Degrees, Minutes, and Seconds.

Reduce the time to minutes, or minutes and seconds; divide by 4, and call the quotient of the minutes, degrees; of the seconds, minutes; and multiply the remainder by 15, for the seconds.

EXAM. 1. Convert 5h. 41m. 10sec. into degrees, &c.

h. m. sec.
5 41 10
60

4) 341 10

85° 17′ 30″

2. Convert 7h. 48m. 50sec. into degrees, &c. *Ans.* 117° 12′ 30″.

3. Convert 11h. 17m. 21sec. into degrees, &c. *Ans.* 169° 20′ 15″.

PROBLEM V.

The Longitudes of two Places, and the Time at one of them being given, to find the corresponding Time at the other.

Express the given time astronomically. Thus, when it is in the morning, add 12 hours, and diminish the number of the day by a unit. When the given time is in the afternoon, it is already in astronomical time.

Find the difference of longitude of the two places, by subtracting the less longitude from the greater, when they are both of the same name, that is, both east or both west; but by adding the two longitudes together when they are of different names. When one of the places is Greenwich, the longitude of the other is the difference of longitude.

Then, if the place, at which the time is required, is to the *east* of the other place, *add* the difference of longitude, in time, to the given time; but if it is to the *west*, *subtract* the difference of longitude from the given time. The sum or remainder is the required time.

Note. The longitudes of the places mentioned in the following examples, are given in table VI.

EXAM. 1. When it is August 8th, 2h. 12m. 17sec. A. M. at Greenwich, what is the time as reckoned at Philadelphia?

	d.	h.	m.	sec.	
Time at Greenwich, August, . . .	7	14	12	17	
Diff. of Long., ...		5	0	40	
Time at Philadelphia, ...	7	9	11	37	P. M.

2. When it is April 11th, 3h. 15m. 20sec. P. M. at New York, what is the corresponding time at Greenwich?

	d.	h.	m.	sec.	
Time at New York, April, ...	11	3	15	20	
Diff. of Long., ...		4	56	4	
Time at Greenwich, ...	11	8	11	24	P. M.

3. When it is Sept. 10th, 3h. 20m. 35sec. P. M. at Paris, what is the time as reckoned at New Haven?

	h.	m.	sec.	
Longitude of Paris, ...	0	9	22	E.
do. of New Haven, ...	4	51	51	W
Diff. of Long., ...	5	1	13	

	d.	h.	m.	sec.
Time at Paris, September, ...	10	3	20	35
Diff. of Long., ...		5	1	13
Time at New Haven, ...	9	22	19	22

Or, Sept. 10th, 10h. 19m. 22sec. A. M.

4. When it is January 15th, 9h. 12m. 10sec. P. M. at Washington, what is the corresponding time at Berlin? *Ans.* Jan. 16th, 3h. 13m. 52sec. A. M.

5. When it is Oct. 5th, 7h. 8m. A. M. at Quebec, what is the time at Richmond? *Ans.* Oct. 5th, 6h. 43m. 18 sec. A. M.

6. When it is noon of the 10th of June at Greenwich, what is the time at Philadelphia? *Ans.* June 10th, 6h. 59m. 20sec. A. M.

PROBLEM VI.

For a given mean time, to find the Sun's Longitude, Semidiameter, Hourly Motion, the apparent Obliquity of the Ecliptic and the Earth's Radius Vector; also the Sun's Right Ascension and Declination and the Apparent Time.

For the Longitude.

When the given time is not for the meridian at Greenwich, reduce it to that meridian by the last problem.

With the mean time at Greenwich, take from Tables XXII., XXIII., and XXIV., the quantities corresponding to the year, month, day, hour, minute, and second, and find their sums.* The sum in the column of mean longitudes will be the *tabular* mean longitude of the sun; the sum in the column of perigee, will be the tabular longitude of the perigee; and the sums in the columns I., II., III., and N., will be the arguments for the small equations of the sun's longitude, and for the equation of the equinoxes, which forms one of them.

Subtract the longitude of the perigee from the sun's mean longitude, borrowing 12 signs when necessary; the remainder is the sun's mean anomaly. With the mean anomaly, take the equation of the sun's centre

* In adding quantities that are expressed in signs, degrees, &c., reject 12 or 24 signs, when the sum exceeds either of these quantities. In adding any arguments, expressed in 100, 1000, &c., parts of the circle, when they are expressed by two figures, reject the hundreds from the sum: when by three figures, the thousands; and when by four figures, the ten thousands.

from table XXVII., and with the arguments I., II., and III., take the corresponding equations from table XXVIII. The equation of the centre, and the three other equations, added to the mean longitude, give the true longitude, reckoned from the mean equinox.

With the argument N, take the equation of the equinoxes, or, which is the same, the nutation in longitude, from table XXX., and apply it, according to its sign, to the true longitude already found, and the result will be the true longitude, from the apparent equinox.

For the Hourly Motion and Semidiameter.

With the sun's mean anomaly, take the hourly motion and semidiameter from tables XXV. and XXVI.

For the apparent Obliquity of the Ecliptic.

To the mean obliquity, taken from table XXIX., apply, according to its sign, the nutation in obliquity, taken from table XXX., with the argument N, and the result will be the Apparent Obliquity.

For the Earth's Radius Vector.

With the sun's mean anomaly and the arguments I., II., and III., take the corresponding quantities from table XXXI., and the small table on the same page, and the sum of these will be the Radius Vector.

For the Right Ascension and Declination.

To the log. cosine of the apparent obliquity of the ecliptic, add the log. tangent of the sun's true longitude, and reject 10 from the index of the sum; the result will be the log. tangent of the Right Ascension, which must always be taken in the same quadrant as the longitude.

To the log. sine of the apparent obliquity, add the log. sine of the longitude, and reject 10 from the index of the sum; the result will be the log. sine of the Declination, which must be taken less than 90°; it will be *north* or *south* according as its sign is *affirmative* or *negative*.

For the Equation of Time and the Apparent Time.

To the sun's tabular mean longitude, increased by 2°, apply, according to its sign, the nutation in right ascension, taken from table XXX., with the argument N, and the result will be the sun's mean longitude from the true equinox. Take the difference* between this longitude and the sun's right ascension, making it *affirmative* or *negative* according as the right ascension is *less* or *greater* than the longitude, and the result will be the Equation of Time in arc. This may be converted into time by Prob. III.

The equation of time, applied, according to its sign, to the *mean* time, gives the *apparent* time.

EXAM. 1. Required the sun's longitude, hourly motion, &c., on the 25th of October, 1836, at 10h. 37m. 10sec. A. M. mean time at Boston.

	d.	h.	m.	sec.
Astron. time at Boston, Oct.	24	22	37	10
Diff. of Long.		4	44	17
Greenwich time	25	3	21	27

	M. Long.				Long. Perigee.				I.	II.	III.	N.
1836	9s	9°	10'	6"	9s	8°	7'	2"	456	517	644	842
Octob.	8	29	4	54				46	250	684	468	40
25 d.		23	39	20				4	810	60	41	4
3 h.			7	23				0	4	0	0	0
21 m.				52								
27 sec.				1	9	8	7	52	520	261	153	886
					7	2	2	36				
Tab. M. Long.	7	2	2	36	9	23	54	44	Mean Anomaly			
Equat. Centre		0	13	9								
I.				9	Sun's hourly motion. . . .						2'	29"
II.				17								
III.				9	" semidiameter						16	8
	7	2	16	20								
Nutation				— 11	M. Obliq. Ecliptic					23°	27'	38"
True Long.	7	2	16	9	Nutation						+	7
					Appar. Obliquity					23	27	45

For Radius Vector.

Arg. Anom. gives	0.99333
" I. "	2
" II. "	4
" III. "	1
Radius Vector	0.99340

* When one of these quantities is near to 0° and the other to 360°, the less must be increased by 360°, and the sum be regarded as its value.

For Right Ascension and Declination.

Obliq.	23° 27′ 45″	l. cos 9.96252	Obliq.	. . .	l. sin 9.60005
Long.	212 16 9	l. tan 9.80032	Long.	. .	l. sin 9.72746n
R. A.	210 4 48	l. tan 9.76284	Dec.	— 12° 16′ 24″	l. sin 9.32751n

For Equation of Time and Appar. Time.

Sun's tab. M. Long. + 2° . .		214° 2′ 36″
Nutation in Right Ascen. . . .		— 10
Sun's M. Long. from true equinox . . .		214 2 26
Sun's true Right Ascension		210 4 48
Equat. of Time, in arc	+	3 57 38
Equat. of Time, in time	+	15m. 50.5sec.

The equation of time, added to the given mean time, gives 10h. 53m. 0.5sec. A. M. for the apparent time.

Exam. 2. Required the sun's longitude, hourly motion, &c., on the 15th of May, 1836, at 8h. 59m. 29sec. A. M., mean time at Philadelphia.

Ans.	Sun's Longitude	54° 42′ 12″
	" hourly mot.	2 25
	" semidiameter	15 50
	App. obliq. of ecliptic	23 27 44
	Right Ascen.	52 20 23
	Declination	18 57 45 N
	Equat. of time	+ 3m. 56sec.
	Appar. Time	9h. 3m. 16sec. A. M.
	Radius Vector	1.01167

PROBLEM VII.

To find the Sidereal Time corresponding to a given Mean Time.

To the sun's mean longitude from the true equinox, found as in the last problem, and converted into time by Prob. III., add the given mean time of the day, expressed astronomically, rejecting 24 hours from the sum, if it exceeds that quantity, and the result will be the sidereal time.

When the sidereal time or right ascension of the zenith is required in arc, and not in time, it is most conveniently obtained by adding the mean time, expressed in arc, to the sun's mean longitude from the true equinox.

Note. When the sidereal time has been found for a given mean time, it may be found for any other time, a few hours later or earlier, by simply adding or subtracting the sum of the quantities corresponding to the hours, minutes and seconds of the interval, taken from table X.

EXAM. 1. What was the sidereal time at Philadelphia, on the 28th of May, 1840, at 3h. 19m. 20sec. P. M., mean time?

Greenw. time	M. Long.	N
1840	9ˢ 9° 11′ 56″	056
May	3 28 16 40	18
28d	26 36 45	4
8h	19 43	
20m.	49	
Tab. M. Long.	2 4 25 53	78

Sun's tab. M. Long. + 2°	66° 25′ 53″
Nut. in Right Ascen.	+ 8
M. Long. from true equinox,	66 26 1
" " " in time	4h. 25m. 44.1sec.
Given mean time,	3 19 20
Sidereal time,	9h. 45m. 4.1sec.

When the sidereal time is required in arc, and not time.

M. Long. from true equinox	66° 26′ 1″
Given mean time, in arc	49 50 0
Sidereal time, in arc	116 16 1

Let now the sidereal time be required at 6h. 40m. 56.5sec. P. M. Then the interval is 3h. 21m. 36.5sec.

Table X., for,

	h. m. sec.
3h. gives	3 0 29.6
21m. "	21 3.4
36.5sec. "	36.6
	3 22 9.6
	7 45 4.1
Sidereal time	11 7 13.7

PROBLEM VIII.

To find the Sidereal Time corresponding to a given Mean Time, using the Nautical Almanac.

In the Greenwich Nautical Almanac, the sidereal time at Greenwich mean noon is given for each day, and that for any other meridian may be found by adding if *west,* and subtracting if *east,* the sidereal acceleration for the difference of longitude expressed in time. The *acceleration* may be found by taking from table X., the equivalents for the hours, minutes and seconds, rejecting the hours, minutes and seconds of the equivalents respectively, and adding the results. Or, by multiplying 9.8565sec. by the difference of longitude expressed in hours.

Then, having found, as directed above, the sidereal time at mean noon, add to it the sidereal equivalents for the hours, minutes and seconds of the given mean time, taken from table X., and the sum will be the sidereal time required.

EXAM. 1. What was the sidereal time at Philadelphia on the 17th of May, 1836, at 8h. 17m. 10.5sec. mean time?

The acceleration for the diff. of Long. is by table X.,

for 5 hours	$49^s.282$
" 40 seconds	.110
	49.392

	h.	m.	sec.
Sid. time at Green. M. N. (Table LXV.)	3	40	51.11
Sid. time at Philada. M. N. . .	3	41	40.502
Sid. equiv. for 8h.	8	1	18.852
" " 17m.		17	2.793
" " 10sec.			10.027
" " 0^s.5sec.			0.501
Sidereal time required . . .	12h.	0m.	12.675sec.

EXAM. 2. Required the sidereal time at St. Petersburg on the 20th of February, 1850, at 11h. 42m. 25sec. mean time, the Berlin sidereal time at mean noon being 22h. 0m. 6.36sec.

	h.	m.	sec.	
St. Petersburg east of Greenwich . .	2	1	16.0	
Berlin " " . .		53	35.5	
St. Petersburg east of Berlin . .	1	7	40.5	$=1^h.128$
Acceleration $= 9^s.8565 \times 1.128 =$			$11^s.117$	
Sid. time at M. N., Berlin . .	22	0	6.36	
" " St. Petersburg	21	59	55.243	
Sid. Equiv. for 11h.	11	1	48.421	
" " 42m. . . .		42	6.900	
" " 25sec. . . .			25.069	
Sidereal time required . .	9h.	44m.	15.633sec.	

EXAM. 3. Required the sidereal time at New Haven on the 10th of May, 1836, at 15h. 3m. 13.5sec., mean time. *Ans.* 18h. 19m. 45.04sec.

PROBLEM IX.

To find the Mean Solar Time corresponding to a given Sidereal Time.

Find, as in the last problem, the sidereal time at mean noon, subtract it from the given sidereal time (adding 24h. if necessary), and the remainder will be the interval of sidereal time from noon. Add together the mean solar equivalents for the hours, minutes and seconds of this interval, taken from table XI., and the sum will be the mean time required.

EXAMPLE. Required the mean solar time at Philadelphia on the 27th of May, 1836, at 1h. 21m. 47.5sec., sidereal time.

	h.	m.	sec.
Sid. time at M. N., Green. . . .	4	20	16.68
Acceleration for Philada. . . .		+	49.39
Sid. time at M. N., Philada. . .	4	21	6.07
Given Sid. time + 24h. . .	25	21	47.5
Sid. interval from noon . . .	21	0	41.43

	h.	m.	sec.
Mean solar equivalent for 21h. . .	20	56	33.579
" " 41sec. .			40.888
" " 0.43sec. .			.429
Mean time required	20	57	14.896

PROBLEM X.

To find, from the Tables, the Moon's Longitude, Latitude, Equatorial Parallax, Semidiameter, and Hourly Motions, in Longitude and Latitude, for a given time.

When the given time is not for the meridian of Greenwich, reduce it to that meridian; and when it is apparent time, reduce it to mean time.

With the mean time at Greenwich, take out from tables XXXII. to XXXVI., the arguments numbered 1, 2, 3, &c., to 20, and find their sums, rejecting the ten thousands in the first nine, and the thousands in the others. The resulting quantities will be the arguments for the first twenty equations of Longitude.

With the same time, and from the same tables, take out the remaining arguments and quantities, entitled Evection, Anomaly, Variation, Longitude, Supplement of the Node, II., V., VI., VII., VIII., IX., and X.; and add the quantities in the column for the Supplement of the Node.

For the Longitude.

With the first twenty arguments of longitude, take, from tables XXXVII. to XLII., the corresponding equations, and place their sum in the column of Evection. Then, the sum of the quantities in this column will be the corrected argument of Evection.

With the corrected argument of Evection, take the Evection from table XLIII., and add it to the sum of the preceding equations. Place the resulting sum in the column of Anomaly. Then, the sum of the quantities in this column will be the corrected Anomaly.

With the corrected Anomaly, take the Equation of the Centre from table XLIV., and add it to the sum of all the preceding equations. Place the resulting sum in the column of variation. Then, the sum of the quantities in this column will be the corrected argument of variation.

With the corrected argument of Variation, take the variation from table XLV., and add it to the sum of all the preceding equations; the result will be the sum of the first twenty-three equations of the Longitude. Place this sum in the column of Longitude. Then, the sum of the quantities in this column will be the Orbit Longitude of the Moon, reckoned from the mean equinox.

Add the Orbit Longitude to the Supplement of the Node. The result will be the argument of the Reduction. It will also be the first argument of Latitude.

With the argument of Reduction, take the reduction from table XLVI., and add it to the Orbit Longitude. Also, with the 19th argument, which is the same as argument N, for the Sun's Longitude, take the Nutation in Longitude, from table XXX., and apply it, according to its signs, to the last sum. The result will be the Moon's true Longitude from the Apparent equinox.

For the Latitude.

Place the sum of the first twenty-three equations of Longitude, taken to the nearest minute, in the column of Arg. II. Then the sum of the quantities in this column will be Arg. II. of Latitude, corrected. The Moon's true Longitude is the 3d argument of Latitude. The 20th argument of Longitude is the 4th argument of Latitude. Convert the degrees and minutes, in the sum of the first twenty-three equations of Longitude, into thousandth parts of the circle, by taking from table L. the number corresponding to them. Place this number in the columns V., VI., VII., VIII., and IX.; but not in column X. Then the sums of the quantities in columns V., VI., VII., VIII., IX., and X., rejecting the thousands, will be the 5th, 6th, 7th, 8th, 9th, and 10th arguments of Latitude.

With the sum of the Supplement of the Node, and the Moon's Orbit Longitude, which is Arg. 1. of Latitude, take the Moon's distance from the North Pole of the Ecliptic, from table XLVII., and with the remaining nine arguments, take the corresponding equations from tables XLVIII., XLIX., and LI. The sum of these ten quantities will be the Moon's true distance from the North pole of the Ecliptic. The difference between this distance and 90°, will be the Moon's true latitude; which will be *north* or *south* according as the distance is *less* or *greater* than 90°.

For the Equatorial Parallax.

With the corrected arguments Evection, Anomaly, and Variation, take the corresponding quantities from tables LII., LIII., and LIV. Their sum will be the Equatorial Parallax.

For the Semidiameter.

With the Equatorial Parallax take the Moon's Semidiameter from table LV.

For the Hourly Motion in Longitude.

With the arguments 2, 3, 4, and 5, of Longitude, rejecting the two right hand figures in each, take the corresponding equations from table LVI. Also, with the correct argument of Evection, take the equation from table LVII.

With the sum of the preceding equations at top, and the correct anomaly at the side, take the equation from table LVIII. Also, with the correct anomaly, take the equation from table LIX.

With the sum of all the preceding equations at the top, and the correct argument of Variation at the side, take the equation from table LX. With the correct argument of Variation, take the equation from table LXI. And, with the argument of Reduction, take the equation from table LXII. These three equations added to the sum of all the preceding ones, will give the Moon's Hourly Motion in Longitude.

For the Hourly Motion in Latitude.

With the 1st and 2nd arguments of Latitude, take the corresponding quantities from table LXIII. and LXIV., and find their sum, attending to the signs. Then 32′ 56″ : the moon's true hourly motion in Longitude : : this sum : the moon's true hourly motion in Latitude. When the sign is affirmative, the moon is tending north; and, when it is negative, she is tending south.

Exam. 1. Required the moon's longitude, latitude, equatorial parallax, semidiameter, and hourly motions in longitude and latitude, on the 6th of August, 1821, at 8h. 46m. 33sec. A. M. mean time at Philadelphia.

	d.	h.	m.	sec.
Mean time at Philadelphia, August,	5	20	46	33
Diff. of Long.		5	0	40
Mean time at Greenwich, August,	6	1	47	13

	1	2	3	4	5	6	7	8	9	10	11	12	13	14	15	16	17	18	19	20
1821	0027	8365	5389	1368	6970	7714	6319	7024	7800	620	917	842	142	979	067	923	331	134	036	036
August	5804	7776	518	838	1134	8874	2233	2742	5148	873	599	938	245	22	471	759	917	503	31	97
6 d.	137	3249	5201	1435	1678	1860	289	1952	121	351	156	352	171	496	153	183	210	130	1	2
1 h.	1	27	43	12	14	16	2	16	1	3	1	3	1	4	1	2	2	1		
47 m. 13 sec.	1	21	34	9	11	12	2	13	1	2	1	2	1	3						
	5970	9438	1185	3662	9807	8476	8845	1747	3071	849	674	137	560	504	692	867	460	768	68	135

	Evection.	Anomaly.	Variation.	Longitude.	Supp. of Node.	II.	V.	VI.	VII.	VIII.	IX.	X.
1821	1s 19° 43′ 47″	8s 9° 54′ 17″	10s 22° 4′ 2″	8s 2° 7′ 41″	0s 13° 3′ 29″	0s 27° 41″	706	711	074	079	637	596
August	7 29 5 59	8 9 46 42	2 4 26 20	9 3 23 46	0 11 13 35	6 24 15	210	371	987	147	597	126
6 d.	1 26 34 57	2 5 19 30	2 0 57 13	2 5 52 55	15 53	1 25 46	170	197	141	169	179	27
1 h.	28 17	32 40	30 29	32 56	8	28	1	2	1	1	1	0
47 min.	22 10	25 35	23 53	25 48	6	22	0	0	0	0	0	
13 sec.	6	7	7	7	0	0	0	0	0	0	0	
Sum of Equa.	36 14	1 47 41	6 1 51	6 32 2		6 32	18	18	18	18	18	
	11 16 51 30	6 27 46 32	3 4 23 35	7 18 55 15	0 24 33 11	9 25 4	105	299	221	414	432	749
			Reduction	. . 3 15	7 18 55 15							
				7 18 58 30	8 13 28 26	Arg. 1 of Latitude.						
			Nutat. in Long.	+ 8								
			Moon's true Longitude	7 18 58 38								

Arg.	Equ.	☽'s Long.	
1	0°	18′	51″
2		1	39
3			20
4		5	42
5		1	36
6		2	39
7		1	39
8			17
9		1	4
10			16
11			14
12			6
13			27
14			20
15			19
16			10
17			8
18			3
19			7
20			17
	0	36	14
Ev.	1	11	27
Sum.	1	47	41
An.	4	13	50
Sum.	6	1	31
Var.		30	31
Sum.	6	32	2

Arg.	Eqs.	☽'s Lat.	
I	94°	43′	57″
II		16	57
☽'s long.			2
20, long.			4
V			45
VI			7
VII			3
VIII			20
IX			18
X			43
	95	3	16
	90	0	0
Moon's Lat.	5	3	16 S.

Arg.	☽'s Eq. Par.	
Ev.	1′	27″
An.	53	0
Var.		4

Moon's Eq. Par. 54 31

Moon's Semidiameter 14′ 51″

Moon's Hourly Motion in Longitude.

Arguments.	Equations.	
2 of long.	0′	5″
3 do.		2
4 do.		1
5 do.		3
Evection	1	19
	1	30
An. & Sum Eqs.		6
Anomaly	27	58
	29	34
Var. & Sum Eqs.		9
Variation		3
Reduction		17
Hourly Motion in Long.	30	3

Moon's Hourly Motion in Latitude.

Arg.	Equations.
I	— 0′ 51″
II	+ 2
Sum.	— 0 49

As 32′ 56″ : 30′ 3″ : : - - 0′ 49″ : — 0′ 45″ = Moon's hourly motion in latitude tending south.

2. Required the moon's longitude, latitude, equatorial parallax, semidiameter, and hourly motions in longitude and latitude, on the 27th of April, 1821, at 9h. 43m. 30sec. P. M. mean time at Baltimore. *Ans.* Long. 11^s 13° 31′ 44″; lat. 6′ 55″ N. equat. par. 60′ 0″; semidiam. 16′ 21″; hor. mot. in long. 36′ 11″; and hor. mot. in lat. 3′ 14″, tending north.

3. What will be the moon's longitude, latitude, equatorial parallax, semidiameter, and hourly motions in longitude and latitude, on the 19th of August, 1822, at 5h. 56m. 14sec. P. M. mean time at Philadelphia? *Ans.* Long. 6^s 3° 7′ 28″; lat. 3° 51′ 35″ S.; equat. par. 56′ 19″; semidiam. 15′ 21″; hor. mot. in long. 32′ 7″; and hor. mot. in lat. 2′ 1″, tending south.

PROBLEM XI.

To find the Moon's Longitude, Latitude, Hourly Motions, Equatorial Parallax, and Semidiameter, for a given Time, from the Nautical Almanac.

Reduce the given time to Mean time at Greenwich. Then,

For the Longitude.

Take from the Nautical Almanac, the two longitudes, for the noon and midnight, or midnight and noon, next preceding the time at Greenwich, and also the two immediately following these, and set them in succession, one under another. Then, having regard to the signs, subtract each longitude from the next following one, and the three remainders will be the *first differences.* Call the middle one A. Subtract each first difference from the following, for the *second differences.* Take the half sum of the second differences, and call it B.

Call the excess of the given time at Greenwich, above the time of the second longitude, T. Then 12h : T : : A : *fourth term,* which must have the same sign as A.

With the time T at the side, take from table LXVI. the quantities corresponding to the minutes, tens of seconds, and seconds of B, at the top, the sum of these, with a *contrary* sign to that of B, will be the *correction of second differences.*

The sum of the second longitude, the fourth term, and the correction of second differences, having regard to the signs, will be the required longitude.

For the Hourly Motion in Longitude.

To the logistical logarithm of $\frac{1}{12}$ of T, add the logistical logarithm of B, and find the quantity corresponding to the sum. Call this quantity E, and prefix to it the same sign as that of B.

Or E may be found without logarithms; thus 12h. : T : : B : E.

Divide the sum of A, ½ B with its sign changed, and E, by 12, and the quotient will be the required hourly motion in longitude.

For the Latitude.

Prefix to *north* latitudes the *affirmative* sign, but to *south* latitudes the *negative* sign, and then proceed in the same manner as for the longitude. The resulting latitude will be *north* or *south*, according as its sign is *affirmative* or *negative.*

Note. The Moon's Declination may be found in the same manner.

For the Hourly Motion in Latitude.

With T, and the values that A and B have, in finding the latitude, find the hourly motion in latitude, in the same manner as directed for finding the hourly motion in longitude. When the resulting hourly motion in latitude is *affirmative*, the moon is tending *north*, and when it is *negative*, she is tending *south.*

For the Semidiameter and Equatorial Parallax.

The moon's semidiameter and equatorial, horizontal parallax, may be taken from the Nautical Almanac with sufficient accuracy by simply pro portioning for the odd time between noon and midnight, or midnight and noon.

EXAM. 1. Required the moon's longitude, latitude, equatorial parallax, semidiameter, and hourly motions in longitude and latitude from the Nautical Almanac, table LXV., on the 4th of May, 1836, at 4h. 30m. 8sec. P. M. mean time at Philadelphia.

		d.	h.	m.	sec.
Mean time at Philadelphia, May, . .		4	4	30	8
Diff. of Longitude	+		5	0	40
Mean time at Greenwich, May, . . .		4	9	30	48

For the Longitude and Hourly Motion in Longitude.

	Longitudes			1st diff.			2d diff.	
3d midn.	260°	32′	39.3″	7	23	46.9	— 2′	34.4″
4th noon	267	56	26.2	7	21	12.5	— 3	15.1
" midn.	275	17	38.7	7	17	57.4		
5th noon	282	35	36.1					
				7	21	12.5	— 2	54.7
					A		B	

h.	T h.	m.	sec.		A		
12 :	9	30	48 : :	7° 21′ 12″.5 :	5° 49′ 46.″8, fourth term.		

Second Longitude	267°	56′	26.2″
Fourth term	5	49	46.8
Cor. for 2d diff.	+		14.4
Moon's true longitude	273	46	27.4

	m.	sec.		
$\frac{1}{12}$ T	47	34	L. L.	1008
B	— 2	54.7	L. L. —	13141
E . . .	— 2	18.5		— 14149

A	7°	21′	12.5″
$\frac{1}{2}$ B, sign changed	+	1	27.3
E	—	2	18.5
	12)7	20	21.3
Hor. mot. in long.		36	41.8

For the Latitude and Hourly Motion in Latitude.

	Latitudes.	1st diff.	2d diff.
3d midn.	— 2° 41′ 22.2″		
		— 33′ 52.6″	
4th noon	— 3 15 14.8		+ 3′ 26.3″
		— 30 26.3	
" midn.	— 3 45 41.1		+ 3 54.9
		— 26 31.4	
5th noon	— 4 12 12.5		
		— 30 26.3	3 40.6
		A	B

h.	T h.	m.	sec.		A	
12 :	9	30	48 : :	— 30′ 26.3″ :	— 24′ 7″.8, fourth term.	

Second latitude	— 3°	15′	14.8″
Fourth term	—	24	7.8
Cor. for 2d diff.	—		18.1
Moon's true latitude	3	39	40.7 S

	m.	sec.			
$\frac{1}{12}$ T	47	34		L. L.	1008
B	3	40.6		L. L.	12127
E	2	54.9			13135

A	—30′	26.3″
$\frac{1}{2}$ B, sign changed	— 1	50.3
E	+ 2	54.9
	12)—29	21.7
Moon's hor. mot. in lat.	— 2	26.8. tending south.

Moon's semidiameter	16′	24.2″
" equatorial parallax	60	11.7

2. Required the moon's longitude, latitude, equatorial parallax, semidiameter, and hourly motions in longitude and latitude, on the 6th of May, 1836, at 1h. 41m. 40sec. P. M. mean time at Greenwich.

Ans. Long. 297° 59′ 57.1″; lat. 4° 54′ 18.6″ S; equat. par. 59′ 15.0″; semidiam. 16′ 8.7″; hor. mot. in long. 35′ 34.9″; hor. mot. in lat. 1′ 13 7″, tending south.

Note 1. When the moon's longitude and latitude are required with great precision, the third and fourth differences should be noticed. To do this, take from the ephemeris, the three longitudes or latitudes, preceding the given time, and the three following it, and find the first, second, third, and fourth differences, as directed in the rule, for the first and second differences. Call the middle first difference, A, the half sum of the two middle second differences, B, the middle third difference, C, and the half sum of the fourth differences, D. Then taking T, equal the excess of the given time above the time of the third longitude or latitude, find the fourth term and the correction for second differences, as directed in the rule.

With the time T, and middle third difference, D, take from table LXVII., the correction for third differences, which, when T is less than 6 hours, must have the same sign as C, but a *contrary* sign, when T is more than 6 hours.

With the time T, and half sum of fourth differences, D, take from table LXVIII., the correction for fourth differences, which must always have the same sign as D.

The sum of the third longitude or latitude, the fourth term and the corrections for second, third, and fourth differences, having regard to the signs of all the quantities, will be the longitude or latitude required.

Note 2. When great precision is required in the moon's parallax and semidiameter, the corrections for second differences should be applied in the same manner as for the longitude or latitude.

Taking the time, as in the first example, let the moon's longitude be required, as corrected for third and fourth differences.

	Longitudes.	1st diff.	2d diff.	3d diff.	4th diff.
3d noon	253° 7′ 3.7″				
		7° 25′ 35.6″			
" midn.	260 32 39.3				
		7 23 46.9	— 1 48.7		
4th noon	267 56 26.2			— 45.7″	
		7 21 12.5	— 2 34.4		+ 5.0″
" midn.	275 17 38.7			— 40.7	
		7 17 57.4	— 3 15.1		+ 6.2
5th noon	282 35 36.1			— 34.5	
		7 14 7.8	— 3 49.6		
" midn.	289 49 43.9				
		7 21 12.5	— 2 54.7	— 40.7	+ 5.6
		A	B	C	D

Third longitude	267° 56′ 26.2″
Fourth term	5 49 46.81
Cor. for 2d diff.	+ 14.31
" 3d "	+ 0.31
" 4th "	+ 0.06
Moon's true longitude	273 46 27.7

PROBLEM XII.

To find the approximate Time of New or Full Moon, for a given Year and Month.

For New Moon.

Take from table XVIII., the mean new moon in January, for the given year, and the arguments I., II., III., and IV. Take from table XIX., as many lunations, and the corresponding arguments I., II., III., and IV., as the number* of the given month exceeds a unit, and add these quantities to the former, rejecting the ten thousands in the first two arguments, and the hundreds in the other two. Take the number of days corresponding to the given month, from the *second* or *third* column of table XX., according as the given year is a *common* or *bissextile* year, and subtract it from

* The numbers for the months are, for January, 1; for February, 2; for March, 3; &c.

the sum, in the column of mean new moon; the remainder will be the tabular time of mean new moon in the given month. If the number of days, taken from table XX., is greater than the sum of the days in the column of mean new moon, as will sometimes be the case, one lunation more than is directed above, with the corresponding arguments, must be added.

With the arguments I., II., III., and IV., take the corresponding equations from table XXI., and add them to the time of mean new moon; the sum will be the *approximate* time of new moon, expressed in mean time at Greenwich.

For Full Moon.

When the time of mean new moon in January of the given year is on or after the 16th, *subtract* from it, and the arguments I., II., III., and IV., a half lunation, with the corresponding arguments, taken from table XIX., increasing when necessary, either or both of the first two of the former by 10,000, and of the two latter by 100; but *add* them, when the time is before the 16th. The result will be the tabular time of mean full moon in January, and the corresponding arguments. Then proceed to find the approximate time of full moon, in the same manner as directed for the new moon.*

EXAM. 1. Required the mean time of new moon in August, 1821, expressed in approximate time at Greenwich.

	M. New Moon.	I.	II.	III.	IV.
	d. h. m.				
1821,	2 17 59	0092	7859	80	78
8 lun.	236 5 52	6468	5737	22	93
	238 23 51	6560	3596	02	71
Days,	212				
August,	26 23 51				
I.	0 54				
II.	2 13				
III.	9				
IV.	10				
August,	27 3 17	Approximate time.			

* When the half lunation and arguments are to be *added*, the addition may be left till the proper number of lunations, with their corresponding arguments, are placed under, and thus make one addition serve.

2. Required the approximate time of full moon in July, 1823, expressed in mean Greenwich time.

	M. New Moon.	I.	II.	III.	IV.
	d. h. m.				
1823,	11 0 20	0304	5787	61	55
½ lun.	14 18 22	404	5359	58	50
6 lun.	177 4 24	4851	4303	92	95
	202 23 6	5559	5449	11	0
Days,	181				
July,	21 23 6				
I.	2 55				
II.	13 7				
III.	5				
IV.	20				
July,	22 15 33	Approximate time.			

3. Required the approximate times of new and full moon in February, 1822, expressed in mean time at Greenwich.

Ans. New moon 21d. 7h. 47m.
Full moon 5d. 17h. 39m.

PROBLEM XIII.

To determine what Eclipses may be expected to occur in any given year, and the times nearly at which they will take place.

For the Eclipses of the Sun.

Take, for the given year, from table XVIII., the time of mean new moon in January, the arguments and the number N.* If the number N differs less than 53, from 0,500, or 1000, an eclipse of the sun may be expected at that new moon. If the difference is less than 37, there *must* be one. When the difference is between 37 and 53, there is a doubt, which can only be removed by calculation.

If an eclipse may or must occur in January, calculate the approximate time of new moon by problem XII., and it will be the time, nearly, at which the eclipse will take place, expressed in mean time at Greenwich. This time may be reduced to the meridian of any other place by problem V.

Look in column N of table XIX., and, excluding the number belonging to the half lunation, seek the first number that, added to the number

* The number N, in this table, designates the sun's mean distance from the moon's ascending node, expressed in thousandth parts of the circle.

N of the given year, will make the sum come within 53, of 0,500, or 1000. Take the corresponding lunations and arguments, and this number N, and add them to the similar quantities for the given year. Take from the *second* or *third* column of table XX., according as the given year is *common* or *bissextile*, the number of days next less than the sum of the days in the column of mean new moon, and subtract it from the time in that column; the remainder will be the tabular time of mean new moon in the month corresponding to the days, taken from table XX. At this new moon an eclipse of the sun may be expected; and if the sum of the numbers N, differs less than 37 from the numbers mentioned above, there *must* be one. Find the time nearly, of the eclipse, by calculating the approximate time of new moon as directed above.

If there are any other numbers in the column N, of table XIX., that, when added to the number N of the given year, will make the sum come within the limit 53, proceed in a similar manner to find the time of the eclipses.

Note. When the time at which an eclipse of the sun will take place is thus found, nearly, and reduced to the meridian of a given place in north latitude, if it comes during the day time, and if the sum of the numbers N, or the number N itself when the eclipse is in January, is a little above 0, or a little less than 500, there is a *probability* that the eclipse will be visible at the given place. When the number N in January, or the sum of the numbers N in other months, is more than 500, the eclipse will seldom be visible in northern latitudes, except near the equator.

For the Eclipses of the Moon.

When the time of new moon in January of the given year is on or after the 16th, *subtract* from it, from the arguments, and the number N, a half lunation, the corresponding arguments, and the number N; but when it is before the 16th, *add* them. The results will be the time of mean full moon in January, and the corresponding arguments, and number N. Proceed to find the times at which eclipses of the moon *may* or *must* occur, exactly as directed for the sun, except that the limits 35 and 25, must be used instead of 53 and 37.

Note. In an eclipse of the moon, when the time is found nearly, and reduced to the meridian of a given place, if it comes in the night, it will be visible at that place.

EXAM. 1. Required the eclipses that may be expected in the year 1822, and the times nearly, at which they will take place.

For the Eclipses of the Sun.

	M. New Moon.	I.	II.	III.	IV.	N.
	d. h. m.					
1822,	21 15 32	0602	7182	78	66	930
1 lun.	29 12 44	808	717	15	96	85
	51 4 16	1410	7899	93	66	15
	31					
Feb.	20 4 16					
I.	7 38					
II.	19 29					
III.	13					
IV.	11					
Feb.	21 7 47	Mean time at Greenwich.				

As the sum of the numbers N comes within 37 of 0, there must be an eclipse.

	M. New Moon.	I.	II.	III.	IV.	N.
	d. h. m.					
1822,	21 15 32	0602	7182	78	66	930
7 lun.	206 17 8	5659	5020	7	94	596
	228 8 40	6261	2202	85	60	526
	212					
August,	16 8 40					
I.	1 24					
II.	0 40					
III.	16					
IV.	14					
August,	16 11 14	Mean time at Greenwich.				

As the sum of the numbers N comes within 37 of 500, there must be an eclipse.

	M. New Moon.	I.	II.	III.	IV.	N
	d. h. m.					
1822.	21 15 32	0602	7182	78	66	930
½ lun.	14 18 22	404	5359	58	50	43
	6 21 10	0198	1823	20	16	887
1 lun.	29 12 44	808	717	15	99	85
	36 9 54	1006	2540	35	15	972
	31					
Feb.	5 9 54					
I.	6 52					
II.	0 20					
III.	4					
IV.	29					
Feb.	5 17 39	Mean time at Greenwich.				

As the sum of the numbers N, although it comes within 35 of 1000, does not come within 25, the eclipse may be considered doubtful. It may, however, be observed, that further calculation by the next problem would show that there will be a small eclipse.

	M. Full Moon.	I.	II.	III.	IV.	N.
	d. h. m.					
1822,	6 21 10	0198	1823	20	16	887
7 lun.	206 17 8	5659	5020	7	94	596
	213 14 18	5857	6843	27	10	483
	212					
August,	1 14 18					
I.	2 14					
II.	19 26					
III.	3					
IV.	26					
August,	2 12 27	Mean time at Greenwich.				

As the sum of the numbers N comes within 25 or 500, there must be an eclipse.

2. Required the eclipses that may be expected in 1823, and the times, nearly, at which they will take place, expressed in mean time at Greenwich. *Ans.* One of the moon on the 26th of January, at 5h. 24m. P. M.; one of the sun on the 11th of February, at 3h. 12m. A. M.; one of the sun on the 8th of July, at 6h. 50m. A. M.; and one of the moon on the 23d of July, at 3h. 33m. A. M.

PROBLEM XIV.

To calculate an Eclipse of the Moon.

Find the approximate time of full moon, by prob. XII., and, for this time, compute the sun's longitude, semidiameter, and hourly motion, and the moon's longitude, latitude, equatorial parallax, semidiameter, and hourly motions in longitude and latitude. Subtract the sun's longitude from the moon's, and call the remainder R. Also, subtract the hourly motion of the sun from that of the moon. Then, as the difference of the hourly motions : the difference between R and VI. signs : : 60 minutes : a correction. The correction, *added* to the approximate time of full moon, when R is *less* than VI. signs, but *subtracted* when it is *greater*, will give the *true* time of full moon for the meridian at Greenwich. Reduce this time to the time at the place for which the calculation is to be made, and call the reduced time T.

For the Semidiameter of the Earth's Shadow.

To the moon's equatorial parallax, add the sun's, which may be taken 9″, and from the sum subtract the semidiameter of the sun. Increase the result by $\frac{1}{50}$ part, and it will be the semidiameter of the earth's shadow, which call S.

For the Inclination of the Moon's Relative Orbit.

To the arithmetical complement of the logarithm of the difference between the hourly motions in longitude of the moon and sun, add the logarithm of the moon's hourly motion in latitude, and the result will be the log. tangent of the inclination, which call I.

Add together the constant logarithm 3.55630, the log. cosine of I., and the arithmetical complement of the log. difference between the hourly motions of the moon and sun, in longitude, rejecting the tens in the index, and call the resulting logarithm R.

For the Time of the Middle of the Eclipse.

Add together the logarithm R, the logarithm of the moon's latitude at the true time of full moon, and the sine of I., rejecting the tens in the index, and the result will be the logarithm of an interval t, in seconds of time, which, *added* to T, when the latitude is *decreasing*, but *subtracted* when its *increasing*, will give the time of the middle of the eclipse.

For the times of Beginning and End.

To the logarithm of the moon's latitude at the true time of full moon, add the log. cosine of I., rejecting the tens in the index, and the result will be the logarithm of an arc, which call c. Call the moon's semidiameter d.

To, and from, the sum of S and d, add and subtract c. Then add together the logarithms of the results, $S + d + c$ and $S + d - c$, divide the sum by 2, and to the quotient add the logarithm R, and the result will be the logarithm of an interval x, in seconds of time, which, subtracted from, and added to, the time of the middle, will give the times of the beginning and end.

Note. If c is equal to, or greater than, the sum of S and d, there cannot be an eclipse.

For the Times of Beginning and End of the Total Eclipse.

To and from the difference of S and d, add and subtract c. Then add together the logarithms of the results, $S - d + c$ and $S - d - c$, divide the sum by 2, and to the quotient add the logarithm R, and the result will be the logarithm of an interval x', in seconds of time, which, subtracted from, and added to, the time of the middle, will give the times of the beginning and end of the total eclipse.

Note. When c is greater than the difference of S and d, the eclipse cannot be total.

For the Quantity of the Eclipse.

Add together the constant logarithm 0.77815, the logarithm of $(S + d - c)$, and the arithmetical complement of the logarithm of d, rejecting the tens in the index, and the result will be the logarithm of the quantity of the eclipse, in digits.

Note 1. In partial eclipses of the moon, the southern part of the moon is eclipsed when the latitude is north, and the northern part when the latitude is south.

2. When the eclipse commences before sunset, the moon rises about the same time the sun sets. To obtain the quantity of the eclipse nearly, at the time the moon rises, take the difference between the time of sunset and the middle of the eclipse. Then, as 1 hour : this difference : : difference between the hourly motion of the moon and sun, in longitude : a fourth term. Add together the squares of this fourth term and of the arc c, both in seconds, and extract the square root of the sum. Use this root

instead of c, in the above rule, and it will give the quantity of the eclipse at the time of the moon's rising very nearly. When the eclipse ends after sunrise in the morning, the quantity at the time of the moon's setting may be found in the same manner, only using sunrise instead of sunset.

3. The relative positions of the earth's shadow and moon, at the time of the eclipse, may be easily represented. Let AB, *Fig.* 42, be a part of the ecliptic, and C the position of the centre of the earth's shadow at the time of full moon. Draw LCK perpendicular to AB, and make CM equal to the moon's latitude at the time of full moon, taken from a scale of equal parts, *above* AB if the latitude is *north*, but *below* if it is *south*. Draw Ma parallel to AB, and make it equal to the difference between the hourly motions of the moon and sun in longitude; and draw ac parallel to LK, *above* Ma, when the latitude is tending *north*, but *below*, when it is tending *south*. Then PQ, drawn through M and c, will represent the moon's relative orbit. Draw CN perpendicular to PQ, meeting it in H. Then will H be the place of the moon's centre at the middle of the eclipse. With the centre C and a radius equal to S, the semidiameter of the earth's shadow, describe the circle LNK, to represent the shadow. With the same centre and a radius equal to $(S + d)$, describe arcs, cutting PQ in D and E, which will be the positions of the moon's centre at the beginning and end of the eclipse. With the centres D, H, and E, and a radius equal to d, the moon's semidiameter, describe circles to represent the moon's disc, at the beginning, middle, and end of the eclipse. When the eclipse is total, describe, with the centre C and a radius equal to $(S - d)$, arcs, cutting PQ in F and G, which will be the positions of the moon's centre at the beginning and end of the total eclipse.

EXAM. 1. Required to calculate, for the meridian of Philadelphia, the eclipse of the moon in July, 1823.

The approximate time of full moon, is July 22, at 15h. 33m.

Sun's longitude at that time, .	3^s	29°	25′	23″	
Do. hourly motion, . . .			2	23	
Do. semidiameter, . . .			15	46	
Moon's longitude, . . .	9	29	24	51	
Do. latitude, . . .			9	10	N.
Do. equatorial parallax, . .			54	1	
Do. semidiameter, . . .	d		14	43	
Do. hor. mot. in long. . .			29	34	
Do. do. in lat. . . .			2	43,	tending north.

		d. h. m. sec.
Approx. time of full moon, July,		22 15 33 0
Correction,		+ 1 11
True time, in mean time at Greenwich,		22 15 34 11
Diff. of Long.		5 0 40
Mean time at Philadelphia,	T	22 10 33 31

As 60m. : 1m. 11sec. : : 2′ 43″ : 3″, the correct. of lat.

Moon's lat. at approx. time,	9′ 10″ N.
Correction,	+ 3
Moon's lat. at true time,	9 13 N.

Moon's equatorial parallax,	54′ 1
Sun's do	9
Sum,	54 10
Sun's semidiameter,	15 46
	38 24
Add	0 46
Semidiam. of earth's shadow,	S 39 10

Moon's hor. mot. less sun's,	1631″	Ar. Co. log. 6.78755
Moon's hor. mot. in lat.	163	log. 2.21219
I	5° 42′	log. tan 8.99974

		3.55630
I	5° 42′	log. cos. 9.99785
Moon's hor. mot. less sun's		Ar. Co. log. 6.78755
		log. R. 0.34170
Moon's lat.	553″	log. 2.74272
I	5° 42′	log. sin 8.99704
t	121 sec. = 2m. 1sec.	log. 2.08146
T	10h. 33m. 31sec.	
Middle,	10h. 31m. 30sec.	

Moon's lat.		log.	2.74272
I		log. cos	9.99785
c . .	$550'' = 9' 10''$ . . .	log.	2.74057
$S + d + c$ . . .	$3783''$ . .	log.	3.57784
$S + d - c$	2683 . . .	log.	3.42862
			2)7.00646
			3.50323
		log. R.	0.34170
$x = \overset{\text{sec.}}{6997} = \overset{\text{h. m. sec.}}{1\ 56\ 37}$		log.	3.84493

	h.	m.	sec.	
Middle, ...	10	31	30	
x	1	56	37	
Beginning, . .	8	34	53	
End, ...	12	28	7	A. M. of 23d day

$S - d + c$ . . .	$2017''$ . . .	log.	3.30471
$S - d - c$ . . .	917 . . .	log.	2.96237
			2)6.26708
			3.13354
		log. R.	0.34170
$x' = \overset{\text{sec.}}{2987} = \overset{\text{m. sec.}}{49\ 47}$		log.	3.47524

	h.	m.	sec.
Middle	10	31	30
x'		40	47
Beginning of the total eclipse,	9	41	43
End do.	11	21	17

			0.77815
$S + d - c$		log.	3.42862
d	$883''$	Ar. Co. log.	7.05404
Digits eclipsed,	18.2	log.	1.26081

2. Required to calculate, for the meridian of Philadelphia, the eclipse of the moon, on the 9th of April, 1838.

	h.	m.
Ans. Beginning at	7	31½
Middle	8	58
End	10	24½

Digits eclipsed 7.2, on northern limb.

PROBLEM XV.

Given the Latitude of a place, to find the logarithms of ρ *cos* ϕ' *and* ρ *sin* ϕ'*, in which* ϕ' *is the Geocentric Latitude and* ρ *the Radius of the Earth, for the given place.*

To the log. cosine and log. sine of the given latitude, add respectively, log. x and log. y, taken from table XVII., with the latitude as the argument, and the sums will be log. ρ cos ϕ' and log. ρ sin ϕ'.

Note. When the logarithms are required to more than five decimal figures, they may be found by the formulæ (App. 51, A and B).

Exam. Required the logarithms of ρ cos ϕ' and ρ sin ϕ', for Philadelphia.*

Lat. 39° 56′ 59″	log. cos 9.88457	log. sin 9.80762
From table VI.,	log. x 0.00060	log. y 9.99770
ρ cos ϕ' . . .	log. 9.88517	ρ sin ϕ' . . log. 9.80532

2. Required the logarithms of ρ cos ϕ' and ρ sin ϕ', for Boston.

Ans. Log. ρ cos $\phi' = 9.86930$

Log. ρ sin $\phi' = 9.82623$

PROBLEM XVI.

To calculate an Eclipse of the Sun for a given place, using the tables of the sun and moon contained in this work.

For quantities independent of the place.

1. Find by prob. XII., the approximate time of new moon, in mean time at Greenwich, and let T represent this time, taken to the nearest whole hour.† Put

---

* The latitudes and longitudes of various places are given in table VI.

† When the approximate time of new moon, if expressed in time at the given place, would be as much as three or four hours *before* or *after* noon, it is generally better to take, for T, the whole hour which is an hour *earlier* or *later*, than the nearest whole hour.

L = moon's longitude,
λ = " latitude, *negative* when *south*,
π = " horizontal parallax,
L' = sun's longitude,
δ' = " semidiameter,
A' = " right ascension,
D' = " declination, *negative* when *south*,
ε = the apparent obliquity of the ecliptic,
E = the equation of time in arc,
H = hour angle at the given place.

2. For the time T, find, by prob. VI., the values of L', δ', ε, A', D', and E, and also the sun's hourly motion. To the value of L' at the time T, add the sun's hourly motion, and the sum will be the value of L' at the time (T + 1 hr.). Also, for the time T, find, by prob. X., the values of L, λ, π, and the moon's hourly motions in longitude and latitude; and then, by means of the hourly motions, find the values of L and λ, for the time (T + 1 hr.).

3. Using the values of the quantities at the time T, to log. A, taken from table IV., add Ar. Co. log. cos D', and call the result log. C. Expressing $(L - L')$ and λ in seconds, to log. C add log. of $(L - L')$, and also, to log. C add log. of λ, and the sums will be respectively the logarithms of two quantities a and b. To log. of a and also to log. of b, add log. tang ε, taken from table III., and log. cos L', and the sums will be respectively the logarithms of two quantities c and d. Attending to the signs of the quantities, subtract d from a, and call the result p; and add b and c together, and call the result q.

To log. B, taken from table IV., add log. tang δ', from table V., and the sum will be the logarithm of a quantity l'. To l' add 2732, and call the sum l. To the logarithm 9.4180 and from it, add and subtract log. sin D', and call the results log. D and log. E. To the logarithm 8.250 add log. cos D', and call the sum log. M.

With the same log. C and log. tang ε, and the values of L, L', and λ, at the time (T + 1 hr.), find, as above, the values of p and q, for this time. Subtract the value of p, at the time T, from its value at the time (T + 1 hr.), and call the remainder p'. Do the same with the values of q, calling the remainder q'. With p' and q', which are the hourly changes of the values of p and q, and which may be regarded as constant during the eclipse, find the values of p and q for the times (T — 1 hr.), (T — 2 hrs.), &c., and for (T + 2 hrs.), &c., by subtracting for the former and adding for the latter, and arrange them in a small table, as in the following example. From the values of p and q thus found for whole hours, their values for any intermediate time may be easily obtained. Multiply 15° by the interval in hours between the time T and noon, the interval being marked *negative* when the time T is in the *forenoon*, and to the product add E. The sum will be the hour angle at Greenwich, at the time T. Call this hour angle H'.

For quantities dependent on the given place.

5. Find, by the last problem, log. ρ cos ϕ' and log. ρ sin ϕ', and, increasing the index of each by 4, call the results log. U and log. V. Then, using the value of D' at the time T, to log. U add log. sin D', and call the sum log. G. To log. V add log. cos D', and the sum will be the logarithm of a quantity f. Add together log. V, log. sin D', and the logarithm 7.668, and the sum will be the logarithm of a quantity a. Subtract a from l, found by Art. 3, and call the result h'. These quantities may be regarded as constant during the eclipse.

6. To H', the hour angle at Greenwich at the time T, add the longitude of the given place, expressed in arc and marked *affirmative* when *east*, but *negative* when *west*, and the sum will be the value of H at the time T. Its value at any other time T', may be found by adding (T' — T). 15°, found either by multiplication or from table XII., to its value at the time T.

To find the approximate time of greatest obscuration.

7. Taking for p, q, and H, their values at the time T, to log. U and log. G add, respectively, log. sin H and log. cos H, and the sums will be the logarithms of two quantities u and g. To log. of u, add log. D, and to log. of g add log. E, and the sums will be the logarithms of two quantities v' and u'. Subtract g from f, and the remainder will be a quantity v.

8. To log. of $(q' - v')$ add Ar. Co. log. of $(p' - u')$, and the sum will be the log. cotangent of an affirmative arc N, less than 180°. To log cot N add log of $(q - v)$, and the sum will be the logarithm of a quantity c. Add together twice log. sin N, log. of $(p - u + c)$, and Ar. Co. log. of $(p - u')$, and the sum will be the logarithm of an interval of time t'. Then will T — t' be the *approximate* time of *greatest obscuration*, in mean time at Greenwich.

To find the true time of greatest obscuration, and approximate times of beginning and end*

9. Taking T′ to represent the approximate time of greatest obscuration or nearly so, find p, q, and H, for this time; and then (Art. 7) find u, v, u', and v'. To log. of u', add log. M, and the sum will be the logarithm of a quantity b. Subtract b from h', and call the remainder h. Find N, as in the last article, and to log. cot N add log. of $(p - u)$, and the sum will be the logarithm of a quantity d. Add together log. sin N, log. of $(d + v - q)$, and Ar. Co. log. of h, and the sum will be the log. cosine of an *affirmative* arc F, less than 180°.

Add together log. cos (N + F), log. of h, and Ar. Co. log. of $(p' - u')$, and the sum will be the logarithm of an interval t. Add together log. cos (N — F), log. of h, and Ar. Co. log. of $(p' - u')$, and the sum will be the logarithm of an interval t'. And add together log. of $(p - u)$ and Ar. Co. log. of $(p' - u')$, and the sum will be the logarithm of an interval t''. Then will $T' - t'' + \frac{1}{2}(t + t')$ be the *true time of greatest obscuration;* $T' - t'' + t$ will be the *approximate time of beginning;* and $T' - t'' + t'$ will be the *approximate time of end.*

To find the quantity of the Eclipse.

10. Add together the constant log. 1.0792, log. of h, Ar. Co. log. of $(h - 2732)$, and twice log. sin ½ F, or twice log. cos ½ F, according as F is *less* or *greater* than 90°, and the sum will be the number of *digits eclipsed;* on the *north* limb when $(d + v - q)$ is *negative*, but on the *south* when it is *affirmative*.

To find the true times of beginning and end.

11. Taking now T′ to represent the approximate time of beginning or nearly so, proceed, as in Art. 9, to find t and t'', omitting the computation of t'. Then will $T' + t - t''$ be the *true time of beginning*, very

* The expression, *true time*, is to be taken here and in the subsequent part of the rule, in a relative sense; as only implying that the time found has an accuracy corresponding with that of the tables, from which the places of the sun and moon have been obtained, and of the number of decimals used in the calculation. With reference to a more exact determination, with more accurate data, they are *near approximate times*. They may frequently be in error to the amount of two or three tenths of a minute; and sometimes, perhaps, to the amount of half a minute.

nearly. Then, taking T′ to represent the approximate time of end, find t' and t'', omitting the computation of t, and $T' + t' - t''$ will be the time of the end.

To find an arc V, *expressing the angular distance, from the sun's vertex, of the point at which the eclipse begins or ends.*

12. With the values of u and v, at the time T′, for beginning, and their hourly changes of value u' and v', at that time, find the value of u and v, at the true time of beginning. Then using these values, to log. of u add Ar. Co. log. of v, and the sum will be the log. tangent of an arc Q, less than 180°, which must have the same sign as u, and which will be numerically *less* or *greater* than 90°, according as v is *affirmative* or *negative.* Then $V = 270° + Q - (N + F)$ will be the distance of the point of beginning from the sun's vertex, reckoned to the *right* or *west* if V is *affirmative,* but in a contrary direction if V is *negative.* Finding, in like manner, u and v, and then Q, for the true time of end, we have $V = 270° + Q - (N - F)$.

If u, v, and Q be found for the true time of greatest obscuration, and we take $V = 270° + Q - N$, when $(d + v - q)$ at the approximate time of greatest obscuration is affirmative, but $V = 90 + Q - N$, when $(d + v - q)$ is negative, then will V express, for the time of greatest obscuration, the angular distance of the moon's centre from the sun's vertex reckoned as before to the *right* or *west.*

ANNULAR OR TOTAL ECLIPSE.

13. If the value of $(d + v - q)$, at the approximate time of greatest obscuration, is numerically less than $(h - 5464)$ or if it is so little greater, that the sum of log. cos N and log. of $(d + v - q)$ is numerically less than log. of $(h - 5464)$, the eclipse will be *total* or *annular; total* when $(h - 5464)$ is *negative,* but *annular,* when it is affirmative.

14. When it is ascertained that the eclipse will be total or annular, take N, F, $(p' - u')$, and t'' as found for the approximate time of greatest obscuration (art. 9), and find t and t', using $(h - 5464)$ instead of h. Then will $T' + t - t''$ and $T' + t' - t''$ be the times at which the eclipse begins and ceases to be total or annular.

Note. The times obtained by the above rules are expressed in mean time at Greenwich. They may be changed to mean time at the given place by Prob. V.

2. The above rule follows from the formulæ for eclipses investigated in the Appendix to Part I.

EXAMPLE.

It is required to calculate for Philadelphia, the eclipse of the sun of May 15th, 1836.

The approx. time of new moon is 15d. 2h. 8m., Greenwich mean time.

At time T = 15d. 2h.

L' = ...	54°	42′	12″
Sun's hourly motion =		2	25
δ' = ...		15	50
ε = ...	23	27	44
A' = ...	52	20	23
D' = ...	+ 18	57	45
E = ...	+	59	3
L = ...	54	38	55
λ = ...	+	25	27
π = ...		54	24
Moon's hor. mot. in long. =		29	57
" " " in lat. =		2	45

At time (T + 1h.) = 15d. 3h.

L' = ...	54°	44′	37″
L = ...	55	8	52
λ = ...	+	28	12

At time T = 2 hrs.

Log. A.	0.44993
D' = 18° 57′ 45″ A. C. cos.	0.02423
Log. C.	0.47416

	Log. C. 0.47416		Log. C. 0.47416
$L - L' = -197''$	log. $2.29447n$	$\lambda = +1527''$	log. 3.18384
$a = -587$	$2.76863n$	$b = +4550$	3.65800
ε . .	log. tang. 9.63752	ε .	log. tang. 9.63752
$L' = 54°\ 42'\ 12''$	log. cos. 9.76178	L .	log. cos. 9.76178
$c = -147$ . .	$2.16793n$	$d = +1141$.	3.05730

$$p = a - d = -1728 \quad ; q = b + c = +4403$$

	Log. B.	5.80184		9.4180	8.250
$\delta' = 15' 50''$	log. tang	7.66330	D′ log. sin	9.5118	D′log. cos 9.976
$l' = 2918$	. .	3.46514	log. D.	8.9298	log. K. 8.226
2732	. . . .		log. E.	9.9062	
$l = 5650$					

At time, (T + 1h.) = 3 hrs.

	Log. C.	0.47416		Log. C.	0.47416
$L - L' = + 1455''$	log.	3.16286	$\lambda = + 1692''$	log.	3.22840
$a = + 4335$		3.63702	$b = + 5042$		3.70256
ε . . .	log. tang	9.63752	ε . . .	log. tang	9.63752
$L' = 54° 44' 37''$	log. cos	9.76136	L′ . . .	log. cos	9.76136
$c = + 1086$		3.03590	$d = + 1263$		3.10144

$p = + 3072$; $q = + 6128$.
$p' = + 4800$; $q' = + 1725$.

d	*h*	*p*	*q*	
15	0	—11328	+ 953	
	1	— 6528	+ 2678	
	2	— 1728	+ 4403	$H' = + 2 \times 15° + E = + 30° 59'$
	3	+ 3072	+ 6128	
	4	+ 7872	+ 7853	

For Philadelphia, Lat. 39° 56′ 59″ N

Log. $\rho \cos \phi' = 9.8852$; Log. $\rho \sin \phi' = 9.8053$
log. U. 3.8852 log. V. 3.8053

	log. U.	3.8852		log. V.	3.8053
D′ . .	log. sin	9.5118	D′	log. cos.	9.9758
	log. G.	3.3970	$f = + 6041$		3.7811
	log. V.	3.805	H′ =		+ 30° 59′
D′ . .	log. sin	9.512	Long. of Phila. =		— 75 10
		7.668	H, at time T =		— 44 11
$a = + 10$ . .		0.985			

$h' = l - a = 5640$

For the *approximate* time of greatest obscuration.

$T = 2$ hrs.; $p = -1728$; $q = +4403$; $H = -44° 11'$

	log. U. 3.8852		log. G. 3.3970
H	log. sin 9.8432 n	H .	log. cos 9.8556
$u = -5350$ . . .	3.7284 n	$g = +1789$	3.2526
	log. D. 8.9298		log. E. 9.9062
$v' = -455$ . . .	2.6582 n	$u' = +1441$	3.1588

$v = f - g = +4252.$

$q' - v' = +2180$	log. 3.3385	N . .	log. sin 9.9237
$p' - u' = +3359$	A. C. " 6.4738	N	log. sin 9.9237
$N = 57° 1'$	log. cot. 9.8123	$p - u + c = +3720$	log. 3.5705
$q - v = +151$	log. 2.1790	$p' - u'$.	A. C. " 6.4738
$c = +98$ . .	1.9913	$t' = +0.78$ . .	9.8917

$T - t' = 1.22$ h. = approx. time of gr. obscur.

For *true* time of greatest obscuration and *approximate* times of beginning and end.

$T' = 1.22$ h.; $p = -5472$; $q = +3057$; $H = -55° 53'$.

	log. U. 3.8852		log. G. 3.3970
H	log. sin 9.9180 n	H . .	log. cos 9.7489
$u = -6356$	3.8032 n	$g = +1399$	3.1459
	log. D. 8.9298 n		log. E. 9.9062
$v' = -541$ . .	2.7330 n	$u' = +1127$	3.0521
			log. K. 8.226
$v = +4642$; $h = 5621$		$b = +19$	1.278

$q' - v' = 2265$	log. 3.3550	N	log. sin 9.9301
$p' - u' = 3673$	A. C. " 6.4349	$d + v - q = 2130$	log. 3.3284
$N = 58° 21'$	log. cot 9.7899	h	A. C. " 6.2502
$p - u = 884$	log. 2.9465	$F = 71° 11'$	log. cos 9.5087
$d = 545$	2.7364		

$N + F = 129° 32'$;	log. cos 9.8038 n	$N - F = -12° 50'$	log. cos 9.9890
h	log. 3.7498	h	log. 3.7498
$p' - u'$	A. C. " 6.4349	$p' - u'$.	A. C. " 6.4349
$t = -0.974$ . .	9.9885 n	$t' = -1.492$ . .	0.1737

$p - u$	log. 2.9465	$T' - t'' + \frac{1}{2}(t + t') =$	$1^{h}.238 = 1^{h}\ 14^{m}.3 =$ *true* time of gr. obs.
$p' - u'$ A.C. "	6.4349	$T' - t'' + t =$	0.005 = *approx.* time of beg.
$t'' = 0.241$	9.3814		
		$T' - t'' + t' =$	2.471 = " " end.

For Quantity of the Eclipse.

		1.0792
h	log.	3.7498
$h - 2732 = 2889$	A. C. "	6.5392
$\frac{1}{2}$ F = 35° 35′ $\frac{1}{2}$	log. sin	9.7649
$\frac{1}{2}$ F	"	9.7649
Digits eclipsed 7.9, on *south* limb,	log.	0.8980

For *true* time of beginning.

$T' = 0^{h}\ 0^{m}$; $p = -11328$; $q = 953$; $H = -74° 11'$

	log. U. 3.8852		log. G. 3.3970
H . . .	log. sin 9.9832 n	H . .	log. cos 9.4355
$u = -7386$	3.8684 n	$g = 680$	2.8325
	log. D. 8.9298		log. E. 9.9062
$v' = -628$	2.7982 n	$u' = 548$	2.7387
			log. K. 8.226
$v = 5361$; $h = 5631$		$b = 9$ . . .	0.965
$q' - v' = 2353$	log. 3.3716	N . .	log. sin 9.9420
$p' - u' = 4252$	A. C. " 6.3714	$d + v - q = 2227$	log. 3.3477
N = 61° 3′	log. cot 9.7430	h . .	A. C. " 6.2494
$p - u = -3942$.	log. 3.5957 n	F = 69° 45′	log. cos 9.5391
$d = -2181$	3.3387 n		
N + F = 130° 48′	log. cos 9.8152 n	$p - u$ . . .	log. 3.5957 n
h . . .	log. 3.7506	$p' - u'$	A. C. " 6.3714
$p' - u'$ . . .	A. C. " 6.3714	$t'' = -0.927$	9.9671 n
$t = -0.865$	9.9372 n		

$T' - t'' + t = 0^{h}.062 = 0^{h}\ 3^{m}.7 =$ true time of beginning.

For the *true* time of end.

$T' = 2\overset{h.}{.}47$; $p = 528$; $q = 5214$; $H = -37° 8'$

	log. U. 3.8852		log. G. 3.3970
H . . .	log. sin 9.7808 n	H . .	log. cos 9.9016
$u = -4634$	. . 3.6660 n	$g = 1989$	3.2986
	log. D. 8.9298		log. E. 9.9062
$v = -394$	. . 2.5958 n	$u' = 1603$	3.2048
			log. K. 8.226
		$b = 27$	1.431

$v = 4052$; $h = 5613$

$q' - v' = 2119$	log. 3.3261	N . .	log. sin. 9.9209
$p' - u' = 3197$ A. C.	" 6.4952	$d + v - q = 2259$	log. 3.3539
$N = 56° 28'$	log. cot. 9.8213	h . .	A. C. " 6.2508
$p - u = 5162$	log. 3.7128	$F = 70° 24'$	log. cos. 9.5256
$d = 3421$.	3.5341		

$N - F = -13° 56'$	log. cos. 9.9870	$p - u$ . . .	log. 3.7128
h	log. 3.7492	$p' - u'$.	A. C. " 6.4952
$p' - u'$ A. C.	" 6.4952	$t'' = 1.614$ . .	0.2080
$t' = 1.704$ . .	0.2314		

$T' - t'' + t' = 2\overset{h.}{.}560 = 2^{h.}\ 33^{m.}.6$ = true time of end.

For V, at beginning.		For V, at end.	
$u = -7352$	log. 3.8664 n	$u = -4490$	log. 3.6522 n
$v = +5322$ A. C.	" 6.2739	$v = +4017$ A. C.	" 6.3961
$Q = -54° 6'$	log. tang 0.1403 n	$Q = -48° 11'$	log. tang. 0.0483 n
$V = 270° + Q - (N + F) = 85° 6'$		$V = 270° + Q - (N - F) = 235° 45'$	

For V, at greatest obscuration.

$u = -6336$ log. 3.8018 n

$v = +4632$ A. C. " 6.3342

$Q = -53° 50'$ log. tang 0.1360

$V = 270° + Q - N = 157° 49'$

Reducing to Philadelphia, mean time, we have,

	h.	m.
Beginning of Eclipse at	7	3.0, A. M.
Greatest obscuration	8	13.6
End . . .	9	32.9

Digits eclipsed 7.9, on south limb.

Eclipse begins	85° 6′,	from vertex to the right
Greatest obscur.	157 49,	" " "
Eclipse ends	124 15,	" " left.

To construct a figure representing the eclipse at the time of greatest obscuration, and showing the points of beginning and end.

With a radius 6, taken from a scale of equal parts, describe a circle VBE, *Fig.* 67, to represent the sun's disc; then, taking a point V, at the top, to represent the vertex, draw the vertical diameter VV′. Make VB, or the angle VSB, equal to the value of the arc V, found for the time of beginning, and B will be the point at which the eclipse commences. Make the arc VBE, equal to the value of V at the end of the eclipse, or, which is the same, make the angle VSE, equal to its supplement to 360°, and E will be the point at which the eclipse ends.

Make the arc VG equal to the value of the arc V, at the time of greatest obscuration, and drawing the diameter GG′, make GD equal to the digits eclipsed, taken from the same scale. Then, as $h - 2732 : 2732 :: 6 :$ a fourth term. Take DM equal to this fourth term, and with the centre M, and radius MD, describe the arc *a*D*b*. Then will *a*D*b*G*a* represent the quantity and position of the part eclipsed at the time of greatest obscuration.

EXAM. 2. It is required to calculate for Philadelphia, an eclipse of the sun that occurred in May, 1854.

	h.	m.
Ans. Beginning at	4	10.8 P. M. mean time
Greatest obscur.	5	26.9
End . .	6	34.0

Digits eclipsed 10¼, on northern limb.

Eclipse begins	147° 52′	from vertex to the right,
Greatest obscur.	117 45	" " " left,
Ends	25 27	" " " "

PROBLEM XVII.

To find a series of places at which an eclipse of the sun will be central.

1. Take for p, q, p', and q' the 10000th parts of the values of these quantities, as found by the last problem, for the time T, and for D′ its value at that time. To log. of q', add Ar. Co. log. of p', and the sum will be the log. cotangent of an affirmative arc N, less than 180°. To log. cot N add log. of p, and the sum will be the logarithm of a quantity d. To log. sin N add log. of $(d - q)$, and the sum* will be the log. cosine of an affirmative arc F, less than 180°.

Find the intervals t, t', and t'' from the formulæ, log. of t = log. cos $(N + F)$ + Ar. Co. log. of p'; log. of t' = log. cos $(N - F)$ + Ar. Co. log. of p'; and log. of t'' = log. of p + Ar. Co. log. of p'. Then, $T - t'' + \frac{1}{2}(t + t')$ will be T′, the time of the middle of the central eclipse for the earth in general; $T - t'' + t$, will be the time of beginning; and $T - t'' + t'$, will be the time of end.

2. To the value of H at the time T, found by the last problem, add $(T' - T) .15°$, and the sum will be the value of H′ at the time T′. From this value of H′, and to it, subtract and add, $\frac{1}{2}(t' - t). 15°$, and the remainder and sum will be the values of H′ at the beginning and end of the central eclipse.

3. To log. sin $(N + F + 180°)$, add log. cos D′, and the sum will be the log. sine of the geocentric latitude ϕ', not exceeding 90°, and *north* or *south*, according as the sine is *affirmative* or *negative*. To ϕ', add the reduction of latitude, taken from table XVI., with ϕ' as the argument, and the sum will be ϕ, the latitude of the place at which the eclipse begins to be central. To log. cot $(N + F)$ add Ar. Co. log. sin D′, and the sum will be the log. tangent of an hour angle H, less than 180°, and *affirmative* when $(N + F)$ is in the *first* or *fourth quadrant*, but *negative* when it is in the *second* or *third*. When H is to be *negative*, it must be taken *greater* than 90°, if its tangent is affirmative, but *less* than 90° if the tangent is *negative*. Subtract H from the value of H′ at the beginning of the central eclipse, and the result will be λ, the longitude of the place at which the eclipse begins to be central. It will be *west* if *affirmative*, but *east* if *negative*.

Using $(N - F)$ instead of $(N + F)$, and taking the value of H′ at the

---

* When this sum, after one 10 has been rejected from the index, is greater than 10, which is the greatest log. cosine, the eclipse cannot be central at any place.

end of the central eclipse, we find, in like manner, the latitude and longitude of the place at which the eclipse ceases to be central.

4. To log. sin D′, add log. tang N, and the sum will be the log. tangent of an arc M′ not exceeding 180°, and with the *same* sign as D′. When M′ is to be *negative*, it must be taken *greater* than 90°, if its tangent is *affirmative*, but *less* than 90°, if the tangent is *negative*. To log. sin D′, add log. cot. N, and the sum will be the log. tangent of an arc N′, less than 90°, and with the same sign as its tangent. Then take S = M′ + N′. Find

log. B′ = log. tang D′ + Ar. Co. log. sin M′

log. C′ = log. cos F + Ar. Co. log. cos D′ + Ar. Co. log. sin N

log. E′ = log. cos D′ + log. sin N + log. cos N + log. of 15 + Ar. Co. log. of p'.

log. G′ = log. sin N + log. sin N + Ar. Co. log. cos N′ + log. of 15 + Ar. Co. log. of p'.

5. Add M′ to the value of H′ at the middle of the central eclipse, and call the sum H″. Add M′ to the value of H at the beginning of the central eclipse and call the sum L′; also add M′ to the value of H at the end of the central eclipse and call the sum L″. The quantities found in the last article and this, may all be regarded as constant.

6. Let L be the value of (H + M′) at any time during the central eclipse. Then, assuming for L any value at pleasure within its limits L and L″, to log. sin L add log. B′, and the sum will be the log. tangent of an arc B, not exceeding 90°, and with the same sign as its tangent. To log. cos B add log. C′, and the sum will be the log. sine of an arc C less than 90° and with the same sign as its sine. Then will B + C be the value of ϕ', the geocentric latitude of the place at which the eclipse is central when L has its assumed value. From ϕ' we find ϕ, as in Art. 3.

Add together log. sin ϕ' and log. E′, and the sum will be the logarithm of an arc E, in degrees and decimals of a degree. Add together log. cos ϕ', log. sin (L — S), and log. G′, and the sum will be the logarithm of another arc G. Then will H″ + E + G — L, be λ, the longitude of the place.

We may thus, by assuming a series of values for L within its limits L′ and L″, find a series of places at which the eclipse will be central.

Note 1. When two assumed values of L differ only in the sign, the corresponding values of B will differ only in the sign, and the values of C will be precisely the same. When two assumed values of L are supplements of each other, the latitudes and values of E will be the same for each. By attention to these circumstances, the computation of a series of places may be considerably shortened.

2. If we subtract M′ from any assumed value of L, the remainder will be the hour angle at the place at which the eclipse is then central.

EXAMPLE.

Required the times and places at which the eclipse of May, 1836, began and ceased to be central; and also a series of other places at which it was central.

From the calculation of the last problem, we have T = 15 d. 2 h. 0 m., D = + 18° 57′.7, and H′ = + 30° 59′; also dividing the values of p and q at the time T, and the values of p' and q', by 10000, we have, $p = -.1728$, $q = +.4403$, $p' = +.4800$, and $q' = +.1725$.

q' . . .	log. 9.2367		N . . .	log. sin 9.9736
p' . . A. C.	" 0.3188		$d - q = -.5024$.	log. 9.7011 n
N = 70° 14′ log. cot	9.5555		F = 118° 13′ .	log. cos 9.6747 n
p . . .	log. 9.2375 n			
$d = -.0621$	8.7930 n		p	log. 9.2375 n
			p' . . A. C.	" 0.3188
			$t'' = -.3600$ . .	9.5563 n

N + F = 188° 27′ log. cos	9.9953 n	N — F = — 47° 59′ log. cos	9.8257
p' . . . A. C. log.	0.3188	p' . . A. C. log.	0.3188
$t = -2.061$ . .	0.3141 n	$t' = 1.395$ . .	0.1445

$T' = T - t'' + \frac{1}{2}(t + t')$ = 2.027 h. = time of middle of central eclipse.
$T - t'' + t$ = 0.299 = " beginning " "
$T - t'' + t'$ = 3.755 = " end " "

H′, at time T, ...	+ 30°	59′
15° × .027	+	24
H′, at middle of centr. eclipse	+ 31	23
15° × 1.728 . . .	25	55
H′, at beginning of centr. eclipse	+ 5	28
H′, " end " "	+ 57	18

For place of beginning.

$N + F + 180° = 8° 27'$.	log. sin 9.1672
D'	log. cos 9.9758
$\phi' = 7° 59'$ N. . . .	9.1430
$N + F = 188° 27'$ . .	log. cot 0.8281
D'	Ar. Co. log. sin 0.4882
$H = - 92° 46'$ . . .	log. tang 1.3163

$\phi = 8° 2'$ N.; $\lambda = H' - H = 98° 14'$ W

For place of end.

$N - F + 180° = 132° 1'$	log. sin 9.8710
D'	log. cos 9.9758
$\phi' = 44° 39'$ N. . .	log. sin 9.8468
$N - F = - 47° 59'$ . .	log. cot 9.9547n
D'	A. C. log. sin 0.4882
$H = 109° 50'$ . . .	log. tang 0.4429n

$\phi = 44° 50'$ N.; $\lambda = - 52° 32'$ E.

For constant quantities.

D' . .	log. sin 9.5118	D' . .	log. sin 9.5118
N' . .	log. tang 0.4445	N . .	log. cot 9.5555
$M' = + 42° 7'$	log. tang 9.9563	$N' = 6° 40'$	log. tang 9.0673

$S = + 48° 47'$

D' .	log. tang 9.5360	F	log. cos 9.6746
M'	A. C. log. sin 0.1735	D' . .	A. C. log. cos 0.0242
	B' 9.7095	N . .	A. C. log. sin 0.0264
			C' 9.7252

D' .	log. cos 9.9758	N .	log. sin 9.9736
N .	log. sin 9.9736	N .	log. sin 9.9736
N .	log. cos 9.5292	N' A. C.	log. cos 0.0029
15 .	log. 1.1761	15 .	log. 1.1761
p' . A. C.	" 0.3188	p' A. C.	" 0.3188
	log. E' 0.9735		log. G' 1.4450

H. at beg. centr. eclipse $= -92° \ 46$
M′ = . . . $+42 \quad 7$

L′ $= -50 \quad 39$

H, at end centr. eclipse $= +109° \ 50$
M = . . . $+ \ 42 \quad 7$

L″ $= 151 \quad 57$

H′, at middle of centr. eclipse $= 31° \ 23'$
M′ = ... $42 \quad 7$

H″ $= 73 \quad 30 = 73°.50$

For places at which the eclipse will be central.

Assume $L = -50°$.

L . log. sin 9.8843n B . log. cos 9.9689
log. B′ 9.7095 log. C′ 9.7252

$B = -21° 26'$ log. tang 9.5938n $C = 29° 38'$ log. sin 9.6941

$\phi' = B + C = 8° \ 12'$ N., $\phi = 8° \ 15'$ N.

ϕ' . . log. sin 9.1542 ϕ' . . . log. cos 9.9955
log. E′ 0.9735 $L - S = -98° 47'$ log. sin 9.9949n
$E = 1°.34$. 0.1277 log. G′ 1.4450

$G = -27°.25$. 1.4354n

$\lambda = H'' + E + G - L = +97°.59 = 97° \ 35'$ W.

Assume $L = +50°$.

Then (*Note*), $B = +21° 26'$, $C = +29° 38'$, $\phi' = 51° 4'$ N., $\phi = 51° 15'$.

$\phi' = +51° \ 4'$ log. sin 9.8909 ϕ' . . log. cos 9.7982
log. E′ 0.9735 $L - S = 1° 13'$ log. sin 8.3270
$E = 7°.32$. 8644 log. G′ 1.4450

$G = 0°.37$ . . 9.5702

$\lambda = 31° \ 11'$ W.

Assume $L = +130°$.

Then (*Note*), $\phi' = 51° \ 4'$ N., $\phi = 51° \ 15'$ N., $E = 7°.32$.

$\phi' = 51° \ 4'$. log. cos 9.7982
$L - S = 81° \ 13'$ log. sin 9.9949
log. G′ 1.4450 $\lambda = -31°.88 = 31° \ 53'$ E.

$G = 17°.30$ . . 1.2381

By assuming for L various other values between its limits L′ and L″, the latitudes and longitudes of a series of places at which the eclipse will be central, as given in the following table, may easily be found. The computation of a part of these may serve as an exercise for the student.

L.		Lat.		Long.	
— 50°	39′	8°	2′ N.	98°	14′ W.
— 50	0	8	15	97	35
— 40	0	12	9	88	14
— 30	0	16	42	80	0
— 20	0	21	44	72	49
— 10	0	27	0	66	30
— 0	0	32	15	60	44
+ 10	0	37	12	55	14
+ 20	0	41	39	49	41
+ 30	0	45	31	43	53
+ 40	0	48	43	37	44
+ 50	0	51	15	31	11
+ 60	0	53	9	24	17
+ 70	0	54	29	17	2
+ 80	0	55	16	9	29
+ 90	0	53	31	1	42 W.
+ 100	0	55	16	6	21 E.
+ 110	0	54	29	14	36
+ 120	0	53	9	23	5
+ 130	0	51	15	31	53
+ 140	0	48	43	41	0
+ 150	0	45	31	50	36
+ 151	57	44	50	52	33 E.

PROBLEM XVIII.

The value of a quantity at three consecutive whole hours, $T - 1$, T, and $T + 1$, being given, to find its value at an intermediate time T', and its hourly variation at that time.

Attending to the signs, subtract the value of the quantity at the time T — 1, from its value at the time T; and its value at the time T, from its value at the time T + 1; and the remainders will be the *first differences.* Subtract the first of these from the second, and the remainder will be the *second difference.* Let $a =$ the value of the quantity at the time T; $b =$ the half sum of the first differences; $c =$ the second difference; and $t =$ the interval between T and T′, expressed in the fraction of an hour, and marked negative when T′ is *earlier* than T. Then the value of the quantity at the time T′, will be

$$a + t.\,b + \frac{t^2}{2}\,c.$$

And the hourly variation of the quantity at the time T′, will be

$$b + t.\,c.$$

EXAMPLE.

Given the moon's declination, on a certain day, as follows: at 10h., D = + 15° 58′ 50″.1; at 11 h., D = 15° 47′ 11″.0; and at 12 h., 15° 35′ 27″.1. Required its value at $10\frac{3}{5}$ h.

	D	1st diff.	2nd diff.
10	+ 15° 58′ 50.1″		
		— 11′ 39.1″	
11	15 47 11.0		— 4.8″
		— 11 43.9	
12	15 35 27.1		

$a = + 15°\ 47'\ 11.0''$, $b = -\ 11'\ 41.5''$, $c = -\ 4.8''$, $t = -\ \frac{3}{5}$.

$t.\,b = +\quad 4\ 40.6$

$\frac{t^2}{2}\,c = -\quad 0.4$

D = + 15 51 51.2, at time T′.

$b = -\ 11'\ 41.5''$

$t.\,c = +\quad 1.9$

— 11 39.6 = hourly variation at time T′.

PROBLEM XIX.

To calculate an Eclipse of the Sun for a given place, obtaining the places, &c., of the moon and sun, from the Nautical Almanac.

1. Let A and A′ represent the right ascensions of the moon and sun, respectively, D and D′, their declinations, marked *negative* when *south*, π and π' their equatorial horizontal parallaxes, and δ the sun's apparent semi-diameter. Also let T be the Greenwich mean time of new moon, taken to the nearest whole hour; other contiguous whole hours, earlier and later being denoted by T — 1, T — 2, &c., and T + 1, T + 2, &c.

For the times T — 2, T — 1, T, T + 1, and T + 2, find the values of A, A′, D, D′, π, and δ'. And for the time T find the value of π', which may be regarded as constant during the eclipse. Also, for the times T — 2, T — 1, &c., find the log. tang δ', log. sin D′, log. cos D, log. A = log. sin D′ + log. tang δ', and log. B = log. cos D′ + log. tang δ'. Arrange the quantities thus found in tables opposite the hours, as in the following example; and find a quantity g', from the formula.

$$\text{log. } g' = \text{Ar. Co. log. sin } (\pi - \pi') + \text{log. sin } \pi + 9.4353665.$$

For the times, T — 2, T — 1, &c. find by the following formulæ the quantities p, q, l, and l'.

$$[+ \log.\sin(D - D'),$$
$$\log. p = \text{Ar. Co.} \log.\sin(\pi - \pi') + \log.\sin(A - A') + \log.\cos D;\ \ \log. a = \text{Ar. Co.} \log.\sin(\pi - \pi')$$
$$\log. b = \log. p + \log.\sin(A - A') + \log.\sin D';\ q = a + \tfrac{1}{2} b. \qquad [+ \log.\text{tang}\ \delta';$$
$$\log. d = \log. p + \log.\sin(A - A') + \log.\cos D';\ \log. c = \text{Ar. Co.} \log.\sin(\pi - \pi') + \log \cos.(D - D')$$
$$l = c - \tfrac{1}{2} d + g';\ l' = c - \tfrac{1}{2} d - g'.$$

From the values of p and q for the whole hours, find by the last problem, their values for the intermediate half hours. The value of l or l', at each half hour may be taken equal to the half sum of its values at the preceding and following whole hours. Arrange the values thus found in a table, placing in columns adjacent to those containing the values of p and q, the differences of their values for each half hour. The values of p and q, for any intermediate time may then be easily obtained by proportion. The values of p' and q', at each of the times T — 1, T, and T + 1, are respectively equal to the sums of the preceding and following differences. Their values at the times T — 2 and T + 2 become known from their hourly changes of value.

Take from the Nautical Almanac, the sidereal time at mean noon at Greenwich, on the day of the eclipse. To this apply the sidereal time corresponding to the interval between Greenwich mean noon and the time T, by *adding* or *subtracting*, according as T is *after* or *before* noon. The result, converted into degrees, will be Z′, the right ascension of the zenith of Greenwich, at the time T. All the preceding quantities are independent of the given place.

2. To the value of Z′ or from it, *add* or *subtract* the longitude of the given place, according as it is *east* or *west*, and the sum or remainder will be Z, the right ascension of the zenith of the given place at the time T. From Z, subtract the value of A′ at the time T, and the remainder will be H, the hour angle at the given place at the time T. Take the difference between the values of A′ at the times T and T + 1, and call it A″. Then the value of H at another time T′ may be found by adding to its value at the time T, $(T' - T) \times 15° + (T' - T) \times (2' 28'' - A'')$. When A″ does not differ more than two or three seconds from 2′ 28″ the second product may be omitted without material error.

Having found for the given place, the logarithms $\rho \cos \phi'$ and $\rho \sin \phi'$, by Prob. XV., the values of u, v, u', v', and h, may be found by the following formulæ :

$$\text{Log. } a = \text{log. } \rho \sin \phi' + \text{log. A}; \qquad \text{log. } f = \text{log. } \rho \sin \phi' + \text{log. cos D}',$$
$$\text{log. } u = \text{log. } \rho \cos \phi' + \text{log. sin H}; \qquad \text{log. G} = \text{log. } \rho \cos \phi' + \text{log. cos H},$$
$$\text{log. } v' = \text{log. } u + \text{log. sin D}' + 9.4180; \text{ log. } g = \text{log. G} + \text{log. sin D}'$$
$$v = f - g,$$
$$\text{log. } u' = \text{log. G} + 9.4180; \qquad \text{log. } b = \text{log. G} + \text{log. B}.$$
$$h = l - (a + b).$$

3. With the values of p, q, u, v, &c., found for the requisite times, make the computation by Arts. 8, 9, &c., of the rule to Prob. XVI., using logarithms to four decimal figures, and natural numbers to three or four decimals. Then, for the times of beginning and end thus found, taken to the nearest hundredth of an hour, repeat the calculation, using logarithms to five or more decimal figures.

When the eclipse is annular or total, the times of its beginning and ceasing to be so, are found in a similar manner, only using l' instead of l.

Note.—The general quantities, whose values are found by the first article, serve not only for calculating the times of beginning, &c., of an eclipse for any place at which it will be visible, but also for the calculations requisite to determine the longitude of a place, from the observed times of beginning and end at that place.*

EXAMPLE.

Let it be required to compute, for Philadelphia, the eclipse of May 15th, 1836.

1. *For quantities independent of the place.*

By the Nautical Almanac, the time of new moon is May 15th, at 2hrs. 6.9m. Taking, therefore, T = 15d. 2h., we find from the Nautical Almanac,† and from tables of sines, tangents, &c., the following quantities:

D. h.	A	A′	A—A′	π	π′	π—π′
15 0	51° 10′ 25″.35	52° 15′ 28″.95	—65′ 3″.60	54′ 25″.60		54′ 17″.12
1	51 40 28 .14	52 17 57 .14	—37 29 .00	54 24 .90		54 16 .42
2	52 10 33 .90	52 20 25 .34	— 9 51 .44	54 24 .21	8.48	54 15 .73
3	52 40 42 .62	52 22 53 .55	+17 49 .07	54 23 .53		54 15 .05
4	53 10 54 .30	52 25 21 .78	+45 32 .52	54 22 .85		54 14 .73

* The values of p, q, u, v, &c., are found by the formulæ in Art. 75 of the Appendix. To find them by the formulæ in (App. 69), the values of a, d, and g, must first be computed (App. 68). For all important eclipses, these and the other general data are given in the Berlin Ephemeris.

† The values of A, A′, D, D′, π, and δ' may be obtained from the part of the Nautical Almanac given in Table LXV. The value of π' is given in a different part of the Almanac.

D. h.	D	D′	D — D′	δ′	log. tang δ′
15 0	19° 1′24.70″N.	18°56′35.90″N.	+ 4′48.80″	15′49.90″	7.6632559
1	19 11 33.01	18 57 11.11	+ 14 21.90	15 49.89	7.6632525
2	19 21 36.10	18 57 46.28	+ 23 49.82	15 49.88	7.6632491
3	19 31 33.96	18 58 21.42	+ 33 12.54	15 49.88	7.6632456
4	19 41 26.60	18 58 56.53	+ 42 30.07	15 49.87	7.6632422

D. h.	log. sin D′	log. cos D′	log. A.	log. B.
15 0	9.51139	9.97582	7.1746	7.6391
1	9.51161	9.97579	7.1749	7.6390
2	9.51182	9.97577	7.1751	7.6390
3	9.51204	9.97574	7.1753	7.6390
4	9.51225	9.97572	7.1755	7.6390

$\pi - \pi'$ Ar. Co. log. sin 1.8017948
π . . log. sin 8.1993348
9.4353665
$g' = +\ 0.273210$ 9.4363961

At time T — 2. or 0hr.

$\pi - \pi'$ Ar. Co. log. sin 1.8016094
A — A′ . log. sin 8.2770142n
D . . . log. cos 9.9756086
$p = -\ 1.13301$ 0.0542322n

$\pi - \pi'$ Ar. Co. log. sin 1.8016094
D — D′ . log. sin 7.1461719
$a = +\ 0.088671$ 8.9477813
$\frac{1}{2}b = +$ 3480
$q = +\ 0.09215$

A — A . . log. sin 8.27701n
D′ . . log. sin 9.51139
$b = +\ 0.006960$ 7.84263

$\pi - \pi'$ Ar. Co. log. sin 1.8016094
D — D′ log. cos 9.9999996
δ' . . log. tang 7.6632559
$c = +\ 0.291652$ 9.4648649
$\frac{1}{2}d = +$ .000046
0.291606
$g' =$ 0.273210
$l =$ 0.56482

p . . . log. 0.0542n
A — A′ . . log. sin 8.2770n
log. B. 7.6391
$d = +\ 0.000093$ 5.9703

Making similar calculations for the other hours, finding the values of p and q for the half hours, by the last problem, and proceeding as directed in the rule, we obtain the following values.

h.m	p.	Diff.	p′	q.	Diff.	q′	l.	l′
0 0	— 1.13301		+ .4807	+ 0.09215		+ .1738	0.56481	0.01840
		+ .24036			+ .08688			
30	— 0.89265			0.17903			.56486	.01844
		.24039			.08681			
1 0	— 0.65226		.4808	0.26584		.1735	.56490	.01848
		.24042			.08674			
30	— 0.41184			0.35258			.56494	.01852
		.24044			.08668			
2 0	— 0.17140		.4809	0.43926		.1733	.56497	.01855
		.24047			.08661			
30	— 0.06907			0.52587			.56500	.01858
		.24049			.08654			
3 0	+ 0.30956		.4810	0.61241		.1730	.56502	.01860
		.24049			.08648			
30	+ 0.55005			0.69889			.56504	.01862
		.24051			.08641			
4 0	+ 0.79056		.4811	0.78530		.1727	.56505	.01863

	h. m. sec.
Sidereal time at mean noon at Greenwich, . . .	3 22 57.98
Add for interval of 2hrs., from tab. X. . .	2 0 16.71
	5 33 17.69
Z′ = ...	83° 19′ 25″.3

For Philadelphia.

Z′ = . . .	83° 19′ 25″.3
Long. of Philada.	— 75 10
Z, at time T, . .	8 9 25.3
A′, " " . .	52 20 25.3
H, at time T, (2hrs.)	— 44 11 0

A″ = 2′ 28″.2

Making now the approximate calculation, the results obtained would be nearly the same as those found in the first example to Prob. XVI. We may, therefore, for finding the times of beginning and end more accurately, take T′ = 0.06h. for the beginning, and T′ = 2.56h. for the end.

For the beginning.

T′ = 0.06 h. = 0 h. 3.6 m.

$p = -1.10417$; $q = 0.10258$; $p' = 0.4807$; $q' = 0.1738$; H = —73°17′0″.

log. ρ sin ϕ' . .	9.8053	log. ρ sin ϕ' . .	9.80532
log. A.	7.1746	D′ . . log. cos	9.97582
a = .00095	6.9799	f = .60414 . .	9.78114
log. ρ cos. ϕ' . .	9.88517	log. ρ cos. ϕ' . .	9.88517
H . log. sin	9.98125 n	H . . log. cos	9.45885
u = — .73522	9.86642 n	log. G.	9.34402
D′ . log. sin	9.5114	D′ . . log. sin	9.51140
	9.4180	g = .07168 . .	8.85542
v' = — .0625	8.7958 n	v = 0.53246	
log. G.	9.3440	log. G.	9.3440
	9.4180	log. B.	7.6391
u' = .0578	8.7620	b = .00096	6.9831

$$h = l - (a + b) = .56290$$

$q' - v' = .2363$ . . . log. 9.37346
$p' - u' = .4229$ . . Ar. Co. " 0.37376

$N = 60° 48' 20''$. log. cot 9.74722
$p - u$ $-.36895$ . . log. $9.56697n$

$d = -.20615$ . . . $9.31419n$

N . . . . . . log. sin 9.94100
$d + v - q = -.22373$. log. 9.34973
h . . . . Ar. Co. " 0.24957

$F = 69° 42' 51''$ . . log. cos 9.54030

$N + F = 130° 31' 11''$. log. cos $9.81272n$
$h =$ . . . . : log. 9.75043
$p' - u'$ . . . Ar. Co. " 0.37376

$t = -.86478$ . . . $9.93691n$

$p - u$ . . . . . log. $9.56697n$
$p' - u'$ . . . Ar. Co. " 0.37376

$t'' = -.87242$. . . $9.94073n$

$T' - t' + t = 0.06764^{h.} = 0^{h.}\ 4^{m.}\ 3.5^{sec.} =$ true time of beginning.

For the end.

$T' = 2.56^{h.} = 2^{h.}.\ 33.6^{m.}$

$p = .09793$; $q = .53625$; $p' = .4810$; $q' = .1731$; $H = -35° 47' 0''$.

log. ρ sin ϕ' . . 9.8053
log. A 7.1752

$a = .00096$. 6.9805

log. ρ cos. ϕ' . . 9.88517
H . . log. sin $9.76695n$

$u = -.44887$. $9.65212n$
D′ . . log. sin 9.5119

9.4180

$v' = -.0382$. $8.5820n$

log. ρ sin ϕ' . . 9.80532
D′ . . log. cos 9.97575

$f = .60404$ 9.78107

log. ρ cos. ϕ' . . 9.88517
H . . log. cos 9.90915

log. G 9.79432
D′ . log. sin 9.51194

$g = .20242$. 9.30626

$v = .40162$

	log. G 9.7943		log. G 9.7943
	9.4180		log. B 7.6390
$u' = .1630$ . .	9.2123	$b = .00271$.	7.4333

$h = .56133.$

$q' - v' = .2113$ . .	log. 9.32490
$p' - u' = .3180$. Ar. Co. "	0.49757
$N = 56° 23' 51''$	log. cot 9.82247
$p - u = .54680$ . .	log. 9.73783
$d = .36333$ . .	9.56030

N	log. sin 9.92059
$d + v - q = .22870$.	log. 9.35927
h . . . Ar. Co. "	0.25078
$F = 70° 9' 49''$.	log. cos 9.53064

$N - F = - 13° 45' 58''$	log. cos 9.98734
h	log. 9.74922
$p' - u'$ . . Ar. Co. "	0.49757
$t' = 1.7145$.	0.23413

$p - u$	log. 9.73783
$p' - u'$ . . . Ar. Co. "	0.49757
$t'' = 1.7195$ . .	0.23540

$T' - t'' + t = 2.555$ h. $= 2$ h. 33 m. 18 sec. $=$ true time of end.

	h.	m.	sec.	
Beginning	7	3	23.5	A. M., Philad'a mean time.
End	9	32	38.0	" " " "

If a more accurate computation of the time of greatest obscuration and of the quantity of the eclipse is desired, let T' = the time before found, taken to the nearest hundredth of an hour, and find the values of p, q, u, v, &c., for this time. The computation may then be made by articles 8 and 10 of the rule to Prob. XVI., using logarithms to five decimal figures, and putting the value of g' found by the first part of the present rule, instead of the number 2732.

PROBLEM XX.

To find the longitude of a place, from the observed mean times of beginning and end of an eclipse of the sun, at that place.

Taking, for the place, an *assumed* longitude as nearly correct as the knowledge of its situation permits, reduce the observed time of beginning to Greenwich time, and for this time, find the equation of time from the Nautical Almanac. Apply the equation of time, according to the direction at the head of its column, to the observed time of beginning, and the result will be the apparent time of beginning. The interval between this time and noon, marked *negative* when the time is *before* noon, will, when converted into degrees, be the hour angle H.

Let T′ = the Greenwich time of beginning, taken to the nearest whole minute; and for the time T′, find, as in the last problem, p, q, p', q', l, u, v, and h, omitting u' and v', which are not required. Then, using p' and q instead of $(p' - u')$ and $(q' - v')$, find the corrected Greenwich time of beginning, as in the last Problem. The difference between this corrected Greenwich time and the observed time of beginning, will be the longitude of the place in time, as deduced from the observed beginning; to the *west* when the observed time is the *earlier* of the two, but to the *east* when it is the *later*. If the longitude thus obtained differs several minutes from the assumed longitude, the calculation should be repeated, taking the longitude obtained for the assumed longitude.

In a similar manner find the longitude from the observed time of the end. The half sum of the two results will the longitude of the place as given by the observations of both beginning and end.

Note 1. When a table of the values of p, q, &c., has not been previously calculated, the values of p, q, and l, may be computed from the formulæ for the time T′ at beginning and the time T′ at the end. The value of p, at the time T′ for beginning, subtracted from its value for the time T′ at the end, will be the change of value during the interval between these two times; from this, the value of p', the hourly change of value of p, may be easily obtained, with sufficient accuracy, by proportion. In the same way, the value of q' may be found from the two computed values of q.

2. When the eclipse has been observed at places whose longitudes are accurately known, corrections of the computed longitude, due to errors in the tables, may be obtained by the method in Art. 82 of the Appendix.

EXAMPLE.

The observed beginning of the eclipse of May 15th, 1836, at Haverford school, latitude 40° 1′ 12″ N., and assumed longitude 5 h. 1 m. 25 sec. W. was at 7 h. 3 m. 24.5 sec. A. M., mean time; and the end, at 9 h. 31 m. 47 sec. Required the longitude.

	d.	h.	m.	sec.
Observed time of beginning . .	14	19	3	24.5
Assumed diff. of long., add . . .		5	1	25
Greenwich time of beginning . .	15	0	4	49.5

	d.	h.	m.	sec.
Observed time of beginning . . .	14	19	3	24.5
Equation of time, add . . .			3	56.05
	14	19	7	20.55
Interval		4	52	39.45

H = — 73° 9′ 52″.

At T′ = 15 d. 0 h. 5 m.

From the table of values in the example in the last problem, we find, $p = -1.09295$; $q = +0.10663$; $p' = +0.4807$; $q' = +0.1738$; $l = .56482$

log. $\rho \sin \phi'$ . . .		9.8059	log. $\rho \sin \phi'$ . .		9.80595
	log. A.	7.1746	D′ . .	log. cos	9.97582
$a = .00096$		6.9805	$f = .60501$.		9.78177
log. ρ cos. ϕ' . . .		9.88473	log. $\rho \cos \phi'$ . .		9.88473
H . .	log. sin	9.98098 n	H .	log. cos	9.46184
$u = -.73402$.		9.86571 n		log. G.	9.34657
			D	log. sin	9.51141
	log. G.	9.3466	$g = .07211$.		8.85798
	log. B.	7.6391	$v = .53290$		
$b = .00097$ . .		6.9857			

$h = .56289$

q'	log.	9.24005	N . .	log. sin	9.97332
p' .	Ar. Co. "	0.31813	$d+v-q = .29649$	log.	9.47201
N = 70° 7′ 19″	log. cot	9.55818	h . .	Ar. Co. "	0.24958
$p-u = -.35893$	log.	9.55501 n	F = 60° 18′ 27″	log. cos	9.69491
$d = -.12978$		9.11319 n			

$N+F=130°\,25'\,46''$	log. cos	$9.81192\,n$	$p-u$	log.	$9.55501\,n$
h	log.	9.75042	p'	Ar. C. "	0.31813
p'	Ar. Co. "	0.31813	$t'=-0.74668$		$9.87314\,n$
$t=-0.7594$		$9.88047\,n$			

	d.	h.	m.	sec.	
$T'-t'+t=$	15	0	4	14.2	= corrected Greenwich time of beginning.
	14	19	3	24.5	= observed time "
		5	0	49.7 W	= longitude, from observed beginning.

	d.	h.	m.	sec.
Observed time of end,	14	21	31	47
Assumed longitude, add		5	1	25
Greenwich time of end	15	2	33	12

	d.	h.	m.	sec.
Observed time of end,	14	21	31	47
Equation of time, add			3	55.96
	14	21	35	42.96
Interval =		— 2	24	17.04
H =		— 36°	4′	16″

At $T'=$ 15 d. 2 h. 33 m.

$p=0.09312$; $q=0.53452$; $p'=0.4810$; $q'=0.1731$; $l=0.56500$

log. $\rho \sin \phi'$		9.8059	log. $\rho \sin \phi'$		9.80595
	log. A.	7.1752	D′	log. cos	9.97575
$a=.0096$		6.9811	$f=.60492$		9.78170

log. $\rho \cos \phi'$		9.88473	log. $\rho \cos \phi'$		9.88473
H	log. sin	$9.76996\,n$	H	log. cos	9.90757
$u=-.45153$		$9.65469\,n$		log. G	9.79230
			D′	log. sin	9.51194
	log. G	9.7923	$g=.20148$		9.30424
	log. B	7.6390	$v=.40344$		
$b=.00270$		7.4313			

$h=.56134$

q' . . . log. 9.23830		N . . log. sin 9.97355	
p' . Ar. Co. " 0.33785		$d+v-q=.06492$ log. 8.81238	
$N=70°\ 12'\ 27''$ log. cot 9.55615		h . . Ar. Co. " 0.25077	
$p-u=.54465$ log. 9.73612		$F=83°\ 45'\ 10''$ log cos. 9.03670	
$d=.19600$ . . 9.29227			

$N-F=-10°\ 32'\ 43''$ log. cos 9.98775	$p-u$. log. 9.73612
h . . . log. 9.74923	p' . A. C. " 0.31785
p' . . A. C. " 0.31785	$t''=1.13232$. 0.05397
$t'=1.13456$ . . 0.05483	

	d.	h.	m.	sec.		
$T'-t''+t=$	15	2	33	8.1		= corrected Greenwich time of end.
	14	21	31	47		= observed time "
		5	1	21.1	W	= longitude, from observed end.
		5	0	49.7	W	= " " beginning.
half sum =		5	1	5.4	W	= longitude from both observations.

Scholium. The longitude thus obtained is subject, however, to the error which results from errors in the tables. But the present eclipse being visible and observed, at many of the European Observatories, as well as in this country, the longitudes of which had been previously ascertained with considerable accuracy, the means have been afforded for correcting this error, by the method in the Appendix (82). C. Rumker, Director of the Hamburg Observatory, computed the principal observations made both in Europe and this country, and thence obtained equations for correcting the errors of the tables. From these, Sears C. Walker, of Philadelphia, has obtained $\varepsilon=-2''.934$, and $\zeta=-7''.198$.* With these values and those of a and b, which are easily found from their expressions (App. 82 y), we obtain (App. 82 z), -15.22 sec., and -4.64 sec., for the corrections to be added to the longitudes found above, from the observed beginning and end respectively. Since the longitudes are west, they are, in accordance with the formula, to be regarded as negative. Hence the corrected longitude deduced from the observed beginning is 5 h. 1 m. 4.92 sec. W., and that from the end, 4 h. 1 m. 25.74 sec. W.;

* Transactions of the American Philosophical Society, vol. VI., new series.

the mean of which, 5h. 1m. 15.33sec. W., is the longitude of Haverford School, as given by the observations.*

The observations of the eclipse made in this country, combined with those made in Europe, afforded favourable means for determining the moon's parallax. The constant of her equatorial horizontal parallax, deduced by S. C. Walker, is 57′ 2″; which, agreeing very nearly with a late determination of its value by Henderson, from an extensive series of meridian observations made at Greenwich, Cambridge, and the Cape of Good Hope, is probably to be regarded as more accurate than 57′ 1″, given in the former part of the work (95).

PROBLEM XXI.

To calculate an Occultation of a fixed star by the moon, for a given place.

1. Let A = moon's right ascension, A′ = star's right ascension, D = moon's declination, D′ = star's declination, A″ = moon's hourly variation in right ascension, D″ moon's hourly variation in declination, π = moon's equatorial horizontal parallax, H′ = star's hour angle for Greenwich, and H = star's hour angle for the given place.

2. Let T = the mean time of conjunction of the moon and star in right ascension, taken to the nearest whole hour; and for the time T, find the quantities p, q, p', and q', from the following formulæ.

$$p = \frac{(A - A') \cos D}{\pi}\,; \quad q = \frac{D - D'}{\pi};$$
$$p' = \frac{A'' \cos D}{\pi}\,; \qquad q' = \frac{D''}{\pi}.$$

The quantities p' and q', which are the hourly variations of p and q, may be regarded as constant. The values of p and q for a time T′, may be found by *adding* to, or *subtracting* from, their values at the time T, the quantities (T ∽ T′). p', and (T ∽ T′). q', respectively, according as T′ is *later* or *earlier* than T.

3. To the logarithm 9.4192, and from it, add and subtract log. sin D′, and the sum and remainder will be respectively two logarithms of D and E,

4. To the sidereal time at mean noon at Greenwich, on the day of the occultation, taken from the Nautical Almanac, add the sidereal time cor-

---

* From the eclipse of September, 1838, Prof. Kendall, who computed the longitudes of various places at which the eclipse was observed, obtained for that of Haverford School, 5h. 1m. 15.0sec. W.—Am. Philos. Trans. vol. VII.

responding to the interval between noon and the time T, taken from table X. From the sum subtract A′, the star's apparent right ascension, and, converting the remainder into degrees, it will be the value of H′, at the time T. To this apply the longitude of the given place by *adding* when it is *east*, but *subtracting* when it is *west*, and the result will be the value of H, at the time T. The value of H at any other time T′, may be found by applying to its value at the time T, the change in the right ascension of the zenith during the interval between T and T′, taken from table XIV.; *adding* if T′ is *later* than T, but *subtracting* if it is *earlier*.

5. Having found the logarithms of $\rho \cos \phi'$ and $\rho \sin \phi'$, for the given place, by Prob. XV., find f and log. G, from the expressions log. f = log. $\rho \sin \phi'$ + log. cos D′, and log. G = log. $\rho \cos \phi'$ + log. sin D′.

6. Taking the values of p, q, and H, at the time T, find the quantities u, v, &c., by the following formulæ.

$$\log. u = \log. \rho \cos \phi' + \log. \sin H; \quad \log. g = \log. G + \log. \cos H;$$
$$\log. v' = \log. u + \log. D \quad ; \quad \log. u' = \log. g + \log. E;$$
$$v = f - g.$$

$$\log. \cot N = \log. (q' - v') + \text{Ar. Co. } \log. (p' - u'),$$
$$\log. d' = \log. \cot N + \log. (p - u); \ \log. \cos F = \log. \sin N + \log. (d + v - q) + 0.5646$$

N and F both to be less than 180°.

$$\log. t = \log. \cos (N + F) + 9.4354 + \text{Ar. Co. } \log. (p' - u');$$
$$\log. t' = \log. \cos (N - F) + 9.4354 + \text{Ar. Co. } \log. (p' - u');$$
$$\log. t'' = \log. (p - u) + \text{Ar. Co. } \log. (p' - u').$$

Then will $T - t'' + t$ = approximate time of immersion,

and $T - t'' + t'$ = " " emersion.

7. Taking T′ to stand for the approximate time of immersion, find p, q, and H, for this time; and proceeding as in the last article, find t and t'', omitting t'. Then $T' - t'' + t$, will be the time of immersion, very nearly. In like manner, finding, for the approximate time of emersion, t' and t'', we have $T' - t'' + t'$ for the time of emersion, very nearly.

8. With the values of u and v at the approximate time of immersion, find Q and V, as in Prob. XVI., art. 12. The arc V will designate the place of immersion, in reference to the moon's vertex, as seen through a telescope that inverts. A similar process will give the place of emersion

EXAMPLE.

Let it be required to calculate for Greenwich, latitude 51° 28′ 39″ N., the occultation of *i*, Leonis, of Jan. 7th, 1836, having given the following data taken from the Nautical Almanac.

	h.	m.	sec.
Mean time of conjunc. in right ascen. . . .	12	12	17
Star's apparent right ascen.	10	23	26.39
Moon's right ascen. at 12h.	10	23	1.27
" " " at 13h.	10	25	3.92
Star's apparent declination, ...	14°	58′	38.8″
Moon's declin. at 12h.	15	35	27.1
" " " 13h.	15	23	38.4
" equat. hor. par. at 12h.		56	3.5

Sidereal time at mean moon, 19h. 4m. 22.41sec.

From these we have, T = 12h.; A — A′ = — 25.12 sec. = (in arc), — 377″; D — D′ = 36′ 48″ = 2208″; A″ = 2 m. 2.65 sec. = 1840″; D″ = —11′ 48″.7 = — 709″.

A — A′ = — 377″	log. 2.5763 *n*		D — D′ = 2208″ .	log. 33440
π = 3363 Ar. Co.	" 6.4733		π . . Ar. Co.	" 6.4732
D = 15° 35′ log. cos	9.9837		*q* = .656 . .	9.8172
p = — .108 .	9.0333 *n*			
A″ = 1840″ .	log. 3.2648		D″ = — 709″	log. 2.8506 *n*
π . . Ar. Co.	" 6.4733		π . . Ar Co.	" 6.4733
D . . . log. cos	9.9837		*q*′ = — .211 .	9.3239 *n*
p′ = .527 . .	9.7218			

		9.4192
D′ = 14° 59′	log. sin	9.4125
	log. D.	8.8317
	log. E.	0.0067

For Greenwich, log. ρ cos φ′ = 9.79526, and log. ρ sin φ′ = 9.89139

log. ρ sin φ′ . .	9.8914	log. ρ cos φ′ . .	9.7953
D′ log. cos	9.9850	D′ log. sin	9.4125
f = .752	9.8764	log. G.	9.2078

At time T = 12h.

	h.	m.	sec.
Sidereal time of mean noon, Greenwich . .	19	4	22.41
Add for interval 12hrs.	12	1	58.28
	7	6	20.69
A′ = ...	10	23	26.39
	— 3	17	5.7
H′ = ...	49°	16′	25.5″
Long. of Greenwich	0	0	0
H, the star's hour ang. at time T, . .	— 49	16′	25.5″

$p = -.108$; $q = .656$; H = — 49° 16′

log. $\rho \cos \phi'$ . . .	9.7953		log. G 9.2078
H .	log. sin 9.8795 n	H .	log. cos 9.8146
$u = -.473$	9.6748 n	$g = .105$ . .	9.0224
	log. D 8.8317		log. E 0.0067
$v' = +.032$.	8.5065 n	$u' = .107$ . .	9.0291
	$v' = .647$		

$q' - v' = -.179$	log. 9.2529 n	N . .	log. sin 9.9638
$p' - u' =$ .420	Ar. Co." 0.3768	$d + v - q = -.165$	log. 9.2175 n
N = 113° 5′	log. cot 9.6297 n		0.5646
$p - u =$ .365	log. 9.5623	F = 123° 51′	log. cos 9.7459 n
$d = -156$	log. 9.1920 n		

N + F = 236° 56′	log. cos 9.7369 n	N — F = — 9° 46′	log. cos 9.9937
	9.4354		9.4354
$p' - u'$	Ar. Co. log. 0.3768	$p' - u'$	Ar. Co. log. 0.3768
$t = -.354$	9.5491 n	$t' = .640$	9.8059

$p - u$ log. 9.5623 $T - t'' + t = 10.78$ h. = *approx.* time of immersion.

$p' - u'$ Ar.Co." 0.3768 $T - t'' + t' = 11.77$ = " " emersion.

$t'' = .869$ 9.9391

At time T′ = 10.78 h.

$p = -.751$; $q = .913$; H = — 67° 37′

log. $\rho \cos \phi'$ . .	9.7953	log. G	9.2078
H . log. sin	9.9660 n	H . . log. cos	9.5807
$u = -.577$	9.7613 n	$g = .061$ . .	8.7885
log. D	8.8317	log. E	0.0067
$v' = -.039$	8.5930 n	$u' = .062$ . .	8.7952

$v = .691$

$q' - v' = -.172$ log.	9.2355 n	N . . log. sin	9.9722
$p' - u' = .465$ Ar. Co. "	0.3325	$d + v - q = -.158$	9.1987 n
N $= 110°\ 18'$ log. cot	9.5680 n	. . .	0.5646
$p - u = -.174$ log.	9.2405 n	F $= 122°\ 57'$ log. cos	9.7355 n
$d = .064$	8.8085		

N + F $= 233°\ 15'$ log. cos	9.7769 n	$p - u$ . . log.	9.2405 n
	9.4354	$p' - u'$ Ar. Co. "	0.3325
$p' - u'$ Ar. Co. log.	0.3325	$t'' = -.374$	9.5730 n
$t' = -.351$	9.5448 n		

$T' - t'' + t = 10^{h}.803 = 10^{h}\ 48^{m}.2$ = time of immersion.

At time $T' = 11.77$ h.

$p = -.229$; $q = .705$; H $= -52°\ 44'$

log. $\rho \cos \phi'$ . .	9.7953	log. G	9.2078
H . log. sin	9.9008 n	H . . log. cos	9.7821
$u = -.497$	9.6961 n	$g = .098$	8.9899
log. D	8.8316	log. E	0.0067
$v' = -.034$	8.5277 n	$u' = .099$	8.9966

$v = .654$

$q' - v' = -.177$ log.	9.2480 n	N . . log. sin	9.9657
$p' - u' = .428$ Ar. Co. "	0.3686	$d + v - q = -.162$ log.	9.2095 n
N $= 112°\ 28'$ log. cot	9.6166 n		0.5646
$p - u = .268$ log.	9.4281	F $= 123°\ 9'$ log. cos	9.7398 n
$d = -.111$	9.0447 n		

N − F $= -10°\ 41'$ log. cos	9.9924	$p - u$ . . log.	9.4281
	9.4354	$p' - u'$ Ar. Co. "	0.3686
$p' - u'$ Ar. Co.	0.3686	$t'' = .626$	9.7967
$t' = .626$	9.7964		

$T' = t'' + t' = 11^{h}.77 = 11^{h}\ 46^{m}.2$ = time of emersion.

At $T' - 10.78$ h.		At $T' = 11.77$ h.	
$u = -.577$ log.	9.7612 n	$u = -.497$ log.	9.6964 n
$v = .691$ Ar. Co. "	0.1605	$v = .654$ Ar. Co. "	0.1844
Q $= -39°\ 52'$ tang	9.9217 n	Q $= -37°\ 14'$ tang	9.8808 n
V $= 3°.1$, to the east.		V $= 116°.5$, to the east.	

PROBLEM XXII.

To find the longitude of a place from an observed occultation of a fixed star by the moon.

1. Let A, A′, &c., be as in the last problem, and $k = .2725$. Using the estimated longitude of the place, reduce the observed mean time of immersion to Greenwich time. Let T stand for this time, and T′ for the same time taken to the nearest tenth of an hour. From the Nautical Almanac, find, for the time T′, by problem XVIII., the values* of A, D, A″, D″, and, by proportion, the value of π; and also take out the values of A′, D′, and the sidereal time of mean noon.

2. With the values of A, D, &c., at the time T′, find the values of p, q, p', and q', from the following formulæ.†

$$p = \frac{(A - A') \cos D}{\pi} \qquad ; c = \frac{D - D'}{\pi}$$

$$\log. B = \log. p + \log. \sin D' + 4.6856$$

$$d = B. (A - A') \qquad ; q = c + \tfrac{1}{2} d$$

$$a' = \frac{A'' \cos D}{\pi} \qquad ; b' = \frac{D''}{\pi}$$

$$c' = B. A'' \qquad ; d' = B. D''$$

$$p' = a' - d' \qquad ; q' = b' + c'$$

3. To the sidereal time at mean noon add the sidereal time corresponding to the interval that T is past noon, taken from table X., and from the sum subtract A′. To the remainder apply the longitude of the place in time, by *adding* if it is *east*, but *subtracting* if it is *west*, and, converting the result into degrees, it will be H, the star's hour angle at the observed time of immersion.

---

* As the values of A and D are given in the almanac for every hour, the values at the time T′ may be obtained, nearly, by proportion; and the values of A″ and D″ by taking the differences between the values of A and D, respectively, at the preceding and following hours. But it is more accurate and but little additional trouble to employ the problem for interpolation.

† By differentiating the expressions for p and q (App. 75), we obtain the following expressions, very nearly, which are in accordance with the rule:

$$p' = \frac{A'' \cos D}{\pi} - \frac{p \sin D'}{206265} .D'';$$

$$q' = \frac{D''}{\pi} + \frac{p \sin D'}{206265} .A''.$$

4. Having found log. $\rho \cos \phi'$ and log. $\rho \sin \phi'$, for the place, by Prob. XV., find u, v, N, F, t and t'' by the following formulæ.

$$f = \rho \sin \phi' \cos D'$$

$$u = \rho \cos \phi' \sin H \quad ; \quad g = \rho \sin \phi' \cos H \sin D'$$

$$v = f - g.$$

$$\cot N = \frac{q'}{p'} \quad ; \quad d = (p - u) \cot N$$

$$\cos F = \frac{(d + v - q) \sin N}{k}$$

$$t = \frac{k \cos (N + F)}{p'} \quad ; \quad t'' = \frac{p - u}{p'}$$

Then will, $T' - t'' + t$, be the corrected Greenwich mean time of the immersion. The difference between this and the observed time, will be the longitude in time; *west,* if the observed time is the *earlier* of the two, but *east,* if it is the *later.*

In a similar manner we deduce the longitude from the observed emersion, except, that, instead of t, we find $t' = \dfrac{k \cos (N - F)}{p'}$. When the immersion and emersion have both been observed, the longitude should be obtained from each, and the mean of the two results be taken.

EXAMPLE.

Suppose the observed immersion of i, Leonis, on Jan. 7th, 1836, at a place in latitude 52° 8′ 28″ N., estimated longitude 0h. 1m. W., was 10h. 45m. 53.3 sec., mean time; required the longitude of the place.

The observed time of immersion reduced to Greenwich time is, T = 10h. 46m. 53.3 sec. Taking T′ = 10.8h. = $10\frac{4}{5}$h., we easily find from the Nautical Almanac,

	h.	m.	sec.		
A =	10	20	33.89	;	D = + 15° 49′ 31.2″
A′ =	10	23	26.39	;	D′ = + 14 58 38.8

A″ = 122.905 sec. = (in arc), 1843.6″ ; D″ = — 700.5″

A — A′ = — 2587.5″ ; D — D′ = 3052.4″ π = 3362.0″

A — A′ . .	log. 3.41288n	D — D′ . .	log. 3.48464
π . Ar. Co.	“ 6.47340	π . Ar. Co.	“ 6.47340
D′ .	log. cos 9.98322	c = .9079	.95804
p = — 7404 .	9.86950n	$\frac{1}{2}$ d = .0012	
D′ . .	log. sin 9.4124	q = .9091	
	4.6856		
	log. B 3.9675n		
A — A′ . .	log. 3.4129n		
d = .0024 .	7.3804		

A″ . . log. 3.26567		D″ . . log. 2.84541n	
π . Ar. Co. " 6.47340		π . Ar. Co. " 6.47340	
D . log. cos 9.98322		$b' = -.2084$ 9.31881n	
$a' = .5276$ 9.72229			

	log. B 3.9675n		log. B 3.9675n
A″ . .	3.2657	D″	2.8454n
$c' = -.0017$	7.2332n	$d' = .0006$.	6.8129

$p' = a' - d' = .5270$; $q' = b' + c' = -.2101$.

	h.	m.	sec.
Sidereal time at mean noon, Greenw. from N. A.,	19	4	22.41
Inter. from noon to time T. { 10h. gives from tab. X. . .	10	1	38.56
46m. " "		46	7.56
53.3.sec. " "			53.45
	5	53	1.98
A′	10	23	26.39
	— 4	30	24.41
Estimated long. W.		1	0.
	— 4	31	24.41
H =	—67°	51′	6″

By Prob. XV., we have log. $\rho \cos \phi' = 9.78888$, log. $\rho \sin \phi' = 9.89538$.

log. $\rho \sin \phi'$		9.89538
D′	log. cos	9.98499
$f = .7592$		9.88037

log $\rho \cos \phi'$ . .		9.78888	log. $\rho \cos \phi'$ . .		9.78888
H	log. sin	9.96671n	H .	log. cos	9.57635
$u = -.5696$		9.75559n	D′ .	log. sin	9.41236
$v = .6993$			$g = .0599$		8.77759

q' . .	log.	9.32243 n	N . . .	log. sin	9.96797
p'	Ar. Co. "	0.27819	$d + v - g = -.1417$	log.	9.15137 n
N = 111° 44′ 10″	log. cot	9.60062 n	k . .	Ar. Co. "	0.56463
$p - u = +.1708$	log.	9.23249 n	F = 118° 53′ 0″	log. cos	9.68397 n
$d = .0681$		8.83311			

N + F = 230° 37′ 10″	log. cos	9.80241 n	$p - u$.	log.	9.23249 n
k	log.	9.43537	p'	Ar. Co. "	0.27819
p'	Ar. Co. "	0.27819	$t'' = -.3241$		9.51068 n
$t = -.3281$		9.51597 n			

	h.	m.	sec.
$T' - t'' + t =$	10	47	45.6
Observed time $=$	10	45	53.3
Long. of place		1	52.3 W.

PROBLEM XXIII.

To find the Heliocentric Longitude and Latitude, and the Radius Vector of Mercury, for a given time.

Reduce the given time to mean time at Greenwich. Take from table LXIX., the mean longitude of Mercury, the longitudes of the aphelion and node, and the arguments II. and III., corresponding to the given year. Under the three former, place the motions for the months, days, hours, minutes, and seconds of the given time, taken from tables LXX. to LXXII.; and under each of the latter, place the number D, in table LXX., corresponding to the given month, and also the number expressing the day of the month, diminished by a unit. Add together the quantities in each column, rejecting 12 signs when either sum in the first three columns admits the rejection, but setting down the whole amount in each of the last two columns. Subtract the resulting longitude of the aphelion from the mean longitude of the planet, and the remainder will be the mean anomaly.

With the mean anomaly as the argument, take the equation of the centre from table LXXIII., and applying it according to its signs to the mean longitude, add to the result the equations II. and III., taken from table LXXIV., with their respective arguments. The sum will be the *orbit longitude* of Mercury.

From the orbit longitude subtract the longitude of the node, and the remainder will be the argument for the latitude; it will also be the argument for the reduction to the ecliptic. With this argument take the reduction from table LXXVI., and apply it according to its sign to the orbit longitude. The result will be the heliocentric ecliptic longitude, reckoned from the mean equinox. With the argument N, found from the solar tables for the given time, take the nutation in longitude from table XXX., and apply it according to its sign to the longitude from the mean equinox, and it will give the longitude from the true equinox.

With the argument of latitude take, from table LXXVIII. the latitude, and also its secular variations. Multiply the secular variation by the number of years,—the given time is subsequent to 1800,—and divide the product by 100. The result added to the latitude taken from the table, will give the correct heliocentric latitude.

With the mean anomaly as the argument, take the radius vector from table LXXV.

EXAM. 1. Required the heliocentric longitude and latitude, and the radius vector of Mercury, on the 13th of June, 1836, at 20h. 12m. 30sec. mean time at Greenwich.

	M. Long.	Aphelion.	Node.	II.	III.
1836.	9s 8° 52′ 11″	8s 14° 53′ 43″	1s 16° 22′ 26″	377	1925
June	8 22 2 29	23	18	152	152
13th.	1 19 6 31	2	1	12	12
20 h.	3 24 37				
12 m.	2 3	8 14 54 8	1 16 22 45	541	2089
30 sec	5	7 23 27 56	8 0 23 32		
	7 23 27 56	11 8 33 48	6 14 0 47		
Eq. Cent.	+ 6 55 27	M. Anom.	Arg. Lat.		
			N		
	8 0 23 23	1836.	842		
Eq. II.	+ 4	June	22		
Eq. III.	+ 5	13 d.	2		
Orbit Long.	8 0 23 32		866		
Reduct.	— 6 16				
	8 0 17 16	Hel. Lat. from table		1° 41′ 29.6″ S.	
Nut.	— 13	Sec. Variation		+ 1.5	
Hel. Long.	8 0 17 3	True Hel. Lat.		1 41 31 S	
		Radius Vector	0.46287.		

2. Required the heliocentric longitude and latitude, and the radius vector of Mercury, on the 17th of November, 1837, at 11h. 29m. 20sec. mean time at Philadelphia. *Ans.* Long. 221° 47′ 1″; Lat. 0° 33′ 55″N. Rad. Vect. 0.44780.

PROBLEM XXIV.

The heliocentric longitude and latitude, and the radius vector of Mercury at a given time being given, to find its geocentric longitude and latitude and its horizontal parallax and semidiameter at that time.

For the Geocentric Longitude.

To the sun's longitude at the given time, found by Prob. VI., add 180° 0′ 20″, and the sum will be the earth's longitude at the time.* Find

* The sun's longitude found from the tables is the apparent longitude as affected by aberration, and is therefore 20″ less than true longitude.

the earth's radius vector, by adding to the radius vector taken from table XXXI., with the sun's mean anomaly as the argument, the perturbations taken from the small table on the same page, with the arguments I., II., and III.

Subtract the longitude of the earth from the heliocentric longitude of the planet; the remainder, if less than 180°, will be the angle of commutation, to be marked *west;* but, if the remainder is greater than 180°, its supplement to 360° will be the angle of commutation, to be marked *east.* Take half the angle of commutation, and subtracting it from 90°, call the remainder A.

Add together the log. cosine of the planet's heliocentric latitude, the logarithm of its radius vector, and the arithmetical complement of the logarithm of the earth's radius vector, rejecting the tens from the index of the sum, and the result will be the log. tangent of the arc B.

To the log. tangent of the difference between the arc B. and 45° add the log. tangent of the arc A, rejecting ten from the index of the sum, and the result will be the log. tangent of an arc C. Subtracting C from A, the remainder will be the angle of elongation, of the same name as the angle of commutation.

If the angle of elongation is *east, add* it to the sun's longitude increased by 20″; but if it is *west, subtract* it from the sun's longitude thus increased; and the sum or remainder will be the true geocentric longitude.

Add together the arcs A and C, and the sum will be the annual parallax. With the elongation, annual parallax, and geocentric latitude as arguments, find the aberration in longitude from table LXXIX., and, applying it to the true longitude, the result will be the apparent longitude.

For the Geocentric Latitude.

Add together the log. tangent of the heliocentric latitude, the log. sine of the elongation, and the arithmetical complement of the log. sine of the commutation, rejecting ten from the index of the sum, and the result will be the log. tangent of the true geocentric latitude, which will be of the same name as the heliocentric latitude.

With the angle of elongation increased by 270°, and the annual parallax and geocentric longitude, each increased by 90°, as arguments, take from table LXXIX., parts I., II., and III., respectively, of the aberration in longitude, and add them together, having regard to their signs. Multiply the sum by the multiplier, taken from a small table at the bottom of page 74, and the product will be the first three parts of the aberration in latitude. Take from the other small table on the same page, part IV. of the aberration in latitude, and add it to the former three, attending to the signs, and the sum will be the aberration in latitude. Considering the true geocentric latitude as affirmative or negative, according as it is north or south, add to it the aberration in latitude, and the result will be the apparent latitude.

For the Horizontal Parallax and Semidiameter.

Add together the constant logarithm 0.93337, the log. sine of the true geocentric latitude, the arithmetical complement of the log. sine of the heliocentric latitude, and the arithmetical complement of the logarithm of Mercury's radius vector, rejecting ten from the index of the sum, and the result will be the logarithm of the horizontal parallax, in seconds.

To the constant logarithm 9.57584 add the logarithm of the horizontal parallax, and the sum, rejecting ten from the index, will be the logarithm of the semidiameter in seconds.

Note 1. The true geocentric longitude and latitude and the horizontal parallax of the planet Venus, may be found in the same manner. The constant logarithm 9.98302 added to the logarithm of the horizontal parallax, and ten rejected from the index of the sum, will give the logarithm of the semidiameter.

2. The geocentric longitude and latitude of a superior planet may also be found in the same manner, except that, in finding the arc B, one ten must be retained in the index of the log. tangent, and the sum of C and A must be taken for the elongation, instead of their difference.

3. At the time of conjunction, the angles of commutation and elongation are each nothing, and, consequently, the geocentric latitude and the parallax cannot be found by the rule. In this case, add the log. cosine of the heliocentric latitude to the logarithm of the planet's radius vector, rejecting ten from the index of the sum, and the result will be the logarithm of the curtate distance of the planet. Take the difference between the curtate distance and the earth's radius vector, or the sum of the two, according as the conjunction is inferior or superior, and add together the arithmetical complement of the logarithm of this difference or sum, the logarithm of the planet's radius vector, and the log. sine of the heliocentric latitude, rejecting ten from the index of the sum, and the result will be the log. tangent of the true geocentric latitude. Also, add the logarithm of the above mentioned difference or sum to the logarithm 0.93337, and the sum will be the logarithm of the parallax.

Exam. 1. Required the geocentric longitude and latitude, the horizontal parallax, and the semidiameter of Mercury at the time given in first example of the last problem.

For the Geocentric Longitude.

Sun's longitude found by Prob. VII. . . .	83° 13′ 14″
Add	180 0 20
Earth's longitude	263 13 34
Sun's anom. 5ˢ 12° 32′ 2″, gives, tab. XXXI., .	1.01592
Arg. I. 999, gives	8
" II. 926, "	4
" III. 924, "	2
Earth's radius vector	1.01606
Heliocentric long. Mercury	240° 17′ 3″
Longitude of Earth	263 13 34
	337 3 29
	360 0 0
Commutation	22 56 31 E
	11 28 15
	90 0 0
	A = 78 31 45

Mercury's hel. lat.	1° 41′ 31″ S	log. cos 9.99981
" rad. vect.	0.46287	log. 9.66546
Earth's " "	1.01606	Ar. Co. " 9.99309
B . . .	24° 28′ 58″	log. tan 9.65836
45° — B	20° 31′ 2″	log. tan 9.57313
A . . .	78 31 45	log. tan 10.69267
C	61 31 52	log. tan 10.26580
Elongation . .	16 59 53 E	

Sun's long. + 20″	83° 13′ 34″
Elongation	16 59 53 E
Mercury's true longitude	100 13 27

Elongation 17° 0′	gives, tab. LXXIX.,	part I.		— 19″
Ann. par. 140 14	"	"	II.	+ 25
Geoc. Long.	"	"	III.	— 6
Aber. in long.				0

Hence the apparent geocentric longitude is in this case the same as the true.

For the Geocentric Latitude.

Hel. lat.	1° 14′ 31″ S	log. tan 8.47038
Elongation	16 59 53	log. sin 9.46589
Commutation	22 56 31	Ar. Co. " 0.40916
True geoc. lat.	1 16 9 S	log. tan 8.34543

Elong.	+ 270° = 287° 0′	Part I.	— 6″
Ann. par.	+ 90 = 230 4	" II.	+ 21
Geoc. long.	+ 90 = 199 13	" III.	+ 3
			+ 18
Multiplier			0.02
			+ 0.36″
Arg. of lat. 194° 1′, gives, part IV.			+ 3.0
Aber. in lat.			+ 3″

True geoc. lat.	— 1° 16′ 9″
Aber. in lat.	+ 0 0 3
Appar. geoc. lat.	1 16 6 S

		C. log.	0.93337
True geoc. lat.	1° 16′ 9″	log. sin	8.34535
Hel. lat.	1 41 31	Ar. Co. "	1.52980
Rad. vect.	0.46287	Ar. Co. log.	0.33454
Hor. parallax	13.9″		1.14306
		C. log.	9.57584
Semidiam.	5.2″		0.71890

2. Required the apparent geocentric longitude and latitude, the horizontal parallax and the semidiameter of Mercury at the time given in the second example of last problem. *Ans.* App. geoc. long. 230° 59′ 53″, app. geoc. lat. 0° 10′ 42″ N.; hor. par. 6.0″; and semidiam. 2.3″.

PROBLEM XXV.

To Calculate a Transit of Mercury.

1. For Greenwich mean noon of the day on which the transit occurs, find the sun's longitude, hourly motion, the apparent obliquity of the ecliptics and Mercury's heliocentric longitude. To Mercury's mean anomaly, add 10′ 14″, the mean hourly motion in anomaly, and the sum will be the

mean anomaly, an hour after noon. With this anomaly take out again the equation of the centre, and adding 10′ 14″ to it, subtract from the sum the equation of the centre at noon. The remainder will be Mercury's hourly motion in longitude, nearly. To the sun's longitude, add 180° 0′ 20″, and the sum will be the earth's longitude. Then, as the difference of the hourly motions of Mercury and the Sun : the difference of their longitudes : : 1 hour : an interval of time. When the earth's longitude is *greater* than Mercury's, *add* this interval to the mean noon at Greenwich, but when it is *less subtract* the interval, and the sum or remainder will be the approximate time of conjunction in longitude.

2. Let T be the approximate time of conjunction, taken to the nearest whole hour, and t an interval of two or three hours. For the times $T - t$, and $T + t$, find the sun's longitude, radius vector and declination, and Mercury's heliocentric longitude, latitude, and the radius vector, and thence the apparent geocentric longitude and latitude. Take half the sum of the sun's mean anomalies at the times $T - t$ and $T + t$, and it will be the mean anomaly at the time T. In like manner find the radius vectors of the earth and Mercury at the time T. With these find the sun's semidiameter and Mercury's equatorial horizontal parallax and semidiameter. Add together the semidiameters of the sun and Mercury, and, expressing the sum in seconds, call it k. To the constant logarithm 7.95071 add the logarithm of the sun's semidiameter in seconds, and the sum, rejecting ten from the index, will be the logarithm of the sun's horizontal parallax.

Take half the sum of the sun's longitude, at the times $T - t$, and $T + t$, and it will be his longitude, at the time T. In like manner, find the sun's tabular mean longitude, at the time T. To this add 2°, and the equation of the equinoxes in right ascension, found from table XXX., with the argument N, at the time $T - t$, and the result will be the sun's mean longitude from the true equinox. With the sun's longitude at the time T, and the obliquity of the ecliptic, find his right ascension. Subtract the right ascension from the corrected mean longitude, and the remainder, converted into time, would be the equation of time.

3. Let L = Mercury's geocentric longitude, λ = her geocentric latitude, π = her horizontal parallax, L' = sun's longitude, A' = his right ascension, D' = his declination, *negative* when south, and ε = apparent obliquity of the ecliptic. Then taking the values of the quantities at the times $T - t$ and $T + t$, respectively, find, for each time, the values of p and q from the following formulæ :

$$\text{Log. } a = \text{log. } (L - L') + \text{log. cos } \varepsilon + \text{Ar. Co. log. cos } D',$$
$$\text{log. } b = \text{log. } \lambda + \text{log. cos } \varepsilon + \text{Ar. Co. log. cos } D',$$
$$\text{log. } c = \text{log. } a + \text{log. tang } \varepsilon + \text{log. cos } L',$$
$$\text{log. } d = \text{log. } b + \text{log. tang. } \varepsilon + \text{log. cos. } L',$$
$$p = a - d;\ q = b + c.$$

Subtract the value of p, at the time $T - t$, from its value at the time $T + t$, and, dividing the result by the number of the hours in $2t$, the quotient will be p'. Do the same with the values of q, to obtain q'.

4. Take T' and T'', to represent the times $T - t$, and $T + t$, respectively, and let $h =$ the difference of the semidiameters of the sun and Mercury. Then, using the values of p and q, at the time T', find N, d, F, t, t', and t'', by the following formulæ; observing that the arc N is to be *negative*, and will, therefore, be between 0° and — 90° when its cotangent is affirmative, but between —90° and —180°, when the cotangent is negative.

$$\text{Log. cot. N} = \text{log. } q' + \text{Ar. Co. log. } p'\text{; log. } d = \text{log. cot N} + \text{log. } p;$$
$$\text{log. cos F} = \text{log. sin N} + \text{log. } (d - q) + \text{Ar. Co. log. } h;$$

The arc F to be affirmative and less than 180°;

$$\text{log. } t = \text{log. cos } (N + F) + \text{log. } h + \text{Ar. Co. log. } p';$$
$$\text{log. } t' = \text{log. cos } (N - F) + \text{log. } h + \text{Ar. Co. log. } p';$$
$$\text{log. } t'' = \text{log. } p + \text{Ar. Co. log. } p'.$$

Then we shall have, in Greenwich mean time,

$T' - t'' + t =$ time of first contact, for the earth's centre,
$T' - t'' + t' =$ " last " " " .

To find auxiliary quantities for computing the effect of parallax on the times of ingress and egress.

5. Let H′ be the hour angle at Greenwich, and find its value for the time of first contact, by adding the equation of time to the mean time, and converting the interval between the resulting apparent time and noon, into degrees. And using the value of D′ at the time T′, find,

$$\text{log. G} = 3.5563 + \text{log. } (\pi - \pi') + \text{log. sin N} + \text{Ar. Co. log. sin F} + \text{Ar. Co. log. } p';$$
$$\text{log. A} = \text{log. G} + \text{log. cos } (N + F)\text{; log. B} = \text{log. G} + \text{log. sin } (N + F)$$
$$\text{log. C} = \text{log. B} + \text{log. cos } D';$$
$$\text{log. tang M}^* = \text{log. B} + \text{Ar. Co. log. A} + \text{log. sin } D'\text{; log. D} = \text{log. A} + \text{Ar. Co. log. cos M};$$
$$m = H' + M.$$

---

* The arc M to be affirmative and less than 180°.

Proceed in a similar manner for the time of last contact, using the same log. G, the value of D′ at the time T″, (N — F) instead of (N + F), and the value of H′, found for the time of last contact.

To find, for a given place, the correction of the time of contact, on account of parallax.

6. Find $\rho \cos \phi'$ and $\rho \sin \phi'$ for the given place, and put $l =$ longitude of the place, *affirmative* if *east*, but *negative* if *west*. Taking the values of the quantities found for the time of ingress, find a, b, and c, from the formulæ,

$$\log. a = \log. \varrho \cos \phi' + \log. D + \log. \sin (l + m)\,;\ \log. b = \log. \varrho \sin \phi' + \log. C\,;$$
$$c = a - b.$$

Then will c be the correction, in seconds, to be *subtracted* from the time of ingress for the earth's centre, to obtain the time at the given place.

In like manner, using the values of the quantities found for the time of egress for the earth's centre, find a correction to be *added* to that time.*

---

* The parallax of Mercury being small, its influence on the time of contact is also small. We may, therefore, with but very slight error, disregard the variations of the quantities u and v, between the times of contact for the earth's centre and a place on the surface.

Taking for p and q their values at the time of contact for the centre, the equations of contact for the centre will be,

$$\left.\begin{aligned} h \cos P &= p \\ h \sin P &= q \end{aligned}\right\} \quad \ldots \quad (A)$$

And taking $c =$ the interval between the times of contact for the centre and a given place on the surface, and taking for distinction P′ instead of P, the equations for a place on the surface will be,

$$\left.\begin{aligned} h \cos P' &= p - u + nc \sin N \\ h \sin P' &= q - v + nc \cos N \end{aligned}\right\} \quad \ldots \quad (B)$$

Multiply the first of equations (A) by cos P, and the second by sin P, and add the products; and, after doing the same with the equations (B), subtract the last sums from the first. We shall thus obtain,

$$h - h \cos (P - P') = u \cos P + v \sin P - nc \sin (P + N).$$

But it is evident that P′ must be very nearly equal to P. We may, therefore, regard $\cos (P - P') = 1$. Hence,

$$nc = \sin (P + N) = u \cos P + v \sin P.$$

Or putting $(P + N) = \mp F$, in which the upper sign is for ingress, and the lower for egress, we have,

$$c = \mp \frac{u \cos (N \pm F) - v \sin (N \pm F)}{n \sin F}.$$

Substituting in this the expressions for u and v, or for their equivalents, x'' and y'' (App. 75 b), and for n, its equal $\frac{p'}{\sin N}$, we obtain, after some reductions of form, the formulæ which are given in the rule for finding the correction on account of parallax.

7. The distance of the place of contact from the north point of the sun's disc, to the *east* if *affirmative*, but to the *west* if *negative*, is, for the ingress, $90° + N + F$, and, for the egress, $90° + N - F$.

If the distance from the vertex is desired, find Q from the following formulæ; in which ϕ is the geographic latitude of the place, and H the hour angle at the place at the time of contact; observing that Q may be taken less than 90° and with the same sign as its tangent, except the value of $(d - f)$ is negative, in which case, the arc Q must be more than 90° and with a sign contrary to that of the tangent.

$$\text{Log. } d = \text{log. tang } \phi + \text{log. cos D}; \text{ log. } f = \text{log. cos H} + \text{log. sin D}';$$
$$\text{log. tang Q} = \text{log. sin H} + \text{Ar. Co. log. } (d - f).$$

The value of Q being found for each contact, we have, for the distance from the vertex at ingress, $V = 90° + N + F - Q$, and, at egress, $V = 90° + N - F - Q$.

Note 1. To find the times of internal contact, take $h =$ the difference between the semidiameters of the Sun and Mercury, and, using this value of h, compute again (art. 4) the values of F, t, and t'. Then will $T' - t'' + t$, and $T' - t'' + t'$, be the times of internal contact for the earth's centre. The corrections on account of parallax will be nearly the sum for the internal, as for the external contacts.

2. A transit of Venus may be computed in a similar manner. Table LXXVII. contains the heliocentric longitudes and latitudes and values of the radius vector at the times of the next two transits.

EXAMPLE.

It is required to calculate the transit of Mercury that will occur on the 8th of May, 1845, and to find the effect of parallax in changing the times of ingress and egress for Philadelphia.

The sun's long. at Greenwich mean noon is	47°	42′	47″
" earth's " " "	227	43	7
Mercury's helio. long.	227	5	7
Sun's hourly motion in long. . . .		2	25
Mercury's		7	19
App. obliq. of ecliptic	23	27	29

$$4' \, 54'' : 38' \, 0'' :: 1^{h} : 7^{h} \, 45^{m}.$$

Hence the time of conjunction in long. is 7h. 45m. P. M., nearly. Taking, therefore, T to denote the 8th day at 8h., let $t = 3$h.

For the time T — t, we find,

Sun's longitude	47° 54′ 51″
" declination	17 11 0 N.
Mercury's app. geoc. longitude . . .	48 6 35
" " latitude . .	7 11 S.
Earth's or sun's radius vector . . .	1.01012
Mercury's " " . .	0.45357

For the time T + t,

Sun's longitude	48° 9′ 20″
" declination	17 15 1 N.
Mercury's app. geoc. longitude . . .	47 57 29
" " " latitude . . .	11 33 S.
Earth's radius vector . .	1.01018
Mercury's " " . . .	0.45423

For the time T, or, without material error, during the transit, we find,

Sun's semidiameter	951.8″
" hor. parallax . . .	8.5
Mercury's semidiameter . . .	5.8
" hor. parallax . . .	15.4
Equation of time . . .	+ 3m. 43sec.

At time T — t'.

L — L′ = 704″	. log.	2.84757	λ = — 431″	. log.	2.63448n
E = 23° 27′ 29″	log. cos.	9.96254	E . .	log. cos.	9.96254
D′ = 17 11	Ar. Co. "	0.01983	D′ .	Ar. Co. "	0.01983
a = 796.36	. .	2.82994	b = — 413.85	.	2.61685n
E . . .	log. tang	9.63743	E . .	log. tang	9.63743
L′ = 47° 54′ 51″	log. cos	9.82623	L′ . .	log. cos	9.82623
c = 196.61	. .	2.29360	d = — 120.37	.	2.08051n

$$p = 796.36\ ;\ q = -217.24$$

At time T + t.

L — L′ = — 711″	. log.	2.85187n	λ = — 693″	. log.	2.84073n
E . . .	log. cos	9.96254	E . .	log. cos	9.96254
D′ = 17° 15′	Ar. Co. "	0.01999	D′ .	Ar. Co. "	0.01999
a = — 682.97	. .	2.83440n	b = — 665.67	.	2.82326n
E . . .	. log. tang	9.63743	E .	log. tang	9.63743
L′ = 48° 9′ 20″	log. cos	9.82419	L′ . .	log. cos	9.82419
c = — 197.70	. .	2.29602n	d = — 192.70		2.28488n

$$p = -490.27\ ;\ q = -863.37$$

Hence we have, $p = -214.44$, and $q = -107.69$

At $T' = T - t' = 5$ hrs.

q'	log. 2.03218n	N . . .	log. sin 9.95116n
p' . Ar. Co.	" 7.66870n	$d - q = 617.18$	log. 2.79042
N = — 116° 39′ 58″	log. cot. 9.70088	$h = 957''.6$	Ar. Co. 7.01882
p	log. 2.90111	F = 125° 10′ 3″	log. cos 9.76040n
$d = 399.94$ . .	2.60199		

N + F = 8° 30′ 5″	log. cos 9.99520	N — F = — 241° 50′ 1″	log. cos 9.67398n
h . .	log. 2.98118	h	log. 2.98118
p' . . Ar. Co.	" 7.66870n	p' . . Ar. Co.	" 7.66870n
$t = -4.4165$.	0.64508n	$t' = 2.1080$ . .	0.32386

p . . log. 2.90111
p' . . Ar. Co. 7.66870
$t'' = -3.7137$ 0.56981

	h.	m.	sec.	
$T' - t'' + t =$	4	17	50	= Greenwich time of first contact.
$T' - t'' + t' =$	10	49	18	= " " last "

For Ingress.

		2.5563
$\pi - \pi' = 6''.9$		0.8388
N	log. sin	9.9512n
F	Ar. Co. "	0.0875
p' . .	" " log.	7.6687n
	log. G	2.1025

	log. G 2.1025		log. G 3.1025
N + F = 8° 30′	log. cos 9.9952	N + F .	log. sin 9.1697
	log. A 2.0977		log. B 1.2722
		D′ = 17° 11′	log. cos. 9.9802
	log. B 1.2722		log. G 1.2524
D′ . .	log. sin 9.4705		
A . .	Ar. Co. log. 7.9023		log. A 2.0977
M = 2° 32′	log. tang 8.6452	M .	Ar. Co. cos 0.0004
H′ = 65 23			log. D 2.0981
m = 67 55			

For Egress.

	log. G 2.1025		log. G 2.1025
$N - F = -241° 50'$	log. cos 9.6740n	$N - F$	log. sin 9.9453
	log. A 1.7765n		log. B 2.0478
		$D' = 17° 15'$	log. cos. 9.9800
			log. C 2.0278
	log. B 2.0478		
D'	log. sin 9.4721		log. A 1.7765n
A . . .	Ar. Co. log. 8.2235n	M .	Ar. Co. cos 0.0581n
$M = 151° 1'$	log. tang 9.7434n		log. D 1.8346
$H' = 163 15$			
$m = 314 16$			

For correction of ingress at Phila.

log. $\rho \cos \phi'$ . .	9.8852	log. $\rho \sin \phi'$.	9.8053
	log. D 2.0981		log. C 1.2524
$l + m = -7° 15'$	log. sin 9.1011n	$b = 11.4$ sec.	1.0577
$a = -12.1$	1.0844n		

$c = a - b = -23$ sec.

	h.	m.	sec.
First contact at Phila.	4	18	13, Greenwich time.

For correction of egress.

log. $\rho \cos \phi'$ . .	9.8852	log. $\rho \sin \phi'$.	9.8053
	log. D 1.8346		log. C 2.0278
$l + m = 239° 6'$	log. sin 9.9335n	$b = 68.1$ sec.	1.8331
$a = -45.1$ sec. .	1.6533n		

	h.	m.	sec.
Last contact at Phila.	10	47	25, Greenwich time.

With the values of H at the times of ingress and egress, which are respectively, $H = -9° 47'$, and $H = 88° 5'$, we easily find by means of the formula, at the ingress $Q = -18° 28'$, and at egress, $Q' = 51° 40'$. With these and the values of N and F, the values of V become known.

Reducing the times found to Philadelphia time, we have,

	h.	m.	sec.
First contact, at	11	17	33 A. M. mean time.
Last "	5	46	45 P. M. " "

First contact	117°,	from vertex, to the east.
Last "	156½	" " " west.

Scholium. The *corrections* of the times of contact on account of parallax, obtained as above, may be regarded as very nearly true. But the times of contact obtained for the earth's centre, and consequently those for a given place, cannot be depended on as equally correct; as an error of three or four seconds in the longitude of the sun or Mercury, may produce an error of a minute in time.

PROBLEM XXVI.

To correct the observed altitude of a heavenly body on account of Refraction.

With the given altitude, take the corresponding mean refraction* from table VII., and subtract it from the altitude. The remainder will be the corrected altitude, very nearly.

If greater accuracy is desired and the states of the barometer and Fahrenheit's thermometer have been observed, take from the table, the numbers corresponding to the given altitude, that are in the two columns following that of the mean refraction. Multiply the first of these by the number of inches in the height of the barometer, *less* 30, and the second by 50, *less* the number of degrees in the height of the thermometer. The products will be the corrections of the refraction in seconds, depending on the states of the barometer and thermometer respectively. Add these, attending to their signs, to the mean refraction, and the result will be the true refraction; which being subtracted from the observed altitude, gives the correct altitude.

EXAMPLES.

1. The observed altitude of a body being 35° 25′ 35″, what is its altitude, corrected for mean refraction?

Observed altitude . . .	35°	25′	35″
Mean refraction from table . .		1	21.7
Corrected altitude . .	35	24	13.3

2. The observed altitude of a star, when the barometer stood at 30.5 inches, and the thermometer, at 62°, was 15° 6′ 30″. Required the corrected altitude.

M. Refrac.	3′ 32.8″	Bar. 7″.15;	Ther.	0″.422
Cor. for bar. +	3.6	+ 5.		— 12
" " ther. —	5.1	+ 3.575		— 5.064
True refrac. 3	31.3			

* The mean refraction is that which corresponds to a height of 30 inches of the barometer and 50° of Fahrenheit's thermometer.

Observed altitude . .	15°	6′	30″
Refraction . . .		3	31.3
Corrected altitude . .	15	2	58.7

PROBLEM XXVII.

From the observed altitude of a star, or the under or upper limb of the sun, to obtain the true altitude.

For a star. The observed altitude, corrected for refraction by the last problem, gives the true altitude of the star.

For the sun. Correct the observed altitude for refraction, by the last problem. Find the sun's semidiameter by Prob. VI., or take it from the Nautical Almanac or other ephemeris in which it is given. Then, if the lower limb was observed, add the semidiameter to the corrected altitude; but if the observation was on the upper limb, subtract the semidiameter; and the result will be the altitude of the centre, corrected for refraction. To this, add the parallax in altitude, taken from table VIII., and the sum will be the true altitude.

EXAMPLES.

1. Suppose the observed altitude of the sun's *lower* limb at a certain time was 18° 48′ 5″; the barometer standing at 29.7 inches, the thermometer at 70°, and the sun's semidiameter being 15′ 47″.4. Required the true altitude.

Observed altitude of lower limb . .	18°	48′	5″
Refraction, found by last prob. .		2	41.1
	18	45	23.9
Sun's semidiameter, add		15	47.4
	19	1	11.3
Sun's parallax in alt. from tab. VIII. .			8.1
True altitude	19	1	19.4

2. The observed altitude of the sun's *upper* limb being 21° 7′ 12″, the barometer 30.3 inches, the thermometer 40°, and the sun's semidiameter 16′ 17″.2; required the true altitude.

Ans. 20° 48′ 28.9″.

PROBLEM XXVIII.

To find the apparent right ascension and declination of any of the stars in the small catalogue, tab. IX., *for a given day.*

1. *To find the Variations in mean right ascension and declination.*

Reduce the months and days of the given time to the decimal of a year, by means of the small table at the foot of the second page of table IX., and annexing it to the years, find the interval between this time and the date of the table, marking the interval negative when the given time is prior to that date. Take from the table the annual variations of the given star, and multiplying each by the interval, the products will be the variations of the mean right ascension and declination, respectively.

2. *To find the Aberrations.*

Find L′, the sun's longitude, for the given day, by Prob. VI., or take it from an ephemeris, and take from tab. IX., the values of ϕ, log. m, θ, and log. n, for the given star. Then

$$\text{log. } (\textit{aber. in right ascen.}) = \text{log. } m + \text{log. sin } (\text{L}' + \phi),$$
$$\text{log. } (\textit{aber. in decl.}) = \text{log. } n + \text{log. sin } (\text{L}' + \theta).$$

3. *To find the Nutations.*

Find N, the mean longitude of the moon's ascending node, for the given day, by taking the supplement of the node, obtained as in Prob. X., from $12^s\ 0°\ 7'$, or take it from an ephemeris. Take from tab. IX., the values of ϕ', log. m', θ', and log. n', for the given star. Then

$$\text{log. } (\textit{nut. in right ascen.}) = \text{log. } m' + \text{log. sin } (\text{N} + \phi'),$$
$$\text{log. } (\textit{nut. in decl.}) = \text{log. } n' + \text{log. sin } (\text{N} + \theta').$$

4. Attending to the signs, add to the mean right ascension of the star, given in the table, the variation, aberration and nutation in right ascension, and the sum will be the apparent right ascension. In like manner, find the apparent declination, observing that the declination is regarded as *negative* when it is *south*, and *positive* when it is *north.*

EXAMPLES.

1. Required the apparent right ascension and declination of α Bootis, (*Arcturus,*) the 1st of May, 1837; the sun's longitude, at that time, being 40° 52′, and the mean longitude of the node 31° 14′.

Given time .	1837, May 1st =	1837.329
		1850.
Interval		— 12.671

	sec.		
An. Var. in R. Asc.	+ 2.733	An. Var. in Decl.	— 18.95″
Multiply by,	— 12.67	Multiply by,	— 12.67
Var. in M. R. Asc.	— 34.63	Var. in M. Decl.	+ 240.1

	log. m 0.1335		log. n 1.0968
L′ + ϕ = 96° 31′	log. sin 9.9972	L′ + θ = 339° 4′	log. sin 9.5530 n
Aber. R. Asc. = + 1.35sec.	0.1307	Aber. in Decl. = — 4″.5	0.6498 n

	log. m' 9.9932		log. n' 1.8810
N + ϕ' = 200° 7′	log. sin 9.5365 n	N + θ' = 351° 8′	log. sin 9.1879 n
Nut. in R. Asc. = — 0.34sec	9.5297 n	Nut. in Decl. = — 1.2″	0.0689 n

	h. m. sec.		
Mean R. Asc. 1840,	14 8 49.22	Mean Decl. 1840,	19° 57′ 56.1″
Var. . . .	— 34.63	Var. . .	+ 4 0.1
Aber.	+ 1.35	Aber. . .	— 4.5
Nut. . . .	— 0.34	Nut. . .	— 1.2
App. R. Ascen.	14 8 15.60	App. Declin.	20 1 50.5

2. Required the apparent right ascension and declination of α Leonis, (*Regulus*), the 19th of August, 1842; the sun's longitude being 146° 2′, and the mean longitude of the moon's ascending node 288° 43′.

h. m. sec.
Ans. App. R. Asc. 9 59 59.02
App. Declin. 12° 44′ 2.3″.

3. Required the apparent right ascension and declination of β Libræ, the 12th of October, 1853.

h. m. sec.
Ans. App. R. Asc. 15 9 6.29
App. Declin. — 8° 50′ 15.9.″

PROBLEM XXIX.

To find the Latitude of a place, having given the corrected altitude of a star, its apparent right ascension and declination, and the mean time of observation.*

Find the sidereal time corresponding to the given mean time, by Prob. VII., or obtain it from an ephemeris. Take the difference between this time and the star's apparent right ascension; and if the difference exceeds 12 hours, subtract it from 24 hours. The result, converted into arc, will be the distance of the star from the meridian. Call this distance H, the star's apparent declination, regarded *affirmative* whether north or south, D, and the corrected altitude A. Then, find two arcs B and C, neither of them exceeding 90°, from the formulæ.

log. tang B = log. cot D + log. cos H, or log. sin (H — 90°) when H exceeds 90°.
log. sin C = Ar. Co. log. sin D + log. sin A + log. cos B.

When H, the star's distance from the meridian, exceeds 90°, the sum of B and C, is the latitude of the place. When the star's declination is of the same name with the latitude of the place and *less* than it, and its position at the time of observation is on the opposite side of the prime vertical, from the elevated pole, the supplement of the sum of B and C, is the latitude. In all other cases the latitude is equal to the difference between B and C.

Note. The observation should not be made so near the prime vertical as to make the side on which the star is situated, doubtful. It is always best, when convenient, to make it near the meridian; as then, a small error in the clock or in the longitude of the place, required in finding the sidereal time, produces but very slight influence on the computed latitude.

Several observations of the altitude and corresponding time should be taken, and the latitude be deduced from each. The mean of these, that is, their sum divided by their number, may be regarded as more accurate than the latitude obtained from a single observation. The probable accuracy of the determination will be still further increased, if, near the same time, the latitude be deduced in like manner from observations on a star† on the opposite side of the zenith, and the half sum of the two latitudes thus obtained, be taken for the latitude. (See Art. 183, 1*st method.*)

---

* The altitude may be taken with a sextant and artificial horizon. For the method of adjusting the instrument and making the observation, the student is referred to Simms' small work on instruments, mentioned in a note on page 26.

† A star whose altitude is within a few degrees of the former, is to be preferred.

EXAMPLES.

1. Given the corrected altitude of α Ursae Minoris 41° 33′ 21.4″, obtained from an observation at a place, long. 5h. 1m. 15sec. W. and lat. about 40° N., on the 25th of November, 1839, at 8h. 34m. 17sec. P. M., mean time, to find the latitude; the apparent right ascension of the star being 1h. 2m. 22.63sec., and declination 88° 27′ 39.3″ N.

The sidereal time corresponding to the given mean time is found to be 0h. 51m. 29.96sec.

	h.	m.	sec.
Star's app. right ascen. . . .	1	2	22.63
Sidereal time	0	51	29.96
Difference		10	52.67
H, the star's dist. from merid. .	2°	43′	10″

D = 88° 27′ 39.3″	log. cot 8.4292440	D		Ar. Co. log. sin 0.0001567
H = 2 43 10	log. cos 9.9995106	A = 41° 33′ 21.4″		log. sin 9.8217434
B = 1 32 14.5	8.4287546	B = 1 32 14.5		log. cos 9.9998436
		C = 41 33 21.5		log. sin 9.8217437
		Lat. = 40 1 7.0		

2. Given for the same place and same night as in the last example, the corrected altitude of β Orionis, 41° 29′ 36″.1, at 12h. 35m. 19sec. P. M., mean time, to find the latitude; the apparent right ascension of the star being 5h. 6m. 52.40sec., and declination 8° 23′ 17″.1 S.

Ans. 40° 0′ 54″.9.

PROBLEM XXX.

To find the Latitude of a place, having given a series of circum-meridian altitudes of a star in the region of the equator, with the times of observation, and the apparent right ascension and declination of the star.

Let A = the meridian altitude of the star,
A′ = an observed altitude, corrected for refraction,
H = the hour angle, which should not exceed 4°,
D = the star's declination, *negative* when it is south,
l = the assumed latitude of the place.

Then, putting $x = A - A'$, the value of x may be computed by the following formulæ :

$$x = Bm - B^2.n \text{ , in which}$$

$$B = \frac{\cos l \cos D}{\cos A'},$$

$$m = \frac{2 \sin^2 \frac{1}{2} H}{\sin 1''},$$

and $$n = \frac{2 \sin^4 \frac{1}{2} H}{\sin 1''}.$$

The values of m and n are given in table LXXXI., for values of H to 17 minutes of time.

Find the mean time of the star's meridian passage by Problem IX. The differences between this time and the times of observation will be the hour angles expressed in mean time. These, increased at the rate of one-sixth of a second for each minute, will be the required hour angles. Take from table LXXXI., the values of m and n for each hour angle.

Take the mean of the observed altitudes, and correct it for refraction. Also, take the means of the values of m and of n; and, using these means for A', m and n, compute x by the above formula. Add x to the mean value of A', and the sum will be A, the meridian altitude of the star resulting from the observations.

To or from the complement of A, add or subtract D, according as the place is in north or south latitude, and the result will be the latitude required.

Note. If the chronometer used is too fast or too slow, its error should be added to, or subtracted from the mean time of the star's transit, and the result, which will be the chronometer time of transit, should be used instead of the mean time.

EXAM. 1. On the 18th of October, 1841, a series of observations were made for the determination of the latitude of a place, whose longitude is 4h. 31m. 7sec. west of Greenwich, and assumed latitude 46° 53′ north. The star observed was α Ceti, its right ascension was 2h. 54m. 2.38sec. and its declination + 3° 28′ 8″.2. The altitudes and times of observation are contained in the first and second columns of the following table. The chronometer used was 4m. 33sec. slow of mean time. The indications of the barometer and thermometer were 28.7 inches and 26°.4 respectively.

Observed Altitudes.			Chron. Times $=t$			T ∽ t		Hour Angles.		m	n
°		′	h.	m.	sec.	m.	sec.	m.	sec.	″	″
46	33	25	12	51	12.0	8	11.7	8	13.0	132.55	0.04
	34	35		54	24.0	4	59.7	5	0.5	49.25	0.01
	35	25		55	20.4	4	3.3	4	4.0	32.47	0.00
	35	25		57	24.2	1	59.5	1	59.8	7.82	0.00
	35	35		59	20.7	0	3.0	0	3.0	0.00	0.00
	35	30	13	1	27.0	2	3.3	2	3.6	8.33	0.00
	35	5		3	15.4	3	51.7	3	52.4	29.46	0.00
	34	45		4	30.4	5	6.7	5	7.5	51.57	0.01
	34	15		5	45.0	6	21.3	6	22.3	79.71	0.02
	33	50		6	45.6	7	21.9	7	23.1	107.08	0.03
	347	30								498.24	0.11
46	34	45								49.82	0.01

To find the mean time of transit of α Ceti, we have,

	h.	m.	sec.
Sid. time at mean noon, Greenwich	13	47	12.35
Sid. acceleration in 4h. 31m. 7sec.			44.54
Sid. time at m. n. at place of observation . .	13	47	56.89
Right asc. of α Ceti + 24 hrs.	26	54	2.38
	13	6	5.49
Mean Solar Equiv. (table XI.) for 13h.	12	57	52.22
" " 6m. . . .		5	59.02
" " 5s.			4.99
" " 0.49 sec. . . .			0.49
Mean time of transit of α Ceti	13	3	56.72
Chronometer slow, subtract		4	33.
Chronometer time of transit . . . . . T =	12	59	23.7

Take the differences between T and the several values of t and insert them in the third column of the above table. These differences are then reduced to sidereal intervals by table X., or by increasing them in the ratio of one-sixth of a second for each minute. Having thus obtained the hour angles, take from table LXXXI., the corresponding values of m and n.

				log. m = 1.6974
Mean Obs. Alt. . .		46 34 45.00		log. cos. l = 9.8347
Refraction . . .		55.73		" " D = 9.9992
	A′ =	46 33 49.27	A. C.	" " A′ = 0.1627
	x =	49.43	. .	log x = 1.6940
Meridian Alt.,	A =	46 34 38.7		
	90° — A =	43 25 21.3		
	D =	+ 3 28 8.2		
Resulting latitude	=	46 53 29.5		

Note. It will be seen by this example, that when the hour angle does not exceed 8 or 10 minutes, it is not necessary to compute the value of $B^2.n$.

EXAM. 2. It is required to find the latitude of the High School Observatory, Philadelphia, from the following data. The mean of eight observed altitudes of α Virginis, taken April 26th, 1839, was 39° 43′ 52″.6, and the times of observation were, 10h. 58m. 3sec., 11h. 0m. 16^s.5, 11h. 3m. 44sec, 11h. 6m. 7sec., 11h. 8m. 12^s.5, 11h. 9m. 12^s.5, 11h. 10m. 30sec., and 11h. 12m. 58sec. Barom., 30.0 inches; and therm., 60°.4. The right ascension of α Virginis was 13h. 16m. 45^s.38; and its declination, 10° 19′ 22″.4^s. The Greenwich sidereal time at mean noon was 2h. 15m. 12^s47. The chronometer used was 7m. 23^s.2 fast of mean time. *Ans.* 39° 57′ 8″.7.

PROBLEM XXXI.

Given the true altitude of the sun, obtained from observation at a given place, and the time of observation as indicated by a clock, to find the time, and the error of the clock.

Find the sun's declination for the given time, and subtracting from 90 when it is of the *same* name with the latitude of the place, but adding when of *a contrary* name, we have his polar distance. Call the polar distance D, the altitude A, the latitude of the place L, and the hour angle or distance of the sun from the meridian H. Add the values of A, D, and L together, and call the sum S.

Add together Ar. Co. log sin D, Ar. Co. log cos L, log cos ½ S, and sin log (½S — A) without rejecting any 10 from the index, and taking half the sum, it will be log sin ½H. When the observation is made in the afternoon, the hour angle H, converted into time, is the apparent time; but when it is made in the forenoon, the difference between this interval and 12 hours, is the apparent time. To the apparent time, apply the equation of time, and we obtain the mean time. The difference between this and the time shown by the clock, is the error of the clock.

Note 1. The observations for finding the time should be made when the sun is several hours from the meridian; the nearer the prime vertical, the better, provided the altitude is not less than 12 or 15°.

2. The time and error of the clock may be obtained in nearly a similar manner, from the corrected altitude of a star. Having found the star's apparent right ascension and declination, and computed the hour angle H, using the star's polar distance and altitude, add it to the right ascension when the observation is made to the west of the meridian, but subtract it from the right ascension when the observation is made to the east. The result will be the sidereal time of the observation. From this, subtract the sidereal time found for mean noon of the given day, and converting the remainder into mean time by means of tab. XI., it will be the mean time of the observation.

EXAMPLE 1. Given the corrected altitude of the sun, 31° 16′ 33.4″, at a place, long. 5h. 1m. 15sec. W., lat. 40° 1′ 12″ N., on the 25th of June, 1842, at 7h. 28m. 50sec. A. M., mean time by the clock, to find the time, and the error of the clock. The sun's declination found for the time shown by the clock, is 23° 24′ 55″.0 N.; which subtracted from 90°, gives D = 66° 35′ 5″.

A =	31°	16′	33.4″	
D =	66	35	5.0	Ar. Co. log. sin 0.0373236
L =	41	1	12.	" " log. cos 0.1223520
S =	138	52	50.4	
½ S =	69	26	25.2	log. cos 9.5455330
½ S − A =	38	9	51.8	log. sin 9.7909321
				2)19.4961407
½ H =	34	2	43.1	log. sin 9.7480703
H =	68	5	26.2	
			4	

	h.	m.	sec.	
	4	32	21.73	
	12	0	0	
App. time	7	27	38.27	A. M.
Equat. of time	+	2	8.94	found from Naut. Alm.
Time of obs.	7	29	47.2	A. M.
Time by clock	7	28	50.	
Error		0	57.2	; clock too *slow*.

EXAMPLE 2. At a place in long. 5h. 0m. 42sec. W., and lat. 39° 57′ 8″ N. On the 7th of September, 1838, the following observations were made:—Time by a mean solar clock, 11h. 20m. 26ˢ.2; altitude of α Arietis (east of the meridian), 39° 48′ 38″; barom., 30.2 inches; therm., 75°. From the Nautical Almanac, the right ascension and declination of α Arietis were, 1h. 58m. 6ˢ.2 and + 22° 41′ 55″.3; and the sidereal time at Greenwich mean noon was, 11h. 4m. 28ˢ.07. Required the error of the clock. *Ans.* 14m. 43.1sec. *too fast.*

EXAMPLE 3. The date, place, and instruments being the same as in the last example, the observed altitude of α Lyræ (west of the meridian) was, 38° 44′ 49″ and clock time 12h. 10m. 45ˢ.2. The right ascension and declination of α Lyræ were, 18h. 31m. 29ˢ.06 and + 38° 38′ 20″.5. Required the error of the clock. *Ans.* 14m. 39.9sec. *too fast.*

Note. The mean of the results obtained by east and by west observations, (like those of the last two examples,) will be nearly independent of any error in the instrument used in measuring the altitudes; for, the hour angles will be both too great or both too small, and since, in one case the hour angle is subtracted and in the other added, to obtain the sidereal time, it follows that one of the resulting times will be too great and the other too small.

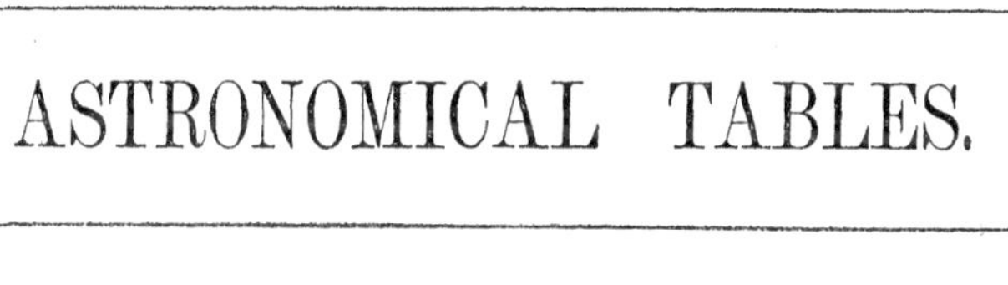

ASTRONOMICAL TABLES.

No.	Log.	D.	No.	Log.	D.
100	.0000	43	160	.2041	27
101	.0043	43	161	.2068	27
102	.0086	42	162	.2095	27
103	.0128	42	163	.2122	26
104	.0170	42	164	.2148	27
105	.0212	41	165	.2175	26
106	.0253	41	166	.2201	26
107	.0294	40	167	.2227	26
108	.0334	40	168	.2253	26
109	.0374	40	169	.2279	25
110	.0414	39	170	.2304	26
111	.0453	39	171	.2330	25
112	.0492	39	172	.2355	25
113	.0531	38	173	.2380	25
114	.0569	38	174	.2405	25
115	.0607	38	175	.2430	25
116	.0645	37	176	.2455	25
117	.0682	37	177	.2480	24
118	.0719	36	178	.2504	25
119	.0755	37	179	.2529	24
120	.0792	36	180	.2553	24
121	.0828	36	181	.2577	24
122	.0864	35	182	.2601	24
123	.0899	35	183	.2625	23
124	.0934	35	184	.2648	24
125	.0969	35	185	.2672	23
126	.1004	34	186	.2695	23
127	.1038	34	187	.2718	24
128	.1072	34	188	.2742	23
129	.1106	33	189	.2765	23
130	.1139	34	190	.2788	22
131	.1173	33	191	.2810	23
132	.1206	33	192	.2833	23
133	.1239	32	193	.2856	22
134	.1271	32	194	.2878	22
135	.1303	32	195	.2900	23
136	.1335	32	196	.2923	22
137	.1367	32	197	.2945	22
138	.1399	31	198	.2967	22
139	.1430	31	199	.2989	21
140	.1461	31	200	.3010	22
141	.1492	31	201	.3032	22
142	.1523	30	202	.3054	21
143	.1553	31	203	.3075	21
144	.1584	30	204	.3096	22
145	.1614	30	205	.3118	21
146	.1644	29	206	.3139	21
147	.1673	30	207	.3160	21
148	.1703	29	208	.3181	20
149	.1732	29	209	.3201	21
150	.1761	29	210	.3222	21
151	.1790	28	211	.3243	20
152	.1818	29	212	.3263	21
153	.1847	28	213	.3284	20
154	.1875	28	214	.3304	20
155	.1903	28	215	.3324	21
156	.1931	28	216	.3345	20
157	.1959	28	217	.3365	20
158	.1987	27	218	.3385	19
159	.2014	27	219	.3404	20
160	.2041		220	.3424	

	43	42	41	40	39	38	37	36	35	34	33	32	31	30	29	28	27	27	26	25	24	23	22	21	20	19
1	4	4	4	4	4	4	4	4	3	3	3	3	3	3	3	3	3	3	3	2	2	2	2	2	2	2
2	9	8	8	8	8	8	7	7	7	7	7	6	6	6	6	6	5	5	5	5	5	5	4	4	4	4
3	13	13	12	12	12	11	11	11	10	10	10	10	9	9	9	8	8	8	8	7	7	7	7	6	6	6
4	17	17	16	16	16	15	15	14	14	14	13	13	12	12	12	11	11	11	10	10	10	9	9	8	8	8
5	21	21	20	20	20	19	18	18	17	17	16	16	15	15	14	14	13	13	13	12	12	11	11	10	10	9
6	26	25	25	24	23	23	22	22	21	20	20	19	19	18	17	17	16	16	16	15	14	14	13	13	12	11
7	30	29	29	28	27	27	26	25	24	24	23	22	22	21	20	20	19	19	18	17	17	16	15	15	14	13
8	34	34	33	32	31	30	30	29	28	27	26	26	25	24	23	22	22	22	21	20	19	18	18	17	16	15
9	39	38	37	36	35	34	33	32	31	31	30	29	28	27	26	25	24	24	23	22	22	21	20	19	18	17

TABLE I. *Logarithms to four decimal figures.*

No.	Log.	D.		No.	Log.	D.		No.	Log.	D.	
220	.3424	20	20: 1 2, 2 4, 3 6, 4 8, 5 10, 6 12, 7 14, 8 16, 9 18	280	.4472	15	16: 1 2, 2 3, 3 5, 4 6, 5 8, 6 10, 7 11, 8 13, 9 14	340	.5315	13	
221	.3444	20		281	.4487	15		341	.5328	12	
222	.3464	19		282	.4502	16		342	.5340	13	
223	.3483	19		283	.4518	15		343	.5353	13	
224	.3502	20		284	.4533	15		344	.5466	12	
225	.3522	19		285	.4548	16		345	.5378	13	
226	.3541	19		286	.4564	15		346	.5391	12	
227	.3560	19		287	.4579	15		347	.5403	13	
228	.3579	19	19: 1 2, 2 4, 3 6, 4 8, 5 9, 6 11, 7 13, 8 15, 9 17	288	.4594	15		348	.5416	12	
229	.3598	19		289	.4609	15		349	.5428	13	13: 1 1, 2 3, 3 4, 4 5, 5 6, 6 8, 7 9, 8 10, 9 12
230	.3617	19		290	.4624	15		350	.5441	12	
231	.3636	19		291	.4639	15	15: 1 1, 2 3, 3 4, 4 6, 5 7, 6 9, 7 10, 8 12, 9 13	351	.5453	12	
232	.3655	19		292	.4654	15		352	.5465	13	
233	.3674	18		293	.4669	14		353	.5478	12	
234	.3692	19		294	.4683	15		354	.5490	12	
235	.3711	18		295	.4698	15		355	.5502	12	
236	.3729	18		296	.4713	15		356	.5514	13	
237	.3747	19		297	.4728	14		357	.5527	12	
238	.3766	18		298	.4742	15		358	.5539	12	
239	.3784	18	18: 1 2, 2 4, 3 5, 4 7, 5 9, 6 11, 7 13, 8 14, 9 16	299	.4757	14		359	.5551	12	
240	.3802	18		300	.4771	15		360	.5563	12	
241	.3820	18		301	.4786	14		361	.5575	12	
242	.3838	18		302	.4800	14		362	.5587	12	
243	.3856	18		303	.4814	15		363	.5599	12	
244	.3874	18		304	.4829	14		364	.5611	12	
245	.3892	17		305	.4843	14		365	.5623	12	
246	.3909	18		306	.4857	14		366	.5635	12	12: 1 1, 2 2, 3 4, 4 5, 5 6, 6 7, 7 8, 8 10, 9 11
247	.3927	18		307	.4871	15		367	.5647	11	
248	.3945	17		308	.4886	14		368	.5658	12	
249	.3962	17		309	.4900	14	14: 1 1, 2 3, 3 4, 4 6, 5 7, 6 8, 7 10, 8 11, 9 13	369	.5670	12	
250	.3979	18		310	.4914	14		370	.5682	12	
251	.3997	17		311	.4928	14		371	.5694	11	
252	.4014	17	17: 1 2, 2 3, 3 5, 4 7, 5 8, 6 10, 7 12, 8 14, 9 15	312	.4942	13		372	.5705	12	
253	.4031	17		313	.4955	14		373	.5717	12	
254	.4048	17		314	.4969	14		374	.5729	11	
255	.4065	17		315	.4983	14		375	.5740	12	
256	.4082	17		316	.4997	14		376	.5752	11	
257	.4099	17		317	.5011	13		377	.5763	12	
258	.4116	17		318	.5024	14		378	.5775	11	
259	.4133	17		319	.5038	14		379	.5786	12	
260	.4150	16		320	.5052	13		380	.5798	11	
261	.4166	17		321	.5065	14		381	.5809	12	
262	.4183	17		322	.5079	13		382	.5821	11	
263	.4200	16		323	.5092	13		383	.5832	11	
264	.4216	16		324	.5105	14		384	.5843	12	
265	.4232	17	16: 1 2, 2 3, 3 5, 4 6, 5 8, 6 10, 7 11, 8 13, 9 14	325	.5119	13		385	.5855	11	
266	.4249	16		326	.5132	13	13: 1 1, 2 3, 3 4, 4 5, 5 6, 6 8, 7 9, 8 10, 9 12	386	.5866	11	11: 1 1, 2 2, 3 3, 4 4, 5 5, 6 7, 7 8, 8 9, 9 10
267	.4265	16		327	.5145	14		387	.5877	11	
268	.4281	17		328	.5159	13		388	.5888	11	
269	.4298	16		329	.5172	13		389	.5899	12	
270	.4314	16		330	.5185	13		390	.5911	11	
271	.4330	16		331	.5198	13		391	.5922	11	
272	.4346	16		332	.5211	13		392	.5933	11	
273	.4362	16		333	.5224	13		393	.5944	11	
274	.4378	15	15: 1 1, 2 3, 3 4, 4 6, 5 7, 6 9, 7 10, 8 12, 9 13	334	.5237	13		394	.5955	11	
275	.4393	16		335	.5250	13		395	.5966	11	
276	.4409	16		336	.5263	13		396	.5977	11	
277	.4425	15		337	.5276	13		397	.5988	11	
278	.4440	16		338	.5289	13		398	.5999	11	
279	.4456	16		339	.5302	13		399	.6010	11	
280	.4472			340	.5315			400	.6021		

TABLE I. *Logarithms to four decimal figures.*

No.	Log.	D.	No.	Log.	D.	No.	Log.	D.	No.	Log.	D.	No.	Log.	D.
400	.6021	10	460	.6628	9	520	.7160	8	580	.7634	8	640	.8062	7
401	.6031	11	461	.6637	9	521	.7168	9	581	.7642	7	641	.8069	6
402	.6042	11	462	.6646	10	522	.7177	8	582	.7649	8	642	.8075	7
403	.6053	11	463	.6656	9	523	.7185	8	583	.7657	7	643	.8082	7
404	.6064	11	464	·6665	10	524	.7193	9	584	.7664	8	644	.8089	7
405	.6075	10	465	.6675	9	525	.7202	8	585	.7672	7	645	.8096	6
406	.6085	11	466	.6684	9	526	.7210	8	586	.7679	7	646	.8102	7
407	.6096	11	467	.6693	9	527	.7218	8	587	.7686	8	647	.8109	7
408	.6107	10	468	.6702	10	528	.7226	9	588	.7694	7	648	.8116	6
409	.6117	11	469	.6712	9	529	.7235	8	589	.7701	8	649	.8122	7
410	.6128	10	470	.6721	9	530	.7243	8	590	.7709	7	650	.8129	7
411	.6138	11	471	.6730	9	531	.7251	8	591	.7716	7	651	.8136	6
412	.6149	11	472	.6739	10	532	.7259	8	592	.7723	8	652	.8142	7
413	.6160	10	473	.6749	9	533	.7267	8	593	.7731	7	653	.8149	7
414	.6170	10	474	.6758	9	534	.7275	9	594	.7738	7	654	.8156	6
415	.6180	11	475	.6767	9	535	.7284	8	595	.7745	7	655	.8162	7
416	.6191	10	476	.6776	9	536	.7292	8	596	.7752	8	656	.8169	7
417	.6201	11	477	.6785	9	537	.7300	8	597	.7760	7	657	.8176	6
418	.6212	10	478	.6794	9	538	.7308	8	598	.7767	7	658	.8182	7
419	.6222	10	479	.6803	9	539	.7316	8	599	.7774	8	659	.8189	6
420	.6232	11	480	.6812	9	540	.7324	8	600	.7782	7	660	.8195	7
421	.6243	10	481	.6821	9	541	.7332	8	601	.7789	7	661	.8202	7
422	.6253	10	482	.6830	9	542	.7340	8	602	.7796	7	662	.8209	6
423	.6263	11	483	.6839	9	543	.7348	8	603	.7803	7	663	.8215	7
424	.6274	10	484	.6848	9	544	.7356	8	604	.7810	8	664	.8222	6
425	.6284	10	485	.6857	9	545	.7364	8	605	.7818	7	665	.8228	7
426	.6294	10	486	.6866	9	546	.7372	8	606	.7825	7	666	.8235	6
427	.6304	10	487	6875	9	547	.7380	8	607	.7832	7	667	.8241	7
428	.6314	11	488	.6884	9	548	.7388	8	608	.7839	7	668	.8248	6
429	.6325	10	489	.6893	9	549	.7396	8	609	.7846	7	669	.8254	7
430	.6335	10	490	.6902	9	550	.7404	8	610	.7853	7	670	.8261	6
431	.6345	10	491	.6911	9	551	.7412	7	611	.7860	8	671	.8267	7
432	.6355	10	492	.6920	8	552	.7419	8	612	.7868	7	672	.8274	6
433	.6365	10	493	.6928	9	553	.7427	8	613	.7875	7	673	.8280	7
434	.6375	10	494	6937	9	554	.7435	8	614	.7882	7	674	.8287	6
435	.6385	10	495	.6946	9	555	.7443	8	615	.7889	7	675	.8293	6
436	.6395	10	496	.6955	9	556	.7451	8	616	.7896	7	676	.8299	7
437	.6405	10	497	.6964	8	557	.7459	7	617	.7903	7	677	.8306	6
438	.6415	10	498	.6972	9	558	.7466	8	618	.7910	7	678	.8312	7
439	.6425	10	499	.6981	9	559	.7474	8	619	.7917	7	679	.8319	6
440	.6435	9	500	.6990	8	560	.7482	8	620	.7924	7	680	.8325	6
441	.6444	10	501	.6998	9	561	.7490	7	621	.7931	7	681	.8331	7
442	.6454	10	502	.7007	9	562	.7497	8	622	.7938	7	682	.8338	6
443	.6464	10	503	.7016	8	563	.7505	8	623	.7945	7	683	.8344	7
444	.6474	10	504	.7024	9	564	.7513	7	624	.7952	7	684	.8351	6
445	.6484	9	505	.7033	9	565	.7520	8	625	.7959	7	685	.8357	6
446	.6493	10	506	.7042	8	566	.7528	8	626	.7966	7	686	.8363	7
447	.6503	10	507	.7050	9	567	.7536	7	627	.7973	7	687	.8370	6
448	.6513	9	508	.7059	8	568	.7543	8	628	.7980	7	688	.8376	6
449	.6522	10	509	.7067	9	569	.7551	8	629	.7987	6	689	.8382	6
450	.6532	10	510	.7076	8	570	.7559	7	630	.7993	7	690	.8388	7
451	.6542	9	511	.7084	9	571	.7566	8	631	.8000	7	691	.8395	6
452	.6551	10	512	.7093	8	572	.7574	8	632	.8007	7	692	.8401	6
453	.6561	10	513	.7101	9	573	.7582	7	633	.8014	7	693	.8407	7
454	.6571	9	514	.7110	8	574	.7589	8	634	.8021	7	694	.8414	6
455	.6580	10	515	.7118	8	575	.7597	7	635	.8028	7	695	.8420	6
456	.6590	9	516	.7126	9	576	.7604	8	636	.8035	6	696	.8426	6
457	.6599	10	517	.7135	8	577	.7612	7	637	.8041	7	697	.8432	7
458	.6609	9	518	.7143	9	578	.7619	8	638	.8048	7	698	.8439	6
459	.6618	10	519	.7152	8	579	.7627	7	639	.8055	7	699	.8445	6
460	.6628		520	.7160		580	.7634		640	.8062		700	.8451	

No.	Log.	D.	No.	Log.	D.	No.	Log.	D.	No.	Log.	D.	No.	Log.	D.
700	.8451	6	760	.8808	6	820	.9138	5	880	.9445	5	940	.9731	5
701	.8457	6	761	.8814	6	821	.9143	6	881	.9450	5	941	.9736	5
702	.8463	7	762	.8820	5	822	.9149	5	882	.9455	5	942	.9741	4
703	.8470	6	763	.8825	6	823	.9154	5	883	.9460	5	943	.9745	5
704	.8476	6	764	.8831	6	824	.9159	6	884	.9465	4	944	.9750	4
705	.8482	6	765	.8837	5	825	.9165	5	885	.9469	5	945	.9754	5
706	.8488	6	766	.8842	6	826	.9170	5	886	.9474	5	946	.9759	4
707	.8494	6	767	.8848	6	827	.9175	5	887	.9479	5	947	.9763	5
708	.8500	6	768	.8854	5	828	.9180	6	888	.9484	5	948	.9768	5
709	.8506	6	769	.8859	6	829	.9186	5	889	.9489	5	949	.9773	4
710	.8513	6	770	.8865	6	830	.9191	5	890	.9494	5	950	.9777	5
711	.8519	6	771	.8871	5	831	.9196	5	891	.9499	5	951	.9782	4
712	.8525	6	772	.8876	6	832	.9201	5	892	.9504	5	952	.9786	5
713	.8531	6	773	.8882	5	833	.9206	6	893	.9509	4	953	.9791	4
714	.8537	6	774	.8887	6	834	.9212	5	894	.9513	5	954	.9795	5
715	.8543	6	775	.8893	6	835	.9217	5	895	.9518	5	955	.9800	5
716	.8549	6	776	.8899	5	836	.9222	5	896	.9523	5	956	.9805	4
717	.8555	6	777	.8904	6	837	.9227	5	897	.9528	5	957	.9809	5
718	.8561	6	778	.8910	5	838	.9232	6	898	.9533	5	958	.9814	4
719	.8567	6	779	.8915	6	839	.9238	5	899	.9538	4	959	.9818	5
720	.8573	6	780	.8921	6	840	.9243	5	900	.9542	5	960	.9823	4
721	.8579	6	781	.8927	5	841	.9248	5	901	.9547	5	961	.9827	5
722	.8585	6	782	.8932	6	842	.9253	5	902	.9552	5	962	.9832	4
723	.8591	6	783	.8938	5	843	.9258	5	903	.9557	5	963	.9836	5
724	.8597	6	784	.8943	6	844	.9263	6	904	.9562	4	964	.9841	4
725	.8603	6	785	.8949	5	845	.9269	5	905	.9566	5	965	.9845	5
726	.8609	6	786	.8954	6	846	.9274	5	906	.9571	5	966	.9850	4
727	.8615	6	787	.8960	5	847	.9279	5	907	.9576	5	967	.9854	5
728	.8621	6	788	.8965	6	848	.9284	5	908	.9581	5	968	.9859	4
729	.8627	6	789	.8971	5	849	.9289	5	909	.9586	4	969	.9863	5
730	.8633	6	790	.8976	6	850	.9294	5	910	.9590	5	970	.9868	4
731	.8639	6	791	.8982	5	851	.9299	5	911	.9595	5	971	.9872	5
732	.8645	6	792	.8987	6	852	.9304	5	912	.9600	5	972	.9877	4
733	.8651	6	793	.8993	5	853	.9309	6	913	.9605	4	973	.9881	5
734	.8657	6	794	.8998	6	854	.9315	5	914	.9609	5	974	.9886	4
735	.8663	6	795	.9004	5	855	.9320	5	915	.9614	5	975	.9890	4
736	.8669	6	796	.9009	6	856	.9325	5	916	.9619	5	976	.9894	5
737	.8675	6	797	.9015	5	857	.9330	5	917	.9624	4	977	.9899	4
738	.8681	5	798	.9020	5	858	.9335	5	918	.9628	5	978	.9903	5
739	.8686	6	799	.9025	6	859	.9340	5	919	.9633	5	979	.9908	4
740	.8692	6	800	.9031	5	860	.9345	5	920	.9638	5	980	.9912	5
741	.8698	6	801	.9036	6	861	.9350	5	921	.9643	4	981	.9917	4
742	.8704	6	802	.9042	5	862	.9355	5	922	.9647	5	982	.9921	5
743	.8710	6	803	.9047	6	863	.9360	5	923	.9652	5	983	.9926	4
744	.8716	6	804	.9053	5	864	.9365	5	924	.9657	4	984	.9930	4
745	.8722	5	805	.9058	5	865	.9370	5	925	.9661	5	985	.9934	5
746	.8727	6	806	.9063	6	866	.9375	5	926	.9666	5	986	.9939	4
747	.8733	6	807	.9069	5	867	.9380	5	927	.9671	4	987	.9943	5
748	.8739	6	808	.9074	5	868	.9385	5	928	.9675	5	988	.9948	4
749	.8745	6	809	.9079	6	869	.9390	5	929	.9680	5	989	.9952	4
750	.8751	5	810	.9085	5	870	.9395	5	930	.9685	4	990	.9956	5
751	.8756	6	811	.9090	6	871	.9400	5	931	.9689	5	991	.9961	4
752	.8762	6	812	.9096	5	872	.9405	5	932	.9694	5	992	.9965	4
753	.8768	6	813	.9101	5	873	.9410	5	933	.9699	4	993	.9969	5
754	.8774	5	814	.9106	6	874	.9415	5	934	.9703	5	994	.9974	4
755	.8779	6	815	.9112	5	875	.9420	5	935	.9708	5	995	.9978	5
756	.8785	6	816	.9117	5	876	.9425	5	936	.9713	4	996	.9983	4
757	.8791	6	817	.9122	6	877	.9430	5	937	.9717	5	997	.9987	4
758	.8797	6	818	.9128	5	878	.9435	5	938	.9722	5	998	.9991	5
759	.8802	6	819	.9133	5	879	.9440	5	939	.9727	4	999	.9996	4
760	.8808		820	.9138		880	.9445		940	.9731		1000	.0000	

	0°				1°				
	Sin.	Cos.	Tang.	Cotang.	Sin.	Cos.	Tang.	Cotang.	
0′		0.0000			8.2419	9.9999	8.2419	1.7581	60′
1	6.4637	0.0000	6.4637	3.5363	8.2490	9.9999	8.2491	1.7509	59
2	6.7648	0.0000	6.7648	3.2352	8.2561	9.9999	8.2562	1.7438	58
3	6.9408	0.0000	6.9408	3.0592	8.2630	9.9999	8.2631	1.7369	57
4	7.0658	0.0000	7.0658	2.9342	8.2699	9.9999	8.2700	1.7300	56
5	7.1627	0.0000	7.1627	2.8373	8.2766	9.9999	8.2767	1.7233	55
6	7.2419	0.0000	7.2419	2.7581	8.2832	9.9999	8.2833	1.7167	54
7	7.3088	0.0000	7.3088	2.6912	8.2898	9.9999	8.2899	1.7101	53
8	7.3668	0.0000	7.3668	2.6332	8.2962	9.9999	8.2963	1.7037	52
9	7.4180	0.0000	7.4180	2.5820	8.3025	9.9999	8.3026	1.6974	51
10	7.4637	0.0000	7.4637	2.5363	8.3088	9.9999	8.3089	1.6911	50
11	7.5051	0.0000	7.5051	2.4949	8.3150	9.9999	8.3150	1.6850	49
12	7.5429	0.0000	7.5429	2.4571	8.3210	9.9999	8.3211	1.6789	48
13	7.5777	0.0000	7.5777	2.4223	8.3270	9.9999	8.3271	1.6729	47
14	7.6099	0.0000	7.6099	2.3901	8.3329	9.9999	8.3330	1.6670	46
15	7.6398	0.0000	7.6398	2.3602	8.3388	9.9999	8.3389	1.6611	45
16	7.6678	0.0000	7.6678	2.3322	8.3445	9.9999	8.3446	1.6554	44
17	7.6942	0.0000	7.6942	2.3058	8.3502	9.9999	8.3503	1.6497	43
18	7.7190	0.0000	7.7190	2.2810	8.3558	9.9999	8.3559	1.6441	42
19	7.7425	0.0000	7.7425	2.2575	8.3613	9.9999	8.3614	1.6386	41
20	7.7648	0.0000	7.7648	2.2352	8.3668	9.9999	8.3669	1.6331	40
21	7.7859	0.0000	7.7860	2.2140	8.3722	9.9999	8.3723	1.6277	39
22	7.8061	0.0000	7.8062	2.1938	8.3775	9.9999	8.3776	1.6224	38
23	7.8255	0.0000	7.8255	2.1745	8.3828	9.9999	8.3829	1.6171	37
24	7.8439	0.0000	7.8439	2.1561	8.3880	9.9999	8.3881	1.6119	36
25	7.8617	0.0000	7.8617	2.1383	8.3931	9.9999	8.3932	1.6068	35
26	7.8787	0.0000	7.8787	2.1213	8.3982	9.9999	8.3983	1.6017	34
27	7.8951	0.0000	7.8951	2.1049	8.4032	9.9999	8.4033	1.5967	33
28	7.9109	0.0000	7.9109	2.0891	8.4082	9.9999	8.4083	1.5917	32
29	7.9261	0.0000	7.9261	2.0739	8.4131	9.9999	8.4132	1.5868	31
30	7.9408	0.0000	7.9409	2.0591	8.4179	9.9999	8.4181	1.5819	30
31	7.9551	0.0000	7.9551	2.0449	8.4227	9.9998	8.4229	1.5771	29
32	7.9689	0.0000	7.9689	2.0311	8.4275	9.9998	8.4276	1.5724	28
33	7.9822	0.0000	7.9823	2.0177	8.4322	9.9998	8.4323	1.5677	27
34	7.9952	0.0000	7.9952	2.0048	8.4368	9.9998	8.4370	1.5630	26
35	8.0078	0.0000	8.0078	1.9922	8.4414	9.9998	8.4416	1.5584	25
36	8.0200	0.0000	8.0200	1.9800	8.4459	9.9998	8.4461	1.5539	24
37	8.0319	0.0000	8.0319	1.9681	8.4504	9.9998	8.4506	1.5494	23
38	8.0435	0.0000	8.0435	1.9565	8.4549	9.9998	8.4551	1.5449	22
39	8.0548	0.0000	8.0548	1.9452	8.4593	9.9998	8.4595	1.5405	21
40	8.0658	0.0000	8.0658	1.9342	8.4637	9.9998	8.4638	1.5362	20
41	8.0765	0.0000	8.0765	1.9235	8.4680	9.9998	8.4682	1.5318	19
42	8.0870	0.0000	8.0870	1.9130	8.4723	9.9998	8.4725	1.5275	18
43	8.0972	0.0000	8.0972	1.9028	8.4765	9.9998	8.4767	1.5233	17
44	8.1072	0.0000	8.1072	1.8928	8.4807	9.9998	8.4809	1.5191	16
45	8.1169	0.0000	8.1170	1.8830	8.4848	9.9998	8.4851	1.5149	15
46	8.1265	0.0000	8.1265	1.8735	8.4890	9.9998	8.4892	1.5108	14
47	8.1358	0.0000	8.1359	1.8641	8.4930	9.9998	8.4933	1.5067	13
48	8.1450	0.0000	8.1450	1.8550	8.4971	9.9998	8.4973	1.5027	12
49	8.1539	0.0000	8.1540	1.8460	8.5011	9.9998	8.5013	1.4987	11
50	8.1627	0.0000	8.1627	1.8373	8.5050	9.9998	8.5053	1.4947	10
51	8.1713	0.0000	8.1713	1.8287	8.5090	9.9998	8.5092	1.4908	9
52	8.1797	0.0000	8.1798	1.8202	8.5129	9.9998	8.5131	1.4869	8
53	8.1880	9.9999	8.1880	1.8120	8.5167	9.9998	8.5170	1.4830	7
54	8.1961	9.9999	8.1962	1.8038	8.5206	9.9998	8.5208	1.4792	6
55	8.2041	9.9999	8.2041	1.7959	8.5243	9.9998	8.5246	1.4754	5
56	8.2119	9.9999	8.2120	1.7880	8.5281	9.9998	8.5283	1.4717	4
57	8.2196	9.9999	8.2196	1.7804	8.5318	9.9997	8.5321	1.4679	3
58	8.2271	9.9999	8.2272	1.7728	8.5355	9.9997	8.5358	1.4642	2
59	8.2346	9.9999	8.2346	1.7654	8.5392	9.9997	8.5394	1.4606	1
60	8.2419	9.9999	8.2419	1.7581	8.5428	9.9997	8.5431	1.4569	0
	Cos.	Sin.	Cotang.	Tang.	Cos.	Sin.	Cotang.	Tang.	
	89°				88°				

	2°				3°				
	Sin.	Cos.	Tang.	Cotang.	Sin.	Cos.	Tang.	Cotang.	
0′	8.5428	9.9997	8.5431	1.4569	8.7188	9.9994	8.7194	1.2806	60′
1	8.5464	9.9997	8.5467	1.4533	8.7212	9.9994	8.7218	1.2782	59
2	8.5500	9.9997	8.5503	1.4497	8.7236	9.9994	8.7242	1.2758	58
3	8.5535	9.9997	8.5538	1.4462	8.7260	9.9994	8.7266	1.2734	57
4	8.5571	9.9997	8.5573	1.4427	8.7283	9.9994	8.7290	1.2710	56
5	8.5605	9.9997	8.5608	1.4392	8.7307	9.9994	8.7313	1.2687	55
6	8.5640	9.9997	8.5643	1.4357	8.7330	9.9994	8.7337	1.2663	54
7	8.5674	9.9997	8.5677	1.4323	8.7354	9.9994	8.7360	1.2640	53
8	8.5708	9.9997	8.5711	1.4289	8.7377	9.9994	8.7383	1.2617	52
9	8.5742	9.9997	8.5745	1.4255	8.7400	9.9993	8.7406	1.2594	51
10	8.5776	9.9997	8.5779	1.4221	8.7423	9.9993	8.7429	1.2571	50
11	8.5809	9.9997	8.5812	1.4188	8.7446	9.9993	8.7452	1.2548	49
12	8.5842	9.9997	8.5845	1.4155	8.7468	9.9993	8.7475	1.2525	48
13	8.5875	9.9997	8.5878	1.4122	8.7491	9.9993	8.7497	1.2503	47
14	8.5907	9.9997	8.5911	1.4089	8.7513	9.9993	8.7520	1.2480	46
15	8.5939	9.9997	8.5943	1.4057	8.7535	9.9993	8.7542	1.2458	45
16	8.5972	9.9997	8.5975	1.4025	8.7557	9.9993	8.7565	1.2435	44
17	8.6003	9.9997	8.6007	1.3993	8.7580	9.9993	8.7587	1.2413	43
18	8.6035	9.9996	8.6038	1.3962	8.7602	9.9993	8.7609	1.2391	42
19	8.6066	9.9996	8.6070	1.3930	8.7623	9.9993	8.7631	1.2369	41
20	8.6097	9.9996	8.6101	1.3899	8.7645	9.9993	8.7652	1.2348	40
21	8.6128	9.9996	8.6132	1.3868	8.7667	9.9993	8.7674	1.2326	39
22	8.6159	9.9996	8.6163	1.3837	8.7688	9.9992	8.7696	1.2304	38
23	8.6189	9.9996	8.6193	1.3807	8.7710	9.9992	8.7717	1.2283	37
24	8.6220	9.9996	8.6223	1.3777	8.7731	9.9992	8.7739	1.2261	36
25	8.6250	9.9996	8.6254	1.3746	8.7752	9.9992	8.7760	1.2240	35
26	8.6279	9.9996	8.6283	1.3717	8.7773	9.9992	8.7781	1.2219	34
27	8.6309	9.9996	8.6313	1.3687	8.7794	9.9992	8.7802	1.2198	33
28	8.6339	9.9996	8.6343	1.3657	8.7815	9.9992	8.7823	1.2177	32
29	8.6368	9.9996	8.6372	1.3628	8.7836	9.9992	8.7844	1.2156	31
30	8.6397	9.9996	8.6401	1.3599	8.7857	9.9992	8.7865	1.2135	30
31	8.6426	9.9996	8.6430	1.3570	8.7877	9.9992	8.7886	1.2114	29
32	8.6454	9.9996	8.6459	1.3541	8.7898	9.9992	8.7906	1.2094	28
33	8.6483	9.9996	8.6487	1.3513	8.7918	9.9992	8.7927	1.2073	27
34	8.6511	9.9996	8.6515	1.3485	8.7939	9.9992	8.7947	1.2053	26
35	8.6539	9.9996	8.6544	1.3456	8.7959	9.9992	8.7967	1.2033	25
36	8.6567	9.9996	8.6571	1.3429	8.7979	9.9991	8.7988	1.2012	24
37	8.6595	9.9995	8.6599	1.3401	8.7999	9.9991	8.8008	1.1992	23
38	8.6622	9.9995	8.6627	1.3373	8.8019	9.9991	8.8028	1.1972	22
39	8.6650	9.9995	8.6654	1.3346	8.8039	9.9991	8.8048	1.1952	21
40	8.6677	9.9995	8.6682	1.3318	8.8059	9.9991	8.8067	1.1933	20
41	8.6704	9.9995	8.6709	1.3291	8.8078	9.9991	8.8087	1.1913	19
42	8.6731	9.9995	8.6736	1.3264	8.8098	9.9991	8.8107	1.1893	18
43	8.6758	9.9995	8.6762	1.3238	8.8117	9.9991	8.8126	1.1874	17
44	8.6784	9.9995	8.6789	1.3211	8.8137	9.9991	8.8146	1.1854	16
45	8.6810	9.9995	8.6815	1.3185	8.8156	9.9991	8.8165	1.1835	15
46	8.6837	9.9995	8.6842	1.3158	8.8175	9.9991	8.8185	1.1815	14
47	8.6863	9.9995	8.6868	1.3132	8.8194	9.9991	8.8204	1.1796	13
48	8.6889	9.9995	8.6894	1.3106	8.8213	9.9990	8.8223	1.1777	12
49	8.6914	9.9995	8.6920	1.3080	8.8232	9.9990	8.8242	1.1758	11
50	8.6940	9.9995	8.6945	1.3055	8.8251	9.9990	8.8261	1.1739	10
51	8.6965	9.9995	8.6971	1.3029	8.8270	9.9990	8.8280	1.1720	9
52	8.6991	9.9995	8.6996	1.3004	8.8289	9.9990	8.8299	1.1701	8
53	8.7016	9.9994	8.7021	1.2979	8.8307	9.9990	8.8317	1.1683	7
54	8.7041	9.9994	8.7046	1.2954	8.8326	9.9990	8.8336	1.1664	6
55	8.7066	9.9994	8.7071	1.2929	8.8345	9.9990	8.8355	1.1645	5
56	8.7090	9.9994	8.7096	1.2904	8.8363	9.9990	8.8373	1.1627	4
57	8.7115	9.9994	8.7121	1.2879	8.8381	9.9990	8.8392	1.1608	3
58	8.7140	9.9994	8.7145	1.2855	8.8400	9.9990	8.8410	1.1590	2
59	8.7164	9.9994	8.7170	1.2830	8.8418	9.9989	8.8428	1.1572	1
60	8.7188	9.9994	8.7194	1.2806	8.8436	9.9989	8.8446	1.1554	0
	Cos.	Sin.	Cotang.	Tang.	Cos.	Sin.	Cotang.	Tang.	
	87°				86°				

	4°				5°				
	Sin.	Cos.	Tang.	Cotang.	Sin.	Cos.	Tang.	Cotang.	
0′	8.8436	9.9989	8.8446	1.1554	8.9403	9.9983	8.9420	1.0580	60′
1	8.8454	9.9989	8.8465	1.1535	8.9417	9.9983	8.9434	1.0566	59
2	8.8472	9.9989	8.8483	1.1517	8.9432	9.9983	8.9449	1.0551	58
3	8.8490	9.9989	8.8501	1.1499	8.9446	9.9983	8.9463	1.0537	57
4	8.8508	9.9989	8.8518	1.1482	8.9460	9.9983	8.9477	1.0523	56
5	8.8525	9.9989	8.8536	1.1464	8.9475	9.9983	8.9492	1.0508	55
6	8.8543	9.9989	8.8554	1.1446	8.9489	9.9983	8.9506	1.0494	54
7	8.8560	9.9989	8.8572	1.1428	8.9503	9.9983	8.9520	1.0480	53
8	8.8578	9.9989	8.8589	1.1411	8.9517	9.9983	8.9534	1.0466	52
9	8.8595	9.9989	8.8607	1.1393	8.9531	9.9982	8.9549	1.0451	51
10	8.8613	9.9989	8.8624	1.1376	8.9545	9.9982	8.9563	1.0437	50
11	8.8630	9.9988	8.8642	1.1358	8.9559	9.9982	8.9577	1.0423	49
12	8.8647	9.9988	8.8659	1.1341	8.9573	9.9982	8.9591	1.0409	48
13	8.8665	9.9988	8.8676	1.1324	8.9587	9.9982	8.9605	1.0395	47
14	8.8682	9.9988	8.8694	1.1306	8.9601	9.9982	8.9619	1.0381	46
15	8.8699	9.9988	8.8711	1.1289	8.9614	9.9982	8.9633	1.0367	45
16	8.8716	9.9988	8.8728	1.1272	8.9628	9.9982	8.9646	1.0354	44
17	8.8733	9.9988	8.8745	1.1255	8.9642	9.9982	8.9660	1.0340	43
18	8.8749	9.9988	8.8762	1.1238	8.9655	9.9981	8.9674	1.0326	42
19	8.8766	9.9988	8.8778	1.1222	8.9669	9.9981	8.9688	1.0312	41
20	8.8783	9.9988	8.8795	1.1205	8.9682	9.9981	8.9701	1.0299	40
21	8.8799	9.9987	8.8812	1.1188	8.9696	9.9981	8.9715	1.0285	39
22	8.8816	9.9987	8.8829	1.1171	8.9709	9.9981	8.9729	1.0271	38
23	8.8833	9.9987	8.8845	1.1155	8.9723	9.9981	8.9742	1.0258	37
24	8.8849	9.9987	8.8862	1.1138	8.9736	9.9981	8.9756	1.0244	36
25	8.8865	9.9987	8.8878	1.1122	8.9750	9.9981	8.9769	1.0231	35
26	8.8882	9.9987	8.8895	1.1105	8.9763	9.9980	8.9782	1.0218	34
27	8.8898	9.9987	8.8911	1.1089	8.9776	9.9980	8.9796	1.0204	33
28	8.8914	9.9987	8.8927	1.1073	8.9789	9.9980	8.9809	1.0191	32
29	8.8930	9.9987	8.8944	1.1056	8.9803	9.9980	8.9823	1.0177	31
30	8.8946	9.9987	8.8960	1.1040	8.9816	9.9980	8.9836	1.0164	30
31	8.8962	9.9986	8.8976	1.1024	8.9829	9.9980	8.9849	1.0151	29
32	8.8978	9.9986	8.8992	1.1008	8.9842	9.9980	8.9862	1.0138	28
33	8.8994	9.9986	8.9008	1.0992	8.9855	9.9980	8.9875	1.0125	27
34	8.9010	9.9986	8.9024	1.0976	8.9868	9.9979	8.9888	1.0112	26
35	8.9026	9.9986	8.9040	1.0960	8.9881	9.9979	8.9901	1.0099	25
36	8.9042	9.9986	8.9056	1.0944	8.9894	9.9979	8.9915	1.0085	24
37	8.9057	9.9986	8.9071	1.0929	8.9907	9.9979	8.9928	1.0072	23
38	8.9073	9.9986	8.9087	1.0913	8.9919	9.9979	8.9940	1.0060	22
39	8.9089	9.9986	8.9103	1.0897	8.9932	9.9979	8.9953	1.0047	21
40	8.9104	9.9986	8.9118	1.0882	8.9945	9.9979	8.9966	1.0034	20
41	8.9119	9.9985	8.9134	1.0866	8.9958	9.9979	8.9979	1.0021	19
42	8.9135	9 9985	8.9150	1.0850	8.9970	9.9978	8.9992	1.0008	18
43	8.9150	9.9985	8.9165	1.0835	8.9983	9.9978	9.0005	0.9995	17
44	8.9166	9.9985	8.9180	1.0820	8.9996	9.9978	9.0017	0.9983	16
45	8.9181	9.9985	8.9196	1.0804	9.0008	9.9978	9.0030	0.9970	15
46	8.9196	9.9985	8.9211	1.0789	9.0021	9.9978	9.0043	0.9957	14
47	8.9211	9.9985	8.9226	1.0774	9.0033	9.9978	9.0055	0.9945	13
48	8.9226	9.9985	8.9241	1.0759	9.0046	9.9978	9.0068	0.9932	12
49	8.9241	9.9985	8.9256	1.0744	9.0058	9.9978	9.0080	0.9920	11
50	8.9256	9.9985	8.9272	1.0728	9.0070	9.9977	9.0093	0.9907	10
51	8.9271	9.9984	8.9287	1.0713	9.0083	9.9977	9.0105	0.9895	9
52	8.9286	9.9984	8.9302	1.0698	9.0095	9.9977	9.0118	0.9882	8
53	8.9301	9.9984	8.9316	1.0684	9.0107	9.9977	9.0130	0.9870	7
54	8.9315	9.9984	8.9331	1.0669	9.0120	9.9977	9.0143	0.9857	6
55	8.9330	9.9984	8.9346	1.0654	9.0132	9.9977	9.0155	0.9845	5
56	8.9345	9.9984	8.9361	1.0639	9.0144	9.9977	9.0167	0.9833	4
57	8.9359	9.9984	8.9376	1.0624	9.0156	9.9977	9.0180	0.9820	3
58	8.9374	9.9984	8.9390	1.0610	9.0168	9.9976	9.0192	0.9808	2
59	8.9388	9.9984	8.9405	1.0595	9.0180	9.9976	9.0204	0.9796	1
60	8.9403	9.9983	8.9420	1.0580	9.0192	9.9976	9.0216	0.9784	0
	Cos.	Sin.	Cotang.	Tang.	Cos.	Sin.	Cotang.	Tang.	
	85°				84°				

	6°				7°				
	Sin.	Cos.	Tang.	Cotang.	Sin.	Cos.	Tang.	Cotang.	
0′	9.0192	9.9976	9.0216	0.9784	9.0859	9.9968	9.0891	0.9109	60′
1	9.0204	9.9976	9.0228	0.9772	9.0869	9.9967	9.0902	0.9098	59
2	9.0216	9.9976	9.0240	0.9760	9.0879	9.9967	9.0912	0.9088	58
3	9.0228	9.9976	9.0253	0.9747	9.0890	9.9967	9.0923	0.9077	57
4	9.0240	9.9976	9.0265	0.9735	9.0900	9.9967	9.0933	0.9067	56
5	9.0252	9.9975	9.0277	0.9723	9.0910	9.9967	9.0943	0.9057	55
6	9.0264	9.9975	9.0289	0.9711	9.0920	9.9967	9.0954	0.9046	54
7	9.0276	9.9975	9.0300	0.9700	9.0930	9.9966	9.0964	0.9036	53
8	9.0287	9.9975	9.0312	0.9688	9.0940	9.9966	9.0974	0.9026	52
9	9.0299	9.9975	9.0324	0.9676	9.0951	9.9966	9.0984	0.9016	51
10	9.0311	9.9975	9.0336	0.9664	9.0961	9.9966	9.0995	0.9005	50
11	9.0323	9.9975	9.0348	0.9652	9.0971	9.9966	9.1005	0.8995	49
12	9.0334	9.9975	9.0360	0.9640	9.0981	9.9966	9.1015	0.8985	48
13	9.0346	9.9974	9.0371	0.9629	9.0991	9.9965	9.1025	0.8975	47
14	9.0357	9.9974	9.0383	0.9617	9.1001	9.9965	9.1035	0.8965	46
15	9.0369	9.9974	9.0395	0.9605	9.1011	9.9965	9.1045	0.8955	45
16	9.0380	9.9974	9.0407	0.9593	9.1020	9.9965	9.1055	0.8945	44
17	9.0392	9.9974	9.0418	0.9582	9.1030	9.9965	9.1066	0.8934	43
18	9.0403	9.9974	9.0430	0.9570	9.1040	9.9965	9.1076	0.8924	42
19	9.0415	9.9974	9.0441	0.9559	9.1050	9.9964	9.1086	0.8914	41
20	9.0426	9.9973	9.0453	0.9547	9.1060	9.9964	9.1096	0.8904	40
21	9.0438	9.9973	9.0464	0.9536	9.1070	9.9964	9.1106	0.8894	39
22	9.0449	9.9973	9.0476	0.9524	9.1080	9.9964	9.1116	0.8884	38
23	9.0460	9.9973	9.0487	0.9513	9.1089	9.9964	9.1125	0.8875	37
24	9.0472	9.9973	9.0499	0.9501	9.1099	9.9964	9.1135	0.8865	36
25	9.0483	9.9973	9.0510	0.9490	9.1109	9.9964	9.1145	0.8855	35
26	9.0494	9.9973	9.0521	0.9479	9.1118	9.9963	9.1155	0.8845	34
27	9.0505	9.9972	9.0533	0.9467	9.1128	9.9963	9.1165	0.8835	33
28	9.0516	9.9972	9.0544	0.9456	9.1138	9.9963	9.1175	0.8825	32
29	9.0527	9.9972	9.0555	0.9445	9.1147	9.9963	9.1185	0.8815	31
30	9.0539	9.9972	9.0567	0.9433	9.1157	9.9963	9.1194	0.8806	30
31	9.0550	9.9972	9.0578	0.9422	9.1167	9.9963	9.1204	0.8796	29
32	9.0561	9.9972	9.0589	0.9411	9.1176	9.9962	9.1214	0.8786	28
33	9.0572	9.9972	9.0600	0.9400	9.1186	9.9962	9.1223	0.8777	27
34	9.0583	9.9971	9.0611	0.9389	9.1195	9.9962	9.1233	0.8767	26
35	9.0594	9.9971	9.0622	0.9378	9.1205	9.9962	9.1243	0.8757	25
36	9.0605	9.9971	9.0633	0.9367	9.1214	9.9962	9.1252	0.8748	24
37	9.0616	9.9971	9.0645	0.9355	9.1224	9.9962	9.1262	0.8738	23
38	9.0626	9.9971	9.0656	0.9344	9.1233	9.9961	9.1272	0.8728	22
39	9.0637	9.9971	9.0667	0.9333	9.1242	9.9961	9.1281	0.8719	21
40	9.0648	9.9971	9.0678	0.9322	9.1252	9.9961	9.1291	0.8709	20
41	9.0659	9.9970	9.0688	0.9312	9.1261	9.9961	9.1300	0.8700	19
42	9.0670	9.9970	9.0699	0.9301	9.1271	9.9961	9.1310	0.8690	18
43	9.0680	9.9970	9.0710	0.9290	9.1280	9.9960	9.1319	0.8681	17
44	9.0691	9.9970	9.0721	0.9279	9.1289	9.9960	9.1329	0.8671	16
45	9.0702	9.9970	9.0732	0.9268	9.1299	9.9960	9.1338	0.8662	15
46	9.0712	9.9970	9.0743	0.9257	9.1308	9.9960	9.1348	0.8652	14
47	9.0723	9.9969	9.0754	0.9246	9.1317	9.9960	9.1357	0.8643	13
48	9.0734	9.9969	9.0764	0.9236	9.1326	9.9960	9.1367	0.8633	12
49	9.0744	9.9969	9.0775	0.9225	9.1336	9.9959	9.1376	0.8624	11
50	9.0755	9.9969	9.0786	0.9214	9.1345	9.9959	9.1385	0.8615	10
51	9.0765	9.9969	9.0796	0.9204	9.1354	9.9959	9.1395	0.8605	9
52	9.0776	9.9969	9.0807	0.9193	9.1363	9.9959	9.1404	0.8596	8
53	9.0786	9.9969	9.0818	0.9182	9.1372	9.9959	9.1413	0.8587	7
54	9.0797	9.9968	9.0828	0.9172	9.1381	9.9959	9.1423	0.8577	6
55	9.0807	9.9968	9.0839	0.9161	9.1390	9.9958	9.1432	0.8568	5
56	9.0818	9.9968	9.0849	0.9151	9.1399	9.9958	9.1441	0.8559	4
57	9.0828	9.9968	9.0860	0.9140	9.1409	9.9958	9.1450	0.8550	3
58	9.0838	9.9968	9.0871	0.9129	9.1418	9.9958	9.1460	0.8540	2
59	9.0849	9.9968	9.0881	0.9119	9.1427	9.9958	9.1469	0.8531	1
60	9.0859	9.9968	9.0891	0.9109	9.1436	9.9958	9.1478	0.8522	0
	Cos.	Sin.	Cotang.	Tang.	Cos.	Sin.	Cotang.	Tang.	
	83°				82°				

	8°				9°				
	Sin.	Cos.	Tang.	Cotang.	Sin.	Cos.	Tang.	Cotang.	
0′	9.1436	9.9958	9.1478	0.8522	9.1943	9.9946	9.1997	0.8003	60′
1	9.1445	9.9957	9.1487	0.8513	9.1951	9.9946	9.2005	0.7995	59
2	9.1453	9.9957	9.1496	0.8504	9.1959	9.9946	9.2013	0.7987	58
3	9.1462	9.9957	9.1505	0.8495	9.1967	9.9946	9.2022	0.7978	57
4	9.1471	9.9957	9.1515	0.8485	9.1975	9.9945	9.2030	0.7970	56
5	9.1480	9.9957	9.1524	0.8476	9.1983	9.9945	9.2038	0.7962	55
6	9.1489	9.9956	9.1533	0.8467	9.1991	9.9945	9.2046	0.7954	54
7	9.1498	9.9956	9.1542	0.8458	9.1999	9.9945	9.2054	0.7946	53
8	9.1507	9.9956	9.1551	0.8449	9.2007	9.9945	9.2062	0.7938	52
9	9.1516	9.9956	9.1560	0.8440	9.2015	9.9944	9.2070	0.7930	51
10	9.1525	9.9956	9.1569	0.8431	9.2022	9.9944	9.2078	0.7922	50
11	9.1533	9.9956	9.1578	0.8422	9.2030	9.9944	9.2086	0.7914	49
12	9.1542	9.9955	9.1587	0.8413	9.2038	9.9944	9.2094	0.7906	48
13	9.1551	9.9955	9.1596	0.8404	9.2046	9.9944	9.2102	0.7898	47
14	9.1560	9.9955	9.1605	0.8395	9.2054	9.9943	9.2110	0.7890	46
15	9.1568	9.9955	9.1613	0.8387	9.2061	9.9943	9.2118	0.7882	45
16	9.1577	9.9955	9.1622	0.8378	9.2069	9.9943	9.2126	0.7874	44
17	9.1586	9.9954	9.1631	0.8369	9.2077	9.9943	9.2134	0.7866	43
18	9.1594	9.9954	9.1640	0.8360	9.2085	9.9943	9.2142	0.7858	42
19	9.1603	9.9954	9.1649	0.8351	9.2092	9.9942	9.2150	0.7850	41
20	9.1612	9.9954	9.1658	0.8342	9.2100	9.9942	9.2158	0.7842	40
21	9.1620	9.9954	9.1667	0.8333	9.2108	9.9942	9.2166	0.7834	39
22	9.1629	9.9954	9.1675	0.8325	9.2115	9.9942	9.2174	0.7826	38
23	9.1637	9.9953	9.1684	0.8316	9.2123	9.9941	9.2181	0.7819	37
24	9.1646	9.9953	9.1693	0.8307	9.2131	9.9941	9.2189	0.7811	36
25	9.1655	9.9953	9.1702	0.8298	9.2138	9.9941	9.2197	0.7803	35
26	9.1663	9.9953	9.1710	0.8290	9.2146	9.9941	9.2205	0.7795	34
27	9.1672	9.9953	9.1719	0.8281	9.2153	9.9941	9.2213	0.7787	33
28	9.1680	9.9952	9.1728	0.8272	9.2161	9.9940	9.2221	0.7779	32
29	9.1689	9.9952	9.1736	0.8264	9.2169	9.9940	9.2228	0.7772	31
30	9.1697	9.9952	9.1745	0.8255	9.2176	9.9940	9.2236	0.7764	30
31	9.1705	9.9952	9.1754	0.8246	9.2184	9.9940	9.2244	0.7756	29
32	9.1714	9.9952	9.1762	0.8238	9.2191	9.9940	9.2252	0.7748	28
33	9.1722	9.9951	9.1771	0.8229	9.2199	9.9939	9.2259	0.7741	27
34	9.1731	9.9951	9.1779	0.8221	9.2206	9.9939	9.2267	0.7733	26
35	9.1739	9.9951	9.1788	0.8212	9.2214	9.9939	9.2275	0.7725	25
36	9.1747	9.9951	9.1797	0.8203	9.2221	9.9939	9.2282	0.7718	24
37	9.1756	9.9951	9.1805	0.8195	9.2229	9.9939	9.2290	0.7710	23
38	9.1764	9.9951	9.1814	0.8186	9.2236	9.9938	9.2298	0.7702	22
39	9.1772	9.9950	9.1822	0.8178	9.2243	9.9938	9.2305	0.7695	21
40	9.1781	9.9950	9.1831	0.8169	9.2251	9.9938	9.2313	0.7687	20
41	9.1789	9.9950	9.1839	0.8161	9.2258	9.9938	9.2321	0.7679	19
42	9.1797	9.9950	9.1848	0.8152	9.2266	9.9937	9.2328	0.7672	18
43	9.1806	9.9950	9.1856	0.8144	9.2273	9.9937	9.2336	0.7664	17
44	9.1814	9.9949	9.1864	0.8136	9.2280	9.9937	9.2343	0.7657	16
45	9.1822	9.9949	9.1873	0.8127	9.2288	9.9937	9.2351	0.7649	15
46	9.1830	9.9949	9.1881	0.8119	9.2295	9.9937	9.2359	0.7641	14
47	9.1838	9.9949	9.1890	0.8110	9.2303	9.9936	9.2366	0.7634	13
48	9.1847	9.9949	9.1898	0.8102	9.2310	9.9936	9.2374	0.7626	12
49	9.1855	9.9948	9.1906	0.8094	9.2317	9.9936	9.2381	0.7619	11
50	9.1863	9.9948	9.1915	0.8085	9.2324	9.9936	9.2389	0.7611	10
51	9.1871	9.9948	9.1923	0.8077	9.2332	9.9936	9.2396	0.7604	9
52	9.1879	9.9948	9.1931	0.8069	9.2339	9.9935	9.2404	0.7596	8
53	9.1887	9.9948	9.1940	0.8060	9.2346	9.9935	9.2411	0.7589	7
54	9.1895	9.9947	9.1948	0.8052	9.2353	9.9935	9.2419	0.7581	6
55	9.1903	9.9947	9.1956	0.8044	9.2361	9.9935	9.2426	0.7574	5
56	9.1911	9.9947	9.1964	0.8036	9.2368	9.9934	9.2434	0.7566	4
57	9.1919	9.9947	9.1973	0.8027	9.2375	9.9934	9.2441	0.7559	3
58	9.1927	9.9947	9.1981	0.8019	9.2382	9.9934	9.2448	0.7552	2
59	9.1935	9.9946	9.1989	0.8011	9.2390	9.9934	9.2456	0.7544	1
60	9.1943	9.9946	9.1997	0.8003	9.2397	9.9934	9.2463	0.7537	0
	Cos.	Sin.	Cotang.	Tang.	Cos.	Sin.	Cotang.	Tang.	
	81°				80°				

	10°				11°				
	Sin.	Cos.	Tang.	Cotang.	Sin.	Cos.	Tang.	Cotang.	
0′	9.2397	9.9934	9.2463	0.7537	9.2806	9.9919	9.2887	0.7113	60′
1	9.2404	9.9933	9.2471	0.7529	9.2812	9.9919	9.2893	0.7107	59
2	9.2411	9.9933	9.2478	0.7522	9.2819	9.9919	9.2900	0.7100	58
3	9.2418	9.9933	9.2485	0.7515	9.2825	9.9919	9.2907	0.7093	57
4	9.2425	9.9933	9.2493	0.7507	9.2832	9.9918	9.2913	0.7087	56
5	9.2432	9.9932	9.2500	0.7500	9.2838	9.9918	9.2920	0.7080	55
6	9.2439	9.9932	9.2507	0.7493	9.2845	9.9918	9.2927	0.7073	54
7	9.2447	9.9932	9.2515	0.7485	9.2851	9.9918	9.2934	0.7066	53
8	9.2454	9.9932	9.2522	0.7478	9.2858	9.9917	9.2940	0.7060	52
9	9.2461	9.9931	9.2529	0.7471	9.2864	9.9917	9.2947	0.7053	51
10	9.2468	9.9931	9.2536	0.7464	9.2870	9.9917	9.2953	0.7047	50
11	9.2475	9.9931	9.2544	0.7456	9.2877	9.9917	9.2960	0.7040	49
12	9.2482	9.9931	9.2551	0.7449	9.2883	9.9916	9.2967	0.7033	48
13	9.2489	9.9931	9.2558	0.7442	9.2890	9.9916	9.2973	0.7027	47
14	9.2496	9.9930	9.2565	0.7435	9.2896	9.9916	9.2980	0.7020	46
15	9.2503	9.9930	9.2573	0.7427	9.2902	9.9916	9.2987	0.7013	45
16	9.2510	9.9930	9.2580	0.7420	9.2909	9.9915	9.2993	0.7007	44
17	9.2517	9.9930	9.2587	0.7413	9.2915	9.9915	9.3000	0.7000	43
18	9.2524	9.9929	9.2594	0.7406	9.2921	9.9915	9.3006	0.6994	42
19	9.2531	9.9929	9.2601	0.7399	9.2928	9.9915	9.3013	0.6987	41
20	9.2538	9.9929	9.2609	0.7391	9.2934	9.9914	9.3020	0.6980	40
21	9.2545	9.9929	9.2616	0.7384	9.2940	9.9914	9.3026	0.6974	39
22	9.2551	9.9929	9.2623	0.7377	9.2947	9.9914	9.3033	0.6967	38
23	9.2558	9.9928	9.2630	0.7370	9.2953	9.9914	9.3039	0.6961	37
24	9.2565	9.9928	9.2637	0.7363	9.2959	9.9913	9.3046	0.6954	36
25	9.2572	9.9928	9.2644	0.7356	9.2965	9.9913	9.3052	0.6948	35
26	9.2579	9.9928	9.2651	0.7349	9.2972	9.9913	9.3059	0.6941	34
27	9.2586	9.9927	9.2658	0.7342	9.2978	9.9913	9.3065	0.6935	33
28	9.2593	9.9927	9.2666	0.7334	9.2984	9.9912	9.3072	0.6928	32
29	9.2600	9.9927	9.2673	0.7327	9.2990	9.9912	9.3078	0.6922	31
30	9.2606	9.9927	9.2680	0.7320	9.2997	9.9912	9.3085	0.6915	30
31	9.2613	9.9926	9.2687	0.7313	9.3003	9.9912	9.3091	0.6909	29
32	9.2620	9.9926	9.2694	0.7306	9.3009	9.9911	9.3098	0.6902	28
33	9.2627	9.9926	9.2701	0.7299	9.3015	9.9911	9.3104	0.6896	27
34	9.2634	9.9926	9.2708	0.7292	9.3021	9.9911	9.3110	0.6890	26
35	9.2640	9.9925	9.2715	0.7285	9.3027	9.9911	9.3117	0.6883	25
36	9.2647	9.9925	9.2722	0.7278	9.3034	9.9910	9.3123	0.6877	24
37	9.2654	9.9925	9.2729	0.7271	9.3040	9.9910	9.3130	0.6870	23
38	9.2661	9.9925	9.2736	0.7264	9.3046	9.9910	9.3136	0.6864	22
39	9.2667	9.9925	9.2743	0.7257	9.3052	9.9910	9.3142	0.6858	21
40	9.2674	9.9924	9.2750	0.7250	9.3058	9.9909	9.3149	0.6851	20
41	9.2681	9.9924	9.2757	0.7243	9.3064	9.9909	9.3155	0.6845	19
42	9.2687	9.9924	9.2764	0.7236	9.3070	9.9909	9.3162	0.6838	18
43	9.2694	9.9924	9.2770	0.7230	9.3077	9.9909	9.3168	0.6832	17
44	9.2701	9.9923	9.2777	0.7223	9.3083	9.9908	9.3174	0.6826	16
45	9.2707	9.9923	9.2784	0.7216	9.3089	9.9908	9.3181	0.6819	15
46	9.2714	9.9923	9.2791	0.7209	9.3095	9.9908	9.3187	0.6813	14
47	9.2721	9.9923	9.2798	0.7202	9.3101	9.9908	9.3193	0.6807	13
48	9.2727	9.9922	9.2805	0.7195	9.3107	9.9907	9.3200	0.6800	12
49	9.2734	9.9922	9.2812	0.7188	9.3113	9.9907	9.3206	0.6794	11
50	9.2740	9.9922	9.2819	0.7181	9.3119	9.9907	9.3212	0.6788	10
51	9.2747	9.9922	9.2825	0.7175	9.3125	9.9906	9.3219	0.6781	9
52	9.2754	9.9921	9.2832	0.7168	9.3131	9.9906	9.3225	0.6775	8
53	9.2760	9.9921	9.2839	0.7161	9.3137	9.9906	9.3231	0.6769	7
54	9.2767	9.9921	9.2846	0.7154	9.3143	9.9906	9.3237	0.6763	6
55	9.2773	9.9921	9.2853	0.7147	9.3149	9.9905	9.3244	0.6756	5
56	9.2780	9.9920	9.2859	0.7141	9.3155	9.9905	9.3250	0.6750	4
57	9.2786	9.9920	9.2866	0.7134	9.3161	9.9905	9.3256	0.6744	3
58	9.2793	9.9920	9.2873	0.7127	9.3167	9.9905	9.3262	0.6738	2
59	9.2799	9.9920	9.2880	0.7120	9.3173	9.9904	9.3269	0.6731	1
60	9.2806	9.9919	9.2887	0.7113	9.3179	9.9904	9.3275	0.6725	0
	Cos.	Sin.	Cotang.	Tang.	Cos.	Sin.	Cotang.	Tang.	
	79°				78°				

	12°				13°				
	Sin.	Cos.	Tang.	Cotang.	Sin.	Cos.	Tang.	Cotang.	
0′	9.3179	9.9904	9.3275	0.6725	9.3521	9.9887	9.3634	0.6366	60′
1	9.3185	9.9904	9.3281	0.6719	9.3526	9.9887	9.3639	0.6361	59
2	9.3191	9.9904	9.3287	0.6713	9.3532	9.9887	9.3645	0.6355	58
3	9.3197	9.9903	9.3293	0.6707	9.3537	9.9886	9.3651	0.6349	57
4	9.3202	9.9903	9.3300	0.6700	9.3543	9.9886	9.3657	0.6343	56
5	9.3208	9.9903	9.3306	0.6694	9.3548	9.9886	9.3662	0.6338	55
6	9.3214	9.9902	9.3312	0.6688	9.3554	9.9885	9.3668	0.6332	54
7	9.3220	9.9902	9.3318	0.6682	9.3559	9.9885	9.3674	0.6326	53
8	9.3226	9.9902	9.3324	0.6676	9.3564	9.9885	9.3680	0.6320	52
9	9.3232	9.9902	9.3330	0.6670	9.3570	9.9885	9.3685	0.6315	51
10	9.3238	9.9901	9.3336	0.6664	9.3575	9.9884	9.3691	0.6309	50
11	9.3244	9.9901	9.3343	0.6657	9.3581	9.9884	9.3697	0.6303	49
12	9.3250	9.9901	9.3349	0.6651	9.3586	9.9884	9.3702	0.6298	48
13	9.3255	9.9901	9.3355	0.6645	9.3591	9.9883	9.3708	0.6292	47
14	9.3261	9.9900	9.3361	0.6639	9.3597	9.9883	9.3714	0.6286	46
15	9.3267	9.9900	9.3367	0.6633	9.3602	9.9883	9.3719	0.6281	45
16	9.3273	9.9900	9.3373	0.6627	9.3608	9.9883	9.3725	0.6275	44
17	9.3279	9.9899	9.3379	0.6621	9.3613	9.9882	9.3731	0.6269	43
18	9.3284	9.9899	9.3385	0.6615	9.3618	9.9882	9.3736	0.6264	42
19	9.3290	9.9899	9.3391	0.6609	9.3624	9.9882	9.3742	0.6258	41
20	9.3296	9.9899	9.3397	0.6603	9.3629	9.9881	9.3748	0.6252	40
21	9.3302	9.9898	9.3403	0.6597	9.3634	9.9881	9.3753	0.6247	39
22	9.3308	9.9898	9.3409	0.6591	9.3640	9.9881	9.3759	0.6241	38
23	9.3313	9.9898	9.3416	0.6584	9.3645	9.9880	9.3764	0.6236	37
24	9.3319	9.9897	9.3422	0.6578	9.3650	9.9880	9.3770	0.6230	36
25	9.3325	9.9897	9.3428	0.6572	9.3655	9.9880	9.3776	0.6224	35
26	9.3331	9.9897	9.3434	0.6566	9.3661	9.9880	9.3781	0.6219	34
27	9.3336	9.9897	9.3440	0.6560	9.3666	9.9879	9.3787	0.6213	33
28	9.3342	9.9896	9.3446	0.6554	9.3671	9.9879	9.3792	0.6208	32
29	9.3348	9.9896	9.3452	0.6548	9.3677	9.9879	9.3798	0.6202	31
30	9.3353	9.9896	9.3458	0.6542	9.3682	9.9878	9.3804	0.6196	30
31	9.3359	9.9896	9.3464	0.6536	9.3687	9.9878	9.3809	0.6191	29
32	9.3365	9.9895	9.3469	0.6531	9.3692	9.9878	9.3815	0.6185	28
33	9.3370	9.9895	9.3475	0.6525	9.3698	9.9877	9.3820	0.6180	27
34	9.3376	9.9895	9.3481	0.6519	9.3703	9.9877	9.3826	0.6174	26
35	9.3382	9.9894	9.3487	0.6513	9.3708	9.9877	9.3831	0.6169	25
36	9.3387	9.9894	9.3493	0.6507	9.3713	9.9876	9.3837	0.6163	24
37	9.3393	9.9894	9.3499	0.6501	9.3719	9.9876	9.3842	0.6158	23
38	9.3399	9.9894	9.3505	0.6495	9.3724	9.9876	9.3848	0.6152	22
39	9.3404	9.9893	9.3511	0.6489	9.3729	9.9876	9.3853	0.6147	21
40	9.3410	9.9893	9.3517	0.6482	9.3734	9.9875	9.3859	0.6141	20
41	9.3416	9.9893	9.3523	0.6477	9.3739	9.9875	9.3864	0.6136	19
42	9.3421	9.9892	9.3529	0.6471	9.3745	9.9875	9.3870	0.6130	18
43	9.3427	9.9892	9.3535	0.6465	9.3750	9.9874	9.3875	0.6125	17
44	9.3432	9.9892	9.3541	0.6459	9.3755	9.9874	9.3881	0.6119	16
45	9.3438	9.9892	9.3546	0.6454	9.3760	9.9874	9.3886	0.6114	15
46	9.3444	9.9891	9.3552	0.6448	9.3765	9.9873	9.3892	0.6108	14
47	9.3449	9.9891	9.3558	0.6442	9.3770	9.9873	9.3897	0.6103	13
48	9.3455	9.9891	9.3564	0.6436	9.3775	9.9873	9.3903	0.6097	12
49	9.3460	9.9890	9.3570	0.6430	9.3781	9.9872	9.3908	0.6092	11
50	9.3466	9.9890	9.3576	0.6424	9.3786	9.9872	9.3914	0.6086	10
51	9.3471	9.9890	9.3581	0.6419	9.3791	9.9872	9.3919	0.6081	9
52	9.3477	9.9890	9.3587	0.6413	9.3796	9.9872	9.3924	0.6076	8
53	9.3482	9.9889	9.3593	0.6407	9.3801	9.9871	9.3930	0.6070	7
54	9.3488	9.9889	9.3599	0.6401	9.3806	9.9871	9.3935	0.6065	6
55	9.3493	9.9889	9.3605	0.6395	9.3811	9.9871	9.3941	0.6059	5
56	9.3499	9.9888	9.3611	0.6389	9.3816	9.9870	9.3946	0.6054	4
57	9.3504	9.9888	9.3616	0.6384	9.3822	9.9870	9.3952	0.6048	3
58	9.3510	9.9888	9.3622	0.6378	9.3827	9.9870	9.3957	0.6043	2
59	9.3515	9.9888	9.3628	0.6372	9.3832	9.9869	9.3962	0.6038	1
60	9.3521	9.9887	9.3634	0.6366	9.3837	9.9869	9.3968	0.6032	0
	Cos.	Sin.	Cotang.	Tang.	Cos.	Sin.	Cotang.	Tang.	
	77°				76°				

	14°				15°				
	Sin.	Cos.	Tang.	Cotang.	Sin.	Cos.	Tang.	Cotang.	
0′	9.3837	9.9869	9.3968	0.6032	9.4130	9.9849	9.4281	0.5719	60′
1	9.3842	9.9869	9.3973	0.6027	9.4135	9.9849	9.4286	0.5714	59
2	9.3847	9.9868	9.3978	0.6022	9.4139	9.9849	9.4291	0.5709	58
3	9.3852	9.9868	9.3984	0.6016	9.4144	9.9848	9.4296	0.5704	57
4	9.3857	9.9868	9.3989	0.6011	9.4149	9.9848	9.4301	0.5699	56
5	9.3862	9.9867	9.3995	0.6005	9.4153	9.9848	9.4306	0.5694	55
6	9.3867	9.9867	9.4000	0.6000	9.4158	9.9847	9.4311	0.5689	54
7	9.3872	9.9867	9.4005	0.5995	9.4163	9.9847	9.4316	0.5684	53
8	9.3877	9.9867	9.4011	0.5989	9.4168	9.9847	9.4321	0.5679	52
9	9.3882	9.9866	9.4016	0.5984	9.4172	9.9846	9.4326	0.5674	51
10	9.3887	9.9866	9.4021	0.5979	9.4177	9.9846	9.4331	0.5669	50
11	9.3892	9.9866	9.4027	0.5973	9.4181	9.9846	9.4336	0.5664	49
12	9.3897	9.9865	9.4032	0.5968	9.4186	9.9845	9.4341	0.5659	48
13	9.3902	9.9865	9.4037	0.5963	9.4191	9.9845	9.4346	0.5654	47
14	9.3907	9.9865	9.4042	0.5958	9.4195	9.9845	9.4351	0.5649	46
15	9.3912	9.9864	9.4048	0.5952	9.4200	9.9844	9.4356	0.5644	45
16	9.3917	9.9864	9.4053	0.5947	9.4205	9.9844	9.4361	0.5639	44
17	9.3922	9.9864	9.4058	0.5942	9.4209	9.9844	9.4366	0.5634	43
18	9.3927	9.9863	9.4064	0.5936	9.4214	9.9843	9.4371	0.5629	42
19	9.3932	9.9863	9.4069	0.5931	9.4219	9.9843	9.4376	0.5624	41
20	9.3937	9.9863	9.4074	0.5926	9.4223	9.9843	9.4381	0.5619	40
21	9.3942	9.9862	9.4079	0.5921	9.4228	9.9842	9.4386	0.5614	39
22	9.3947	9.9862	9.4085	0.5915	9.4232	9.9842	9.4390	0.5610	38
23	9.3952	9.9862	9.4090	0.5910	9.4237	9.9842	9.4395	0.5605	37
24	9.3957	9.9861	9.4095	0.5905	9.4242	9.9841	9.4400	0.5600	36
25	9.3961	9.9861	9.4100	0.5900	9.4246	9.9841	9.4405	0.5595	35
26	9.3966	9.9861	9.4106	0.5894	9.4251	9.9841	9.4410	0.5590	34
27	9.3971	9.9860	9.4111	0.5889	9.4255	9.9840	9.4415	0.5585	33
28	9.3976	9.9860	9.4116	0.5884	9.4260	9.9840	9.4420	0.5580	32
29	9.3981	9.9860	9.4121	0.5879	9.4264	9.9839	9.4425	0.5575	31
30	9.3986	9.9859	9.4127	0.5873	9.4269	9.9839	9.4430	0.5570	30
31	9.3991	9.9859	9.4132	0.5868	9.4274	9.9839	9.4435	0.5565	29
32	9.3996	9.9859	9.4137	0.5863	9.4278	9.9838	9.4440	0.5560	28
33	9.4001	9.9858	9.4142	0.5858	9.4283	9.9838	9.4445	0.5555	27
34	9.4005	9.9858	9.4147	0.5853	9.4287	9.9838	9.4449	0.5551	26
35	9.4010	9.9858	9.4153	0.5847	9.4292	9.9837	9.4454	0.5546	25
36	9.4015	9.9857	9.4158	0.5842	9.4296	9.9837	9.4459	0.5541	24
37	9.4020	9.9857	9.4163	0.5837	9.4301	9.9837	9.4464	0.5536	23
38	9.4025	9.9857	9.4168	0.5832	9.4305	9.9836	9.4469	0.5531	22
39	9.4030	9.9856	9.4173	0.5827	9.4310	9.9836	9.4474	0.5526	21
40	9.4035	9.9856	9.4178	0.5822	9.4314	9.9836	9.4479	0.5521	20
41	9.4039	9.9856	9.4184	0.5816	9.4319	9.9835	9.4484	0.5516	19
42	9.4044	9.9855	9.4189	0.5811	9.4323	9.9835	9.4488	0.5512	18
43	9.4049	9.9855	9.4194	0.5806	9.4328	9.9835	9.4493	0.5507	17
44	9.4054	9.9855	9.4199	0.5801	9.4332	9.9834	9.4498	0.5502	16
45	9.4059	9.9854	9.4204	0.5796	9.4337	9.9834	9.4503	0.5497	15
46	9.4063	9.9854	9.4209	0.5791	9.4341	9.9833	9.4508	0.5492	14
47	9.4068	9.9854	9.4214	0.5786	9.4346	9.9833	9.4513	0.5487	13
48	9.4073	9.9853	9.4220	0.5780	9.4350	9.9833	9.4517	0.5483	12
49	9.4078	9.9853	9.4225	0.5775	9.4355	9.9832	9.4522	0.5478	11
50	9.4083	9.9853	9.4230	0.5770	9.4359	9.9832	9.4527	0.5473	10
51	9.4087	9.9852	9.4235	0.5765	9.4364	9.9832	9.4532	0.5468	9
52	9.4092	9.9852	9.4240	0.5760	9.4368	9.9831	9.4537	0.5463	8
53	9.4097	9.9852	9.4245	0.5755	9.4372	9.9831	9.4541	0.5459	7
54	9.4102	9.9851	9.4250	0.5750	9.4377	9.9831	9.4546	0.5454	6
55	9.4106	9.9851	9.4255	0.5745	9.4381	9.9830	9.4551	0.5449	5
56	9.4111	9.9851	9.4260	0.5740	9.4386	9.9830	9.4556	0.5444	4
57	9.4116	9.9850	9.4265	0.5735	9.4390	9.9830	9.4561	0.5439	3
58	9.4121	9.9850	9.4270	0.5730	9.4395	9.9829	9.4565	0.5435	2
59	9.4125	9.9850	9.4275	0.5725	9.4399	9.9829	9.4570	0.5430	1
60	9.4130	9.9849	9.4281	0.5719	9.4403	9.9828	9.4575	0.5425	0
	Cos.	Sin.	Cotang.	Tang.	Cos.	Sin.	Cotang.	Tang.	
	75°				74°				

	16°				17°				
	Sin.	Cos.	Tang.	Cotang.	Sin.	Cos.	Tang.	Cotang.	
0′	9.4403	9.9828	9.4575	0.5425	9.4659	9.9806	9.4853	0.5147	60′
1	9.4408	9.9828	9.4580	0.5420	9.4663	9.9806	9.4858	0.5142	59
2	9.4412	9.9828	9.4584	0.5416	9.4668	9.9805	9.4862	0.5138	58
3	9.4417	9.9827	9.4589	0.5411	9.4672	9.9805	9.4867	0.5133	57
4	9.4421	9.9827	9.4594	0.5406	9.4676	9.9804	9.4871	0.5129	56
5	9.4425	9.9827	9.4599	0.5401	9.4680	9.9804	9.4876	0.5124	55
6	9.4430	9.9826	9.4603	0.5397	9.4684	9.9804	9.4880	0.5120	54
7	9.4434	9.9826	9.4608	0.5392	9.4688	9.9803	9.4885	0.5115	53
8	9.4438	9.9826	9.4613	0.5387	9.4692	9.9803	9.4889	0.5111	52
9	9.4443	9.9825	9.4618	0.5382	9.4696	9.9802	9.4894	0.5106	51
10	9.4447	9.9825	9.4622	0.5378	9.4700	9.9802	9.4898	0.5102	50
11	9.4452	9.9824	9.4627	0.5373	9.4705	9.9802	9.4903	0.5097	49
12	9.4456	9.9824	9.4632	0.5368	9.4709	9.9801	9.4907	0.5093	48
13	9.4460	9.9824	9.4637	0.5363	9.4713	9.9801	9.4912	0.5088	47
14	9.4465	9.9823	9.4641	0.5359	9.4717	9.9801	9.4916	0.5084	46
15	9.4469	9.9823	9.4646	0.5354	9.4721	9.9800	9.4921	0.5079	45
16	9.4473	9.9822	9.4651	0.5349	9.4725	9.9800	9.4925	0.5075	44
17	9.4478	9.9822	9.4655	0.5345	9.4729	9.9799	9.4930	0.5070	43
18	9.4482	9.9822	9.4660	0.5340	9.4733	9.9799	9.4934	0.5066	42
19	9.4486	9.9821	9.4665	0.5335	9.4737	9.9799	9.4939	0.5061	41
20	9.4491	9.9821	9.4669	0.5331	9.4741	9.9798	9.4943	0.5057	40
21	9.4495	9.9821	9.4674	0.5326	9.4745	9.9798	9.4947	0.5053	39
22	9.4499	9.9820	9.4679	0.5321	9.4749	9.9797	9.4952	0.5048	38
23	9.4503	9.9820	9.4683	0.5317	9.4753	9.9797	9.4956	0.5044	37
24	9.4508	9.9820	9.4688	0.5312	9.4757	9.9797	9.4961	0.5039	36
25	9.4512	9.9819	9.4693	0.5307	9.4761	9.9796	9.4965	0.5035	35
26	9.4516	9.9819	9.4697	0.5303	9.4765	9.9796	9.4970	0.5030	34
27	9.4521	9.9818	9.4702	0.5298	9.4769	9.9795	9.4974	0.5026	33
28	9.4525	9.9818	9.4707	0.5293	9.4773	9.9795	9.4978	0.5022	32
29	9.4529	9.9818	9.4711	0.5289	9.4777	9.9795	9.4983	0.5017	31
30	9.4533	9.9817	9.4716	0.5284	9.4781	9.9794	9.4987	0.5013	30
31	9.4538	9.9817	9.4721	0.5279	9.4785	9.9794	9.4992	0.5008	29
32	9.4542	9.9817	9.4725	0.5275	9.4789	9.9793	9.4996	0.5004	28
33	9.4546	9.9816	9.4730	0.5270	9.4793	9.9793	9.5000	0.5000	27
34	9.4550	9.9816	9.4735	0.5265	9.4797	9.9793	9.5005	0.4995	26
35	9.4555	9.9815	9.4739	0.5261	9.4801	9.9792	9.5009	0.4991	25
36	9.4559	9.9815	9.4744	0.5256	9.4805	9.9792	9.5014	0.4986	24
37	9.4563	9.9815	9.4748	0.5252	9.4809	9.9791	9.5018	0.4982	23
38	9.4567	9.9814	9.4753	0.5247	9.4813	9.9791	9.5022	0.4978	22
39	9.4572	9.9814	9.4758	0.5242	9.4817	9.9791	9.5027	0.4973	21
40	9.4576	9.9814	9.4762	0.5238	9.4821	9.9790	9.5031	0.4969	20
41	9.4580	9.9813	9.4767	0.5233	9.4825	9.9790	9.5035	0.4965	19
42	9.4584	9.9813	9.4771	0.5229	9.4829	9.9789	9.5040	0.4960	18
43	9.4588	9.9813	9.4776	0.5224	9.4833	9.9789	9.5044	0.4956	17
44	9.4593	9.9812	9.4781	0.5219	9.4837	9.9789	9.5049	0.4951	16
45	9.4597	9.9812	9.4785	0.5215	9.4841	9.9788	9.5053	0.4947	15
46	9.4601	9.9811	9.4790	0.5210	9.4845	9.9788	9.5057	0.4943	14
47	9.4605	9.9811	9.4794	0.5206	9.4849	9.9787	9.5062	0.4938	13
48	9.4609	9.9811	9.4799	0.5201	9.4853	9.9787	9.5066	0.4934	12
49	9.4614	9.9810	9.4803	0.5197	9.4857	9.9787	9.5070	0.4930	11
50	9.4618	9.9810	9.4808	0.5192	9.4861	9.9786	9.5075	0.4925	10
51	9.4622	9.9809	9.4813	0.5187	9.4865	9.9786	9.5079	0.4921	9
52	9.4626	9.9809	9.4817	0.5183	9.4869	9.9785	9.5083	0.4917	8
53	9.4630	9.9809	9.4822	0.5178	9.4873	9.9785	9.5088	0.4912	7
54	9.4634	9.9808	9.4826	0.5174	9.4876	9.9785	9.5092	0.4908	6
55	9.4639	9.9808	9.4831	0.5169	9.4880	9.9784	9.5096	0.4904	5
56	9.4643	9.9808	9.4835	0.5165	9.4884	9.9784	9.5101	0.4899	4
57	9.4647	9.9807	9.4840	0.5160	9.4888	9.9783	9.5105	0.4895	3
58	9.4651	9.9807	9.4844	0.5156	9.4892	9.9783	9.5109	0.4891	2
59	9.4655	9.9806	9.4849	0.5151	9.4896	9.9782	9.5113	0.4887	1
60	9.4659	9.9806	9.4853	0.5147	9.4900	9.9782	9.5118	0.4882	0
	Cos.	Sin.	Cotang.	Tang.	Cos.	Sin.	Cotang.	Tang.	
	73°				72°				

	18°				19°				
	Sin.	Cos.	Tang.	Cotang.	Sin.	Cos.	Tang.	Cotang.	
0′	9.4900	9.9782	9.5118	0.4882	9.5126	9.9757	9.5370	0.4630	60′
1	9.4904	9.9782	9.5122	0.4878	9.5130	9.9756	9.5374	0.4626	59
2	9.4908	9.9781	9.5126	0.4874	9.5134	9.9756	9.5378	0.4622	58
3	9.4911	9.9781	9.5131	0.4869	9.5137	9.9755	9.5382	0.4618	57
4	9.4915	9.9780	9.5135	0.4865	9.5141	9.9755	9.5386	0.4614	56
5	9.4919	9.9780	9.5139	0.4861	9.5145	9.9755	9.5390	0.4610	55
6	9.4923	9.9780	9.5143	0.4857	9.5148	9.9754	9.5394	0.4606	54
7	9.4927	9.9779	9.5148	0.4852	9.5152	9.9754	9.5398	0.4602	53
8	9.4931	9.9779	9.5152	0.4848	9.5156	9.9753	9.5402	0.4598	52
9	9.4935	9.9778	9.5156	0.4844	9.5159	9.9753	9.5407	0.4593	51
10	9.4939	9.9778	9.5161	0.4839	9.5163	9.9752	9.5411	0.4589	50
11	9.4942	9.9778	9.5165	0.4835	9.5167	9.9752	9.5415	0.4585	49
12	9.4946	9.9777	9.5169	0.4831	9.5170	9.9751	9.5419	0.4581	48
13	9.4950	9.9777	9.5173	0.4827	9.5174	9.9751	9.5423	0.4577	47
14	9.4954	9.9776	9.5178	0.4822	9.5177	9.9751	9.5427	0.4573	46
15	9.4958	9.9776	9.5182	0.4818	9.5181	9.9750	9.5431	0.4569	45
16	9.4962	9.9775	9.5186	0.4814	9.5185	9.9750	9.5435	0.4565	44
17	9.4965	9.9775	9.5190	0.4810	9.5188	9.9749	9.5439	0.4561	43
18	9.4969	9.9775	9.5195	0.4805	9.5192	9.9749	9.5443	0.4557	42
19	9.4973	9.9774	9.5199	0.4801	9.5196	9.9748	9.5447	0.4553	41
20	9.4977	9.9774	9.5203	0.4797	9.5199	9.9748	9.5451	0.4549	40
21	9.4981	9.9773	9.5207	0.4793	9.5203	9.9747	9.5455	0.4545	39
22	9.4984	9.9773	9.5212	0.4788	9.5206	9.9747	9.5459	0.4541	38
23	9.4988	9.9773	9.5216	0.4784	9.5210	9.9747	9.5463	0.4537	37
24	9.4992	9.9772	9.5220	0.4780	9.5213	9.9746	9.5467	0.4533	36
25	9.4996	9.9772	9.5224	0.4776	9.5217	9.9746	9.5471	0.4529	35
26	9.5000	9.9771	9.5228	0.4772	9.5221	9.9745	9.5475	0.4525	34
27	9.5003	9.9771	9.5233	0.4767	9.5224	9.9745	9.5479	0.4521	33
28	9.5007	9.9770	9.5237	0.4763	9.5228	9.9744	9.5483	0.4517	32
29	9.5011	9.9770	9.5241	0.4759	9.5231	9.9744	9.5487	0.4513	31
30	9.5015	9.9770	9.5245	0.4755	9.5235	9.9743	9.5491	0.4509	30
31	9.5019	9.9769	9.5249	0.4751	9.5239	9.9743	9.5495	0.4505	29
32	9.5022	9.9769	9.5254	0.4746	9.5242	9.9743	9.5500	0.4500	28
33	9.5026	9.9768	9.5258	0.4742	9.5246	9.9742	9.5504	0.4496	27
34	9.5030	9.9768	9.5262	0.4738	9.5249	9.9742	9.5508	0.4492	26
35	9.5034	9.9767	9.5266	0.4734	9.5253	9.9741	9.5512	0.4488	25
36	9.5037	9.9767	9.5270	0.4730	9.5256	9.9741	9.5516	0.4484	24
37	9.5041	9.9767	9.5275	0.4725	9.5260	9.9740	9.5520	0.4480	23
38	9.5045	9.9766	9.5279	0.4721	9.5263	9.9740	9.5524	0.4476	22
39	9.5049	9.9766	9.5283	0.4717	9.5267	9.9739	9.5528	0.4472	21
40	9.5052	9.9765	9.5287	0.4713	9.5270	9.9739	9.5531	0.4469	20
41	9.5056	9.9765	9.5291	0.4709	9.5274	9.9739	9.5535	0.4465	19
42	9.5060	9.9764	9.5295	0.4705	9.5278	9.9738	9.5539	0.4461	18
43	9.5064	9.9764	9.5300	0.4700	9.5281	9.9738	9.5543	0.4457	17
44	9.5067	9.9764	9.5304	0.4696	9.5285	9.9737	9.5547	0.4453	16
45	9.5071	9.9763	9.5308	0.4692	9.5288	9.9737	9.5551	0.4449	15
46	9.5075	9.9763	9.5312	0.4688	9.5292	9.9736	9.5555	0.4445	14
47	9.5078	9.9762	9.5316	0.4684	9.5295	9.9736	9.5559	0.4441	13
48	9.5082	9.9762	9.5320	0.4680	9.5299	9.9735	9.5563	0.4437	12
49	9.5086	9.9761	9.5324	0.4676	9.5302	9.9735	9.5567	0.4433	11
50	9.5090	9.9761	9.5329	0.4671	9.5306	9.9734	9.5571	0.4429	10
51	9.5093	9.9761	9.5333	0.4667	9.5309	9.9734	9.5575	0.4425	9
52	9.5097	9.9760	9.5337	0.4663	9.5313	9.9734	9.5579	0.4421	8
53	9.5101	9.9760	9.5341	0.4659	9.5316	9.9733	9.5583	0.4417	7
54	9.5104	9.9759	9.5345	0.4655	9.5320	9.9733	9.5587	0.4413	6
55	9.5108	9.9759	9.5349	0.4651	9.5323	9.9732	9.5591	0.4409	5
56	9.5112	9.9758	9.5353	0.4647	9.5327	9.9732	9.5595	0.4405	4
57	9.5115	9.9758	9.5357	0.4643	9.5330	9.9731	9.5599	0.4401	3
58	9.5119	9.9758	9.5362	0.4638	9.5334	9.9731	9.5603	0.4397	2
59	9.5123	9.9757	9.5366	0.4634	9.5337	9.9730	9.5607	0.4393	1
60	9.5126	9.9757	9.5370	0.4630	9.5341	9.9730	9.5611	0.4389	0
	Cos.	Sin.	Cotang.	Tang.	Cos.	Sin.	Cotang.	Tang.	
	71°				70°				

	20°				21°				
	Sin.	Cos.	Tang.	Cotang.	Sin.	Cos.	Tang.	Cotang.	
0′	9.5341	9.9730	9.5611	0.4389	9.5543	9.9702	9.5842	0.4158	60′
1	9.5344	9.9729	9.5615	0.4385	9.5547	9.9701	9.5846	0.4154	59
2	9.5347	9.9729	9.5619	0.4381	9.5550	9.9701	9.5849	0.4151	58
3	9.5351	9.9728	9.5622	0.4378	9.5553	9.9700	9.5853	0.4147	57
4	9.5354	9.9728	9.5626	0.4374	9.5556	9.9700	9.5857	0.4143	56
5	9.5358	9.9728	9.5630	0.4370	9.5560	9.9699	9.5861	0.4139	55
6	9.5361	9.9727	9.5634	0.4366	9.5563	9.9699	9.5864	0.4136	54
7	9.5365	9.9727	9.5638	0.4362	9.5566	9.9698	9.5868	0.4132	53
8	9.5368	9.9726	9.5642	0.4358	9.5570	9.9698	9.5872	0.4128	52
9	9.5372	9.9726	9.5646	0.4354	9.5573	9.9697	9.5876	0.4124	51
10	9.5375	9.9725	9.5650	0.4350	9.5576	9.9697	9.5879	0.4121	50
11	9.5379	9.9725	9.5654	0.4346	9.5579	9.9696	9.5883	0.4117	49
12	9.5382	9.9724	9.5658	0.4342	9.5583	9.9696	9.5887	0.4113	48
13	9.5385	9.9724	9.5662	0.4338	9.5586	9.9695	9.5891	0.4109	47
14	9.5389	9.9723	9.5665	0.4335	9.5589	9.9695	9.5894	0.4106	46
15	9.5392	9.9723	9.5669	0.4331	9.5592	9.9694	9.5898	0.4102	45
16	9.5396	9.9722	9.5673	0.4327	9.5596	9.9694	9.5902	0.4098	44
17	9.5399	9.9722	9.5677	0.4323	9.5599	9.9693	9.5906	0.4094	43
18	9.5402	9.9722	9.5681	0.4319	9.5602	9.9693	9.5909	0.4091	42
19	9.5406	9.9721	9.5685	0.4315	9.5605	9.9692	9.5913	0.4087	41
20	9.5409	9.9721	9.5689	0.4311	9.5609	9.9692	9.5917	0.4083	40
21	9.5413	9.9720	9.5693	0.4307	9.5612	9.9691	9.5921	0.4079	39
22	9.5416	9.9720	9.5696	0.4304	9.5615	9.9691	9.5924	0.4076	38
23	9.5420	9.9719	9.5700	0.4300	9.5618	9.9690	9.5928	0.4072	37
24	9.5423	9.9719	9.5704	0.4296	9.5621	9.9690	9.5932	0.4068	36
25	9.5426	9.9718	9.5708	0.4292	9.5625	9.9689	9.5935	0.4065	35
26	9.5430	9.9718	9.5712	0.4288	9.5628	9.9689	9.5939	0.4061	34
27	9.5433	9.9717	9.5716	0.4284	9.5631	9.9688	9.5943	0.4057	33
28	9.5436	9.9717	9.5720	0.4280	9.5634	9.9688	9.5947	0.4053	32
29	9.5440	9.9716	9.5724	0.4276	9.5638	9.9687	9.5950	0.4050	31
30	9.5443	9.9716	9.5727	0.4273	9.5641	9.9687	9.5954	0.4046	30
31	9.5447	9.9715	9.5731	0.4269	9.5644	9.9686	9.5958	0.4042	29
32	9.5450	9.9715	9.5735	0.4265	9.5647	9.9686	9.5961	0.4039	28
33	9.5453	9.9714	9.5739	0.4261	9.5650	9.9685	9.5965	0.4035	27
34	9.5457	9.9714	9.5743	0.4257	9.5654	9.9685	9.5969	0.4031	26
35	9.5460	9.9714	9.5747	0.4253	9.5657	9.9684	9.5972	0.4028	25
36	9.5463	9.9713	9.5750	0.4250	9.5660	9.9684	9.5976	0.4024	24
37	9.5467	9.9713	9.5754	0.4246	9.5663	9.9683	9.5980	0.4020	23
38	9.5470	9.9712	9.5758	0.4242	9.5666	9.9683	9.5984	0.4016	22
39	9.5474	9.9712	9.5762	0.4238	9.5670	9.9682	9.5987	0.4013	21
40	9.5477	9.9711	9.5766	0.4234	9.5673	9.9682	9.5991	0.4009	20
41	9.5480	9.9711	9.5770	0.4230	9.5676	9.9681	9.5995	0.4005	19
42	9.5484	9.9710	9.5773	0.4227	9.5679	9.9681	9.5998	0.4002	18
43	9.5487	9.9710	9.5777	0.4223	9.5682	9.9680	9.6002	0.3998	17
44	9.5490	9.9709	9.5781	0.4219	9.5685	9.9680	9.6006	0.3994	16
45	9.5494	9.9709	9.5785	0.4215	9.5689	9.9679	9.6009	0.3991	15
46	9.5497	9.9708	9.5789	0.4211	9.5692	9.9679	9.6013	0.3987	14
47	9.5500	9.9708	9.5792	0.4208	9.5695	9.9678	9.6017	0.3983	13
48	9.5504	9.9707	9.5796	0.4204	9.5698	9.9678	9.6020	0.3980	12
49	9.5507	9.9707	9.5800	0.4200	9.5701	9.9677	9.6024	0.3976	11
50	9.5510	9.9706	9.5804	0.4196	9.5704	9.9677	9.6028	0.3972	10
51	9.5514	9.9706	9.5808	0.4192	9.5708	9.9676	9.6031	0.3969	9
52	9.5517	9.9705	9.5811	0.4189	9.5711	9.9676	9.6035	0.3965	8
53	9.5520	9.9705	9.5815	0.4185	9.5714	9.9675	9.6039	0.3961	7
54	9.5523	9.9704	9.5819	0.4181	9.5717	9.9675	9.6042	0.3958	6
55	9.5527	9.9704	9.5823	0.4177	9.5720	9.9674	9.6046	0.3954	5
56	9.5530	9.9703	9.5827	0.4173	9.5723	9.9674	9.6050	0.3950	4
57	9.5533	9.9703	9.5830	0.4170	9.5726	9.9673	9.6053	0.3947	3
58	9.5537	9.9702	9.5834	0.4166	9.5729	9.9673	9.6057	0.3943	2
59	9.5540	9.9702	9.5838	0.4162	9.5733	9.9672	9.6060	0.3940	1
60	9.5543	9.9702	9.5842	0.4158	9.5736	9.9672	9.6064	0.3936	0
	Cos.	Sin.	Cotang.	Tang.	Cos.	Sin.	Cotang.	Tang.	
	69°				68°				

	22°				23°				
	Sin.	Cos.	Tang.	Cotang.	Sin.	Cos.	Tang.	Cotang.	
0′	9.5736	9.9672	9.6064	0.3936	9.5919	9.9640	9.6279	0.3721	60′
1	9.5739	9.9671	9.6068	0.3932	9.5922	9.9640	9.6282	0.3718	59
2	9.5742	9.9671	9.6071	0.3929	9.5925	9.9639	9.6286	0.3714	58
3	9.5745	9.9670	9.6075	0.3925	9.5928	9.9639	9.6289	0.3711	57
4	9.5748	9.9670	9.6079	0.3921	9.5931	9.9638	9.6293	0.3707	56
5	9.5751	9.9669	9.6082	0.3918	9.5934	9.9638	9.6296	0.3704	55
6	9.5754	9.9669	9.6086	0.3914	9.5937	9.9637	9.6300	0.3700	54
7	9.5758	9.9668	9.6090	0.3910	9.5940	9.9637	9.6303	0.3697	53
8	9.5761	9.9668	9.6093	0.3907	9.5943	9.9636	9.6307	0.3693	52
9	9.5764	9.9667	9.6097	0.3903	9.5945	9.9635	9.6310	0.3690	51
10	9.5767	9.9667	9.6100	0.3900	9.5948	9.9635	9.6314	0.3686	50
11	9.5770	9.9666	9.6104	0.3896	9.5951	9.9634	9.6317	0.3683	49
12	9.5773	9.9666	9.6108	0.3892	9.5954	9.9634	9.6321	0.3679	48
13	9.5776	9.9665	9.6111	0.3889	9.5957	9.9633	9.6324	0.3676	47
14	9.5779	9.9664	9.6115	0.3885	9.5960	9.9633	9.6328	0.3672	46
15	9.5782	9.9664	9.6118	0.3882	9.5963	9.9632	9.6331	0.3669	45
16	9.5785	9.9663	9.6122	0.3878	9.5966	9.9632	9.6334	0.3666	44
17	9.5789	9.9663	9.6126	0.3874	9.5969	9.9631	9.6338	0.3662	43
18	9.5792	9.9662	9.6129	0.3871	9.5972	9.9631	9.6341	0.3659	42
19	9.5795	9.9662	9.6133	0.3867	9.5975	9.9630	9.6345	0.3655	41
20	9.5798	9.9661	9.6136	0.3864	9.5978	9.9629	9.6348	0.3652	40
21	9.5801	9.9661	9.6140	0.3860	9.5981	9.9629	9.6352	0.3648	39
22	9.5804	9.9660	9.6144	0.3856	9.5984	9.9628	9.6355	0.3645	38
23	9.5807	9.9660	9.6147	0.3853	9.5987	9.9628	9.6359	0.3641	37
24	9.5810	9.9659	9.6151	0.3849	9.5990	9.9627	9.6362	0.3638	36
25	9.5813	9.9659	9.6154	0.3846	9.5992	9.9627	9.6366	0.3634	35
26	9.5816	9.9658	9.6158	0.3842	9.5995	9.9626	9.6369	0.3631	34
27	9.5819	9.9658	9.6162	0.3838	9.5998	9.9626	9.6373	0.3627	33
28	9.5822	9.9657	9.6165	0.3835	9.6001	9.9625	9.6376	0.3624	32
29	9.5825	9.9657	9.6169	0.3831	9.6004	9.9625	9.6380	0.3620	31
30	9.5828	9.9656	9.6172	0.3828	9.6007	9.9624	9.6383	0.3617	30
31	9.5831	9.9656	9.6176	0.3824	9.6010	9.9623	9.6386	0.3614	29
32	9.5834	9.9655	9.6179	0.3821	9.6013	9.9623	9.6390	0.3610	28
33	9.5838	9.9655	9.6183	0.3817	9.6016	9.9622	9.6393	0.3607	27
34	9.5841	9.9654	9.6187	0.3813	9.6019	9.9622	9.6397	0.3603	26
35	9.5844	9.9654	9.6190	0.3810	9.6021	9.9621	9.6400	0.3600	25
36	9.5847	9.9653	9.6194	0.3806	9.6024	9.9621	9.6404	0.3596	24
37	9.5850	9.9652	9.6197	0.3803	9.6027	9.9620	9.6407	0.3593	23
38	9.5853	9.9652	9.6201	0.3799	9.6030	9.9620	9.6411	0.3589	22
39	9.5856	9.9651	9.6204	0.3796	9.6033	9.9619	9.6414	0.3586	21
40	9.5859	9.9651	9.6208	0.3792	9.6036	9.9618	9.6417	0.3583	20
41	9.5862	9.9650	9.6211	0.3789	9.6039	9.9618	9.6421	0.3579	19
42	9.5865	9.9650	9.6215	0.3785	9.6042	9.9617	9.6424	0.3576	18
43	9.5868	9.9649	9.6219	0.3781	9.6045	9.9617	9.6428	0.3572	17
44	9.5871	9.9649	9.6222	0.3778	9.6047	9.9616	9.6431	0.3569	16
45	9.5874	9.9648	9.6226	0.3774	9.6050	9.9616	9.6435	0.3565	15
46	9.5877	9.9648	9.6229	0.3771	9.6053	9.9615	9.6438	0.3562	14
47	9.5880	9.9647	9.6233	0.3767	9.6056	9.9615	9.6441	0.3559	13
48	9.5883	9.9647	9.6236	0.3764	9.6059	9.9614	9.6445	0.3555	12
49	9.5886	9.9646	9.6240	0.3760	9.6062	9.9613	9.6448	0.3552	11
50	9.5889	9.9646	9.6243	0.3757	9.6065	9.9613	9.6452	0.3548	10
51	9.5892	9.9645	9.6247	0.3753	9.6068	9.9612	9.6455	0.3545	9
52	9.5895	9.9645	9.6250	0.3750	9.6070	9.9612	9.6459	0.3541	8
53	9.5898	9.9644	9.6254	0.3746	9.6073	9.9611	9.6462	0.3538	7
54	9.5901	9.9643	9.6257	0.3743	9.6076	9.9611	9.6465	0.3535	6
55	9.5904	9.9643	9.6261	0.3739	9.6079	9.9610	9.6469	0.3531	5
56	9.5907	9.9642	9.6264	0.3736	9.6082	9.9610	9.6472	0.3528	4
57	9.5910	9.9642	9.6268	0.3732	9.6085	9.9609	9.6476	0.3524	3
58	9.5913	9.9641	9.6271	0.3729	9.6087	9.9608	9.6479	0.3521	2
59	9.5916	9.9641	9.6275	0.3725	9.6090	9.9608	9.6482	0.3518	1
60	9.5919	9.9640	9.6279	0.3721	9.6093	9.9607	9.6486	0.3514	0
	Cos.	Sin.	Cotang.	Tang.	Cos.	Sin.	Cotang.	Tang.	
	67°				66°				

	24°				25°				
	Sin.	Cos.	Tang.	Cotang.	Sin.	Cos.	Tang.	Cotang.	
0′	9.6093	9.9607	9.6486	0.3514	9.6259	9.9573	9.6687	0.3313	60′
1	9.6096	9.9607	9.6489	0.3511	9.6262	9.9572	9.6690	0.3310	59
2	9.6099	9.9606	9.6493	0.3507	9.6265	9.9572	9.6693	0.3307	58
3	9.6102	9.9606	9.6496	0.3504	9.6268	9.9571	9.6697	0.3303	57
4	9.6104	9.9605	9.6499	0.3501	9.6270	9.9570	9.6700	0.3300	56
5	9.6107	9.9604	9.6503	0.3497	9.6273	9.9570	9.6703	0.3297	55
6	9.6110	9.9604	9.6506	0.3494	9.6276	9.9569	9.6706	0.3294	54
7	9.6113	9.9603	9.6509	0.3491	9.6278	9.9569	9.6710	0.3290	53
8	9.6116	9.9603	9.6513	0.3487	9.6281	9.9568	9.6713	0.3287	52
9	9.6119	9.9602	9.6516	0.3484	9.6284	9.9567	9.6716	0.3284	51
10	9.6121	9.9602	9.6520	0.3480	9.6286	9.9567	9.6720	0.3280	50
11	9.6124	9.9601	9.6523	0.3477	9.6289	9.9566	9.6723	0 3277	49
12	9.6127	9.9601	9.6527	0.3473	9.6292	9.9566	9.6726	0.3274	48
13	9.6130	9.9600	9.6530	0.3470	9.6295	9.9565	9.6729	0.3271	47
14	9.6133	9.9599	9.6533	0.3467	9.6297	9.9564	9.6733	0.3267	46
15	9.6135	9.9599	9.6537	0.3463	9.6300	9.9564	9.6736	0.3264	45
16	9.6138	9.9598	9.6540	0.3460	9.6303	9.9563	9.6739	0.3261	44
17	9.6141	9.9598	9.6543	0.3457	9.6305	9.9563	9.6743	0.3257	43
18	9.6144	9.9597	9.6547	0.3453	9.6308	9.9562	9.6746	0.3254	42
19	9.6147	9.9597	9.6550	0.3450	9.6311	9.9561	9.6749	0.3251	41
20	9.6149	9.9596	9.6553	0.3447	9.6313	9.9561	9.6752	0.3248	40
21	9.6152	9.9595	9.6557	0.3443	9.6316	9.9560	9.6756	0.3244	39
22	9.6155	9.9595	9.6560	0.3440	9.6319	9.9560	9.6759	0.3241	38
23	9.6158	9.9594	9.6564	0.3436	9.6321	9.9559	9.6762	0.3238	37
24	9.6161	9.9594	9.6567	0.3433	9.6324	9.9558	9.6765	0.3235	36
25	9.6163	9.9593	9.6570	0.3430	9.6327	9.9558	9.6769	0.3231	35
26	9.6166	9.9593	9.6574	0.3426	9.6329	9.9557	9.6772	0.3228	34
27	9.6169	9.9592	9.6577	0.3423	9.6332	9.9557	9.6775	0.3225	33
28	9.6172	9.9591	9.6580	0.3420	9.6335	9.9556	9.6778	0.3222	32
29	9.6174	9.9591	9.6584	0.3416	9.6337	9.9555	9.6782	0.3218	31
30	9.6177	9.9590	9.6587	0.3413	9.6340	9.9555	9.6785	0.3215	30
31	9.6180	9.9590	9.6590	0.3410	9.6342	9.9554	9.6788	0.3212	29
32	9.6183	9.9589	9.6594	0.3406	9.6345	9.9554	9.6791	0.3209	28
33	9.6186	9.9589	9.6597	0.3403	9.6348	9.9553	9.6795	0.3205	27
34	9.6188	9.9588	9.6600	0.3400	9.6350	9.9552	9.6798	0.3202	26
35	9.6191	9.9587	9.6604	0.3396	9.6353	9.9552	9.6801	0.3199	25
36	9.6194	9.9587	9.6607	0.3393	9.6356	9.9551	9.6804	0.3196	24
37	9.6197	9.9586	9.6610	0.3390	9.6358	9.9551	9.6808	0.3192	23
38	9.6199	9.9586	9.6614	0.3386	9.6361	9.9550	9.6811	0.3189	22
39	9.6202	9.9585	9.6617	0.3383	9.6364	9.9549	9.6814	0.3186	21
40	9.6205	9.9584	9.6620	0.3380	9.6366	9.9549	9.6817	0.3183	20
41	9.6208	9.9584	9.6624	0.3376	9.6369	9.9548	9.6821	0.3179	19
42	9.6210	9.9583	9.6627	0.3373	9.6371	9.9548	9.6824	0.3176	18
43	9.6213	9.9583	9.6630	0.3370	9.6374	9.9547	9.6827	0.3173	17
44	9.6216	9.9582	9.6634	0.3366	9.6377	9.9546	9.6830	0.3170	16
45	9.6219	9.9582	9.6637	0.3363	9.6379	9.9546	9.6834	0.3166	15
46	9.6221	9.9581	9.6640	0.3360	9.6382	9.9545	9.6837	0.3163	14
47	9.6224	9.9580	9.6644	0.3356	9.6385	9.9545	9.6840	0.3160	13
48	9.6227	9.9580	9.6647	0.3353	9.6387	9.9544	9.6843	0.3157	12
49	9.6230	9.9579	9.6650	0.3350	9.6390	9.9543	9.6846	0.3154	11
50	9.6232	9.9579	9.6654	0.3346	9.6392	9.9543	9.6850	0.3150	10
51	9.6235	9.9578	9.6657	0.3343	9.6395	9.9542	9.6853	0.3147	9
52	9.6238	9.9577	9.6660	0.3340	9.6398	9.9542	9.6856	0.3144	8
53	9.6240	9.9577	9.6664	0.3636	9.6400	9.9541	9.6859	0.3141	7
54	9.6243	9.9576	9.6667	0.3333	9.6403	9.9540	9.6863	0.3137	6
55	9.6246	9.9576	9.6670	0.3330	9.6405	9.9540	9.6866	0.3134	5
56	9.6249	9.9575	9.6674	0.3326	9.6408	9.9539	9.6869	0.3131	4
57	9.6251	9.9575	9.6677	0.3323	9.6411	9.9538	9.6872	0.3128	3
58	9.6254	9.9574	9.6680	0.3320	9.6413	9.9538	9.6875	0.3125	2
59	9.6257	9.9573	9.6683	0.3317	9.6416	9.9537	9.6879	0.3121	1
60	9.6259	9.9573	9.6687	0.3313	9.6418	9.9537	9.6882	0.3118	0
	Cos.	Sin.	Cotang.	Tang.	Cos.	Sin.	Cotang.	Tang.	
	65°				64°				

	26°				27°				
	Sin.	Cos.	Tang.	Cotang.	Sin.	Cos.	Tang.	Cotang.	
0′	9.6418	9.9537	9.6882	0.3118	9.6570	9.9499	9.7072	0.2928	60′
1	9.6421	9.9536	9.6885	0.3115	9.6573	9.9498	9.7075	0.2925	59
2	9.6424	9.9535	9.6888	0.3112	9.6575	9.9497	9.7078	0.2922	58
3	9.6426	9.9535	9.6891	0.3109	9.6578	9.9497	9.7081	0.2919	57
4	9.6429	9.9534	9.6895	0.3105	9.6580	9.9496	9.7084	0.2916	56
5	9.6431	9.9534	9.6898	0.3102	9.6583	9.9496	9.7087	0.2913	55
6	9.6434	9.9533	9.6901	0.3099	9.6585	9.9495	9.7090	0.2910	54
7	9.6437	9.9532	9.6904	0.3096	9.6588	9.9494	9.7093	0.2907	53
8	9.6439	9.9532	9.6907	0.3093	9.6590	9.9494	9.7097	0.2903	52
9	9.6442	9.9531	9.6911	0.3089	9.6593	9.9493	9.7100	0.2900	51
10	9.6444	9.9530	9.6914	0.3086	9.6595	9.9492	9.7103	0.2897	50
11	9.6447	9.9530	9.6917	0.3083	9.6598	9.9492	9.7106	0.2894	49
12	9.6449	9.9529	9.6920	0.3080	9.6600	9.9491	9.7109	0.2891	48
13	9.6452	9.9529	9.6923	0.3077	9.6603	9.9490	9.7112	0.2888	47
14	9.6454	9.9528	9.6927	0.3073	9.6605	9.9490	9.7115	0.2885	46
15	9.6457	9.9527	9.6930	0.3070	9.6607	9.9489	9.7118	0.2882	45
16	9.6460	9.9527	9.6933	0.3067	9.6610	9.9488	9.7121	0.2879	44
17	9.6462	9.9526	9.6936	0.3064	9.6612	9.9488	9.7125	0.2875	43
18	9.6465	9.9525	9.6939	0.3061	9.6615	9.9487	9.7128	0.2872	42
19	9.6467	9.9525	9.6942	0.3058	9.6617	9.9486	9.7131	0.2869	41
20	9.6470	9.9524	9.6946	0.3054	9.6620	9.9486	9.7134	0.2866	40
21	9.6472	9.9524	9.6949	0.3051	9.6622	9.9485	9.7137	0.2863	39
22	9.6475	9.9523	9.6952	0.3048	9.6625	9.9485	9.7140	0.2860	38
23	9.6477	9.9522	9.6955	0.3045	9.6627	9.9484	9.7143	0.2857	37
24	9.6480	9.9522	9.6958	0.3042	9.6629	9.9483	9.7146	0.2854	36
25	9.6483	9.9521	9.6962	0.3038	9.6632	9.9483	9.7149	0.2851	35
26	9.6485	9.9520	9.6965	0.3035	9.6634	9.9482	9.7152	0.2848	34
27	9.6488	9.9520	9.6968	0.3032	9.6637	9.9481	9.7156	0.2844	33
28	9.6490	9.9519	9.6971	0.3029	9.6639	9.9481	9.7159	0.2841	32
29	9.6493	9.9519	9.6974	0.3026	9.6642	9.9480	9.7162	0.2838	31
30	9.6495	9.9518	9.6977	0.3023	9.6644	9.9479	9.7165	0.2835	30
31	9.6498	9.9517	9.6981	0.3019	9.6646	9.9478	9.7168	0.2832	29
32	9.6500	9.9517	9.6984	0.3016	9.6649	9.9478	9.7171	0.2829	28
33	9.6503	9.9516	9.6987	0.3013	9.6651	9.9477	9.7174	0.2826	27
34	9.6505	9.9515	9.6990	0.3010	9.6654	9.9477	9.7177	0.2823	26
35	9.6508	9.9515	9.6993	0.3007	9.6656	9.9476	9.7180	0.2820	25
36	9.6510	9.9514	9.6996	0.3004	9.6659	9.9475	9.7183	0.2817	24
37	9.6513	9.9513	9.6999	0.3001	9.6661	9.9475	9.7186	0.2814	23
38	9.6515	9.9513	9.7003	0.2997	9.6663	9.9474	9.7189	0.2811	22
39	9.6518	9.9512	9.7006	0.2994	9.6666	9.9473	9.7192	0.2808	21
40	9.6521	9.9512	9.7009	0.2991	9.6668	9.9473	9.7196	0.2804	20
41	9.6523	9.9511	9.7012	0.2988	9.6671	9.9472	9.7199	0.2801	19
42	9.6526	9.9510	9.7015	0.2985	9.6673	9.9471	9.7202	0.2798	18
43	9.6528	9.9510	9.7018	0.2982	9.6675	9.9471	9.7205	0.2795	17
44	9.6531	9.9509	9.7022	0.2978	9.6678	9.9470	9.7208	0.2792	16
45	9.6533	9.9508	9.7025	0.2975	9.6680	9.9469	9.7211	0.2789	15
46	9.6536	9.9508	9.7028	0.2972	9.6683	9.9469	9.7214	0.2786	14
47	9.6538	9.9507	9.7031	0.2969	9.6685	9.9468	9.7217	0.2783	13
48	9.6541	9.9507	9.7034	0.2966	9.6687	9.9467	9.7220	0.2780	12
49	9.6543	9.9506	9.7037	0.2963	9.6690	9.9467	9.7223	0.2777	11
50	9.6546	9.9505	9.7040	0.2960	9.6692	9.9466	9.7226	0.2774	10
51	9.6548	9.9505	9.7043	0.2957	9.6695	9.9465	9.7229	0.2771	9
52	9.6551	9.9504	9.7047	0.2953	9.6697	9.9465	9.7232	0.2768	8
53	9.6553	9.9503	9.7050	0.2950	9.6699	9.9464	9.7235	0.2765	7
54	9.6556	9.9503	9.7053	0.2947	9.6702	9.9463	9.7238	0.2762	6
55	9.6558	9.9502	9.7056	0.2944	9.6704	9.9463	9.7241	0.2759	5
56	9.6561	9.9501	9.7059	0.2941	9.6707	9.9462	9.7245	0.2755	4
57	9.6563	9.9501	9.7062	0.2938	9.6709	9.9461	9.7248	0.2752	3
58	9.6566	9.9500	9.7065	0.2935	9.6711	9.9461	9.7251	0.2749	2
59	9.6568	9.9499	9.7069	0.2931	9.6714	9.9460	9.7254	0.2746	1
60	9.6570	9.9499	9.7072	0.2928	9.6716	9.9459	9.7257	0.2743	0
	Cos.	Sin.	Cotang.	Tang.	Cos.	Sin.	Cotang.	Tang.	
	63°				62°				

	28°				29°				
	Sin.	Cos.	Tang.	Cotang.	Sin.	Cos.	Tang.	Cotang.	
0′	9.6716	9.9459	9.7257	0.2743	9.6856	9.9418	9.7438	0.2562	60′
1	9.6718	9.9459	9.7260	0.2740	9.6858	9.9417	9.7440	0.2560	59
2	9.6721	9.9458	9.7263	0.2737	9.6860	9.9417	9.7443	0.2557	58
3	9.6723	9.9457	9.7266	0.2734	9.6863	9.9416	9.7446	0.2554	57
4	9.6726	9.9457	9.7269	0.2731	9.6865	9.9415	9.7449	0.2551	56
5	9.6728	9.9456	9.7271	0.2728	9.6867	9.9415	9.7452	0.2548	55
6	9.6730	9.9455	9.7275	0.2725	9.6869	9.9414	9.7455	0.2545	54
7	9.6733	9.9455	9.7278	0.2722	9.6872	9.9413	9.7458	0.2542	53
8	9.6735	9.9454	9.7281	0.2719	9.6874	9.9413	9.7461	0.2539	52
9	9.6737	9.9453	9.7284	0.2716	9.6876	9.9412	9.7464	0.2536	51
10	9.6740	9.9453	9.7287	0.2713	9.6878	9.9411	9.7467	0.2533	50
11	9.6742	9.9452	9.7290	0.2710	9.6881	9.9410	9.7470	0.2530	49
12	9.6744	9.9451	9.7293	0.2707	9.6883	9.9410	9.7473	0.2527	48
13	9.6747	9.9451	9.7296	0.2704	9.6885	9.9409	9.7476	0.2524	47
14	9.6749	9.9450	9.7299	0.2701	9.6887	9.9408	9.7479	0.2521	46
15	9.6752	9.9449	9.7302	0.2698	9.6890	9.9408	9.7482	0.2518	45
16	9.6754	9.9449	9.7305	0.2695	9.6892	9.9407	9.7485	0.2515	44
17	9.6756	9.9448	9.7308	0.2692	9.6894	9.9406	9.7488	0.2512	43
18	9.6759	9.9447	9.7311	0.2689	9.6896	9.9406	9.7491	0.2509	42
19	9.6761	9.9447	9.7314	0.2686	9.6899	9.9405	9.7494	0.2506	41
20	9.6763	9.9446	9.7317	0.2683	9.6901	9.9404	9.7497	0.2503	40
21	9.6766	9.9445	9.7320	0.2680	9.6903	9.9403	9.7500	0.2500	39
22	9.6768	9.9444	9.7324	0.2676	9.6905	9.9403	9.7503	0.2497	38
23	9.6770	9.9444	9.7327	0.2673	9.6908	9.9402	9.7506	0.2494	37
24	9.6773	9.9443	9.7330	0.2670	9.6910	9.9401	9.7509	0.2491	36
25	9.6775	9.9442	9.7333	0.2667	9.6912	9.9401	9.7512	0.2488	35
26	9.6777	9.9442	9.7336	0.2664	9.6914	9.9400	9.7515	0.2485	34
27	9.6780	9.9441	9.7339	0.2661	9.6917	9.9399	9.7518	0.2482	33
28	9.6782	9.9440	9.7342	0.2658	9.6919	9.9398	9.7521	0.2479	32
29	9.6784	9.9440	9.7345	0.2655	9.6921	9.9398	9.7523	0.2477	31
30	9.6787	9.9439	9.7348	0.2652	9.6923	9.9397	9.7526	0.2474	30
31	9.6789	9.9438	9.7351	0.2649	9.6926	9.9396	9.7529	0.2471	29
32	9.6791	9.9438	9.7354	0.2646	9.6928	9.9396	9.7532	0.2468	28
33	9.6794	9.9437	9.7357	0.2643	9.6930	9.9395	9.7535	0.2465	27
34	9.6796	9.9436	9.7360	0.2640	9.6932	9.9394	9.7538	0.2462	26
35	9.6798	9.9436	9.7363	0.2637	9.6935	9.9393	9.7541	0.2459	25
36	9.6801	9.9435	9.7366	0.2634	9.6937	9.9393	9.7544	0.2456	24
37	9.6803	9.9434	9.7369	0.2631	9.6939	9.9392	9.7547	0.2453	23
38	9.6805	9.9433	9.7372	0.2628	9.6941	9.9391	9.7550	0.2450	22
39	9.6808	9.9433	9.7375	0.2625	9.6943	9.9391	9.7553	0.2447	21
40	9.6810	9.9432	9.7378	0.2622	9.6946	9.9390	9.7556	0.2444	20
41	9.6812	9.9431	9.7381	0.2619	9.6948	9.9389	9.7559	0.2441	19
42	9.6814	9.9431	9.7384	0.2616	9.6950	9.9388	9.7562	0.2438	18
43	9.6817	9.9430	9.7387	0.2613	9.6952	9.9388	9.7565	0.2435	17
44	9.6819	9.9429	9.7390	0.2610	9.6955	9.9387	9.7568	0.2432	16
45	9.6821	9.9429	9.7393	0.2607	9.6957	9.9386	9.7571	0.2429	15
46	9.6824	9.9428	9.7396	0.2604	9.6959	9.9385	9.7573	0.2427	14
47	9.6826	9.9427	9.7399	0.2601	9.6961	9.9385	9.7576	0.2424	13
48	9.6828	9.9427	9.7402	0.2598	9.6963	9.9384	9.7579	0.2421	12
49	9.6831	9.9426	9.7405	0.2595	9.6966	9.9383	9.7582	0.2418	11
50	9.6833	9.9425	9.7408	0.2592	9.6968	9.9383	9.7585	0.2415	10
51	9.6835	9.9424	9.7411	0.2589	9.6970	9.9382	9.7588	0.2412	9
52	9.6837	9.9424	9.7414	0.2586	9.6972	9.9381	9.7591	0.2409	8
53	9.6840	9.9423	9.7417	0.2583	9.6974	9.9380	9.7594	0.2406	7
54	9.6842	9.9422	9.7420	0.2580	9.6977	9.9380	9.7597	0.2403	6
55	9.6844	9.9422	9.7423	0.2577	9.6979	9.9379	9.7600	0.2400	5
56	9.6847	9.9421	9.7426	0.2574	9.6981	9.9378	9.7603	0.2397	4
57	9.6849	9.9420	9.7429	0.2571	9.6983	9.9377	9.7606	0.2394	3
58	9.6851	9.9420	9.7432	0.2568	9.6985	9.9377	9.7609	0.2391	2
59	9.6853	9.9419	9.7435	0.2565	9.6988	9.9376	9.7611	0.2389	1
60	9.6856	9.9418	9.7438	0.2562	9.6990	9.9375	9.7614	0.2386	0
	Cos.	Sin.	Cotang.	Tang.	Cos.	Sin.	Cotang.	Tang.	
	61°				60°				

	30°				31°				
	Sin.	Cos.	Tang.	Cotang.	Sin.	Cos.	Tang.	Cotang.	
0′	9.6990	9.9375	9.7614	0.2386	9.7118	9.9331	9.7788	0.2212	60′
1	9.6992	9.9375	9.7617	0.2383	9.7120	9.9330	9.7791	0.2209	59
2	9.6994	9.9374	9.7620	0.2380	9.7123	9.9329	9.7793	0.2207	58
3	9.6996	9.9373	9.7623	0.2377	9.7125	9.9328	9.7796	0.2204	57
4	9.6998	9.9372	9.7626	0.2374	9.7127	9.9328	9.7799	0.2201	56
5	9.7001	9.9372	9.7629	0.2371	9.7129	9.9327	9.7802	0.2198	55
6	9.7003	9.9371	9.7632	0.2368	9.7131	9.9326	9.7805	0.2195	54
7	9.7005	9.9370	9.7635	0.2365	9.7133	9.9325	9.7808	0.2192	53
8	9.7007	9.9369	9.7638	0.2362	9.7135	9.9325	9.7811	0.2189	52
9	9.7009	9.9369	9.7641	0.2359	9.7137	9.9324	9.7813	0.2187	51
10	9.7012	9.9368	9.7644	0.2356	9.7139	9.9323	9.7816	0.2184	50
11	9.7014	9.9367	9.7646	0.2354	9.7141	9.9322	9.7819	0.2181	49
12	9.7016	9.9367	9.7649	0.2351	9.7144	9.9322	9.7822	0.2178	48
13	9.7018	9.9366	9.7652	0.2348	9.7146	9.9321	9.7825	0.2175	47
14	9.7020	9.9365	9.7655	0.2345	9.7148	9.9320	9.7828	0.2172	46
15	9.7022	9.9364	9.7658	0.2342	9.7150	9.9319	9.7831	0.2169	45
16	9.7025	9.9364	9.7661	0.2339	9.7152	9.9318	9.7833	0.2167	44
17	9.7027	9.9363	9.7664	0.2336	9.7154	9.9318	9.7836	0.2164	43
18	9.7029	9.9362	9.7667	0.2333	9.7156	9.9317	9.7839	0.2161	42
19	9.7031	9.9361	9.7670	0.2330	9.7158	9.9316	9.7842	0.2158	41
20	9.7033	9.9361	9.7673	0.2327	9.7160	9.9315	9.7845	0.2155	40
21	9.7035	9.9360	9.7675	0.2325	9.7162	9.9315	9.7848	0.2152	39
22	9.7037	9.9359	9.7678	0.2322	9.7164	9.9314	9.7850	0.2150	38
23	9.7040	9.9358	9.7681	0.2319	9.7166	9.9313	9.7853	0.2147	37
24	9.7042	9.9358	9.7684	0.2316	9.7168	9.9312	9.7856	0.2144	36
25	9.7044	9.9357	9.7687	0.2313	9.7171	9.9312	9.7859	0.2141	35
26	9.7046	9.9356	9.7690	0.2310	9.7173	9.9311	9.7862	0.2138	34
27	9.7048	9.9355	9.7693	0.2307	9.7175	9.9310	9.7865	0.2135	33
28	9.7050	9.9355	9.7696	0.2304	9.7177	9.9309	9.7868	0.2132	32
29	9.7053	9.9354	9.7699	0.2301	9.7179	9.9308	9.7870	0.2130	31
30	9.7055	9.9353	9.7701	0.2299	9.7181	9.9308	9.7873	0.2127	30
31	9.7057	9.9352	9.7704	0.2296	9.7183	9.9307	9.7876	0.2124	29
32	9.7059	9.9352	9.7707	0.2293	9.7185	9.9306	9.7879	0.2121	28
33	9.7061	9.9351	9.7710	0.2290	9.7187	9.9305	9.7882	0.2118	27
34	9.7063	9.9350	9.7713	0.2287	9.7189	9.9305	9.7885	0.2115	26
35	9.7065	9.9349	9.7716	0.2284	9.7191	9.9304	9.7887	0.2113	25
36	9.7068	9.9349	9.7719	0.2281	9.7193	9.9303	9.7890	0.2110	24
37	9.7070	9.9348	9.7722	0.2278	9.7195	9.9302	9.7893	0.2107	23
38	9.7072	9.9347	9.7725	0.2275	9.7197	9.9301	9.7896	0.2104	22
39	9.7074	9.9346	9.7727	0.2273	9.7199	9.9301	9.7899	0.2101	21
40	9.7076	9.9346	9.7730	0.2270	9.7201	9.9300	9.7902	0.2098	20
41	9.7078	9.9345	9.7733	0.2267	9.7203	9.9299	9.7904	0.2096	19
42	9.7080	9.9344	9.7736	0.2264	9.7205	9.9298	9.7907	0.2093	18
43	9.7082	9.9343	9.7739	0.2261	9.7208	9.9298	9.7910	0.2090	17
44	9.7085	9.9343	9.7742	0.2258	9.7210	9.9297	9.7913	0.2087	16
45	9.7087	9.9342	9.7745	0.2255	9.7212	9.9296	9.7916	0.2084	15
46	9.7089	9.9341	9.7748	0.2252	9.7214	9.9295	9.7918	0.2082	14
47	9.7091	9.9340	9.7750	0.2250	9.7216	9.9294	9.7921	0.2079	13
48	9.7093	9.9340	9.7753	0.2247	9.7218	9.9294	9.7924	0.2076	12
49	9.7095	9.9339	9.7756	0.2244	9.7220	9.9293	9.7927	0.2073	11
50	9.7097	9.9338	9.7759	0.2241	9.7222	9.9292	9.7930	0.2070	10
51	9.7099	9.9337	9.7762	0.2238	9.7224	9.9291	9.7933	0.2067	9
52	9.7102	9.9337	9.7765	0.2235	9.7226	9.9291	9.7935	0.2065	8
53	9.7104	9.9336	9.7768	0.2232	9.7228	9.9290	9.7938	0.2062	7
54	9.7106	9.9335	9.7771	0.2229	9.7230	9.9289	9.7941	0.2059	6
55	9.7108	9.9334	9.7773	0.2227	9.7232	9.9288	9.7944	0.2056	5
56	9.7110	9.9334	9.7776	0.2224	9.7234	9.9287	9.7947	0.2053	4
57	9.7112	9.9333	9.7779	0.2221	9.7236	9.9287	9.7949	0.2051	3
58	9.7114	9.9332	9.7782	0.2218	9.7238	9.9286	9.7952	0.2048	2
59	9.7116	9.9331	9.7785	0.2215	9.7240	9.9285	9.7955	0.2045	1
60	9.7118	9.9331	9.7788	0.2212	9.7242	9.9284	9.7958	0.2042	0
	Cos.	Sin.	Cotang.	Tang.	Cos.	Sin.	Cotang.	Tang.	
	59°				58°				

	32°				33°				
	Sin.	Cos.	Tang.	Cotang.	Sin.	Cos.	Tang.	Cotang.	
0′	9.7242	9.9284	9.7958	0.2042	9.7361	9.9236	9.8125	0.1875	60′
1	9.7244	9.9283	9.7961	0.2039	9.7363	9.9235	9.8128	0.1872	59
2	9.7246	9.9283	9.7964	0.2036	9.7365	9.9234	9.8131	0.1869	58
3	9.7248	9.9282	9.7966	0.2034	9.7367	9.9233	9.8133	0.1867	57
4	9.7250	9.9281	9.7969	0.2031	9.7369	9.9233	9.8136	0.1864	56
5	9.7252	9.9280	9.7972	0.2028	9.7371	9.9232	9.8139	0.1861	55
6	9.7254	9.9279	9.7975	0.2025	9.7373	9.9231	9.8142	0.1858	54
7	9.7256	9.9279	9.7978	0.2022	9.7375	9.9230	9.8145	0.1855	53
8	9.7258	9.9278	9.7980	0.2020	9.7377	9.9229	9.8147	0.1853	52
9	9.7260	9.9277	9.7983	0.2017	9.7379	9.9229	9.8150	0.1850	51
10	9.7262	9.9276	9.7986	0.2014	9.7380	9.9228	9.8153	0.1847	50
11	9.7264	9.9275	9.7989	0.2011	9.7382	9.9227	9.8156	0.1844	49
12	9.7266	9.9275	9.7992	0.2008	9.7384	9.9226	9.8158	0.1842	48
13	9.7268	9.9274	9.7994	0.2006	9.7386	9.9225	9.8161	0.1839	47
14	9.7270	9.9273	9.7997	0.2003	9.7388	9.9224	9.8164	0.1836	46
15	9.7272	9.9272	9.8000	0.2000	9.7390	9.9224	9.8167	0.1833	45
16	9.7274	9.9272	9.8003	0.1997	9.7392	9.9223	9.8169	0.1831	44
17	9.7276	9.9271	9.8006	0.1994	9.7394	9.9222	9.8172	0.1828	43
18	9.7278	9.9270	9.8008	0.1992	9.7396	9.9221	9.8175	0.1825	42
19	9.7280	9.9269	9.8011	0.1989	9.7398	9.9220	9.8178	0.1822	41
20	9.7282	9.9268	9.8014	0.1986	9.7400	9.9219	9.8180	0.1820	40
21	9.7284	9.9268	9.8017	0.1983	9.7402	9.9219	9.8183	0.1817	39
22	9.7286	9.9267	9.8020	0.1980	9.7404	9.9218	9.8186	0.1814	38
23	9.7288	9.9266	9.8022	0.1978	9.7406	9.9217	9.8189	0.1811	37
24	9.7290	9.9265	9.8025	0.1975	9.7407	9.9216	9.8191	0.1809	36
25	9.7292	9.9264	9.8028	0.1972	9.7409	9.9215	9.8194	0.1806	35
26	9.7294	9.9264	9.8031	0.1969	9.7411	9.9214	9.8197	0.1803	34
27	9.7296	9.9263	9.8034	0.1966	9.7413	9.9214	9.8200	0.1800	33
28	9.7298	9.9262	9.8036	0.1964	9.7415	9.9213	9.8202	0.1798	32
29	9.7300	9.9261	9.8039	0.1961	9.7417	9.9212	9.8205	0.1795	31
30	9.7302	9.9260	9.8042	0.1958	9.7419	9.9211	9.8208	0.1792	30
31	9.7304	9.9259	9.8045	0.1955	9.7421	9.9210	9.8211	0.1789	29
32	9.7306	9.9259	9.8047	0.1953	9.7423	9.9209	9.8213	0.1787	28
33	9.7308	9.9258	9.8050	0.1950	9.7425	9.9209	9.8216	0.1784	27
34	9.7310	9.9257	9.8053	0.1947	9.7427	9.9208	9.8219	0.1781	26
35	9.7312	9.9256	9.8056	0.1944	9.7428	9.9207	9.8222	0.1778	25
36	9.7314	9.9255	9.8059	0.1941	9.7430	9.9206	9.8224	0.1776	24
37	9.7316	9.9255	9.8061	0.1939	9.7432	9.9205	9.8227	0.1773	23
38	9.7318	9.9254	9.8064	0.1936	9.7434	9.9204	9.8230	0.1770	22
39	9.7320	9.9253	9.8067	0.1933	9.7436	9.9204	9.8233	0.1767	21
40	9.7322	9.9252	9.8070	0.1930	9.7438	9.9203	9.8235	0.1765	20
41	9.7324	9.9251	9.8072	0.1928	9.7440	9.9202	9.8238	0.1762	19
42	9.7326	9.9251	9.8075	0.1925	9.7442	9.9201	9.8241	0.1759	18
43	9.7328	9.9250	9.8078	0.1922	9.7444	9.9200	9.8243	0.1757	17
44	9.7330	9.9249	9.8081	0.1919	9.7445	9.9199	9.8246	0.1754	16
45	9.7332	9.9248	9.8084	0.1916	9.7447	9.9198	9.8249	0.1751	15
46	9.7334	9.9247	9.8086	0.1914	9.7449	9.9198	9.8252	0.1748	14
47	9.7336	9.9247	9.8089	0.1911	9.7451	9.9197	9.8254	0.1746	13
48	9.7338	9.9246	9.8092	0.1908	9.7453	9.9196	9.8257	0.1743	12
49	9.7340	9.9245	9.8095	0.1905	9.7455	9.9195	9.8260	0.1740	11
50	9.7342	9.9244	9.8097	0.1903	9.7457	9.9194	9.8263	0.1737	10
51	9.7344	9.9243	9.8100	0.1900	9.7459	9.9193	9.8265	0.1735	9
52	9.7345	9.9242	9.8103	0.1897	9.7461	9.9193	9.8268	0.1732	8
53	9.7347	9.9242	9.8106	0.1894	9.7462	9.9192	9.8271	0.1729	7
54	9.7349	9.9241	9.8109	0.1891	9.7464	9.9191	9.8274	0.1726	6
55	9.7351	9.9240	9.8111	0.1889	9.7466	9.9190	9.8276	0.1724	5
56	9.7353	9.9239	9.8114	0.1886	9.7468	9.9189	9.8279	0.1721	4
57	9.7355	9.9238	9.8117	0.1883	9.7470	9.9188	9.8282	0.1718	3
58	9.7357	9.9238	9.8120	0.1880	9.7472	9.9187	9.8284	0.1716	2
59	9.7359	9.9237	9.8122	0.1878	9.7474	9.9187	9.8287	0.1713	1
60	9.7361	9.9236	9.8125	0.1875	9.7476	9.9186	9.8290	0.1710	0
	Cos.	Sin.	Cotang.	Tang.	Cos.	Sin.	Cotang.	Tang.	
	57°				56°				

	34°				35°				
	Sin.	Cos.	Tang.	Cotang.	Sin.	Cos.	Tang.	Cotang.	
0′	9.7476	9.9186	9.8290	0.1710	9.7586	9.9134	9.8452	0.1548	60′
1	9.7477	9.9185	9.8293	0.1707	9.7588	9.9133	9.8455	0.1545	59
2	9.7479	9.9184	9.8295	0.1705	9.7590	9.9132	9.8458	0.1542	58
3	9.7481	9.9183	9.8298	0.1702	9.7591	9.9131	9.8460	0.1540	57
4	9.7483	9.9182	9.8301	0.1699	9.7593	9.9130	9.8463	0.1537	56
5	9.7485	9.9181	9.8303	0.1697	9.7595	9.9129	9.8466	0.1534	55
6	9.7487	9.9181	9.8306	0.1694	9.7597	9.9128	9.8468	0.1532	54
7	9.7489	9.9180	9.8309	0.1691	9.7599	9.9127	9.8471	0.1529	53
8	9.7491	9.9179	9.8312	0.1688	9.7600	9.9127	9.8474	0.1526	52
9	9.7492	9.9178	9.8314	0.1686	9.7602	9.9126	9.8476	0.1524	51
10	9.7494	9.9177	9.8317	0.1683	9.7604	9.9125	9.8479	0.1521	50
11	9.7496	9.9176	9.8320	0.1680	9.7606	9.9124	9.8482	0.1518	49
12	9.7498	9.9175	9.8323	0.1677	9.7607	9.9123	9.8484	0.1516	48
13	9.7500	9.9175	9.8325	0.1675	9.7609	9.9122	9.8487	0.1513	47
14	9.7502	9.9174	9.8328	0.1672	9.7611	9.9121	9.8490	0.1510	46
15	9.7504	9.9173	9.8331	0.1669	9.7613	9.9120	9.8493	0.1507	45
16	9.7505	9.9172	9.8333	0.1667	9.7615	9.9119	9.8495	0.1505	44
17	9.7507	9.9171	9.8336	0.1664	9.7616	9.9119	9.8498	0.1502	43
18	9.7509	9.9170	9.8339	0.1661	9.7618	9.9118	9.8501	0.1499	42
19	9.7511	9.9169	9.8342	0.1658	9.7620	9.9117	9.8503	0.1497	41
20	9.7513	9.9169	9.8344	0.1656	9.7622	9.9116	9.8506	0.1494	40
21	9.7515	9.9168	9.8347	0.1653	9.7624	9.9115	9.8509	0.1491	39
22	9.7517	9.9167	9.8350	0.1650	9.7625	9.9114	9.8511	0.1489	38
23	9.7518	9.9166	9.8352	0.1648	9.7627	9.9113	9.8514	0.1486	37
24	9.7520	9.9165	9.8355	0.1645	9.7629	9.9112	9.8517	0.1483	36
25	9.7522	9.9164	9.8358	0.1642	9.7631	9.9111	9.8519	0.1481	35
26	9.7524	9.9163	9.8361	0.1639	9.7632	9.9110	9.8522	0.1478	34
27	9.7526	9.9163	9.8363	0.1637	9.7634	9.9110	9.8525	0.1475	33
28	9.7528	9.9162	9.8366	0.1634	9.7636	9.9109	9.8527	0.1473	32
29	9.7529	9.9161	9.8369	0.1631	9.7638	9.9108	9.8530	0.1470	31
30	9.7531	9.9160	9.8371	0.1629	9.7640	9.9107	9.8533	0.1467	30
31	9.7533	9.9159	9.8374	0.1626	9.7641	9.9106	9.8535	0.1465	29
32	9.7535	9.9158	9.8377	0.1623	9.7643	9.9105	9.8538	0.1462	28
33	9.7537	9.9157	9.8379	0.1621	9.7645	9.9104	9.8541	0.1459	27
34	9.7539	9.9156	9.8382	0.1618	9.7647	9.9103	9.8543	0.1457	26
35	9.7540	9.9156	9.8385	0.1615	9.7649	9.9102	9.8546	0.1454	25
36	9.7542	9.9155	9.8388	0.1612	9.7650	9.9101	9.8549	0.1451	24
37	9.7544	9.9154	9.8390	0.1610	9.7652	9.9101	9.8551	0.1449	23
38	9.7546	9.9153	9.8393	0.1607	9.7654	9.9100	9.8554	0.1446	22
39	9.7548	9.9152	9.8396	0.1604	9.7655	9.9099	9.8557	0.1443	21
40	9.7550	9.9151	9.8398	0.1602	9.7657	9.9098	9.8559	0.1441	20
41	9.7551	9.9150	9.8401	0.1599	9.7659	9.9097	9.8562	0.1438	19
42	9.7553	9.9149	9.8404	0.1596	9.7661	9.9096	9.8565	0.1435	18
43	9.7555	9.9149	9.8406	0.1594	9.7662	9.9095	9.8567	0.1433	17
44	9.7557	9.9148	9.8409	0.1591	9.7664	9.9094	9.8570	0.1430	16
45	9.7559	9.9147	9.8412	0.1588	9.7666	9.9093	9.8573	0.1427	15
46	9.7561	9.9146	9.8415	0.1585	9.7668	9.9092	9.8575	0.1425	14
47	9.7562	9.9145	9.8417	0.1583	9.7669	9.9091	9.8578	0.1422	13
48	9.7564	9.9144	9.8420	0.1580	9.7671	9.9091	9.8581	0.1419	12
49	9.7566	9.9143	9.8423	0.1577	9.7673	9.9090	9.8583	0.1417	11
50	9.7568	9.9142	9.8425	0.1575	9.7675	9.9089	9.8586	0.1414	10
51	9.7570	9.9142	9.8428	0.1572	9.7676	9.9088	9.8589	0.1411	9
52	9.7571	9.9141	9.8431	0.1569	9.7678	9.9087	9.8591	0.1409	8
53	9.7573	9.9140	9.8433	0.1567	9.7680	9.9086	9.8594	0.1406	7
54	9.7575	9.9139	9.8436	0.1564	9.7682	9.9085	9.8597	0.1403	6
55	9.7577	9.9138	9.8439	0.1561	9.7683	9.9084	9.8599	0.1401	5
56	9.7579	9.9137	9.8442	0.1558	9.7685	9.9083	9.8602	0.1398	4
57	9.7580	9.9136	9.8444	0.1556	9.7687	9.9082	9.8605	0.1395	3
58	9.7582	9.9135	9.8447	0.1553	9.7689	9.9081	9.8607	0.1393	2
59	9.7584	9.9135	9.8450	0.1550	9.7690	9.9080	9.8610	0.1390	1
60	9.7586	9.9134	9.8452	0.1548	9.7692	9.9080	9.8613	0.1387	0
	Cos.	Sin.	Cotang.	Tang.	Cos.	Sin.	Cotang.	Tang.	
	55°				54°				

	36°				37°				
	Sin.	Cos.	Tang.	Cotang.	Sin.	Cos.	Tang.	Cotang.	
0′	9.7692	9.9080	9.8613	0.1387	9.7795	9.9023	9.8771	0.1229	60′
1	9.7694	9.9079	9.8615	0.1385	9.7796	9.9023	9.8774	0.1226	59
2	9.7696	9.9078	9.8618	0.1382	9.7798	9.9022	9.8776	0.1224	58
3	9.7697	9.9077	9.8621	0.1379	9.7800	9.9021	9.8779	0.1221	57
4	9.7699	9.9076	9.8623	0.1377	9.7801	9.9020	9.8782	0.1218	56
5	9.7701	9.9075	9.8626	0.1374	9.7803	9.9019	9.8784	0.1216	55
6	9.7703	9.9074	9.8629	0.1371	9.7805	9.9018	9.8787	0.1213	54
7	9.7704	9.9073	9.8631	0.1369	9.7806	9.9017	9.8790	0.1210	53
8	9.7706	9.9072	9.8634	0.1366	9.7808	9.9016	9.8792	0.1208	52
9	9.7708	9.9071	9.8637	0.1363	9.7810	9.9015	9.8795	0.1205	51
10	9.7710	9.9070	9.8639	0.1361	9.7811	9.9014	9.8797	0.1203	50
11	9.7711	9.9069	9.8642	0.1358	9.7813	9.9013	9.8800	0.1200	49
12	9.7713	9.9069	9.8644	0.1356	9.7815	9.9012	9.8803	0.1197	48
13	9.7715	9.9068	9.8647	0.1353	9.7816	9.9011	9.8805	0.1195	47
14	9.7716	9.9067	9.8650	0.1350	9.7818	9.9010	9.8808	0.1192	46
15	9.7718	9.9066	9.8652	0.1348	9.7820	9.9009	9.8811	0.1189	45
16	9.7720	9.9065	9.8655	0.1345	9.7821	9.9008	9.8813	0.1187	44
17	9.7722	9.9064	9.8658	0.1342	9.7823	9.9007	9.8816	0.1184	43
18	9.7723	9.9063	9.8660	0.1340	9.7825	9.9006	9.8818	0.1182	42
19	9.7725	9.9062	9.8663	0.1337	9.7826	9.9005	9.8821	0.1179	41
20	9.7727	9.9061	9.8666	0.1334	9.7828	9.9004	9.8824	0.1176	40
21	9.7728	9.9060	9.8668	0.1332	9.7830	9.9003	9.8826	0.1174	39
22	9.7730	9.9059	9.8671	0.1329	9.7831	9.9002	9.8829	0.1171	38
23	9.7732	9.9058	9.8674	0.1326	9.7833	9.9001	9.8831	0.1169	37
24	9.7734	9.9057	9.8676	0.1324	9.7835	9.9000	9.8834	0.1166	36
25	9.7735	9.9056	9.8679	0.1321	9.7836	9.9000	9.8837	0.1163	35
26	9.7737	9.9056	9.8682	0.1318	9.7838	9.8999	9.8839	0.1161	34
27	9.7739	9.9055	9.8684	0.1316	9.7840	9.8998	9.8842	0.1158	33
28	9.7740	9.9054	9.8687	0.1313	9.7841	9.8997	9.8845	0.1155	32
29	9.7742	9.9053	9.8689	0.1311	9.7843	9.8996	9.8847	0.1153	31
30	9.7744	9.9052	9.8692	0.1308	9.7844	9.8995	9.8850	0.1150	30
31	9.7746	9.9051	9.8695	0.1305	9.7846	9.8994	9.8852	0.1148	29
32	9.7747	9.9050	9.8697	0.1303	9.7848	9.8993	9.8855	0.1145	28
33	9.7749	9.9049	9.8700	0.1300	9.7849	9.8992	9.8858	0.1142	27
34	9.7751	9.9048	9.8703	0.1297	9.7851	9.8991	9.8860	0.1140	26
35	9.7752	9.9047	9.8705	0.1295	9.7853	9.8990	9.8863	0.1137	25
36	9.7754	9.9046	9.8708	0.1292	9.7854	9.8989	9.8865	0.1135	24
37	9.7756	9.9045	9.8711	0.1289	9.7856	9.8988	9.8868	0.1132	23
38	9.7758	9.9044	9.8713	0.1287	9.7858	9.8987	9.8871	0.1129	22
39	9.7759	9.9043	9.8716	0.1284	9.7859	9.8986	9.8873	0.1127	21
40	9.7761	9.9042	9.8718	0.1282	9.7861	9.8985	9.8876	0.1124	20
41	9.7763	9.9041	9.8721	0.1279	9.7863	9.8984	9.8879	0.1121	19
42	9.7764	9.9041	9.8724	0.1276	9.7864	9.8983	9.8881	0.1119	18
43	9.7766	9.9040	9.8726	0.1274	9.7366	9.8982	9.8884	0.1116	17
44	9.7768	9.9039	9.8729	0.1271	9.7867	9.8981	9.8886	0.1114	16
45	9.7769	9.9038	9.8732	0.1268	9.7869	9.8980	9.8889	0.1111	15
46	9.7771	9.9037	9.8734	0.1266	9.7871	9.8979	9.8892	0.1108	14
47	9.7773	9.9036	9.8737	0.1263	9.7872	9.8978	9.8894	0.1106	13
48	9.7774	9.9035	9.8740	0.1260	9.7874	9.8977	9.8897	0.1103	12
49	9.7776	9.9034	9.8742	0.1258	9.7876	9.8976	9.8899	0.1101	11
50	9.7778	9.9033	9.8745	0.1255	9.7877	9.8975	9.8902	0.1098	10
51	9.7780	9.9032	9.8747	0.1253	9.7879	9.8974	9.8905	0.1095	9
52	9.7781	9.9031	9.8750	0.1250	9.7880	9.8973	9.8907	0.1093	8
53	9.7783	9.9030	9.8753	0.1247	9.7882	9.8972	9.8910	0.1090	7
54	9.7785	9.9029	9.8755	0.1245	9.7884	9.8971	9.8912	0.1088	6
55	9.7786	9.9028	9.8758	0.1242	9.7885	9.8970	9.8915	0.1085	5
56	9.7788	9.9027	9.8761	0.1239	9.7887	9.8969	9.8918	0.1082	4
57	9.7790	9.9026	9.8763	0.1237	9.7889	9.8968	9.8920	0.1080	3
58	9.7791	9.9025	9.8766	0.1234	9.7890	9.8967	9.8923	0.1077	2
59	9.7793	9.9024	9.8769	0.1231	9.7892	9.8966	9.8925	0.1075	1
60	9.7795	9.9023	9.8771	0.1229	9.7893	9.8965	9.8928	0.1072	0
	Cos.	Sin.	Cotang.	Tang.	Cos.	Sin.	Cotang.	Tang.	
	53°				52°				

	38°				39°				
	Sin.	Cos.	Tang.	Cotang.	Sin.	Cos.	Tang.	Cotang.	
0′	9.7893	9.8965	9.8928	0.1072	9.7989	9.8905	9.9084	0.0916	60′
1	9.7895	9.8964	9.8931	0.1069	9.7990	9.8904	9.9086	0.0914	59
2	9.7897	9.8963	9.8933	0.1067	9.7992	9.8903	9.9089	0.0911	58
3	9.7898	9.8962	9.8936	0.1064	9.7993	9.8902	9.9091	0.0909	57
4	9.7900	9.8961	9.8939	0.1061	9.7995	9.8901	9.9094	0.0906	56
5	9.7901	9.8960	9.8941	0.1059	9.7997	9.8900	9.9097	0.0903	55
6	9.7903	9.8959	9.8944	0.1056	9.7998	9.8899	9.9099	0.0901	54
7	9.7905	9.8958	9.8946	0.1054	9.8000	9.8898	9.9102	0.0898	53
8	9.7906	9.8957	9.8949	0.1051	9.8001	9.8897	9.9104	0.0896	52
9	9.7908	9.8956	9.8952	0.1048	9.8003	9.8896	9.9107	0.0893	51
10	9.7910	9.8955	9.8954	0.1046	9.8004	9.8895	9.9110	0.0890	50
11	9.7911	9.8954	9.8957	0.1043	9.8006	9.8894	9.9112	0.0888	49
12	9.7913	9.8953	9.8959	0.1041	9.8007	9.8893	9.9115	0.0885	48
13	9.7914	9.8952	9.8962	0.1038	9.8009	9.8892	9.9117	0.0883	47
14	9.7916	9.8951	9.8965	0.1035	9.8010	9.8891	9.9120	0.0880	46
15	9.7918	9.8950	9.8967	0.1033	9.8012	9.8890	9.9122	0.0878	45
16	9.7919	9.8949	9.8970	0.1030	9.8014	9.8889	9.9125	0.0875	44
17	9.7921	9.8948	9.8972	0.1028	9.8015	9.8888	9.9128	0.0872	43
18	9.7922	9.8947	9.8975	0.1025	9.8017	9.8887	9.9130	0.0870	42
19	9.7924	9.8946	9.8978	0.1022	9.8018	9.8885	9.9133	0.0867	41
20	9.7926	9.8945	9.8980	0.1020	9.8020	9.8884	9.9135	0.0865	40
21	9.7927	9.8944	9.8983	0.1017	9.8021	9.8883	9.9138	0.0862	39
22	9.7929	9.8943	9.8985	0.1015	9.8023	9.8882	9.9140	0.0860	38
23	9.7930	9.8942	9.8988	0.1012	9.8024	9.8881	9.9143	0.0857	37
24	9.7932	9.8941	9.8990	0.1010	9.8026	9.8880	9.9146	0.0854	36
25	9.7934	9.8940	9.8993	0.1007	9.8027	9.8879	9.9148	0.0852	35
26	9.7935	9.8939	9.8996	0.1004	9.8029	9.8878	9.9151	0.0849	34
27	9.7937	9.8938	9.8998	0.1002	9.8031	9.8877	9.9153	0.0847	33
28	9.7938	9.8937	9.9001	0.0999	9.8032	9.8876	9.9156	0.0844	32
29	9.7940	9.8936	9.9003	0.0997	9.8034	9.8875	9.9158	0.0842	31
30	9.7941	9.8935	9.9006	0.0994	9.8035	9.8874	9.9161	0.0839	30
31	9.7943	9.8934	9.9009	0.0991	9.8037	9.8873	9.9164	0.0836	29
32	9.7945	9.8933	9.9011	0.0989	9.8038	9.8872	9.9166	0.0834	28
33	9.7946	9.8932	9.9014	0.0986	9.8040	9.8871	9.9169	0.0831	27
34	9.7948	9.8931	9.9016	0.0984	9.8041	9.8870	9.9171	0.0829	26
35	9.7949	9.8930	9.9019	0.0981	9.8043	9.8869	9.9174	0.0826	25
36	9.7951	9.8929	9.9022	0.0978	9.8044	9.8868	9.9176	0.0824	24
37	9.7953	9.8928	9.9024	0.0976	9.8046	9.8867	9.9179	0.0821	23
38	9.7954	9.8927	9.9027	0.0973	9.8047	9.8866	9.9182	0.0818	22
39	9.7956	9.8926	9.9029	0.0971	9.8049	9.8865	9.9184	0.0816	21
40	9.7957	9.8925	9.9032	0.0968	9.8050	9.8864	9.9187	0.0813	20
41	9.7959	9.8924	9.9035	0.0965	9.8052	9.8863	9.9189	0.0811	19
42	9.7960	9.8923	9.9037	0.0963	9.8053	9.8862	9.9192	0.0808	18
43	9.7962	9.8922	9.9040	0.0960	9.8055	9.8860	9.9194	0.0806	17
44	9.7964	9.8921	9.9042	0.0958	9.8056	9.8859	9.9197	0.0803	16
45	9.7965	9.8920	9.9045	0.0955	9.8058	9.8858	9.9200	0.0800	15
46	9.7967	9.8919	9.9047	0.0953	9.8060	9.8857	9.9202	0.0798	14
47	9.7968	9.8918	9.9050	0.0950	9.8061	9.8856	9.9205	0.0795	13
48	9.7970	9.8917	9.9053	0.0947	9.8063	9.8855	9.9207	0.0793	12
49	9.7972	9.8916	9.9055	0.0945	9.8064	9.8854	9.9210	0.0790	11
50	9.7973	9.8915	9.9058	0.0942	9.8066	9.8853	9.9212	0.0788	10
51	9.7975	9.8914	9.9060	0.0940	9.8067	9.8852	9.9215	0.0785	9
52	9.7976	9.8913	9.9063	0.0937	9.8069	9.8851	9.9218	0.0782	8
53	9.7978	9.8912	9.9066	0.0934	9.8070	9.8850	9.9220	0.0780	7
54	9.7979	9.8911	9.9068	0.0932	9.8072	9.8849	9.9223	0.0777	6
55	9.7981	9.8910	9.9071	0.0929	9.8073	9.8848	9.9225	0.0775	5
56	9.7982	9.8909	9.9073	0.0927	9.8075	9.8847	9.9228	0.0772	4
57	9.7984	9.8908	9.9076	0.0924	9.8076	9.8846	9.9230	0.0770	3
58	9.7986	9.8907	9.9079	0.0921	9.8078	9.8845	9.9233	0.0767	2
59	9.7987	9.8906	9.9081	0.0919	9.8079	9.8844	9.9236	0.0764	1
60	9.7989	9.8905	9.9084	0.0916	9.8081	9.8843	9.9238	0.0762	0
	Cos.	Sin.	Cotang.	Tang.	Cos.	Sin.	Cotang.	Tang.	
	51°				50°				

	40°				41°				
	Sin.	Cos.	Tang.	Cotang.	Sin.	Cos.	Tang.	Cotang.	
0′	9.8081	9.8843	9.9238	0.0762	9.8169	9.8778	9.9392	0.0608	60′
1	9.8082	9.8841	9.9241	0.0759	9.8171	9.8777	9.9394	0.0606	59
2	9.8084	9.8840	9.9243	0.0757	9.8172	9.8776	9.9397	0.0603	58
3	9.8085	9.8839	9.9246	0.0754	9.8174	9.8775	9.9399	0.0601	57
4	9.8087	9.8838	9.9248	0.0752	9.8175	9.8773	9.9402	0.0598	56
5	9.8088	9.8837	9.9251	0.0749	9.8177	9.8772	9.9404	0.0596	55
6	9.8090	9.8836	9.9254	0.0746	9.8178	9.8771	9.9407	0.0593	54
7	9.8091	9.8835	9.9256	0.0744	9.8180	9.8770	9.9409	0.0591	53
8	9.8093	9.8834	9.9259	0.0741	9.8181	9.8769	9.9412	0.0588	52
9	9.8094	9.8833	9.9261	0.0739	9.8182	9.8768	9.9415	0.0585	51
10	9.8096	9.8832	9.9264	0.0736	9.8184	9.8767	9.9417	0.0583	50
11	9.8097	9.8831	9.9266	0.0734	9.8185	9.8766	9.9420	0.0580	49
12	9.8099	9.8830	9.9269	0.0731	9.8187	9.8765	9.9422	0.0578	48
13	9.8100	9.8829	9.9271	0.0729	9.8188	9.8763	9.9425	0.0575	47
14	9.8102	9.8828	9.9274	0.0726	9.8190	9.8762	9.9427	0.0573	46
15	9.8103	9.8827	9.9277	0.0723	9.8191	9.8761	9.9430	0.0570	45
16	9.8105	9.8825	9.9279	0.0721	9.8193	9.8760	9.9432	0.0568	44
17	9.8106	9.8824	9.9282	0.0718	9.8194	9.8759	9.9435	0.0565	43
18	9.8108	9.8823	9.9284	0.0716	9.8195	9.8758	9.9438	0.0562	42
19	9.8109	9.8822	9.9287	0.0713	9.8197	9.8757	9.9440	0.0560	41
20	9.8111	9.8821	9.9289	0.0711	9.8198	9.8756	9.9443	0.0557	40
21	9.8112	9.8820	9.9292	0.0708	9.8200	9.8755	9.9445	0.0555	39
22	9.8114	9.8819	9.9295	0.0705	9.8201	9.8753	9.9448	0.0552	38
23	9.8115	9.8818	9.9297	0.0703	9.8203	9.8752	9.9450	0.0550	37
24	9.8117	9.8817	9.9300	0.0700	9.8204	9.8751	9.9453	0.0547	36
25	9.8118	9.8816	9.9302	0.0698	9.8205	9.8750	9.9455	0.0545	35
26	9.8120	9.8815	9.9305	0.0695	9.8207	9.8749	9.9458	0.0542	34
27	9.8121	9.8814	9.9307	0.0693	9.8208	9.8748	9.9460	0.0540	33
28	9.8122	9.8813	9.9310	0.0690	9.8210	9.8747	9.9463	0.0537	32
29	9.8124	9.8812	9.9312	0.0688	9.8211	9.8746	9.9466	0.0534	31
30	9.8125	9.8810	9.9315	0.0685	9.8213	9.8745	9.9468	0.0532	30
31	9.8127	9.8809	9.9318	0.0682	9.8214	9.8743	9.9471	0.0529	29
32	9.8128	9.8808	9.9320	0.0680	9.8216	9.8742	9.9473	0.0527	28
33	9.8130	9.8807	9.9323	0.0677	9.8217	9.8741	9.9476	0.0524	27
34	9.8131	9.8806	9.9325	0.0675	9.8218	9.8740	9.9478	0.0522	26
35	9.8133	9.8805	9.9328	0.0672	9.8220	9.8739	9.9481	0.0519	25
36	9.8134	9.8804	9.9330	0.0670	9.8221	9.8738	9.9483	0.0517	24
37	9.8136	9.8803	9.9333	0.0667	9.8223	9.8737	9.9486	0.0514	23
38	9.8137	9.8802	9.9335	0.0665	9.8224	9.8736	9.9488	0.0512	22
39	9.8139	9.8801	9.9338	0.0662	9.8225	9.8734	9.9491	0.0509	21
40	9.8140	9.8800	9.9341	0.0659	9.8227	9.8733	9.9494	0.0506	20
41	9.8142	9.8799	9.9343	0.0657	9.8228	9.8732	9.9496	0.0504	19
42	9.8143	9.8797	9.9346	0.0654	9.8230	9.8731	9.9499	0.0501	18
43	9.8145	9.8796	9.9348	0.0652	9.8231	9.8730	9.9501	0.0499	17
44	9.8146	9.8795	9.9351	0.0649	9.8233	9.8729	9.9504	0.0496	16
45	9.8148	9.8794	9.9353	0.0647	9.8234	9.8728	9.9506	0.0494	15
46	9.8149	9.8793	9.9356	0.0644	9.8235	9.8727	9.9509	0.0491	14
47	9.8150	9.8792	9.9358	0.0642	9.8237	9.8725	9.9511	0.0489	13
48	9.8152	9.8791	9.9361	0.0639	9.8238	9.8724	9.9514	0.0486	12
49	9.8153	9.8790	9.9364	0.0636	9.8240	9.8723	9.9516	0.0484	11
50	9.8155	9.8789	9.9366	0.0634	9.8241	9.8722	9.9519	0.0481	10
51	9.8156	9.8788	9.9369	0.0631	9.8242	9.8721	9.9522	0.0478	9
52	9.8158	9.8787	9.9371	0.0629	9.8244	9.8720	9.9524	0.0476	8
53	9.8159	9.8785	9.9374	0.0626	9.8245	9.8719	9.9527	0.0473	7
54	9.8161	9.8784	9.9376	0.0624	9.8247	9.8718	9.9529	0.0471	6
55	9.8162	9.8783	9.9379	0.0621	9.8248	9.8716	9.9532	0.0468	5
56	9.8164	9.8782	9.9381	0.0619	9.8249	9.8715	9.9534	0.0466	4
57	9.8165	9.8781	9.9384	0.0616	9.8251	9.8714	9.9537	0.0463	3
58	9.8167	9.8780	9.9387	0.0613	9.8252	9.8713	9.9539	0.0461	2
59	9.8168	9.8779	9.9389	0.0611	9.8254	9.8712	9.9542	0.0458	1
60	9.8169	9.8778	9.9392	0.0608	9.8255	9.8711	9.9544	0.0456	0
	Cos.	Sin.	Cotang.	Tang.	Cos.	Sin.	Cotang.	Tang.	
	49°				48°				

	42°				43°				
	Sin.	Cos.	Tang.	Cotang.	Sin.	Cos.	Tang.	Cotang.	
0′	9.8255	9.8711	9.9544	0.0456	9.8338	9.8641	9.9697	0.0303	60′
1	9.8257	9.8710	9.9547	0.0453	9.8339	9.8640	9.9699	0.0301	59
2	9.8258	9.8708	9.9549	0.0451	9.8341	9.8639	9.9702	0.0298	58
3	9.8259	9.8707	9.9552	0.0448	9.8342	9.8638	9.9704	0.0296	57
4	9.8261	9.8706	9.9555	0.0445	9.8343	9.8637	9.9707	0.0293	56
5	9.8262	9.8705	9.9557	0.0443	9.8345	9.8635	9.9709	0.0291	55
6	9.8264	9.8704	9.9560	0.0440	9.8346	9.8634	9.9712	0.0288	54
7	9.8265	9.8703	9.9562	0.0438	9.8347	9.8633	9.9714	0.0286	53
8	9.8266	9.8702	9.9565	0.0435	9.8349	9.8632	9.9717	0.0283	52
9	9.8268	9.8700	9.9567	0.0433	9.8350	9.8631	9.9719	0.0281	51
10	9.8269	9.8699	9.9570	0.0430	9.8351	9.8629	9.9722	0.0278	50
11	9.8270	9.8698	9.9572	0.0428	9.8353	9.8628	9.9724	0.0276	49
12	9.8272	9.8697	9.9575	0.0425	9.8354	9.8627	9.9727	0.0273	48
13	9.8273	9.8696	9.9577	0.0423	9.8355	9.8626	9.9729	0.0271	47
14	9.8275	9.8695	9.9580	0.0420	9.8357	9.8625	9.9732	0.0268	46
15	9.8276	9.8694	9.9582	0.0418	9.8358	9.8624	9.9735	0.0265	45
16	9.8277	9.8692	9.9585	0.0415	9.8359	9.8622	9.9737	0.0263	44
17	9.8279	9.8691	9.9588	0.0412	9.8361	9.8621	9.9740	0.0260	43
18	9.8280	9.8690	9.9590	0.0410	9.8362	9.8620	9.9742	0.0258	42
19	9.8282	9.8689	9.9593	0.0407	9.8365	9.8619	9.9745	0.0255	41
20	9.8283	9.8688	9.9595	0.0405	9.8363	9.8618	9.9747	0.0253	40
21	9.8284	9.8687	9.9598	0.0402	9.8366	9.8616	9.9750	0.0250	39
22	9.8286	9.8686	9.9600	0.0400	9.8367	9.8615	9.9752	0.0248	38
23	9.8287	9.8684	9.9603	0.0397	9.8369	9.8614	9.9755	0.0245	37
24	9.8289	9.8683	9.9605	0.0395	9.8370	9.8613	9.9757	0.0243	36
25	9.8290	9.8682	9.9608	0.0392	9.8371	9.8612	9.9760	0.0240	35
26	9.8291	9.8681	9.9610	0.0390	9.8373	9.8610	9.9762	0.0238	34
27	9.8293	9.8680	9.9613	0.0387	9.8374	9.8609	9.9765	0.0235	33
28	9.8294	9.8679	9.9615	0.0385	9.8375	9.8608	9.9767	0.0233	32
29	9.8295	9.8677	9.9618	0.0382	9.8377	9.8607	9.9770	0.0230	31
30	9.8297	9.8676	9.9621	0.0379	9.8378	9.8606	9.9772	0.0228	30
31	9.8298	9.8675	9.9623	0.0377	9.8379	9.8604	9.9775	0.0225	29
32	9.8300	9.8674	9.9626	0.0374	9.8381	9.8603	9.9778	0.0222	28
33	9.8301	9.8673	9.9628	0.0372	9.8382	9.8602	9.9780	0.0220	27
34	9.8302	9.8672	9.9631	0.0369	9.8383	9.8601	9.9783	0.0217	26
35	9.8304	9.8671	9.9633	0.0367	9.8385	9.8600	9.9785	0.0215	25
36	9.8305	9.8669	9.9636	0.0364	9.8386	9.8598	9.9788	0.0212	24
37	9.8306	9.8668	9.9638	0.0362	9.8387	9.8597	9.9790	0.0210	23
38	9.8308	9.8667	9.9641	0.0359	9.8389	9.8596	9.9793	0.0207	22
39	9.8309	9.8666	9.9643	0.0357	9.8390	9.8595	9.9795	0.0205	21
40	9.8311	9.8665	9.9646	0.0354	9.8391	9.8594	9.9798	0.0202	20
41	9.8312	9.8664	9.9648	0.0352	9.8393	9.8592	9.9800	0.0200	19
42	9.8313	9.8662	9.9651	0.0349	9.8394	9.8591	9.9803	0.0197	18
43	9.8315	9.8661	9.9653	0.0347	9.8395	9.8590	9.9805	0.0195	17
44	9.8316	9.8660	9.9656	0.0344	9.8397	9.8589	9.9808	0.0192	16
45	9.8317	9.8659	9.9659	0.0341	9.8398	9.8588	9.9810	0.0190	15
46	9.8319	9.8658	9.9661	0.0339	9.8399	9.8586	9.9813	0.0187	14
47	9.8320	9.8657	9.9664	0.0336	9.8401	9.8685	9.9816	0.0184	13
48	9.8322	9.8655	9.9666	0.0334	9.8402	9.8584	9.9818	0.0182	12
49	9.8323	9.8654	9.9669	0.0331	9.8403	9.8583	9.9821	0.0179	11
50	9.8324	9.8653	9.9671	0.0329	9.8405	9.8582	9.9823	0.0177	10
51	9.8326	9.8652	9.9674	0.0326	9.8406	9.8580	9.9826	0.0174	9
52	9.8327	9.8651	9.9676	0.0324	9.8407	9.8579	9.9828	0.0172	8
53	9.8328	9.8649	9.9679	0.0321	9.8409	9.8578	9.9831	0.0169	7
54	9.8330	9.8648	9.9681	0.0319	9.8410	9.8577	9.9833	0.0167	6
55	9.8331	9.8647	9.9684	0.0316	9.8411	9.8575	9.9836	0.0164	5
56	9.8332	9.8646	9.9686	0.0314	9.8412	9.8574	9.9838	0.0162	4
57	9.8334	9.8645	9.9689	0.0311	9.8414	9.8573	9.9341	0.0159	3
58	9.8335	9.8644	9.9691	0.0309	9.8415	9.8572	9.9843	0.0157	2
59	9.8336	9.8642	9.9694	0.0306	9.8416	9.8571	9.9846	0.0154	1
60	9.8338	9.8641	9.9697	0.0303	9.8418	9.8569	9.9848	0.0152	0
	Cos.	Sin.	Cotang.	Tang.	Cos.	Sin.	Cotang.	Tang.	
	47°				46°				

TABLE II. *Log. Sines and Tangents.*

	44°				
	Sin.	Cos.	Tang.	Cotang.	
0′	9.8418	9.8569	9.9848	0.0152	60′
1	9.8419	9.8568	9.9851	0.0149	59
2	9.8420	9.8567	9.9853	0.0147	58
3	9.8422	9.8566	9.9856	0.0144	57
4	9.8423	9.8564	9.9858	0.0142	56
5	9.8424	9.8563	9.9861	0.0139	55
6	9.8426	9.8562	9.9864	0.0136	54
7	9.8427	9.8561	9.9866	0.0134	53
8	9.8428	9.8560	9.9869	0.0131	52
9	9.8429	9.8558	9.9871	0.0129	51
10	9.8431	9.8557	9.9874	0.0126	50
11	9.8432	9.8556	9.9876	0.0124	49
12	9.8433	9.8555	9.9879	0.0121	48
13	9.8435	9.8553	9.9881	0.0119	47
14	9.8436	9.8552	9.9884	0.0116	46
15	9.8437	9.8551	9.9886	0.0114	45
16	9.8439	9.8550	9.9889	0.0111	44
17	9.8440	9.8548	9.9891	0.0109	43
18	9.8441	9.8547	9.9894	0.0106	42
19	9.8442	9.8546	9.9896	0.0104	41
20	9.8444	9.8545	9.9899	0.0101	40
21	9.8445	9.8544	9.9901	0.0099	39
22	9.8446	9.8542	9.9904	0.0096	38
23	9.8448	9.8541	9.9907	0.0093	37
24	9.8449	9.8540	9.9909	0.0091	36
25	9.8450	9.8539	9.9912	0.0088	35
26	9.8451	9.8537	9.9914	0.0086	34
27	9.8453	9.8536	9.9917	0.0083	33
28	9.8454	9.8535	9.9919	0.0081	32
29	9.8455	9.8534	9.9922	0.0078	31
30	9.8457	9.8532	9.9924	0.0076	30
31	9.8458	9.8531	9.9927	0.0073	29
32	9.8459	9.8530	9.9929	0.0071	28
33	9.8460	9.8529	9.9932	0.0068	27
34	9.8462	9.8527	9.9934	0.0066	26
35	9.8463	9.8526	9.9937	0.0063	25
36	9.8464	9.8525	9.9939	0.0061	24
37	9.8466	9.8524	9.9942	0.0058	23
38	9.8467	9.8522	9.9944	0.0056	22
39	9.8468	9.8521	9.9947	0.0053	21
40	9.8469	9.8520	9.9949	0.0051	20
41	9.8471	9.8519	9.9952	0.0048	19
42	9.8472	9.8517	9.9955	0.0045	18
43	9.8473	9.8516	9.9957	0.0043	17
44	9.8475	9.8515	9.9960	0.0040	16
45	9.8476	9.8514	9.9962	0.0038	15
46	9.8477	9.8512	9.9965	0.0035	14
47	9.8478	9.8511	9.9967	0.0033	13
48	9.8480	9.8510	9.9970	0.0030	12
49	9.8481	9.8509	9.9972	0.0028	11
50	9.8482	9.8507	9.9975	0.0025	10
51	9.8483	9.8506	9.9977	0.0023	9
52	9.8485	9.8505	9.9980	0.0020	8
53	9.8486	9.8504	9.9982	0.0018	7
54	9.8487	9.8502	9.9985	0.0015	6
55	9.8489	9.8501	9.9987	0.0013	5
56	9.8490	9.8500	9.9990	0.0010	4
57	9.8491	9.8499	9.9992	0.0008	3
58	9.8492	9.8497	9.9995	0.0005	2
59	9.8494	9.8496	9.9997	0.0003	1
60	9.8495	9.8495	0.0000	0.0000	0
	Cos.	Sin.	Cotang.	Tang.	
	45°				

TABLE III.

Log. Tangent of Obliquity of Ecliptic.

° ′ ″	Tang.
23 27 0	9.63726
1	9.63727
2	9.63728
3	9.63728
4	9.63729
5	9.63729
6	9.63730
7	9.63730
8	9.63731
9	9.63732
10	9.63732
11	9.63733
12	9.63733
13	9.63734
14	9.63735
15	9.63735
16	9.63736
17	9.63736
18	9.63737
19	9.63737
20	9.63738
21	9.63739
22	9.63739
23	9.63740
24	9.63740
25	9.63741
26	9.63741
27	9.63742
28	9.63743
29	9.63743
30	9.63744
31	9.63744
32	9.63745
33	9.63745
34	9.63746
35	9.63747
36	9.63747
37	9.63748
38	9.63748
39	9.63749
40	9.63750
41	9.63750
42	9.63751
43	9.63751
44	9.63752
45	9.63752
46	9.63753
47	9.63754
48	9.63754
49	9.63755
50	9.63755
51	9.63756
52	9.63756
53	9.63757
54	9.63758
55	9.63758
56	9.63759
57	9.63759
58	9.63760
59	9.63760
23 28 0	9.63761

ARGUMENT. Moon's Equatorial Parallax.

Par.	Log. A	Log. B	Par.	Log. A	Log. B	Par.	Log. A	Log. B
′ ″			′ ″			′ ″		
53 50	0.45449	5.80640	54 50	0.44647	5.79838	55 50	0.43860	5.79052
51	0.45435	5.80626	51	0.44634	5.79825	51	0.43847	5.79039
52	0.45422	5.80613	52	0.44621	5.79812	52	0.43834	5.79026
53	0.45408	5.80599	53	0.44608	5.79799	53	0.43821	5.79013
54	0.45395	5.80586	54	0.44594	5.79786	54	0.43808	5.79000
53 55	0.45382	5.80572	54 55	0.44581	5.79772	55 55	0.43795	5.78987
56	0.45368	5.80559	56	0.44568	5.79759	56	0.43782	5.78974
57	0.45355	5.80546	57	0.44555	5.79746	57	0.43769	5.78961
58	0.45341	5.80532	58	0.44542	5.79733	58	0.43756	5.78948
59	0.45328	5.80519	59	0.44528	5.79719	59	0.43744	5.78935
54 0	0.45314	5.80505	55 0	0.44515	5.79706	56 0	0.43731	5.78922
1	0.45301	5.80492	1	0.44502	5.79693	1	0.43718	5.78909
2	0.45287	5.80478	2	0.44489	5.79680	2	0.43705	5.78896
3	0.45274	5.80465	3	0.44476	5.79667	3	0.43692	5.78883
4	0.45261	5.80451	4	0.44462	5.79654	4	0.43679	5.78870
54 5	0.45247	5.80438	55 5	0.44449	5.79640	56 5	0.43666	5.78857
6	0.45234	5.80425	6	0.44436	5.79627	6	0.43653	5.78844
7	0.45220	5.80411	7	0.44423	5.79614	7	0.43640	5.78831
8	0.45207	5.80398	8	0.44410	5.79601	8	0.43627	5.78818
9	0.45193	5.80384	9	0.44397	5.79588	9	0.43614	5.78805
54 10	0.45180	5.80371	55 10	0.44383	5.79575	56 10	0.43601	5.78792
11	0.45167	5.80358	11	0.44370	5.79561	11	0.43588	5.78779
12	0.45153	5.80344	12	0.44357	5.79548	12	0.43575	5.78767
13	0.45140	5.80331	13	0.44344	5.79535	13	0.43562	5.78754
14	0.45127	5.80317	14	0.44331	5.79522	14	0.43549	5.78741
54 15	0.45113	5.80304	55 15	0.44318	5.79509	56 15	0.43536	5.78728
16	0.45100	5.80291	16	0.44305	5.79496	16	0.43524	5.78715
17	0.45086	5.80277	17	0.44291	5.79483	17	0.43511	5.78702
18	0.45073	5.80264	18	0.44278	5.79469	18	0.43498	5.78689
19	0.45060	5.80251	19	0.44265	5.79456	19	0.43485	5.78676
54 20	0.45046	5.80237	55 20	0.44252	5.79443	56 20	0.43472	5.78663
21	0.45033	5.80224	21	0.44239	5.79430	21	0.43459	5.78651
22	0.45020	5.80211	22	0.44226	5.79417	22	0.43446	5.78638
23	0.45006	5.80197	23	0.44213	5.79404	23	0.43433	5.78625
24	0.44993	5.80184	24	0.44200	5.79391	24	0.43421	5.78612
54 25	0.44980	5.80171	55 25	0.44187	5.79378	56 25	0.43408	5.78599
26	0.44966	5.80157	26	0.44173	5.79365	26	0.43395	5.78586
27	0.44953	5.80144	27	0.44160	5.79352	27	0.43382	5.78573
28	0.44940	5.80131	28	0.44147	5.79338	28	0.43369	5.78560
29	0.44926	5.80117	29	0.44134	5.79325	29	0.43356	5.78548
54 30	0.44913	5.80104	55 30	0.44121	5.79312	56 30	0.43343	5.78535
31	0.44900	5.80091	31	0.44108	5.79299	31	0.43331	5.78522
32	0.44886	5.80077	32	0.44095	5.79286	32	0.43318	5.78509
33	0.44873	5.80064	33	0.44082	5.79273	33	0.43305	5.78496
34	0.44860	5.80051	34	0.44069	5.79260	34	0.43292	5.78483
54 35	0.44846	5.80037	55 35	0.44056	5.79247	56 35	0.43279	5.78471
36	0.44833	5.80024	36	0.44043	5.79234	36	0.43266	5.78458
37	0.44820	5.80011	37	0.44030	5.79221	37	0.43254	5.78445
38	0.44807	5.79998	38	0.44017	5.79208	38	0.43241	5.78432
39	0.44793	5.79984	39	0.44004	5.79195	39	0.43228	5.78419
54 40	0.44780	5.79971	55 40	0.43991	5.79182	56 40	0.43215	5.78407
41	0.44767	5.79958	41	0.43978	5.79169	41	0.43202	5.78394
42	0.44753	5.79944	42	0.43964	5.79156	42	0.43190	5.78381
43	0.44740	5.79931	43	0.43951	5.79143	43	0.43177	5.78368
44	0.44727	5.79918	44	0.43938	5.79130	44	0.43164	5.78355
54 45	0.44714	5.79905	55 45	0.43925	5.79117	56 45	0.43151	5.78343
46	0.44700	5.79891	46	0.43912	5.79104	46	0.43138	5.78330
47	0.44687	5.79878	47	0.43899	5.79091	47	0.43126	5.78317
48	0.44674	5.79865	48	0.43886	5.79078	48	0.43113	5.78304
49	0.44661	5.79852	49	0.43873	5.79065	49	0.43100	5.78291

ARGUMENT. Moon's Equatorial Parallax.

Par.	Log. A	Log. B	Par.	Log. A	Log. B	Par.	Log. A	Log. B
′ ″			′ ″			′ ″		
56 50	0.43087	5.78279	57 50	0.42328	5.77519	58 50	0.41581	5.76773
51	0.43075	5.78266	51	0.42315	5.77507	51	0.41569	5.76761
52	0.43062	5.78253	52	0.42303	5.77494	52	0.41557	5.76748
53	0.43049	5.78240	53	0.42290	5.77482	53	0.41544	5.76736
54	0.43036	5.78228	54	0.42278	5.77469	54	0.41532	5.76724
56 55	0.43024	5.78215	57 55	0.42265	5.77457	58 55	0.41520	5.76711
56	0.43011	5.78202	56	0.42253	5.77444	56	0.41507	5.76699
57	0.42998	5.78189	57	0.42240	5.77432	57	0.41495	5.76687
58	0.42985	5.78177	58	0.42228	5.77419	58	0.41483	5.76675
59	0.42973	5.78164	59	0.42215	5.77407	59	0.41471	5.76662
57 0	0.42960	5.78151	58 0	0.42203	5.77394	59 0	0.41458	5.76650
1	0.42947	5.78139	1	0.42190	5.77382	1	0.41446	5.76638
2	0.42934	5.78126	2	0.42178	5.77369	2	0.41434	5.76625
3	0.42922	5.78113	3	0.42165	5.77357	3	0.41421	5.76613
4	0.42909	5.78100	4	0.42153	5.77344	4	0.41409	5.76601
57 5	0.42896	5.78088	58 5	0.42140	5.77332	59 5	0.41397	5.76589
6	0.42884	5.78075	6	0.42128	5.77319	6	0.41384	5.76576
7	0.42871	5.78062	7	0.42115	5.77307	7	0.41372	5.76564
8	0.42858	5.78050	8	0.42103	5.77294	8	0.41360	5.76552
9	0.42845	5.78037	9	0.42090	5.77282	9	0.41348	5.76539
57 10	0.42833	5.78024	58 10	0.42078	5.77269	59 10	0.41335	5.76527
11	0.42820	5.78011	11	0.42065	5.77257	11	0.41323	5.76515
12	0.42807	5.77999	12	0.42053	5.77244	12	0.41311	5.76503
13	0.42795	5.77986	13	0.42040	5.77232	13	0.41299	5.76490
14	0.42782	5.77973	14	0.42028	5.77219	14	0.41286	5.76478
57 15	0.42769	5.77961	58 15	0.42015	5.77207	59 15	0.41274	5.76466
16	0.42757	5.77948	16	0.42003	5.77194	16	0.41262	5.76454
17	0.42744	5.77935	17	0.41990	5.77182	17	0.41250	5.76441
18	0.42731	5.77923	18	0.41978	5.77170	18	0.41237	5.76429
19	0.42719	5.77910	19	0.41965	5.77157	19	0.41225	5.76417
57 20	0.42706	5.77897	58 20	0.41953	5.77145	59 20	0.41213	5.76405
21	0.42693	5.77885	21	0.41941	5.77132	21	0.41201	5.76392
22	0.42681	5.77872	22	0.41928	5.77120	22	0.41188	5.76380
23	0.42668	5.77859	23	0.41916	5.77107	23	0.41176	5.76368
24	0.42655	5.77847	24	0.41903	5.77095	24	0.41164	5.76356
57 25	0.42643	5.77834	58 25	0.41891	5.77083	59 25	0.41152	5.76344
26	0.42630	5.77822	26	0.41878	5.77070	26	0.41140	5.76331
27	0.42617	5.77809	27	0.41866	5.77058	27	0.41127	5.76319
28	0.42605	5.77796	28	0.41854	5.77045	28	0.41115	5.76307
29	0.42592	5.77784	29	0.41841	5.77033	29	0.41103	5.76295
57 30	0.42580	5.77771	58 30	0.41829	5.77020	59 30	0.41091	5.76283
31	0.42567	5.77758	31	0.41816	5.77008	31	0.41079	5.76270
32	0.42554	5.77746	32	0.41804	5.76996	32	0.41066	5.76258
33	0.42542	5.77733	33	0.41792	5.76983	33	0.41054	5.76246
34	0.42529	5.77721	34	0.41779	5.76971	34	0.41042	5.76234
57 35	0.42516	5.77708	58 35	0.41767	5.76959	59 35	0.41030	5.76222
36	0.42504	5.77695	36	0.41754	5.76946	36	0.41018	5.76209
37	0.42491	5.77683	37	0.41742	5.76934	37	0.41006	5.76197
38	0.42479	5.77670	38	0.41730	5.76921	38	0.40993	5.76185
39	0.42466	5.77658	39	0.41717	5.76909	39	0.40981	5.76173
57 40	0.42454	5.77645	58 40	0.41705	5.76897	59 40	0.40969	5.76161
41	0.42441	5.77632	41	0.41693	5.76884	41	0.40957	5.76149
42	0.42428	5.77620	42	0.41680	5.76872	42	0.40945	5.76137
43	0.42416	5.77607	43	0.41668	5.76860	43	0.40933	5.76124
44	0.42403	5.77595	44	0.41655	5.76847	44	0.40920	5.76112
57 45	0.42391	5.77582	58 45	0.41643	5.76835	59 45	0.40908	5.76100
46	0.42378	5.77570	46	0.41631	5.76822	46	0.40896	5.76088
47	0.42366	5.77557	47	0.41618	5.76810	47	0.40884	5.76076
48	0.42353	5.77544	48	0.41606	5.76798	48	0.40872	5.76064
49	0.42340	5.77532	49	0.41594	5.76785	49	0.40860	5.76052

 TABLE IV. *Logarithms A and B.*

ARGUMENT. Moon's Equatorial Parallax.

Par.	Log. A	Log. B	Par.	Log. A	Log. B
′ ″			′ ″		
59 50	0.40848	5.76039	60 50	0.40126	5.75318
51	0.40835	5.76027	51	0.40114	5.75306
52	0.40823	5.76015	52	0.40102	5.75294
53	0.40811	5.76003	53	0.40090	5.75282
54	0.40799	5.75991	54	0.40078	5.75270
59 55	0.40787	5.75979	60 55	0.40066	5.75258
56	0.40775	5.75967	56	0.40054	5.75246
57	0.40763	5.75955	57	0.40043	5.75234
58	0.40751	5.75943	58	0.40031	5.75223
59	0.40739	5.75930	59	0.40019	5.75211
60 0	0.40726	5.75918	61 0	0.40007	5.75199
1	0.40714	5.75906	1	0.39995	5.75187
2	0.40702	5.75894	2	0.39983	5.75175
3	0.40690	5.75882	3	0.39971	5.75163
4	0.40678	5.75870	4	0.39959	5.75151
60 5	0.40666	5.75858	61 5	0.39947	5.75139
6	0.40654	5.75846	6	0.39936	5.75127
7	0.40642	5.75834	7	0.39924	5.75116
8	0.40630	5.75822	8	0.39912	5.75104
9	0.40618	5.75810	9	0.39900	5.75092
60 10	0.40606	5.75798	61 10	0.39888	5.75080
11	0.40594	5.75786	11	0.39876	5.75068
12	0.40582	5.75773	12	0.39864	5.75056
13	0.40570	5.75761	13	0.39852	5.75044
14	0.40557	5.75749	14	0.39841	5.75033
60 15	0.40545	5.75737	61 15	0.39829	5.75021
16	0.40533	5.75725	16	0.39817	5.75009
17	0.40521	5.75713	17	0.39805	5.74997
18	0.40509	5.75701	18	0.39793	5.74985
19	0.40497	5.75689	19	0.39781	5.74973
60 20	0.40485	5.75677	61 20	0.39770	5.74962
21	0.40473	5.75665	21	0.39758	5.74950
22	0.40461	5.75653	22	0.39746	5.74938
23	0.40449	5.75641	23	0.39734	5.74926
24	0.40437	5.75629	24	0.39722	5.74914
60 25	0.40425	5.75617	61 25	0.39710	5.74902
26	0.40413	5.75605	26	0.39699	5.74891
27	0.40401	5.75593	27	0.39687	5.74879
28	0.40389	5.75581	28	0.39675	5.74867
29	0.40377	5.75569	29	0.39663	5.74855
60 30	0.40365	5.75557	61 30	0.39651	5.74843
31	0.40353	5.75545	31	0.39640	5.74832
32	0.40341	5.75533	32	0.39628	5.74820
33	0.40329	5.75521	33	0.39616	5.74808
34	0.40317	5.75509	34	0.39604	5.74796
60 35	0.40305	5.75497			
36	0.40293	5.75485			
37	0.40281	5.75473			
38	0.40269	5.75461			
39	0.40257	5.75449			
60 40	0.40245	5.75437			
41	0.40233	5.75425			
42	0.40221	5.75413			
43	0.40210	5.75401			
44	0.40198	5.75389			
60 45	0.40186	5.75378			
46	0.40174	5.75366			
47	0.40162	5.75354			
48	0.40150	5.75342			
49	0.40138	5.75330			

TABLE V.

Log. Tangent of Sun's Semidiameter

Semidiam.		Tang.
15′	40″	7.65871
	41	7.65917
	42	7.65963
	43	7.66009
	44	7.66055
15	45	7.66101
	46	7.66147
	47	7.66193
	48	7.66239
	49	7.66284
15	50	7.66330
	51	7.66376
	52	7.66422
	53	7.66467
	54	7.66513
15	55	7.66558
	56	7.66604
	57	7.66649
	58	7.66694
	59	7.66740
16	0	7.66785
	1	7.66830
	2	7.66875
	3	7.66920
	4	7.66966
16	5	7.67011
	6	7.67056
	7	7.67100
	8	7.67145
	9	7.67190
16	10	7.67235
	11	7.67280
	12	7.67324
	13	7.67369
	14	7.67414
16	15	7.67458
	16	7.67503
	17	7.67547
	18	7.67592
	19	7.67636
	20	7.67680

Latitudes and Longitudes from the Meridian of Greenwich, of some Cities, and other conspicuous places.

Names of Places.		Latitude.		Longitude in Time.		Longitude in Degrees.
		° ′ ″		h. m. sec.		° ′ ″
Albany, *Capitol,*	New York,	42 39 3	N.	4 54 59	W.	73 44 45
Altona, *Obs.,*	Denmark,	53 32 45	N.	0 39 47	E.	9 56 45
Amsterdam,	Holland,	52 22 30	N.	0 19 33	E.	4 53 16
Baltimore, *B. Mon't.,*	Maryland,	39 17 13	N.	5 6 31	W.	76 37 50
Berlin, *Obs.,*	Germany,	52 31 13	N.	0 53 35.5	E.	13 23 52
Boston, *State House,*	Mass'ts,	42 21 15	N.	4 44 16.6	W.	71 4 9
Brest, *Obs.,*	France,	48 23 32	N.	0 17 58	W.	4 29 25
Canton,	China,	23 8 9	N.	7 33 8	E.	113 16 54
Cape G. Hope, *Obs.,*	Africa,	33 56 3	S.	1 13 55	E.	18 28 45
Charleston, *Coll.,*	S. Carolina,	32 47 0	N.	5 20 3	W.	80 0 52
Charlottesville, *Univers.,*	Virginia,	38 2 3	N.	5 14 6	W.	78 31 29
Cincinnati,	Ohio,	39 5 54	N.	5 37 36	W.	84 24 0
Copenhagen, *Obs.,*	Denmark,	55 40 53	N.	0 50 19.8	E.	12 34 57
Dorpat, *Obs.,*	Russia,	58 22 47	N.	1 46 55	E.	26 43 45
Dublin, *Obs.,*	Ireland,	53 23 13	N.	0 25 22	W.	6 20 30
Edinburgh, *Obs.,*	Scotland,	55 57 20	N.	0 12 43.6	W.	3 10 54
Gotha, *Seeberg Obs.,*	Germany,	50 56 5	N.	0 42 56.4	E.	10 44 6
Göttingen, *Obs.,*	Germany,	51 31 48	N.	0 39 46.5	E.	9 56 37
Greenwich, *Obs.,*	England,	51 28 39	N.	0 0 0		0 0 0
Hudson, *Obs.,*	Ohio,	41 14 37	N.	5 25 42	W.	81 25 30
Königsberg, *Obs.,*	Prussia,	54 42 50	N.	1 22 0.5	E.	20 30 7
Lancaster,	Penn.,	40 2 36	N.	5 5 22	W.	76 20 33
London, *St. Paul's Ch.,*	England,	51 30 49	N.	0 0 23	W.	0 5 47
Marseilles, *Obs.,*	France,	43 17 50	N.	0 21 29	E.	5 22 15
Milan, *Obs.,*	Italy,	45 28 1	N.	0 36 47	E.	9 11 48
Naples, *Obs.,*	Italy,	40 51 47	N.	0 57 0	E.	14 15 4
New Haven, *Coll.,*	Connecticut,	41 17 58	N.	4 51 51	W.	72 57 46
New Orleans, *C. Hall,*	Louisiana,	29 57 45	N.	6 0 27	W.	90 6 49
New York, *C. Hall,*	New York,	40 42 40	N.	4 56 4	W.	74 1 8
Palermo, *Obs.,*	Italy,	38 6 44	N.	0 53 25.6	E.	13 21 24
Paramatta, *Obs.,*	New Hol.,	33 48 50	S.	10 4 6	E.	151 1 34
Paris, *Obs.,*	France,	48 50 13	N.	0 9 21.6	E.	2 20 24
Petersburgh, *Obs.,*	Russia,	59 56 31	N.	2 1 16	E.	30 19 0
Philadelphia, *Ind'ce H.,*	Penn.,	39 56 59	N.	5 0 40	W.	75 10 0
Pittsburgh,	Penn.,	40 26 15	N.	5 19 52	W.	79 58 6
Point Venus,	Otaheite,	17 29 21	S.	9 57 56	W.	149 28 55
Princeton, *Coll.,*	New Jersey,	40 22	N.	4 58 20	W.	74 35
Providence, *Univers.,*	Rhode Isl.,	41 49 25	N.	4 45 44	W.	71 25 56
Pulkowa, *Obs.,*	Russia,	59 46 18.7	N.	2 1 24.7	E.	30 26 10
Quebec, *Castle,*	L. Canada,	46 49 12	N.	4 45 4	W.	71 16 0
Richmond, *Capitol,*	Virginia,	37 32 17	N.	5 9 46	W.	77 26 28
Rome, *St. Peter's Ch.,*	Italy,	41 54 8	N.	0 49 48	E.	12 27 5
Savannah, *Exch.,*	Georgia,	32 4 56	N.	5 24 29	W.	81 7 9
Stockholm, *Obs.,*	Sweden,	59 20 31	N.	1 12 14	E.	18 3 34
Turin, *Obs.,*	Italy,	45 4 6	N.	0 30 48.4	E.	7 42 6
Vienna, *Obs.,*	Austria,	48 12 35	N.	1 5 32	E.	16 23 0
Wardhus,	Lapland,	70 22 36	N.	2 4 32	E.	31 7 54
Washington, *Obs.,*	Dist. Colum.,	38 53 33	N.	5 8 15	W.	77 3 39

TABLE VII.

Astronomical Refractions.

App. Alt.	Mean Refr.	Diff for + 1 Bar.	Diff for — 1° Fah.	App. Alt.	Mean Refr.	Diff for + 1 Bar.	Diff for — 1° Fah.	App. Alt.	Mean Refr.	Diff for + 1 Bar.	Diff for — 1° Fah.
° ′	′ ″	″	″	° ′	′ ″	″	″	° ′	′ ″	″	
0 0	33 51	74	8.1	4 0	11 52	24.1	1.70	12 0	4 28.1	9.00	0.556
5	32 53	71	7.6	10	11 30	23.4	1.64	10	4 24.4	8.86	.548
10	31 58	69	7.3	20	11 10	22.7	1.58	20	4 20.8	8.74	.541
15	31 5	67	7.0	30	10 50	22.0	1.53	30	4 17.3	8.63	.533
20	30 13	65	6.7	40	10 32	21.3	1.48	40	4 13.9	8.51	.524
25	29 24	63	6.4	50	10 15	20.7	1.43	50	4 10.7	8.41	.517
30	28 37	61	6.1	5 0	9 58	20.1	1.38	13 0	4 7.5	8.30	.509
35	27 51	59	5.9	10	9 42	19.6	1.34	10	4 4.4	8.20	.503
40	27 6	58	5.6	20	9 27	19.1	1.30	20	4 1.4	8.10	.496
45	26 24	56	5.4	30	9 11	18.6	1.26	30	3 58.4	8.00	.490
50	25 43	55	5.1	40	8 58	18.1	1.22	40	3 55.5	7.89	.482
55	25 3	53	4.9	50	8 45	17.6	1.19	50	3 52.6	7.79	.476
1 0	24 25	52	4.7	6 0	8 32	17.2	1.15	14 0	3 49.9	7.70	.469
5	23 48	50	4.6	10	8 20	16.8	1.11	10	3 47.1	7.61	.464
10	23 13	49	4.5	20	8 9	16.4	1.09	20	3 44.4	7.52	.458
15	22 40	48	4.4	30	7 58	16.0	1.06	30	3 41.8	7.43	.453
20	22 8	46	4.2	40	7 47	15.7	1.03	40	3 39.2	7.34	.448
25	21 37	45	4.0	50	7 37	15.3	1.00	50	3 36.7	7.26	.444
30	21 7	44	3.9	7 0	7 27	15.0	0.98	15 0	3 34.3	7.18	.439
35	20 38	43	3.8	10	7 17	14.6	.95	30	3 27.3	6.95	.424
40	20 10	42	3.6	20	7 8	14.3	.93	16 0	3 20.6	6.73	.411
45	19 43	40	3.5	30	6 59	14.1	.91	30	3 14.4	6.51	.399
50	19 17	39	3.4	40	6 51	13.8	.89	17 0	3 8.5	6.31	.386
55	18 52	39	3.3	50	6 43	13.5	.87	30	3 2.9	6.12	.374
2 0	18 29	38	3.2	8 0	6 35	13.3	.85	18 0	2 57.6	5.94	.362
5	18 5	37	3.1	10	6 28	13.1	.83	19 0	2 47.7	5.61	.340
10	17 43	36	3.0	20	6 21	12.8	.82	20 0	2 38.7	5.31	.322
15	17 21	36	2.9	30	6 14	12.6	.80	21 0	2 30.5	5.04	.305
20	17 0	35	2.8	40	6 7	12.3	.79	22 0	2 23.2	4.79	.290
25	16 40	34	2.8	50	6 0	12.1	.77	23 0	2 16.5	4.57	.276
30	16 21	33	2.7	9 0	5 54	11.9	.76	24 0	2 10.1	4.35	.264
35	16 2	33	2.7	10	5 47	11.7	.74	25 0	2 4.2	4.16	.252
40	15 43	32	2.6	20	5 41	11.5	.73	26 0	1 58.8	3.97	.241
45	15 25	32	2.5	30	5 36	11.3	.72	27 0	1 53.8	3.81	.230
50	15 8	31	2.4	40	5 30	11.1	.71	28 0	1 49.1	3.65	.219
55	14 51	30	2.3	50	5 25	11.0	.70	29 0	1 44.7	3.50	.209
3 0	14 35	30	2.3	10 0	5 20	10.8	.69	30 0	1 40.5	3.36	.201
5	14 19	29	2.2	10	5 15	10.6	.67	31 0	1 36.6	3.23	.193
10	14 4	29	2.2	20	5 10	10.4	.65	32 0	1 33.0	3.11	.186
15	13 50	28	2.1	30	5 5	10.2	.64	33 0	1 29.5	2.99	.179
20	13 35	28	2.1	40	5 0	10.1	.63	34 0	1 26.1	2.88	.173
25	13 21	27	2.0	50	4 56	9.9	.62	35 0	1 23.0	2.78	.167
30	13 7	27	2.0	11 0	4 51	9.8	.60	36 0	1 20.0	2.68	.161
35	12 53	26	2.0	10	4 47	9.6	.59	37 0	1 17.1	2.58	.155
40	12 41	26	1.9	20	4 43	9.5	.58	38 0	1 14.4	2.49	.149
45	12 28	25	1.9	30	4 39	9.4	.57	39 0	1 11.8	2.40	.144
50	12 16	25	1.9	40	4 35	9.2	.56	40 0	1 9.3	2.32	.139
55	12 3	25	1.8	50	4 31	9.1	.55	41 0	1 6.9	2.24	.134
4 0	11 52	24	1.7	12 0	4 28	9.0	.55	42 0	1 4.6	2.16	.130

Astronomical Refractions.

App. Alt.	Mean Refr.	Diff. for + 1 Bar.	Diff. for — 1° Fah.	App. Alt.	Mean Refr.	Diff. for + 1 Bar.	Diff. for — 1° Fah.	App. Alt.	Mean Refr.	Diff. for + 1 Bar.	Diff. for — 1° Fah.
°	′ ″	″	″	°	″	″	″	°	″	″	″
42	1 4.6	2.16	0.130	58	36.4	1.22	0.073	74	16.6	0.56	0.033
43	1 2.4	2.09	.125	59	35.0	1.17	.070	75	15.5	.52	.031
44	1 0.3	2.02	.120	60	33.6	1.12	.067	76	14.4	.48	029
45	0 58.1	1.95	.116	61	32.3	1.08	.065	77	13.4	.45	.027
46	0 56.1	1.88	.112	62	31.0	1.04	.062	78	12.3	.41	.025
47	0 54.2	1.81	.108	63	29.7	.99	.060	79	11.2	.38	.023
48	0 52.3	1.75	.104	64	28.4	.95	.057	80	10.2	.34	.021
49	50.5	1.69	.101	65	27.2	.91	.055	81	9.2	.31	.018
50	48.8	1.63	.097	66	25.9	.87	.052	82	8.2	.27	.016
51	47.1	1.58	.094	67	24.7	.83	.050	83	7.1	.24	.014
52	45.4	1.52	.090	68	23.5	.79	.047	84	6.1	.20	.012
53	43.8	1.47	.088	69	22.4	.75	.045	85	5.1	.17	.010
54	0 42.2	1.41	.085	70	21.2	.71	.043	86	4.1	.14	.008
55	40.8	1.36	.082	71	19.9	.67	.040	87	3.1	.10	.006
56	39.3	1.31	.079	72	18.8	.63	.038	88	2.1	.07	.004
57	37.8	1.26	.076	73	17.7	.59	.036	89	1.0	.03	.002
58	36.4	1.22	.073	74	16.6	.56	.033	90	0.0	.00	.000

TABLE VIII.

Sun's Parallax in Altitude.

ARGUMENTS. Sun's Semi-diameter at top and Altitude at side.

	15′ 40″	15′ 50″	16′ 0″	16′ 10″	16′ 20″
°	″	″	″	″	″
0	8.4	8.5	8.6	8.7	8.7
5	8.4	8.4	8.5	8.6	8.7
10	8.3	8.4	8.4	8.5	8.6
15	8.1	8.2	8.3	8.4	8.4
20	7.9	8.0	8.1	8.1	8.2
25	7.6	7.7	7.8	7.8	7.9
30	7,3	7.3	7.4	7.5	7.6
35	6.9	6.9	7.0	7.1	7.2
40	6.4	6.5	6.6	6.6	6.7
45	5.9	6.0	6.1	6.1	6.2
50	5.4	5.5	5.5	5.6	5.6
55	4.8	4.9	4.9	5.0	5.0
60	4.2	4.2	4.3	4.3	4.4
65	3.5	3.6	3.6	3.7	3.7
70	2.9	2.9	2.9	3.0	3.0
75	2.2	2.2	2.2	2.2	2.3
80	1.5	1.5	1.5	1.5	1.5
85	0.7	0.7	0.7	0.8	0.8
90	0.0	0.0	0.0	0.0	0.0

Mean Right Ascensions and Declinations of 30 *principal fixed Stars for January* 1, 1850.

No.	Star's Name.	Mag.	Right Ascen.	Ann. Var.	Declination.	Ann. Var.
			h m s.	s.	° ′ ″	″
1	α Andromedæ	2	0 0 38.51	+ 3.082	+28 15 43.8	+19.90
2	α Urs. Min. (*Polaris*)....	2	1 5 1.30	+17.554	+88 30 34.9	+19.25
3	α Arietis	2	1 58 43.60	+ 3.362	+22 45 1.7	+17.28
4	α Ceti.....................	2.3	2 54 26.52	+ 3.126	+ 3 29 51.0	+14.40
5	α Tauri (*Aldebaran*)	1	4 27 19.04	+ 3.433	+16 12 10.8	+ 7.72
6	α Aurigæ (*Capella*)	1	5 5 36.93	+ 4.418	+45 50 20.7	+ 4.29
7	β Orionis (*Rigel*)	1	5 7 19.81	+ 2.880	— 8 22 46.2	+ 4.54
8	β Tauri	2	5 16 48.80	+ 3.788	+28 28 29.3	+ 3.55
9	α Columbæ................	2	5 34 13.17	+ 2.177	—34 9 24.5	+ 2.25
10	α Orionis	1	5 47 3.11	+ 3.246	+ 7 22 26.6	+ 1.13
11	α Canis Maj. (*Sirius*)....	1	6 38 32.15	+ 2.644	—16 30 53.3	— 4.60
12	α Canis Min. (*Procyon*).	1	7 31 26.83	+ 3.146	+ 5 36 16.3	— 8.86
13	β Geminor (*Pollux*)	2	7 36 7.73	+ 3.682	+28 23 0.7	— 8.23
14	α Hydræ	2	9 20 12.83	+ 2.947	— 8 0 41.1	—15.35
15	α Leonis (*Regulus*)	1	10 0 22.64	+ 3.202	+12 41 53.2	—17.38
16	β Leonis....................	2	11 41 24.22	+ 3.065	+15 24 37.4	—20.09
17	α Virginis (*Spica*)........	1	13 17 17.80	+ 3.149	—10 22 37.8	—18.98
18	α Bootis (*Arcturus*)......	1	14 8 49.22	+ 2.733	+19 57 56.1	—18.95
19	α^2 Libræ....................	2.3	14 42 35.34	+ 3.306	—15 24 54.9	—15.28
20	β Ursæ Minoris..........	2.3	14 51 11.96	— 0.271	+74 46 5.4	—14.76
21	β Libræ	2	15 8 56.39	+ 3.220	— 8 49 33.0	—13.61
22	α Coronæ Borealis.......	2	15 28 20.24	+ 2.537	+27 13 21.3	—12.39
23	β^1 Scorpii....................	2	15 56 43.27	+ 3.478	—19 23 25.6	—10.29
24	α Scorpii (*Antares*)......	1	16 20 13.11	+ 3.666	—26 5 40.3	— 8.50
25	α Lyræ (*Vega*)...........	1	18 31 51.58	+ 2.030	+38 38 47.9	+ 3.05
26	α Aquilæ (*Altair*)........	1.2	19 43 27.85	+ 2.928	+ 8 28 32.5	+ 9.12
27	α Cygni.....................	1.2	20 36 19.13	+ 2.042	+44 44 46.4	+12.63
28	α Aquarii..................	3	21 58 4.66	+ 3.083	— 1 2 48.6	+17.26
29	α Pis. Aus. (*Fomalhaut*)	1	22 49 21.25	+ 3.335	—30 25 2.9	+18.88
30	α Pegasi (*Markab*)	2	22 57 17.52	+ 2.982	+14 23 57.3	+19.29

		Year.	Right Ascension.			Ann. Var.	Declination.	Ann. Var.
			h	m	s.	s.	° ′ ″	″
2	Polaris	1840	1	2	10.68	+16.501	88 27 21.9	+19.32
		1850	1	5	1.30	17.554	88 30 34.9	19.25
		1860	1	8	1.73	18.784	88 33 47.6	19.16

Year.	φ	Log. M	θ	Log. N	φ′	Log. M′	θ′	Log. N′
	° ′		° ′		° ′		° ′	
1840	253 8	1.6679	166 17	1.3050	255 41	1.3557	159 31	0.8487
1850	252 22	1.6833	165 36	1.3048	255 14	1.3704	158 37	0.8498
1860	251 34	1.6995	164 53	1.3045	254 45	1.3855	157 40	0.8510

Constants for the Aberration and Nutation in Right Ascension and Declination.

No.	Aberration.				Nutation.			
	φ	Log. m	θ	Log. n.	φ′	Log. m′	θ′	Log. n′
	° ′		° ′		° ′		° ′	
1	269 58	0.1501	216 34	1.0785	197 22	0.0434	180 3	0.8366
2	253 8	1.6679	166 17	1.3050	255 41	1.3557	159 31	0.8487
3	238 17	0.1400	209 51	0.8966	191 1	0.0685	142 43	0.8754
4	224 3	0.1146	263 3	0.8675	181 26	0.0310	128 8	0.9067
5	201 34	0.1448	233 13	0.5741	183 26	9.0714	107 47	0.9490
6	192 41	0.2876	115 17	0.9108	185 41	0.1817	100 21	0.9592
7	192 13	0.1355	273 40	1.0291	178 48	9.9952	99 58	0.9595
8	190 4	0.1874	138 48	0.3893	182 48	0.1138	98 12	0.9612
9	186 0	0.2146	274 22	1.2325	176 21	9.8736	94 53	0.9635
10	183 6	0.1362	268 27	0.7539	180 14	0.0469	92 31	0.9644
11	171 15	0.1500	265 47	1.1153	181 52	9.9646	82 53	0.9621
12	158 59	0.1296	276 54	0.8078	178 47	0.0401	72 41	0.9498
13	157 54	0.1827	14 14	0.6069	174 0	0.1097	71 45	0.9483
14	132 31	0.1158	257 26	0.9971	183 43	0.0066	48 29	0.8991
15	122 13	0.1161	303 39	0.8462	173 47	0.0465	37 50	0.8765
16	95 12	0.1115	306 11	0.9621	170 57	0.0328	6 24	0.8378
17	69 13	0.1067	243 22	0.8859	185 35	0.0363	334 57	0.8547
18	55 39	0.1335	298 12	1.0968	168 53	9.9932	319 54	0.8810
19	47 2	0.1276	228 16	0.7898	186 27	0.0583	341 4	0.9001
20	44 43	0.6950	345 5	1.3087	86 52	0.2210	338 47	0.9052
21	40 27	0.1212	250 19	0.7985	183 21	0.0447	334 41	0.9145
22	35 39	0.1703	292 23	1.1785	167 21	9.9496	300 12	0.9245
23	28 50	0.1487	213 57	0.6219	185 19	0.0783	294 5	0.9375
24	23 15	0.1730	177 43	0.5796	185 48	0.1017	289 14	0.9466
25	352 46	0.2393	264 22	1.2545	185 37	9.8423	264 3	0.9629
26	336 8	0.1308	262 56	1.0241	182 17	9.9974	250 15	0.9457
27	323 24	0.2679	240 34	1.2635	208 36	9.9031	238 55	0.9226
28	302 49	0.1057	272 25	0.8992	179 27	0.0249	218 28	0.8778
29	289 17	0.1635	337 25	1.0272	163 9	0.0747	203 20	0.8524
30	287 9	0.1120	241 59	1.0143	188 25	0.0145	200 49	0.8492

Months.	Com.	Bis.		Days.		Days.		Days.	
January	.000	.000		1	.000	13	.033	25	.066
February	.085	.085		2	.003	14	.036	26	.068
March	.162	.164		3	.005	15	.038	27	.071
April	.246	.249		4	.008	16	.041	28	.074
May	.329	.331		5	.011	17	.044	29	.077
June	.413	.416		6	.014	18	.047	30	.079
July	.496	.498		7	.016	19	.049	31	.082
August	.580	.583		8	.019	20	.052		
September	.665	.668		9	.022	21	.055		
October	.747	.750		10	.025	22	.057		
November	.832	.835		11	.027	23	.060		
December	.914	.917		12	.030	24	.063		

For converting Intervals *of* Mean Solar *Time into equivalent* Intervals *of* Sidereal *Time.*

Hours.		Minutes.				Seconds.			
Hours of Mean Time.	Equivalents in Sidereal Time.	Minutes of Mean Time.	Equivs. in Sid. Time.	Minutes of Mean Time.	Equivs. in Sid. Time.	Seconds of Mean Time.	Equivs. in S. Time.	Seconds of Mean Time.	Equivs. in S. Time.
	h. m. s.		m. s.		m. s.		s.		s.
1	1 0 9.856	1	1 0.164	31	31 5.092	1	1.003	31	31.085
2	2 0 19.713	2	2 0.329	32	32 5.257	2	2.005	32	32.088
3	3 0 29.569	3	3 0.493	33	33 5.421	3	3.008	33	33.090
4	4 0 39.426	4	4 0.657	34	34 5.585	4	4.011	34	34.093
5	5 0 49.282	5	5 0.821	35	35 5.750	5	5.014	35	35.096
6	6 0 59.139	6	6 0.986	36	36 5.914	6	6.016	36	36.099
7	7 1 8.995	7	7 1.150	37	37 6.078	7	7.019	37	37.101
8	8 1 18.852	8	8 1.314	38	38 6.242	8	8.022	38	38.104
9	9 1 28.708	9	9 1.478	39	39 6.407	9	9.025	39	39.107
10	10 1 38.565	10	10 1.643	40	40 6.571	10	10.027	40	40.110
11	11 1 48.421	11	11 1.807	41	41 6.735	11	11.030	41	41.112
12	12 1 58.278	12	12 1.971	42	42 6.900	12	12.033	42	42.115
13	13 2 8.134	13	13 2.136	43	43 7.064	13	13.036	43	43.118
14	14 2 17.991	14	14 2.300	44	44 7.228	14	14.038	44	44.120
15	15 2 27.847	15	15 2.464	45	45 7.392	15	15.041	45	45.123
16	16 2 37.704	16	16 2.628	46	46 7.557	16	16.044	46	46.126
17	17 2 47.560	17	17 2.793	47	47 7.721	17	17.047	47	47.129
18	18 2 57.417	18	18 2.957	48	48 7.885	18	18.049	48	48.131
19	19 3 7.273	19	19 3.121	49	49 8.049	19	19.052	49	49.134
20	20 3 17.129	20	20 3.285	50	50 8.214	20	20.055	50	50.137
21	21 3 26.986	21	21 3.450	51	51 8.378	21	21.057	51	51.140
22	22 3 36.842	22	22 3.614	52	52 8.542	22	22.060	52	52.142
23	23 3 46.699	23	23 3.778	53	53 8.707	23	23.063	53	53.145
24	24 3 56.555	24	24 3.943	54	54 8.871	24	24.066	54	54.148
		25	25 4.107	55	55 9.035	25	25.069	55	55.151
		26	26 4.271	56	56 9.199	26	26.071	56	56.153
		27	27 4.435	57	57 9.364	27	27.074	57	57.156
		28	28 4.600	58	58 9.528	28	28.077	58	58.159
		29	29 4.764	59	59 9.692	29	29.079	59	59.162
		30	30 4.928	60	60 9.856	30	30.082	60	60.164

For converting Intervals *of* Sidereal *Time into equivalent* Intervals *of* Mean Solar *Time.*

Hours.		Minutes.				Seconds.			
Hours of Sid. Time.	Equivalents in Mean Time.	Minutes of Sid. Time.	Equivs. in Mean Time.	Minutes of Sid. Time.	Equivs. in Mean Time.	Seconds of Sid. Time.	Equivs. in M.Time.	Seconds of Sid. Time.	Equivs. in M.Time.
	h. m. s.		m. s.		m. s.		s.		s.
1	0 59 50.170	1	0 59.836	31	30 54.921	1	0.997	31	30.915
2	1 59 40.341	2	1 59.672	32	31 54.758	2	1.994	32	31.913
3	2 59 30.511	3	2 59.509	33	32 54.594	3	2.992	33	32.910
4	3 59 20.682	4	3 59.345	34	33 54.430	4	3.989	34	33.907
5	4 59 10.852	5	4 59.181	35	34 54.266	5	4.986	35	34.904
6	5 59 1.023	6	5 59.017	36	35 54.102	6	5.984	36	35.902
7	6 58 51.193	7	6 58.853	37	36 53.938	7	6.981	37	36.899
8	7 58 41.364	8	7 58.689	38	37 53.775	8	7.978	38	37.896
9	8 58 31.534	9	8 58.526	39	38 53.611	9	8.975	39	38.894
10	9 58 21.704	10	9 58.362	40	39 53.447	10	9.973	40	39.891
11	10 58 11.875	11	10 58.198	41	40 53.283	11	10.970	41	40.888
12	11 58 2.045	12	11 58.034	42	41 53.119	12	11.967	42	41.885
13	12 57 52.216	13	12 57.870	43	42 52.955	13	12.965	43	42.883
14	13 57 42.386	14	13 57.706	44	43 52.792	14	13.962	44	43.880
15	14 57 32.557	15	14 57.543	45	44 52.629	15	14.959	45	44.877
16	15 57 22.727	16	15 57.379	46	45 52.464	16	15.956	46	45.874
17	16 57 12.898	17	16 57.215	47	46 52.300	17	16.954	47	46.872
18	17 57 3.068	18	17 57.051	48	47 52.136	18	17.951	48	47.869
19	18 56 53.238	19	18 56.887	49	48 51.973	19	18.948	49	48.866
20	19 56 43.409	20	19 56.723	50	49 51.809	20	19.945	50	49.863
21	20 56 33.579	21	20 56.560	51	50 51.645	21	20.943	51	50.861
22	21 56 23.750	22	21 56.396	52	51 51.481	22	21.940	52	51.858
23	22 56 13.920	23	22 56.232	53	52 51.317	23	22.937	53	52.855
24	23 56 4.091	24	23 56.068	54	53 51.153	24	23.934	54	53.853
		25	24 55.904	55	54 50.990	25	24.932	55	54.850
		26	25 55.741	56	55 50 826	26	25.929	56	55.847
		27	26 55.577	57	56 50.662	27	26.926	57	56.844
		28	27 55.413	58	57 50.498	28	27.924	58	57.842
		29	28 55.249	59	58 50.334	29	28.921	59	58.839
		30	29 55.085	60	59 50.170	30	29.918	60	59.836

TABLE XII.

Values of c, *the change of hour angle, corresponding to intervals of mean solar time.*

Int.	c	Int.	c	Int.	c	Int.	c
hr.	° '	hr.	° '	hr.	° '	hr.	° '
0.01	0 9	0.61	9 9	1.21	18 9	1.81	27 9
.02	0 18	.62	9 18	1.22	18 18	1.82	27 18
.03	0 27	.63	9 27	1.23	18 27	1.83	27 27
.04	0 36	.64	9 36	1.24	18 36	1.84	27 36
.05	0 45	.65	9 45	1.25	18 45	1.85	27 45
.06	0 54	.66	9 54	1.26	18 54	1.86	27 54
.07	1 3	.67	10 3	1.27	19 3	1.87	28 3
.08	1 12	.68	10 12	1.28	19 12	1.88	28 12
.09	1 21	.69	10 21	1.29	19 21	1.89	28 21
.10	1 30	.70	10 30	1.30	19 30	1.90	28 30
.11	1 39	.71	10 39	1.31	19 39	1.91	28 39
.12	1 48	.72	10 48	1.32	19 48	1.92	28 48
.13	1 57	.73	10 57	1.33	19 57	1.93	28 57
.14	2 6	.74	11 6	1.34	20 6	1.94	29 6
.15	2 15	.75	11 15	1.35	20 15	1.95	29 15
.16	2 24	.76	11 24	1.36	20 24	1.96	29 24
.17	2 33	.77	11 33	1.37	20 33	1.97	29 33
.18	2 42	.78	11 42	1.38	20 42	1.98	29 42
.19	2 51	.79	11 51	1.39	20 51	1.99	29 51
.20	3 0	.80	12 0	1.40	21 0	2.00	30 0
.21	3 9	.81	12 9	1.41	21 9	2.01	30 9
.22	3 18	.82	12 18	1.42	21 18	2.02	30 18
.23	3 27	.83	12 27	1.43	21 27	2.03	30 27
.24	3 36	.84	12 36	1.44	21 36	2.04	30 36
.25	3 45	.85	12 45	1.45	21 45	2.05	30 45
.26	3 54	.86	12 54	1.46	21 54	2.06	30 54
.27	4 3	.87	13 3	1.47	22 3	2.07	31 3
.28	4 12	.88	13 12	1.48	22 12	2.08	31 12
.29	4 21	.89	13 21	1.49	22 21	2.09	31 21
.30	4 30	.90	13 30	1.50	22 30	2.10	31 30
.31	4 39	.91	13 39	1.51	22 39	2.11	31 39
.32	4 48	.92	13 48	1.52	22 48	2.12	31 48
.33	4 57	.93	13 57	1.53	22 57	2.13	31 57
.34	5 6	.94	14 6	1.54	23 6	2.14	32 6
.35	5 15	.95	14 15	1.55	23 15	2.15	32 15
.36	5 24	.96	14 24	1.56	23 24	2.16	32 24
.37	5 33	.97	14 33	1.57	23 33	2.17	32 33
.38	5 42	.98	14 42	1.58	23 42	2.18	32 42
.39	5 51	.99	14 51	1.59	23 51	2.19	32 51
.40	6 0	1.00	15 0	1.60	24 0	2.20	33 0
.41	6 9	1.01	15 9	1.61	24 9	2.21	33 9
.42	6 18	1.02	15 18	1.62	24 18	2.22	33 18
.43	6 27	1.03	15 27	1.63	24 27	2.23	33 27
.44	6 36	1.04	15 36	1.64	24 36	2.24	33 36
.45	6 45	1.05	15 45	1.65	24 45	2.25	33 45
.46	6 54	1.06	15 54	1.66	24 54	2.26	33 54
.47	7 3	1.07	16 3	1.67	25 3	2.27	34 3
.48	7 12	1.08	16 12	1.68	25 12	2.28	34 12
.49	7 21	1.09	16 21	1.69	25 21	2.29	34 21
.50	7 30	1.10	16 30	1.70	25 30	2.30	34 30
.51	7 39	1.11	16 39	1.71	25 39	2.31	34 39
.52	7 48	1.12	16 48	1.72	25 48	2.32	34 48
.53	7 57	1.13	16 57	1.73	25 57	2.33	34 57
.54	8 6	1.14	17 6	1.74	26 6	2.34	35 6
.55	8 15	1.15	17 15	1.75	26 15	2.35	35 15
.56	8 24	1.16	17 24	1.76	26 24	2.36	35 24
.57	8 33	1.17	17 33	1.77	26 33	2.37	35 33
.58	8 42	1.18	17 42	1.78	26 42	2.38	35 42
.59	8 51	1.19	17 51	1.79	26 51	2.39	35 51
.60	9 0	1.20	18 0	1.80	27 0	2.40	36 0

TABLE XIII.

$$\text{Log. C} = \text{Log. } \frac{2 \sin \frac{1}{2}tc}{t}.$$

Int.	Log. C
hr.	
0.0	9.4180
0.1	9.4180
0.2	9.4179
0.3	9.4179
0.4	9.4178
0.5	9.4177
0.6	9.4175
0.7	9.4174
0.8	9.4172
0.9	9.4170
1.0	9.4167
1.1	9.4165
1.2	9.4162
1.3	9.4159
1.4	9.4155
1.5	9.4152
1.6	9.4148
1.7	9.4144
1.8	9.4140
1.9	9.4135
2.0	9.4130
2.1	9.4125
2.2	9.4120
2.3	9.4114
2.4	9.4108
2.5	9.4102

TABLE XIV.

Values of c′, *the change of Right Ascension of Zenith, corresponding to intervals of mean solar time.*

Int.	c′			Int.	c′			Int.	c′		
hr.	°	′	″	hr.	°	′	″	hr.	°	′	″
0.01	0	9	1.5	0.61	9	10	30.2	1.21	18	11	58.9
.02	0	18	3.0	.62	9	19	31.7	1.22	18	21	0.4
.03	0	27	4.4	.63	9	28	33.1	1.23	18	30	1.9
.04	0	36	5.9	.64	9	37	34.6	1.24	18	39	3.3
.05	0	45	7.4	.65	9	46	36.1	1.25	18	48	4.8
.06	0	54	8.9	.66	9	55	37.6	1.26	18	57	6.3
.07	1	3	10.3	.67	10	4	39.1	1.27	19	6	7.8
.08	1	12	11.8	.68	10	13	40.5	1.28	19	15	9.2
.09	1	21	13.3	.69	10	22	42.0	1.29	19	24	10.7
.10	1	30	14.8	.70	10	31	43.5	1.30	19	33	12.2
.11	1	39	16.3	.71	10	40	45.0	1.31	19	42	13.7
.12	1	48	17.7	.72	10	49	46.4	1.32	19	51	15.2
.13	1	57	19.2	.73	10	58	47.9	1.33	20	0	16.6
.14	2	6	20.7	.74	11	7	49.4	1.34	20	9	18.1
.15	2	15	22.2	.75	11	16	50.9	1.35	20	18	19.6
.16	2	24	23.7	.76	11	25	52.4	1.36	20	27	21.1
.17	2	33	25.1	.77	11	34	53.8	1.37	20	36	22.6
.18	2	42	26.6	.78	11	43	55.3	1.38	20	45	24.0
.19	2	51	28.1	.79	11	52	56.8	1.39	20	54	25.5
.20	3	0	29.6	.80	12	1	58.3	1.40	21	3	27.0
.21	3	9	31.0	.81	12	10	59.8	1.41	21	12	28.5
.22	3	18	32.5	.82	12	20	1.2	1.42	21	21	29.9
.23	3	27	34.0	.83	12	29	2.7	1.43	21	30	31.4
.24	3	36	35.5	.84	12	38	4.2	1.44	21	39	32.9
.25	3	45	37.0	.85	12	47	5.7	1.45	21	48	34.4
.26	3	54	38.4	.86	12	56	7.1	1.46	21	57	35.9
.27	4	3	40.0	.87	13	5	8.6	1.47	22	6	37.3
.28	4	12	41.4	.88	13	14	10.1	1.48	22	15	38.8
.29	4	21	42.9	.89	13	23	11.6	1.49	22	24	40.3
.30	4	30	44.4	.90	13	32	13.1	1.50	22	33	41.8
.31	4	39	45.8	.91	13	41	14.5	1.51	22	42	43.2
.32	4	48	47.3	.92	13	50	16.0	1.52	22	51	44.7
.33	4	57	48.8	.93	13	59	17.5	1.53	23	0	46.2
.34	5	6	50.3	.94	14	8	19.0	1.54	23	9	47.7
.35	5	15	51.7	.95	14	17	20.5	1.55	23	18	49.2
.36	5	24	53.2	.96	14	26	21.9	1.56	23	27	50.6
.37	5	33	54.7	.97	14	35	23.4	1.57	23	36	52.1
.38	5	42	56.2	.98	14	44	24.9	1.58	23	45	53.6
.39	5	51	57.7	.99	14	53	26.4	1.59	23	54	55.1
.40	6	0	59.1	1.00	15	2	27.8	1.60	24	3	56.6
.41	6	10	0.6	1.01	15	11	29.3	1.61	24	12	58.0
.42	6	19	2.1	1.02	15	20	30.8	1.62	24	21	59.5
.43	6	28	3.6	1.03	15	29	32.3	1.63	24	31	1.0
.44	6	37	5.1	1.04	15	38	33.8	1.64	24	40	2.5
.45	6	46	6.5	1.05	15	47	35.2	1.65	24	49	3.9
.46	6	55	8.0	1.06	15	56	36.7	1.66	24	58	5.4
.47	7	4	9.5	1.07	16	5	38.2	1.67	25	7	6.9
.48	7	13	11.0	1.08	16	14	39.7	1.68	25	16	8.4
.49	7	22	12.4	1.09	16	23	41.2	1.69	25	25	9.9
.50	7	31	13.9	1.10	16	32	42.6	1.70	25	34	11.3
.51	7	40	15.4	1.11	16	41	44.1	1.71	25	43	12.8
.52	7	49	16.9	1.12	16	50	45.6	1.72	25	52	14.3
.53	7	58	18.4	1.13	16	59	47.1	1.73	26	1	15.8
.54	8	7	19.8	1.14	17	8	48.5	1.74	26	10	17.3
.55	8	16	21.3	1.15	17	17	50.0	1.75	26	19	18.7
.56	8	25	22.8	1.16	17	26	51.5	1.76	26	28	20.2
.57	8	34	24.3	1.17	17	35	53.0	1.77	26	37	21.7
.58	8	43	25.8	1.18	17	44	54.5	1.78	26	46	23.2
.59	8	52	27.2	1.19	17	53	55.9	1.79	26	55	24.6
.60	9	1	28.7	1.20	18	2	57.4	1.80	27	4	26.1

TABLE XV.

$$\text{Log. } C' = \text{Log. } \frac{2 \sin \tfrac{1}{2} t c'}{t}.$$

Int.	Log. C′
hr.	
0.0	9.4192
0.1	9.4191
0.2	9.4191
0.3	9.4190
0.4	9.4190
0.5	9.4188
0.6	9.4187
0.7	9.4185
0.8	9.4184
0.9	9.4181
1.0	9.4179
1.1	9.4176
1.2	9.4174
1.3	9.4170
1.4	9.4167
1.5	9.4163
1.6	9.4160
1.7	9.4155
1.8	9.4151
1.9	9.4146
2.0	9.4142

TABLE XVI.

Reduction of Lat. and Moon's Hor. Par., and the logarithm of ρ, *for compression* $\frac{1}{300}$.

ARGUMENT. Geographic Latitude.

Arg.	Reduction of Lat.	Reduc. of Hor. Par. 53′	57′	61′	Log. ρ
°	′ ″	″	″	″	
0	0 0.0	0.0	0.0	0.0	0.00000
2	0 47.9	0.0	0.0	0.0	0.00000
4	1 35.5	0.1	0.1	0.1	9.99999
6	2 22.7	0.1	0.1	0.1	9.99998
8	3 9.2	0.2	0.2	0.2	9.99997
10	3 54.8	0.3	0.3	0.4	9.99996
12	4 39.3	0.5	0.5	0.5	9.99994
14	5 22.4	0.6	0.7	0.7	9.99992
16	6 3.9	0.8	0.9	0.9	9.99989
18	6 43.7	1.0	1.1	1.2	9.99986
20	7 21.6	1.2	1.3	1.4	9.99983
22	7 57.3	1.5	1.6	1.7	9.99980
24	8 30.7	1.8	1.9	2.0	9.99976
26	9 1.6	2.0	2.2	2.3	9.99972
28	9 29.9	2.3	2.5	2.7	9.99968
30	9 55.4	2.7	2.9	3.1	9.99964
32	10 18.1	3.0	3.2	3.4	9.99960
34	10 37.8	3.3	3.6	3.8	9.99955
36	10 54.3	3.7	3.9	4.2	9.99950
38	11 7.7	4.0	4.3	4.6	9.99945
40	11 17.9	4.4	4.7	5.0	9.99940
42	11 24.7	4.7	5.1	5.5	9.99935
44	11 28.2	5.1	5.5	5.9	9.99930
46	11 28.4	5.5	5.9	6.3	9.99925
48	11 25.2	5.9	6.3	6.7	9.99920
50	11 18.6	6.2	6.7	6.2	9.99915
52	11 8.8	6.6	7.1	7.6	0.99910
54	10 55.7	6.9	7.5	8.0	9.99905
56	10 39.4	7.3	7.8	8.4	9.99901
58	10 19.9	7.6	8.2	8.8	9.99896
60	9 57.4	7.9	8.5	9.1	9.99892
62	9 32.0	8.3	8.9	9.5	9.99887
64	9 3.8	8.6	9.2	9.9	9.99883
66	8 32.9	8.8	9.5	10.2	9.99879
68	7 59.6	9.1	9.8	10.5	9.99876
70	7 23.8	9.4	10.1	10.8	9.99872
72	6 45.9	9.6	10.3	11.0	9.99869
74	6 6.0	9.8	10.5	11.3	9.09866
76	5 24.3	10.0	10.7	11.5	9.99864
78	4 41.0	10.1	10.9	11.7	9.99861
80	3 56.3	10.3	11.1	11.8	9.99859
		10.4	11.2	12.0	
82	3 10.4	10.5	11.3	12.1	9.99858
84	2 23.7	10.5	11.3	12.1	9.99857
86	1 36.2	10.5	11.3	12.1	9.99856
88	0 48.2	10.6	11.4	12.2	9.99855
90	0 0.0	10.6	11.4	12.2	9.99855

TABLE XVII.

Logarithms x and y.

ARG. Geog. Latitude.

Arg.	Log. x	Log. y
0°	0.00000	9.99710
2	0.00000	9.99710
4	0.00001	9.99711
6	0.00002	9.99712
8	0.00003	9.99713
10	0.00004	9.99714
12	0.00006	9.99716
14	0.00008	9.99718
16	0.00011	9.99721
18	0.00014	9.99724
20	0.00017	9.99727
22	0.00020	9.99730
24	0.00024	9.99734
26	0.00028	9.99738
28	0.00032	9.99742
30	0.00036	9.99746
32	0.00041	9.99751
34	0.00045	9.99755
36	0.00050	9.99760
38	0.00055	9.99765
40	0.00060	9.99770
42	0.00065	9.99775
44	0.00070	9.99780
46	0.00075	9.99785
48	0.00080	9.99790
50	0.00085	9.99795
52	0.00090	9.99800
54	0.00095	9.99805
56	0.00100	9.99810
58	0.00105	9.99815
60	0.00109	9.99819
62	0.00113	9.99823
64	0.00117	9.99827
66	0.00121	9.99831
68	0.00125	9.99835
70	0.00128	9.99838
72	0.00131	9.99841
74	0.00134	9.99844
76	0.00137	9.99847
78	0.00139	9.99849
80	0.00141	9.98851
82	0.00143	9.99853
84	0.00143	9.99853
86	0.00144	9.99854
88	0.00145	9.99855
90	0.00145	9.99855

Mean New Moons and arguments, in January.

A. D.	Mean New Moon in January. D. H. M.	I.	II.	III.	IV.	N.
1821	2 17 59	0092	7859	80	78	823
1822	21 15 32	0602	7182	78	66	930
1823	11 0 20	0304	5787	61	55	953
1824 B.	29 21 53	0814	5110	59	43	060
1825	18 6 41	0516	3716	42	32	083
1826	7 15 30	0218	2321	25	21	105
1827	26 13 3	0728	1644	24	09	213
1828 B.	15 21 51	0430	0250	07	98	235
1829	4 6 40	0131	8855	90	87	257
1830	23 4 12	0642	8178	88	75	365
1831	12 13 1	0343	6784	71	64	387
1832 B.	1 21 50	0045	5389	54	53	409
1833	19 19 22	0555	4712	53	42	517
1834	9 4 11	0257	3318	36	31	539
1835	28 1 43	0768	2641	34	19	647
1836 B.	17 10 32	0469	1246	17	08	669
1837	5 19 20	0171	9852	00	97	692
1838	24 16 53	0681	9175	99	85	799
1839	14 1 42	0383	7780	82	74	822
1840 B.	3 10 30	0085	6386	65	63	844
1841	21 8 3	0595	5709	63	51	951
1842	10 16 51	0297	4314	46	40	974
1843	29 14 24	0807	3637	44	28	081
1844 B.	18 23 13	0509	2243	28	17	104
1845	7 8 1	0211	0848	11	06	126
1846	26 5 34	0721	0171	09	94	234
1847	15 14 22	0423	8777	92	84	256
1848 B.	4 23 11	0125	7382	75	73	278
1849	22 20 43	0635	6705	73	61	386
1850	12 5 32	0337	5311	56	50	408
1851	1 14 21	0038	3916	40	39	431
1852 B.	20 11 53	0549	3239	38	27	538
1853	8 20 42	0251	1845	21	16	560
1854	27 18 14	0761	1168	19	04	668
1855	17 3 3	0463	9773	02	93	690
1856 B.	6 11 51	0164	8379	85	82	713
1857	24 9 24	0675	7702	84	70	820
1858	13 18 13	0376	6307	67	59	843
1859	3 3 1	0078	4913	50	48	865
1860 B.	22 0 34	0588	4236	48	36	972
1861	10 9 22	0290	2840	31	25	995
1862	29 6 55	0800	2163	30	14	102
1863	18 15 44	0504	0769	13	03	125
1864 B.	8 0 32	0204	9374	96	92	147
1865	25 22 5	0714	8698	94	80	256
1866	15 6 53	0416	7303	77	69	277
1867	4 15 42	0118	5909	60	58	299
1868 B.	23 13 14	0628	5231	59	46	407
1869	11 22 3	0330	3837	42	35	429
1870	1 6 51	0032	2442	25	24	451

TABLE XIX.

Mean Lunations and Change in Arguments.

Num.	Lunations. D. H. M.	I.	II.	III.	IV.	N.
½	14 18 22	404	5359	58	50	43
1	29 12 44	808	717	15	99	85
2	59 1 28	1617	1434	31	98	170
3	88 14 12	2425	2151	46	97	256
4	118 2 56	3234	2869	61	96	341
5	147 15 40	4042	3586	76	95	426
6	177 4 24	4851	4303	92	95	511
7	206 17 8	5659	5020	7	94	596
8	236 5 52	6468	5737	22	93	682
9	265 18 36	7276	6454	37	92	767
10	295 7 20	8085	7171	53	91	852
11	324 20 5	8893	7889	68	90	937
12	354 8 49	9702	8606	83	89	22
13	383 21 33	510	9323	98	88	108

TABLE XX.

Number of Days from the commencement of the year to the first of each month.

Months.	Com. Days.	Bis. Days.
January ...	0	0
February..	31	31
March......	59	60
April........	90	91
May.........	120	121
June........	151	152
July	181	182
August.....	212	213
September	243	244
October....	273	274
November.	304	305
December.	334	335

Equations for New and Full Moon.

Arg.	I.	II.	Arg.	I.	II.	Arg.	III.	IV.	Arg.
	H. M.	H. M.		H. M.	H. M.		M.	M.	
000	4 20	10 10	5000	4 20	10 10	25	3	31	25
100	4 36	9 36	5100	4 5	10 50	26	3	31	24
200	4 52	9 2	5200	3 49	11 30	27	3	30	23
300	5 8	8 28	5300	3 34	12 9	28	3	30	22
400	5 24	7 55	5400	3 19	12 48	29	3	30	21
500	5 40	7 22	5500	3 4	13 26	30	3	30	20
600	5 55	6 49	5600	2 49	14 3	31	3	30	19
700	6 10	6 17	5700	2 35	14 39	32	4	30	18
800	6 24	5 46	5800	2 21	15 13	33	4	29	17
900	6 38	5 15	5900	2 8	15 46	34	4	29	16
1000	6 51	4 46	6000	1 55	16 18	35	4	29	15
1100	7 4	4 17	6100	1 42	16 48	36	5	28	14
1200	7 15	3 50	6200	1 31	17 16	37	5	28	13
1300	7 27	3 24	6300	1 19	17 42	38	5	27	12
1400	7 37	2 59	6400	1 9	18 6	39	5	27	11
1500	7 47	2 35	6500	0 59	18 28	40	6	26	10
1600	7 55	2 14	6600	0 50	18 48	41	6	26	9
1700	8 3	1 53	6700	0 42	19 6	42	7	25	8
1800	8 10	1 35	6800	0 34	19 21	43	7	25	7
1900	8 16	1 18	6900	0 28	19 33	44	7	24	6
2000	8 21	1 3	7000	0 22	19 44	45	8	23	5
2100	8 25	0 51	7100	0 17	19 52	46	8	23	4
2200	8 29	0 40	7200	0 14	19 57	47	9	22	3
2300	8 31	0 32	7300	0 11	20 0	48	9	21	2
2400	8 32	0 25	7400	0 9	20 1	49	10	21	1
2500	8 32	0 21	7500	0 8	19 59	50	10	20	0
2600	8 31	0 19	7600	0 8	19 55	51	10	19	99
2700	8 29	0 20	7700	0 9	19 48	52	11	19	98
2800	8 26	0 23	7800	0 11	19 40	53	11	18	97
2900	8 23	0 28	7900	0 15	19 29	54	12	17	96
3000	8 18	0 36	8000	0 19	19 17	55	12	17	95
3100	8 12	0 47	8100	0 24	19 2	56	13	16	94
3200	8 6	0 59	8200	0 30	18 45	57	13	15	93
3300	7 58	1 14	8300	0 37	18 27	58	13	15	92
3400	7 50	1 32	8400	0 45	18 6	59	14	14	91
3500	7 41	1 52	8500	0 53	17 45	60	14	14	90
3600	7 31	2 14	8600	1 3	17 21	61	15	13	89
3700	7 21	2 38	8700	1 13	16 56	62	15	13	88
3800	7 9	3 4	8800	1 25	16 30	63	15	12	87
3900	6 58	3 32	8900	1 36	16 3	64	15	12	86
4000	6 45	4 2	9000	1 49	15 34	65	16	11	85
4100	6 32	4 34	9100	2 2	15 5	66	16	11	84
4200	6 19	5 7	9200	2 16	14 34	67	16	11	83
4300	6 5	5 41	9300	2 30	14 3	68	16	10	82
4400	5 51	6 17	9400	2 45	13 31	69	17	10	81
4500	5 36	6 54	9500	3 0	12 58	70	17	10	80
4600	5 21	7 32	9600	3 16	12 25	71	17	10	79
4700	5 6	8 11	9700	3 32	11 52	72	17	10	78
4800	4 51	8 50	9800	3 48	11 18	73	17	10	77
4900	4 35	9 30	9900	4 4	10 44	74	17	9	76
5000	4 20	10 10	10000	4 20	10 10	75	17	9	75

Sun's Epochs.

Years.	M. Long.	Long. Perig.	I.	II.	III.	N.
1821	9ˢ 8°48′ 19″	9ˢ 7°50′ 43″	920	782	260	036
1822	9 8 34 0	9 7 51 45	280	697	886	090
1823	9 8 19 40	9 7 52 47	640	612	511	143
1824 B.	9 9 4 29	9 7 53 49	034	530	138	197
1825	9 8 50 9	9 7 54 51	394	445	763	251
1826	9 8 35 49	9 7 55 52	754	360	388	304
1827	9 8 21 30	9 7 56 54	114	275	013	358
1828 B.	9 9 6 18	9 7 57 56	508	192	640	412
1829	9 8 51 59	9 7 58 58	868	107	265	466
1830	9 8 37 39	9 8 0 0	228	022	890	519
1831	9 8 23 19	9 8 1 2	588	937	515	573
1832 B.	9 9 8 8	9 8 2 4	982	855	142	627
1833	9 8 53 49	9 8 3 6	342	770	767	681
1834	9 8 39 29	9 8 4 8	702	684	392	734
1835	9 8 25 9	9 8 5 10	062	600	017	788
1836 B.	9 9 10 6	9 8 7 2	456	517	644	842
1837	9 8 55 46	9 8 8 3	816	432	269	895
1838	9 8 41 27	9 8 9 4	176	347	894	949
1839	9 8 27 7	9 8 10 6	536	262	519	003
1840 B.	9 9 11 56	9 8 11 8	930	180	146	056
1841	9 8 57 37	9 8 12 9	290	095	771	110
1842	9 8 43 17	9 8 13 11	650	009	397	164
1843	9 8 28 9	9 8 14 12	010	925	021	218
1844 B.	9 9 13 47	9 8 15 14	404	843	648	272
1845	9 8 59 27	9 8 16 16	764	757	273	325
1846	9 8 45 8	9 8 17 17	124	673	897	379
1847	9 8 30 48	9 8 18 19	484	588	623	433
1848 B.	9 9 15 37	9 8 19 20	878	505	151	487
1849	9 9 1 17	9 8 20 22	238	420	775	540
1850	9 8 46 58	9 8 21 23	598	336	400	594
1851	9 8 32 39	9 8 22 24	958	250	025	648
1852 B.	9 9 17 27	9 8 23 26	353	168	653	701
1853	9 9 3 8	9 8 24 27	713	083	277	755
1854	9 8 48 48	9 8 25 29	073	998	902	809
1855	9 8 34 29	9 8 26 30	433	913	527	863
1856 B.	9 9 19 18	9 8 27 32	827	832	153	916
1857	9 9 4 58	9 8 28 34	187	746	779	970
1858	9 8 50 39	9 8 29 35	547	661	404	024
1859	9 8 36 19	9 8 30 37	907	576	029	078
1860 B.	9 9 21 8	9 8 31 38	301	494	656	131
1861	9 9 6 49	9 8 32 39	661	409	281	185
1862	9 8 52 29	9 8 33 41	021	324	906	239
1863	9 8 38 10	9 8 34 42	381	239	530	292
1864 B.	9 9 22 58	9 8 35 44	775	157	157	346
1865	9 9 8 39	9 8 36 45	135	072	783	400
1866	9 8 54 20	9 8 37 47	495	985	408	453
1867	9 8 40 0	9 8 38 49	855	902	033	507
1868 B.	9 9 24 49	9 8 39 50	249	820	659	561
1869	9 9 10 30	9 8 40 52	609	734	285	615
1870	9 8 56 10	9 8 41 53	969	649	910	668
1882	9 9 1 41	9 8 54 10	391	638	416	313

Sun's Motions for Months, Days, and Hours.

Months.	Longitude.	Per.	I.	II.	III.	N.
Jan.......... Com..	0s 0° 0′ 0″	0″	0	0	0	0
Jan.......... Bis....	11 29 0 52	0	966	997	998	0
Feb.......... Com..	1 0 33 18	5	47	78	53	4
Feb.......... Bis....	0 29 34 10	5	13	75	51	4
March..................	1 28 9 11	10	993	148	101	9
April....................	2 28 42 30	15	42	226	154	13
May......................	3 28 16 40	20	59	301	206	18
June.....................	4 28 49 58	26	110	379	259	22
July......................	5 28 24 8	31	129	454	310	27
August................	6 28 57 26	36	182	531	363	31
September............	7 29 30 44	41	233	609	416	36
October................	8 29 4 54	46	250	684	468	40
November.............	9 29 38 12	52	300	762	521	45
December.............	10 29 12 22	57	313	837	572	49

Days.	Longitude.	Per.	I.	II.	III.	N.	Hours.	Long.	I.
1	0° 0′ 0″	0′	0	0	0	0	1	2′ 28″	1
2	0 59 8	0	34	3	2	0	2	4 56	3
3	1 58 17	0	68	5	3	0	3	7 23	4
4	2 57 25	0	101	8	5	0	4	9 51	6
5	3 56 33	1	135	10	7	1	5	12 19	7
6	4 55 42	1	169	13	9	1	6	14 47	8
7	5 54 50	1	203	15	10	1	7	17 15	10
8	6 53 58	1	236	18	12	1	8	19 43	11
9	7 53 7	1	270	20	14	1	9	22 11	13
10	8 52 15	1	304	23	15	1	10	24 38	14
11	9 51 23	2	338	25	17	1	11	27 6	16
12	10 50 32	2	371	28	19	2	12	29 34	17
13	11 49 40	2	405	30	21	2	13	32 2	18
14	12 48 48	2	439	33	22	2	14	34 30	20
15	13 47 57	2	473	35	24	2	15	36 58	21
16	14 47 5	3	506	38	26	2	16	39 26	23
17	15 46 13	3	540	40	27	2	17	41 53	24
18	16 45 22	3	574	43	29	2	18	44 21	25
19	17 44 30	3	608	45	31	3	19	46 49	27
20	18 43 38	3	641	48	33	3	20	49 17	28
21	19 42 47	3	675	50	34	3	21	51 45	30
22	20 41 55	4	709	53	36	3	22	54 13	31
23	21 41 3	4	743	55	38	3	23	56 40	32
24	22 40 12	4	777	58	39	3	24	59 8	34
25	23 39 20	4	810	60	41	4			
26	24 38 28	4	844	63	43	4			
27	25 37 37	4	878	65	45	4			
28	26 36 45	5	912	68	46	4			
29	27 35 53	5	945	70	48	4			
30	28 35 2	5	979	73	50	4			
31	29 34 10	5	13	75	51	4			

Sun's Motions for Minutes and Seconds.

Min.	Long.	Min.	Long.	Min.	Long.	Sec.	Long.
1	0′ 2″	21	0′ 52″	41	1′ 41″	1	0″
2	0 5	22	0 54	42	1 43	to	
3	0 7	23	0 57	43	1 46	12	0
4	0 10	24	0 59	44	1 48		
5	0 12	25	1 2	45	1 51	13	1
						to	
6	0 15	26	1 4	46	1 53	36	1
7	0 17	27	1 7	47	1 56		
8	0 20	28	1 9	48	1 58	37	2
9	0 22	29	1 11	49	2 1	to	
10	0 25	30	1 14	50	2 3	60	2
11	0 27	31	1 16	51	2 6		
12	0 30	32	1 19	52	2 8		
13	0 32	33	1 21	53	2 11		
14	0 34	34	1 24	54	2 13		
15	0 37	35	1 26	55	2 16		
16	0 39	36	1 29	56	2 18		
17	0 42	37	1 31	57	2 20		
18	0 44	38	1 34	58	2 23		
19	0 47	39	1 36	59	2 25		
20	0 49	40	1 39	60	2 28		

TABLE XXV.

Sun's Horary Motion.

ARGUMENT. Sun's Mean Anomaly.

	0s	Is	IIs	IIIs	IVs	Vs	
0°	2′ 33″	2′ 32″	2′ 30″	2′ 28″	2′ 25″	2′ 24″	30°
10	2 33	2 32	2 29	2 27	2 25	2 23	20
20	2 33	2 31	2 29	2 26	2 24	2 23	10
30	2 32	2 30	2 28	2 25	2 24	2 23	0
	XIs	Xs	IXs	VIIIs	VII.	VIs	

TABLE XXVI.

Sun's Semi Diameter.

ARGUMENT. Sun's Mean Anomaly.

	0s	Is	IIs	IIIs	IVs	Vs	
0°	16′ 18″	16′ 15″	16′ 9″	16′ 1″	15′ 53″	15′ 48″	30°
10	16 18	16 14	16 7	15 58	15 51	15 46	20
20	16 17	16 12	16 4	15 56	15 49	15 46	10
30	16 15	16 9	16 1	15 53	15 48	15 45	0
	XIs	Xs	IXs	VIIIs	VIIs	VIs	

Equation of the Sun's Centre.

ARGUMENT. Sun's Mean Anomaly.

	0s.	Is.	IIs.	IIIs.	IVs.	Vs.
0°	1°59′ 30″	2°58′ 15″	3°40′ 27″	3°54′ 50″	3°38′ 21″	2°56′ 9″
1	2 1 33	3 0 0	3 41 25	3 54 47	3 37 18	2 54 25
2	2 3 37	3 1 44	3 42 21	3 54 41	3 36 14	2 52 40
3	2 5 40	3 3 27	3 43 15	3 54 33	3 35 8	2 50 54
4	2 7 43	3 5 9	3 44 8	3 54 23	3 34 1	2 49 8
5	2 9 46	3 6 49	3 44 58	3 54 11	3 32 51	2 47 20
6	2 11 49	3 8 28	3 45 47	3 53 57	3 31 41	2 45 32
7	2 13 51	3 10 6	3 46 33	3 53 41	3 30 28	2 43 43
8	2 15 54	3 11 43	3 47 17	3 53 23	3 29 14	2 41 53
9	2 17 56	3 13 18	3 48 0	3 53 3	3 27 58	2 40 3
10	2 19 57	3 14 51	3 48 40	3 52 40	3 26 41	2 38 11
11	2 21 58	3 16 24	3 49 18	3 52 16	3 25 22	2 36 19
12	2 23 59	3 17 54	3 49 55	3 51 50	3 24 2	2 34 27
13	2 25 59	3 19 24	3 50 29	3 51 21	3 22 40	2 32 34
14	2 27 59	3 20 51	3 51 1	3 50 51	3 21 17	2 30 40
15	2 29 58	3 22 18	3 51 31	3 50 18	3 19 52	2 28 46
16	2 31 57	3 23 42	3 51 59	3 49 44	3 18 26	2 26 52
17	2 33 55	3 25 5	3 52 25	3 49 7	3 16 58	2 24 56
18	2 35 52	3 26 26	3 52 49	3 48 29	3 15 30	2 23 0
19	2 37 49	3 27 46	3 53 10	3 47 49	3 14 0	2 21 4
20	2 39 45	3 29 4	3 53 30	3 47 7	3 12 28	2 19 8
21	2 41 40	3 30 20	3 53 47	3 46 22	3 10 55	2 17 11
22	2 43 34	3 31 35	3 54 3	3 45 36	3 9 22	2 15 14
23	2 45 28	3 32 48	3 54 16	3 44 48	3 7 46	2 13 16
24	2 47 20	3 33 59	3 54 27	3 43 58	3 6 10	2 11 19
25	2 49 12	3 35 8	3 54 36	3 43 7	3 4 33	2 9 21
26	2 51 2	3 36 16	3 54 43	3 42 13	3 2 54	2 7 23
27	2 52 52	3 37 21	3 54 48	3 41 18	3 1 14	2 5 25
28	2 54 41	3 38 25	3 54 51	3 40 21	2 59 33	2 3 27
29	2 56 28	3 39 27	3 54 52	3 39 22	2 57 52	2 1 28
30	2 58 15	3 40 27	3 54 50	3 38 21	2 56 9	1 59 30

Equations of the Sun's Centre.

ARGUMENT. Sun's Mean Anomaly.

	VI.	VII.	VIII.	IX.	X.	XI.
0°	1°59′ 30″	1° 2′ 51″	0°20′ 39″	0° 4′ 10″	0°18′ 33″	1° 0′ 45″
1	1 57 32	1 1 8	0 19 38	0 4 8	0 19 33	1 2 32
2	1 55 33	0 59 27	0 18 39	0 4 9	0 20 35	1 4 19
3	1 53 35	0 57 46	0 17 42	0 4 12	0 21 39	1 6 8
4	1 51 37	0 56 6	0 16 47	0 4 17	0 22 44	1 7 58
5	1 49 39	0 54 27	0 15 53	0 4 24	0 23 52	1 9 48
6	1 47 41	0 52 49	0 15 2	0 4 33	0 25 1	1 11 40
7	1 45 44	0 51 13	0 14 12	0 4 44	0 26 12	1 13 32
8	1 43 46	0 49 38	0 13 24	0 4 57	0 27 25	1 15 26
9	1 41 49	0 48 5	0 12 38	0 5 13	0 28 40	1 17 20
10	1 39 52	0 46 32	0 11 53	0 5 30	0 29 56	1 19 15
11	1 37 56	0 45 0	0 11 11	0 5 50	0 31 14	1 21 11
12	1 36 0	0 43 30	0 10 31	0 6 11	0 32 34	1 23 8
13	1 34 4	0 42 1	0 9 53	0 6 35	0 33 55	1 25 5
14	1 32 9	0 40 34	0 9 16	0 7 1	0 35 18	1 27 3
15	1 30 14	0 39 8	0 8 42	0 7 29	0 36 42	1 29 2
16	1 28 20	0 37 43	0 8 9	0 7 59	0 38 9	1 31 1
17	1 26 26	0 36 20	0 7 39	0 8 31	0 39 36	1 33 1
18	1 24 33	0 34 58	0 7 10	0 9 5	0 41 6	1 35 1
19	1 22 41	0 33 38	0 6 44	0 9 42	0 42 36	1 37 1
20	1 20 49	0 32 19	0 6 20	0 10 20	0 44 9	1 39 3
21	1 18 57	0 31 2	0 5 57	0 11 0	0 45 42	1 41 4
22	1 17 7	0 29 46	0 5 37	0 11 43	0 47 17	1 43 6
23	1 15 17	0 28 32	0 5 19	0 12 27	0 48 54	1 45 9
24	1 13 28	0 27 19	0 5 3	0 13 13	0 50 32	1 47 11
25	1 11 40	0 26 9	0 4 49	0 14 2	0 52 11	1 49 14
26	1 9 52	0 24 59	0 4 37	0 14 52	0 53 51	1 51 17
27	1 8 6	0 23 52	0 4 27	0 15 45	0 55 33	1 53 20
28	1 6 20	0 22 46	0 4 19	0 16 39	0 57 16	1 55 23
29	1 4 35	0 21 41	0 4 13	0 17 35	0 59 0	1 57 27
30	1 2 51	0 20 39	0 4 10	0 18 33	1 0 45	1 59 30

TABLE XXVIII.

Small Equations of Sun's Longitude.

Arg.	I.	II.	III.	Arg.	I.	II.	III.
0	10″	10″	10″	500	10″	10″	10′
10	10	11	9	510	10	10	9
20	11	11	9	520	9	10	8
30	11	12	8	530	9	10	7
40	11	13	8	540	9	10	7
50	12	14	7	550	8	10	6
60	12	14	7	560	8	9	5
70	12	15	7	570	8	9	4
80	13	15	7	580	7	9	3
90	13	16	7	590	7	9	3
100	13	16	7	600	7	9	2
110	14	17	7	610	6	8	1
120	14	17	7	620	6	8	1
130	14	18	8	630	6	8	1
140	15	18	8	640	5	7	0
150	15	18	9	650	5	7	0
160	15	18	9	660	5	6	0
170	15	18	10	670	5	6	1
180	15	18	10	680	5	6	1
190	16	18	11	690	4	5	2
200	16	18	11	700	4	5	2
210	16	18	12	710	4	4	3
220	16	18	12	720	4	4	3
230	16	18	13	730	4	4	4
240	16	17	14	740	4	3	5
250	16	17	14	750	4	3	6
260	16	17	15	760	4	3	6
270	16	16	16	770	4	2	7
280	16	16	17	780	4	2	8
290	16	16	17	790	4	2	8
300	16	15	18	800	4	2	9
310	16	15	18	810	4	2	9
320	15	14	19	820	5	2	10
330	15	14	19	830	5	2	10
340	15	14	20	840	5	2	11
350	15	13	20	850	5	2	11
360	15	13	20	860	5	2	12
370	14	12	19	870	6	2	12
380	14	12	19	880	6	3	13
390	14	12	19	890	6	3	13
400	13	11	18	900	7	4	13
410	13	11	17	910	7	4	13
420	13	11	17	920	7	5	13
430	12	11	16	930	8	5	13
440	12	11	15	940	8	6	13
450	12	10	14	950	8	6	13
460	11	10	13	960	9	7	12
470	11	10	13	970	9	8	12
480	11	10	12	980	9	9	11
490	10	10	11	990	10	9	11
500	10	10	10	1000	10	10	10

TABLE XXIX.

Mean Obliquity of the Ecliptic.

Years.	M. Obliqu.
1821	23°27′ 46″
1822	23 27 46
1823	23 27 45
1824	23 27 44
1825	23 27 44
1826	23 27 43
1827	23 27 43
1828	23 27 42
1829	23 27 42
1830	23 27 41
1831	23 27 41
1832	23 27 40
1833	23 27 40
1834	23 27 39
1835	23 27 39
1836	23 27 38
1837	23 27 38
1838	23 27 37
1839	23 27 37
1840	23 27 37
1841	23 27 36
1842	23 27 36
1843	23 27 35
1844	23 27 35
1845	23 27 34
1846	23 27 34
1847	23 27 33
1848	23 27 33
1849	23 27 32
1850	23 27 32
1851	23 27 31
1852	23 27 31
1853	23 27 31
1854	23 27 30
1855	23 27 30
1856	23 27 29
1857	23 27 29
1858	23 27 28
1859	23 27 28
1860	23 27 27
1861	23 27 27
1862	23 27 26
1863	23 27 26
1864	23 27 26
1865	23 27 25
1866	23 27 25
1867	23 27 24
1868	23 27 24
1869	23 27 23
1870	23 27 23
1882	23 27 17

TABLE XXX.

Nutations.

ARGUMENT. Supplement of the Node, or N.

N.	Long.	R. Asc.	Obliq.	N.	Long.	R. Asc.	Obliq.
0	+ 0″	+ 0″	+ 10″	500	— 0″	— 0″	— 10″
10	1	1	10	510	1	1	10
20	2	2	10	520	2	2	9
30	3	3	9	530	3	3	9
40	4	4	9	540	4	4	9
50	+ 6	+ 5	+ 9	550	— 6	— 5	— 9
60	7	6	9	560	7	6	9
70	8	7	9	570	8	7	8
80	9	8	8	580	9	8	8
90	10	9	8	590	10	9	8
100	+ 11	+ 10	+ 8	600	— 11	— 10	— 8
110	11	10	7	610	11	10	7
120	12	11	7	620	12	11	7
130	13	12	7	630	13	12	7
140	14	13	6	640	14	13	6
150	+ 15	+ 13	+ 6	650	— 15	— 13	— 6
160	15	14	5	660	15	14	5
170	16	14	5	670	16	14	5
180	16	15	4	680	16	15	4
190	17	15	3	690	17	15	3
200	+ 17	+ 16	+ 3	700	— 17	— 16	— 3
210	17	16	2	710	17	16	2
220	18	16	2	720	18	16	2
230	18	16	1	730	18	16	1
240	18	16	1	740	18	16	1
250	+ 18	+ 16	+ 0	750	— 18	— 16	— 0
260	18	16	— 1	760	18	16	+ 1
270	18	16	1	770	18	16	1
280	18	16	2	780	18	16	2
290	17	16	2	790	17	16	2
300	+ 17	+ 16	— 3	800	— 17	— 16	+ 3
310	17	15	3	810	17	15	3
320	16	15	4	820	16	15	4
330	16	14	5	830	16	14	5
340	15	14	5	840	15	14	5
350	+ 15	+ 13	— 6	850	— 15	— 13	+ 6
360	14	13	6	860	14	13	6
370	13	12	7	870	13	12	7
380	12	11	7	880	12	11	7
390	11	10	7	890	11	10	7
400	+ 11	+ 10	— 8	900	— 11	— 10	+ 8
410	10	9	8	910	10	9	8
420	9	8	8	920	9	8	8
430	8	7	9	930	8	7	9
440	7	6	9	940	7	6	9
450	+ 6	+ 5	— 9	950	— 6	— 5	+ 9
460	4	4	9	960	4	4	9
470	3	3	9	970	3	3	9
480	2	2	10	980	2	2	10
490	1	1	10	990	1	1	10
500	+ 0	+ 0	— 10	1000	— 0	— 0	+ 10

Earth's Radius Vector.

ARGUMENT. Sun's Mean Anomaly.

	0s	Is	IIs	IIIs	IVs	Vs	
0°	0.98313	0.98545	0.99173	1.00018	1.00850	1.01450	30°
1	0.98313	0.98560	0.99199	1.00047	1.00874	1.01464	29
2	0.98314	0.98576	0.99225	1.00077	1.00899	1.01477	28
3	0.98315	0.98592	0.99251	1.00106	1.00923	1.01490	27
4	0.98317	0.98608	0.99278	1.00135	1.00947	1.01503	26
5	0.98319	0.98625	0.99304	1.00164	1.00971	1.01515	25
6	0.98322	0.98643	0.99331	1.00193	1.00994	1.01527	24
7	0.98326	0.98661	0.99359	1.00222	1.01017	1.01538	23
8	0.98330	0.98679	0.99386	1.00251	1.01040	1.01549	22
9	0.98334	0.98698	0.99414	1.00280	1.01062	1.01560	21
10	0.98339	0.98717	0.99441	1.00308	1.01084	1.01569	20
11	0.98344	0.98736	0.99469	1.00337	1.01106	1.01579	19
12	0.98350	0.98756	0.99497	1.00366	1.01128	1.01588	18
13	0.98357	0.98777	0.99526	1.00394	1.01149	1.01596	17
14	0.98364	0.98797	0.99554	1.00422	1.01170	1.01604	16
15	0.98372	0.98818	0.99582	1.00450	1.01190	1.01612	15
16	0.98380	0.98840	0.99611	1.00478	1.01210	1.01619	14
17	0.98388	0.98861	0.99640	1.00506	1.01230	1.01626	13
18	0.98397	0.98883	0.99668	1.00534	1.01249	1.01632	12
19	0.98407	0.98906	0.99697	1.00561	1.01268	1.01638	11
20	0.98417	0.98929	0.99726	1.00588	1.01286	1.01643	10
21	0.98428	0.98952	0.99755	1.00615	1.01304	1.01647	9
22	0.98439	0.98975	0.99784	1.00642	1.01322	1.01652	8
23	0.98450	0.98999	0.99813	1.00669	1.01340	1.01655	7
24	0.98462	0.99023	0.99843	1.00695	1.01357	1.01659	6
25	0.98475	0.99047	0.99872	1.00722	1.01373	1.01661	5
26	0.98486	0.99072	0.99901	1.00748	1.01389	1.01663	4
27	0.98501	0.99096	0.99930	1.00774	1.01405	1.01665	3
28	0.98515	0.99122	0.99960	1.00799	1.01420	1.01666	2
29	0.98530	0.99147	0.99989	1.00825	1.01435	1.01667	1
30	0.98545	0.99173	1.00018	1.00850	1.01450	1.01667	0
	XIs	Xs	IXs	VIIIs	VIIs	VIs	

Perturbations of Earth's Radius Vector.

Arg.	I.	II.	III.	Arg.	I.	II.	III.
0	8	4	3	500	2	0	4
50	8	4	3	550	2	1	4
100	7	4	2	600	3	1	3
150	7	4	1	650	3	2	2
200	6	4	0	700	4	3	1
250	5	4	0	750	5	4	0
300	4	3	1	800	6	4	0
350	3	2	2	850	7	4	1
400	3	1	3	900	7	4	2
450	2	1	4	950	8	4	3
500	2	0	4	1000	8	4	3

TABLE XXXI.

Moon's Epochs.

Years.	1	2	3	4	5	6	7	8	9
1821	0027	8365	5389	1368	6970	7714	6319	7024	7800
1822	0020	5573	5054	6112	9441	3512	7380	9481	6664
1823	0012	2782	4720	0856	1913	9309	8440	1938	5528
1824 B.	0033	0640	5426	5887	4720	5478	9559	4787	4417
1825	0026	7849	5092	0631	7192	1276	0619	7243	3280
1826	0018	5057	4758	5375	9663	7073	1680	9701	2144
1827	0011	2265	4424	0119	2135	2871	2740	2158	1008
1828 B.	0032	0124	5129	5150	4942	9040	3859	5007	9896
1829	0024	7332	4795	9894	7414	4837	4919	7463	8760
1830	0017	4541	4461	4638	9885	0635	5979	9921	7623
1831	0010	1749	4127	9381	2357	6432	7040	2378	6487
1832 B.	0030	9607	4833	4412	5164	2601	8158	5226	5376
1833	0023	6816	4499	9156	7636	8399	9219	7683	4239
1834	0016	4024	4164	3900	0107	4196	0279	0140	3103
1835	0009	1232	3830	8644	2579	9993	1340	2598	1967
1836 B.	0029	9091	4536	3675	5386	6163	2418	5446	0856
1837	0022	6299	4202	8419	7858	1960	3518	7903	9719
1838	0015	3508	3868	3163	0329	7757	4579	0360	8583
1839	0008	0716	3534	7907	2801	3555	5639	2818	7447
1840 B.	0028	8575	4239	2938	5608	9724	6758	5666	6335
1841	0021	5783	3906	7682	8080	5522	7818	8123	5199
1842	0014	2991	3571	2425	0551	1319	8879	0580	4062
1843	0007	0200	3237	7169	3023	7116	9939	3038	2926
1844 B.	0027	8058	3943	2200	5830	3286	1058	5886	1815
1845	0020	5266	3609	6944	8302	9083	2118	8343	0678
1846	0013	2475	3275	1688	0773	4880	3179	0800	9542
1847	0006	9683	2941	6432	3245	0678	4239	3257	8406
1848 B.	0026	7542	3646	1463	6052	6847	5358	6106	7295
1849	0019	4750	3312	6207	8524	2644	6418	8563	6158
1850	0012	1958	2978	0951	0995	8442	7479	1020	5022
1851	0005	9167	2644	5695	3467	4239	8539	3477	3885
1852 B.	0025	7025	3350	0726	6274	0408	9658	6326	2774
1853	0018	4233	3016	5469	8746	6206	0718	8782	1637
1854	0011	1442	2681	0213	1217	2003	1778	1240	0501
1855	0004	8650	2347	4957	3689	7801	2839	3697	9365
1856 B.	0024	6509	3053	9988	6496	3970	3957	6546	8254
1857	0017	3717	2719	4732	8968	9767	5018	9002	7117
1858	0010	0925	2385	9476	1439	5565	6078	1460	5981
1859	0003	8134	2051	4220	3911	1362	7139	3917	4845
1860 B.	0023	5992	2756	9251	6718	7531	8257	6765	3734
1861	0016	3200	2423	3995	9190	3329	9317	9222	2597
1862	0009	0409	2088	8739	1661	9126	0378	1679	1461
1863	0002	7617	1754	3483	4133	4923	1438	4137	0324
1864 B.	0022	5476	2460	8514	6941	1093	2557	6984	9212
1865	0015	2684	2126	3257	9412	6890	3617	9442	8076
1866	0008	9893	1792	8001	1883	2687	4678	1899	6940
1867	0001	7101	1457	2745	4355	8485	5738	4357	5804
1868 B.	0021	4959	2163	7776	7163	4654	6857	7204	4692
1869	0014	2168	1829	2520	9634	0452	7917	9662	3556
1870	0007	9376	1495	7264	2105	6249	8978	2119	2420

TABLE XXXII.

Moon's Epochs.

Years.	10	11	12	13	14	15	16	17	18	19	20
1821	620	917	842	142	979	067	923	331	134	036	036
1822	226	278	562	615	172	208	282	684	609	090	202
1823	833	639	281	088	366	348	641	036	084	143	369
1824 B.	509	030	070	595	659	519	037	431	585	197	537
1825	116	391	790	068	853	659	397	783	060	251	703
1826	722	752	510	541	047	800	756	136	536	304	869
1827	329	113	229	014	241	940	115	488	011	358	036
1828 B.	005	505	019	521	533	111	511	883	512	412	204
1829	612	866	738	994	727	251	871	235	987	466	370
1830	219	226	458	468	921	392	230	588	462	519	536
1831	825	587	177	940	115	532	589	940	937	573	703
1832 B.	502	979	967	447	408	704	985	335	438	627	871
1833	108	340	687	920	602	844	345	688	913	681	037
1834	715	701	406	393	796	984	704	040	388	734	203
1835	321	061	125	866	989	124	063	393	863	788	370
1836 B.	998	453	915	373	282	296	459	787	364	842	538
1837	605	814	635	846	476	436	819	140	840	895	704
1838	211	175	354	319	670	576	178	492	315	949	870
1839	818	536	074	792	864	716	537	845	790	003	037
1840 B.	494	927	863	299	157	888	933	239	291	056	205
1841	101	288	583	772	351	028	293	592	766	110	371
1842	707	649	302	245	544	168	652	944	241	164	537
1843	314	010	022	718	738	308	012	297	716	218	704
1844 B.	990	402	811	225	031	480	407	691	217	272	872
1845	597	763	531	698	225	620	767	044	692	325	038
1846	203	123	250	171	419	760	126	396	167	379	204
1847	810	484	970	644	613	901	486	749	643	433	371
1848 B.	486	876	759	151	905	072	881	143	144	487	539
1849	093	237	479	624	099	212	241	496	619	540	705
1850	700	597	199	097	293	352	600	848	094	594	871
1851	306	958	918	570	487	493	960	201	569	648	038
1852 B.	983	350	707	077	780	664	355	595	070	701	206
1853	589	711	427	550	974	804	715	948	545	755	372
1854	196	072	147	023	168	944	074	300	020	809	539
1855	802	432	866	496	361	085	434	653	495	863	705
1856 B.	479	824	656	003	654	256	829	047	996	916	873
1857	086	185	375	476	848	396	189	400	471	970	039
1858	692	546	095	949	042	537	548	752	947	024	206
1859	299	907	814	422	236	677	908	105	422	078	372
1860 B.	975	298	604	929	529	848	303	499	923	131	540
1861	581	659	323	402	723	988	662	852	398	185	706
1862	187	020	042	875	916	129	021	204	873	239	873
1863	794	381	761	348	110	269	381	557	348	292	039
1864 B.	470	773	551	855	403	440	777	951	849	346	207
1865	077	134	271	328	597	580	136	304	324	400	373
1866	684	494	990	801	791	721	495	657	799	453	540
1867	290	855	710	274	985	861	855	009	274	507	707
1868 B.	967	247	500	781	277	032	251	404	775	561	874
1869	573	608	219	254	471	172	610	756	251	615	040
1870	180	968	939	727	665	313	969	109	726	668	207

TABLE XXXI.

Moon's Epochs.

Years.	Evection.	Anomaly.	Variation.	Longitude.
1821	1ˢ 19° 43′ 47″	8ˢ 9° 54′ 17″	10ˢ 22° 4′ 2″	8ˢ 2° 7′ 41″
1822	7 10 15 16	11 8 37 37	3 1 41 27	0 11 30 46
1823	1 0 46 45	2 7 20 57	7 11 18 51	4 20 53 51
1824 B.	7 2 37 15	5 19 8 11	0 3 7 43	9 13 27 31
1825	0 23 8 44	8 17 51 31	4 12 45 7	1 22 50 37
1826	6 13 40 14	11 16 34 50	8 22 22 32	6 2 13 42
1827	0 4 11 44	2 15 18 10	1 1 59 56	10 11 36 47
1828 B.	6 6 2 13	5 27 5 24	5 23 48 49	3 4 10 27
1829	11 26 33 43	8 25 48 44	10 3 26 13	7 13 33 32
1830	5 17 5 12	11 24 32 4	2 13 3 38	11 22 56 37
1831	11 7 36 41	2 23 16 24	6 22 41 3	4 2 19 42
1832 B.	5 9 27 11	6 5 2 38	11 14 29 54	8 24 53 23
1833	10 29 58 40	9 3 45 58	3 24 7 20	1 4 16 28
1834	4 20 30 11	0 2 29 18	8 3 44 44	5 13 39 33
1835	10 11 1 40	3 1 12 38	0 13 22 9	9 23 2 38
1836 B.	4 12 52 9	6 12 59 52	5 5 11 0	2 15 36 19
1837	10 3 23 39	9 11 43 12	9 14 48 26	6 24 59 24
1838	3 23 55 9	0 10 26 32	1 24 25 50	11 4 22 29
1839	9 14 26 38	3 9 9 53	6 4 3 15	3 13 45 35
1840 B.	3 16 17 8	6 20 57 7	10 25 52 8	8 6 19 15
1841	9 6 48 37	9 19 40 27	3 5 29 32	0 15 42 21
1842	2 27 20 7	0 18 23 47	7 15 6 57	4 25 5 26
1843	8 17 51 37	3 17 7 7	11 24 44 22	9 4 28 31
1844 B.	2 19 42 7	6 28 54 22	4 16 33 14	1 27 2 12
1845	8 10 13 36	9 27 37 42	8 26 10 39	6 6 25 17
1846	2 0 45 6	0 26 21 2	1 5 48 4	10 15 48 23
1847	7 21 16 35	3 25 4 23	5 15 25 29	2 25 11 28
1848 B.	1 23 7 5	7 6 51 37	10 7 14 21	7 17 45 8
1849	7 13 38 35	10 5 34 57	2 16 51 46	11 27 8 14
1850	1 4 10 4	1 4 18 18	6 26 29 11	4 6 31 20
1851	6 24 41 35	4 3 1 38	11 6 6 36	8 15 54 25
1852 B.	0 26 32 5	7 14 48 53	3 27 55 29	1 8 28 6
1853	6 17 3 34	10 13 32 13	8 7 32 53	5 17 51 11
1854	0 7 35 4	1 12 15 34	0 17 10 19	9 27 14 17
1855	5 28 6 33	4 10 58 54	4 26 47 43	2 6 37 22
1856 B.	11 29 57 3	7 22 46 9	9 18 36 36	6 29 11 3
1857	5 20 28 33	10 21 29 29	1 28 14 1	11 8 34 9
1858	11 11 0 2	1 20 12 50	6 7 51 26	3 17 57 14
1859	5 1 31 33	4 18 56 10	10 17 28 52	7 27 20 20
1860 B.	11 3 22 3	8 0 43 25	3 9 17 44	0 19 54 0
1861	4 23 53 33	10 29 26 45	7 18 55 9	4 29 17 6
1862	10 14 25 3	1 28 10 6	11 28 32 34	9 8 40 12
1863	4 4 56 33	4 26 53 27	4 8 10 0	1 18 3 18
1864 B.	10 6 47 2	8 8 40 41	8 29 58 51	6 10 36 58
1865	3 27 18 32	11 7 24 2	1 9 36 17	10 20 0 4
1866	9 17 50 2	2 6 7 23	5 19 13 42	2 29 23 10
1867	3 8 21 32	5 4 50 43	9 28 51 8	7 8 46 15
1868 B.	9 10 12 2	8 16 37 58	2 20 40 0	0 1 19 56
1869	3 0 43 33	11 15 21 19	7 0 17 25	4 10 43 2
1870	8 21 15 2	2 14 4 40	11 9 54 50	8 20 6 8

TABLE XXXII.

Moon's Epochs.

Years.	Supp. of Node.	II.	V.	VI.	VII.	VIII.	IX.	X.
1821	0s 13° 3′ 29″	0s 27° 41′	706	711	074	079	637	596
1822	1 2 23 11	4 18 13	120	124	382	386	717	536
1823	1 21 42 56	8 8 45	533	536	689	692	796	475
1824 B.	2 11 5 47	0 10 26	981	988	026	032	912	420
1825	3 0 25 29	4 0 58	395	401	333	338	992	359
1826	3 19 45 11	7 21 30	809	813	641	645	072	299
1827	4 9 4 53	11 12 2	223	225	949	951	151	238
1828 B.	4 28 27 46	3 13 43	670	677	285	291	267	182
1829	5 17 47 29	7 4 15	084	090	592	597	347	122
1830	6 7 7 11	10 24 47	498	502	900	904	427	062
1831	6 26 26 53	2 15 19	912	914	208	210	506	001
1832 B.	7 15 49 46	6 17 0	360	366	545	550	622	945
1833	8 5 9 28	10 7 32	774	779	852	856	702	885
1834	8 24 29 11	1 28 4	187	191	159	163	782	825
1835	9 13 48 53	5 18 36	601	603	467	469	861	764
1836 B.	10 3 11 46	9 20 18	048	055	804	809	977	708
1837	10 22 31 28	1 10 50	463	468	111	116	057	648
1838	11 11 51 10	5 1 22	876	880	419	423	137	588
1839	0 1 10 52	8 21 54	290	292	726	729	217	527
1840 B.	0 20 33 45	0 23 35	738	744	063	069	332	471
1841	1 9 53 28	4 14 7	152	157	370	375	412	411
1842	1 29 13 10	8 4 39	566	569	678	682	492	350
1843	2 18 32 52	11 25 11	980	980	986	988	572	290
1844 B.	3 7 55 45	3 26 52	427	433	322	328	687	234
1845	3 27 15 27	7 17 24	840	846	629	634	767	174
1846	4 16 35 9	11 7 56	254	258	937	941	847	113
1847	5 5 54 52	2 28 38	668	670	245	247	927	053
1848 B.	5 25 17 45	7 0 9	116	122	582	587	042	997
1849	6 14 37 27	10 20 41	531	535	889	893	122	937
1850	7 3 57 9	2 11 13	944	947	196	200	202	876
1851	7 23 16 51	6 1 45	358	359	504	506	282	816
1852 B.	8 12 39 44	10 3 27	806	811	841	846	398	760
1853	9 1 59 26	1 23 59	220	223	148	152	477	700
1854	9 21 19 9	5 14 31	634	636	456	459	557	639
1855	10 10 38 51	9 5 3	047	048	763	765	637	579
1856 B.	11 0 1 44	1 6 44	495	500	100	105	753	523
1857	11 19 21 26	4 27 16	909	912	407	411	832	463
1858	0 8 41 8	8 17 48	323	325	715	718	912	402
1859	0 28 0 51	0 8 20	736	737	023	024	992	342
1860 B.	1 17 23 43	4 10 1	184	189	359	364	108	286
1861	2 6 43 27	8 0 33	598	601	666	670	187	226
1862	2 26 3 9	11 21 5	012	014	974	977	267	165
1863	3 15 23 11	3 11 37	426	426	282	283	347	105
1864 B.	4 4 45 44	7 13 18	873	878	618	623	463	049
1865	4 24 5 46	11 3 50	287	291	926	929	542	989
1866	5 13 25 28	2 24 22	701	703	233	236	622	928
1867	6 2 45 10	6 14 54	115	115	541	542	702	868
1868 B.	6 22 7 43	10 16 36	563	567	877	882	818	812
1869	7 11 27 46	2 7 8	977	980	185	188	897	752
1870	8 0 47 28	5 27 40	390	392	493	495	977	691

Moon's Motions for Months.

Months.	1	2	3	4	5	6	7	8	9
January { Com.	0000	0000	0000	0000	0000	0000	0000	0000	0000
January { Bis.	9973	9350	8960	9713	9664	9628	9942	9610	9976
February { Com.	849	146	2246	8896	402	1533	1789	2099	753
February { Bis.	821	9497	1205	8609	66	1161	1731	1709	729
March	1615	8343	1371	6931	9797	1951	3404	3027	1433
April	2464	8490	3616	5827	199	3484	5193	5126	2186
May	3285	7986	4822	4436	265	4646	6924	6835	2914
June	4134	8133	7067	3332	666	6179	8713	8934	3667
July	4955	7629	8273	1942	732	7341	444	643	4396
August	5804	7776	518	838	1134	8874	2233	2742	5148
September	6653	7922	2764	9734	1536	408	4021	4842	5901
October	7474	7419	3969	8343	1602	1569	5752	6550	6630
November	8323	7565	6215	7239	2004	3102	7541	8649	7382
December	9144	7062	7420	5848	2070	4264	9272	358	8111

TABLE XXXIII.

Moon's Motions for Months.

Months.	Evection.	Anomaly.	Variation.	M. Longitude.
January { Com.	0ˢ 0° 0′ 0″	0ˢ 0° 0′ 0″	0ˢ 0° 0′ 0″	0ˢ 0° 0′ 0″
January { Bis.	11 18 41 1	11 16 56 6	11 17 48 33	11 16 49 25
February { Com.	11 20 48 42	1 15 0 53	0 17 54 48	1 18 28 6
February { Bis.	11 9 29 43	1 1 56 59	0 5 43 21	1 5 17 31
March	10 7 40 26	1 20 50 4	11 29 15 15	1 27 24 27
April	9 28 29 8	3 5 50 57	0 17 10 3	3 15 52 32
May	9 7 58 51	4 7 47 56	0 22 53 24	4 21 10 3
June	8 28 47 33	5 22 48 49	1 10 48 11	6 9 38 9
July	8 8 17 16	6 24 45 48	1 16 31 32	7 14 55 40
August	7 29 5 59	8 9 46 42	2 4 26 20	9 3 23 46
September	7 19 54 41	9 24 47 35	2 22 21 7	10 21 51 52
October	6 29 24 24	10 26 44 34	2 28 4 28	11 27 9 22
November	6 20 13 6	0 11 45 27	3 15 59 16	1 15 37 28
December	5 29 42 49	1 13 42 26	3 21 42 37	2 20 54 59

TABLE XXXIII.

Moon's Motions for Months.

Months.		10	11	12	13	14	15	16	17	18	19	20
January	Com.	000	000	000	000	000	000	000	000	000	000	000
	Bis.	930	969	930	966	901	969	963	958	974	000	000
February	Com.	175	965	184	59	74	946	135	304	805	5	14
	Bis.	105	934	114	25	975	916	98	262	779	5	14
March		139	836	157	16	851	801	159	482	532	9	27
April		314	801	342	76	925	747	294	786	336	13	41
May		419	735	456	101	899	663	392	47	115	18	55
June		593	700	640	160	973	609	527	351	920	22	69
July		698	634	754	185	948	525	625	613	699	27	83
August		873	599	938	245	22	471	759	917	503	31	97
September		48	563	123	304	96	417	894	221	308	36	111
October		152	497	237	329	71	333	992	483	87	40	125
November		327	462	421	388	145	279	127	787	892	45	139
December		432	396	535	414	120	194	225	49	670	49	153

TABLE XXXIII.

Moon's Motions for Months.

Months.		Supp. of Node.	II.	V.	VI.	VII.	VIII.	IX.	X.
January	Com.	0ˢ 0° 0′ 0″	0ˢ 0° 0′	000	000	000	000	000	000
	Bis.	11 29 56 49	11 18 51	966	961	972	966	964	995
February	Com.	0 1 38 30	11 15 43	54	224	875	45	111	165
	Bis.	0 1 35 19	11 4 34	20	185	847	11	75	159
March		0 3 7 27	9 27 59	7	330	666	989	114	313
April		0 4 45 57	9 13 42	61	554	542	34	225	478
May		0 6 21 16	8 18 15	81	738	389	46	300	638
June		0 7 59 46	8 3 58	136	962	264	91	411	802
July		0 9 35 5	7 8 32	156	147	112	103	486	962
August		0 11 13 35	6 24 15	210	371	987	147	597	126
September		0 12 52 5	6 9 58	265	595	862	193	708	291
October		0 14 27 24	5 14 32	285	780	710	204	783	451
November		0 16 5 53	5 0 15	339	4	585	250	894	615
December		0 17 41 13	4 4 49	359	188	432	261	969	775

Moon's Motions for Days.

Days.	1	2	3	4	5	6	7	8	9
1	0000	0000	0000	0000	0000	0000	0000	0000	0000
2	27	650	1040	287	336	372	58	390	24
3	55	1300	2080	574	671	744	115	781	49
4	82	1950	3121	861	1007	1116	173	1171	73
5	109	2600	4161	1148	1342	1488	231	1561	97
6	137	3249	5201	1435	1678	1860	289	1952	121
7	164	3899	6241	1722	2013	2232	346	2342	146
8	192	4549	7281	2009	2349	2604	404	2732	170
9	219	5199	8321	2296	2684	2976	462	3122	194
10	246	5849	9362	2583	3020	3348	519	3513	219
11	274	6499	402	2870	3355	3720	577	3903	243
12	301	7149	1442	3157	3691	4093	635	4293	267
13	328	7799	2482	3444	4026	4465	692	4684	291
14	356	8449	3522	3731	4362	4837	750	5074	316
15	383	9098	4563	4018	4698	5209	808	5464	340
16	411	9748	5603	4305	5033	5581	866	5854	364
17	438	398	6643	4592	5369	5953	923	6245	389
18	465	1048	7683	4878	5704	6325	981	6635	413
19	493	1698	8723	5165	6040	6697	1039	7025	437
20	520	2348	9763	5452	6375	7069	1096	7416	461
21	548	2998	804	5739	6711	7441	1154	7806	486
22	575	3648	1844	6026	7046	7813	1212	8196	510
23	602	4298	2884	6313	7382	8185	1269	8586	534
24	630	4947	3924	6600	7717	8557	1327	8977	559
25	657	5597	4964	6887	8053	8929	1385	9367	583
26	684	6247	6005	7174	8389	9301	1443	9757	607
27	712	6897	7045	7461	8724	9673	1500	148	631
28	739	7547	8085	7748	9060	45	1558	538	656
29	767	8197	9125	8035	9395	417	1616	928	680
30	794	8847	165	8322	9731	789	1673	1319	704
31	821	9497	1205	8609	66	1161	1731	1709	729

Moon's Motions for Days.

Days.	10	11	12	13	14	15	16	17	18	19	20
1	000	000	000	000	000	000	000	000	000	000	000
2	70	31	70	34	99	31	37	42	26	0	0
3	140	62	141	68	198	61	73	84	52	0	1
4	210	93	211	103	297	92	110	126	78	0	1
5	281	125	282	137	397	122	146	168	104	1	2
6	351	156	352	171	496	153	183	210	130	1	2
7	421	187	423	205	595	183	220	252	156	1	3
8	491	218	493	239	694	214	256	294	182	1	3
9	561	249	564	273	793	244	293	336	208	1	4
10	631	280	634	308	892	275	329	379	234	1	4
11	702	311	705	342	992	305	366	421	260	1	5
12	772	342	775	376	91	336	403	463	286	2	5
13	842	374	845	410	190	366	439	505	312	2	5
14	912	405	916	444	289	397	476	547	337	2	6
15	982	436	986	478	388	427	512	589	363	2	6
16	52	467	57	513	487	458	549	631	389	2	7
17	122	498	127	547	587	488	586	673	415	2	7
18	193	529	198	581	686	519	622	715	441	2	8
19	263	560	268	615	785	549	659	757	467	3	8
20	333	591	339	649	884	580	695	799	493	3	9
21	403	623	409	683	983	611	732	841	519	3	9
22	473	654	480	718	82	641	769	883	545	3	10
23	543	685	550	752	182	672	805	925	571	3	10
24	614	716	621	786	281	702	842	967	597	3	11
25	684	747	691	820	380	733	878	9	623	4	11
26	754	778	762	854	479	763	915	52	649	4	11
27	824	809	832	888	578	794	952	94	675	4	12
28	894	840	903	923	677	824	988	136	701	4	12
29	964	872	973	957	777	855	25	178	727	4	13
30	34	903	43	991	876	885	61	220	753	4	13
31	105	934	114	25	975	916	98	262	779	4	14

Moon's Motions for Days.

Days.	Evection.	Anomaly.	Variation.	M. Long.
1	0s 0° 0′ 0″	0s 0° 0′ 0″	0s 0° 0′ 0″	0s 0° 0′ 0″
2	0 11 18 59	0 13 3 54	0 12 11 27	0 13 10 35
3	0 22 37 59	0 26 7 48	0 24 22 53	0 26 21 10
4	1 3 56 58	1 9 11 42	1 6 34 20	1 9 31 45
5	1 15 15 58	1 22 15 36	1 18 45 47	1 22 42 20
6	1 26 34 57	2 5 19 30	2 0 57 13	2 5 52 55
7	2 7 53 57	2 18 23 24	2 13 8 40	2 19 3 30
8	2 19 12 56	3 1 27 18	2 25 20 7	3 2 14 5
9	3 0 31 55	3 14 31 12	3 7 31 34	3 15 24 40
10	3 11 50 55	3 27 35 6	3 19 43 0	3 28 35 15
11	3 23 9 54	4 10 39 0	4 1 54 27	4 11 45 50
12	4 4 28 54	4 23 42 54	4 14 5 54	4 24 56 25
13	4 15 47 53	5 6 46 48	4 26 17 20	5 8 7 0
14	4 27 6 53	5 19 50 42	5 8 28 47	5 21 17 35
15	5 8 25 52	6 2 54 36	5 20 40 14	6 4 28 10
16	5 19 44 51	6 15 58 29	6 2 51 40	6 17 38 45
17	6 1 3 51	6 29 2 23	6 15 3 7	7 0 49 20
18	6 12 22 50	7 12 6 17	6 27 14 34	7 13 59 55
19	6 23 41 50	7 25 10 11	7 9 26 1	7 27 10 30
20	7 5 0 49	8 8 14 5	7 21 37 27	8 10 21 5
21	7 16 19 49	8 21 17 59	8 3 48 54	8 23 31 40
22	7 27 38 48	9 4 21 53	8 16 0 21	9 6 42 16
23	8 8 57 47	9 17 25 47	8 28 11 47	9 19 52 51
24	8 20 16 47	10 0 29 41	9 10 23 14	10 3 3 26
25	9 1 35 46	10 13 33 35	9 22 34 41	10 16 14 1
26	9 12 54 46	10 26 37 29	10 4 46 7	10 29 24 36
27	9 24 13 45	11 9 41 23	10 16 57 34	11 12 35 11
28	10 5 32 45	11 22 45 17	10 29 9 1	11 25 45 46
29	10 16 51 44	0 5 49 11	11 11 20 28	0 8 56 21
30	10 28 10 43	0 18 53 5	11 23 31 54	0 22 6 56
31	11 9 29 43	1 1 56 59	0 5 43 21	1 5 17 31

Moon's Motions for Days.

Days.	Sup. of Node.	II.	V.	VI.	VII.	VIII.	IX.	X.
1	0° 0′ 0″	0ˢ 0° 0′	000	000	000	000	000	000
2	0 3 11	11 9	34	39	28	34	36	5
3	0 6 21	22 18	68	79	56	67	72	11
4	0 9 32	1 3 27	102	118	85	101	108	16
5	0 12 52	1 14 37	136	158	113	135	143	21
6	0 15 53	1 25 46	170	197	141	169	179	27
7	0 19 4	2 6 55	204	237	169	202	215	32
8	0 22 14	2 18 4	238	276	198	236	251	37
9	0 25 25	2 29 13	272	316	226	270	287	43
10	0 28 36	3 10 22	306	355	254	303	323	48
11	0 31 46	3 21 31	340	395	282	337	358	53
12	0 34 57	4 2 40	374	434	311	371	394	58
13	0 38 7	4 13 50	408	474	339	405	430	64
14	0 41 18	4 24 59	442	513	367	438	466	69
15	0 44 29	5 6 8	476	553	395	472	502	74
16	0 47 39	5 17 17	510	592	424	506	538	80
17	0 50 50	5 28 26	544	632	452	539	573	85
18	0 54 1	6 9 35	578	671	480	573	609	90
19	0 57 11	6 20 44	612	711	508	607	645	96
20	1 0 22	7 1 53	646	750	537	641	681	101
21	1 3 33	7 13 3	680	790	565	674	717	106
22	1 6 43	7 24 12	714	829	593	708	753	112
23	1 9 54	8 5 21	748	869	621	742	788	117
24	1 13 5	8 16 30	782	908	650	775	824	122
25	1 16 15	8 27 39	816	948	678	809	860	128
26	1 19 26	9 8 48	850	987	706	843	896	133
27	1 22 36	9 19 57	884	027	734	877	932	138
28	1 25 47	10 1 6	918	066	762	910	968	143
29	1 28 58	10 12 16	952	106	791	944	003	149
30	1 32 8	10 23 25	986	145	819	978	039	154
31	1 35 19	11 4 34	020	185	947	011	075	159

Moon's Motions for Hours.

Hours.	1	2	3	4	5	6	7	8	9
1	1	27	43	12	14	16	2	16	1
2	2	54	87	24	28	31	5	33	2
3	3	81	130	36	42	47	7	49	3
4	5	108	173	48	56	62	10	65	4
5	6	135	217	60	70	78	12	81	5
6	7	162	260	72	84	93	14	98	6
7	8	190	303	84	98	109	17	114	7
8	9	217	347	96	112	124	19	130	8
9	10	244	390	108	126	140	22	146	9
10	11	271	433	120	140	155	24	163	10
11	12	298	477	131	154	171	26	179	11
12	14	325	520	143	168	186	29	195	12
13	15	352	563	155	182	202	31	211	13
14	16	379	607	167	196	217	34	228	14
15	17	406	650	179	210	233	36	244	15
16	18	433	693	191	224	248	38	260	16
17	19	460	737	203	238	264	41	276	17
18	20	487	780	215	252	279	43	293	18
19	22	515	823	227	266	295	46	309	19
20	23	542	867	239	280	310	48	325	20
21	24	569	910	251	294	326	50	341	21
22	25	596	953	263	308	341	53	358	22
23	26	623	997	275	322	357	55	374	23
24	27	650	1040	287	336	372	58	390	24

Hours.	Evection.	Anomaly.	Variation.	Longitude.
1	0°28′ 17″	0°32′ 40″	0°30′ 29″	0°32′ 56″
2	0 56 35	1 5 19	1 0 57	1 5 53
3	1 24 52	1 37 59	1 31 26	1 38 49
4	1 53 10	2 10 39	2 1 54	2 11 46
5	2 21 27	2 43 19	2 32 23	2 44 42
6	2 49 45	3 15 58	3 2 52	3 17 39
7	3 18 2	3 48 38	3 33 20	3 50 35
8	3 46 20	4 21 18	4 3 49	4 23 32
9	4 14 37	4 53 58	4 34 17	4 56 28
10	4 42 55	5 26 37	5 4 46	5 29 25
11	5 11 12	5 59 17	5 35 15	6 2 21
12	5 39 30	6 31 57	6 5 43	6 35 17
13	6 7 47	7 4 37	6 36 12	7 8 14
14	6 36 5	7 37 16	7 6 40	7 41 10
15	7 4 22	8 9 56	7 37 9	8 14 7
16	7 32 40	8 42 36	8 7 38	8 47 3
17	8 0 57	9 15 16	8 38 6	9 20 0
18	8 29 15	9 47 55	9 8 35	9 52 56
19	8 57 32	10 20 35	9 39 3	10 25 53
20	9 25 50	10 53 15	10 9 32	10 58 49
21	9 54 7	11 25 55	10 40 1	11 31 46
22	10 22 24	11 58 34	11 10 29	12 4 42
23	10 50 42	12 31 14	11 40 58	12 37 39
24	11 18 59	13 3 54	12 11 27	13 10 35

Moon's Motions for Hours.

Hours.	10	11	12	13	14	15	16	17	18
1	3	1	3	1	4	1	2	2	1
2	6	3	6	3	8	3	3	4	2
3	9	4	9	4	12	4	5	5	3
4	12	5	12	6	16	5	6	7	4
5	15	6	15	7	21	6	8	9	5
6	18	8	18	9	25	8	9	11	6
7	20	9	20	10	29	9	11	12	8
8	23	10	23	11	33	10	12	14	9
9	26	12	26	13	37	11	14	16	10
10	29	13	29	14	41	13	15	18	11
11	32	14	32	16	45	14	17	19	12
12	35	16	35	17	49	15	18	21	13
13	38	17	38	18	54	16	20	23	14
14	41	18	41	20	58	18	21	25	15
15	44	19	44	21	62	19	23	26	16
16	47	21	47	23	66	20	25	28	17
17	50	22	50	24	70	21	26	30	18
18	53	23	53	25	74	23	28	32	19
19	56	25	56	27	78	24	29	33	21
20	58	26	58	28	83	25	31	35	22
21	61	27	61	30	87	26	32	37	23
22	64	28	64	31	91	28	34	39	24
23	67	30	67	33	95	29	35	40	25
24	70	31	70	34	99	31	37	42	26

Hours.	Sup. of Node.	II.	V.	VI.	VII.	VIII.	IX.	X.
1	0′ 8″	0°28″	1	2	1	1	1	0
2	0 16	0 56	3	3	2	3	3	0
3	0 24	1 24	4	5	4	4	4	1
4	0 32	1 52	6	7	5	6	6	1
5	0 40	2 19	7	8	6	7	7	1
6	0 48	2 47	9	10	7	9	9	1
7	0 56	3 15	10	12	8	10	10	2
8	1 4	3 43	11	13	9	11	12	2
9	1 11	4 11	13	15	11	13	13	2
10	1 19	4 39	14	16	12	14	15	2
11	1 27	5 7	16	18	13	15	16	2
12	1 35	5 35	17	20	14	17	18	3
13	1 43	6 2	18	21	15	18	19	3
14	1 51	6 30	20	23	16	19	21	3
15	1 59	6 58	21	25	18	21	22	3
16	2 7	7 26	23	26	19	22	24	4
17	2 15	7 54	24	28	20	24	25	4
18	2 23	8 22	26	29	21	25	27	4
19	2 31	8 50	27	31	22	27	28	4
20	2 39	9 18	28	32	24	28	30	4
21	2 47	9 45	30	34	25	29	31	5
22	2 55	10 13	31	36	26	31	33	5
23	3 3	10 41	33	38	27	32	34	5
24	3 11	11 9	34	39	28	34	36	5

TABLE XXXVI.

Moon's Motions for Minutes and Seconds.

Min.	1	2	3	4	5	6	7	8	9	10	11	12	13	14
1	0	0	1	0	0	0	0	0	0	0	0	0	0	0
2	0	1	1	0	0	1	0	1	0	0	0	0	0	0
3	0	1	2	1	1	1	0	1	0	0	0	0	0	0
4	0	2	3	1	1	1	0	1	0	0	0	0	0	0
5	0	2	4	1	1	1	0	1	0	0	0	0	0	0
6	0	3	4	1	1	2	0	2	0	0	0	0	0	0
7	0	3	5	1	2	2	0	2	0	0	0	0	0	0
8	0	4	6	2	2	2	0	2	0	0	0	0	0	1
9	0	4	6	2	2	2	0	2	0	0	0	0	0	1
10	0	5	7	2	2	3	0	3	0	0	0	0	0	1
11	0	5	8	2	3	3	0	3	0	1	0	1	0	1
12	0	5	9	2	3	3	0	3	0	1	0	1	0	1
13	0	6	9	3	3	3	1	4	0	1	0	1	0	1
14	0	6	10	3	3	4	1	4	0	1	0	1	0	1
15	0	7	11	3	3	4	1	4	0	1	0	1	0	1
16	0	7	12	3	4	4	1	4	0	1	0	1	0	1
17	0	8	12	3	4	4	1	5	0	1	0	1	0	1
18	0	8	13	4	4	5	1	5	0	1	0	1	0	1
19	0	9	14	4	4	5	1	5	0	1	0	1	0	1
20	0	9	14	4	5	5	1	5	0	1	0	1	0	1
21	0	10	15	4	5	5	1	6	0	1	0	1	0	1
22	0	10	16	4	5	6	1	6	0	1	0	1	1	2
23	0	10	17	5	5	6	1	6	0	1	0	1	1	2
24	0	11	17	5	6	6	1	7	0	1	1	1	1	2
25	0	11	18	5	6	6	1	7	0	1	1	1	1	2
26	0	12	19	5	6	7	1	7	0	1	1	1	1	2
27	1	12	19	5	6	7	1	7	0	1	1	1	1	2
28	1	13	20	6	7	7	1	8	0	1	1	1	1	2
29	1	13	21	6	7	7	1	8	0	1	1	1	1	2
30	1	14	22	6	7	8	1	8	0	1	1	1	1	2
31	1	14	22	6	7	8	1	8	0	1	1	1	1	2
32	1	14	23	6	7	8	1	9	1	2	1	2	1	2
33	1	15	24	7	8	9	1	9	1	2	1	2	1	2
34	1	15	25	7	8	9	1	9	1	2	1	2	1	2
35	1	16	25	7	8	9	1	10	1	2	1	2	1	2
36	1	16	26	7	8	9	1	10	1	2	1	2	1	3
37	1	17	27	7	9	10	1	10	1	2	1	2	1	3
38	1	17	27	8	9	10	2	10	1	2	1	2	1	3
39	1	18	28	8	9	10	2	11	1	2	1	2	1	3
40	1	18	29	8	9	10	2	11	1	2	1	2	1	3
41	1	19	30	8	10	11	2	11	1	2	1	2	1	3
42	1	19	30	8	10	11	2	11	1	2	1	2	1	3
43	1	19	31	9	10	11	2	12	1	2	1	2	1	3
44	1	20	32	9	10	11	2	12	1	2	1	2	1	3
45	1	20	32	9	10	12	2	12	1	2	1	2	1	3
46	1	21	33	9	11	12	2	12	1	2	1	2	1	3
47	1	21	34	9	11	12	2	13	1	2	1	2	1	3
48	1	22	35	10	11	12	2	13	1	2	1	2	1	3
49	1	22	35	10	11	13	2	13	1	2	1	2	1	3
50	1	23	36	10	11	13	2	13	1	2	1	2	1	3
51	1	23	37	10	12	13	2	14	1	2	1	2	1	4
52	1	24	38	10	12	13	2	14	1	3	1	3	1	4
53	1	24	38	11	12	14	2	14	1	3	1	3	1	4
54	1	24	39	11	12	14	2	14	1	3	1	3	1	4
55	1	25	40	11	13	14	2	15	1	3	1	3	1	4
56	1	25	40	11	13	14	2	15	1	3	1	3	1	4
57	1	26	41	11	13	15	2	15	1	3	1	3	1	4
58	1	26	42	12	13	15	2	16	1	3	1	3	1	4
59	1	27	43	12	14	15	2	16	1	3	1	3	1	4
60	1	27	43	12	14	15	2	16	1	3	1	3	1	4

Moon's Motions for Minutes and Seconds.

Min.	Evec.	Anom.	Variat.	Long.	Sup. nod.	II.	Sec.	Ev.	An.	Var.	Lon.
1	0′ 28″	0′ 33″	0′ 30″	0′ 33″	0″	0′	1	0	0″	1″	0″
2	0 57	1 5	1 1	1 6	0	1	2	1	1	1	1
3	1 25	1 38	1 31	1 39	0	1	3	1	2	2	2
4	1 53	2 11	2 2	2 12	0	2	4	2	2	2	2
5	2 21	2 43	2 32	2 45	1	2	5	2	3	3	3
6	2 50	3 16	3 3	3 18	1	3	6	3	3	3	3
7	3 18	3 49	3 33	3 51	1	3	7	3	4	4	4
8	3 46	4 21	4 4	4 23	1	4	8	4	4	4	4
9	4 15	4 54	4 34	4 56	1	4	9	4	5	5	5
10	4 43	5 27	5 5	5 29	1	5	10	5	5	5	5
11	5 11	5 59	5 35	6 2	1	5	11	5	6	6	6
12	5 40	6 32	6 6	6 35	2	6	12	6	6	6	7
13	6 8	7 5	6 36	7 8	2	6	13	6	7	7	7
14	6 36	7 37	7 7	7 41	2	7	14	7	8	7	8
15	7 4	8 10	7 37	8 14	2	7	15	7	8	8	8
16	7 33	8 43	8 8	8 47	2	7	16	8	9	8	9
17	8 1	9 15	8 38	9 20	2	8	17	8	9	9	9
18	8 29	9 48	9 9	9 53	2	8	18	9	10	9	10
19	8 58	10 21	9 39	10 26	2	9	19	9	10	10	10
20	9 26	10 53	10 10	10 59	3	9	20	9	11	10	11
21	9 54	11 26	10 40	11 32	3	10	21	10	11	11	11
22	10 22	11 59	11 11	12 5	3	10	22	10	12	11	12
23	10 51	12 31	11 41	12 38	3	11	23	11	12	12	13
24	11 19	13 4	12 12	13 11	3	11	24	11	13	12	13
25	11 47	13 37	12 42	13 43	3	12	25	12	14	13	14
26	12 16	14 9	13 13	14 16	3	12	26	12	14	13	14
27	12 44	14 42	13 43	14 49	4	13	27	13	15	14	15
28	13 12	15 15	14 13	15 22	4	13	28	13	15	14	15
29	13 40	15 47	14 44	15 55	4	13	29	14	16	15	16
30	14 9	16 20	15 14	16 28	4	14	30	14	16	15	16
31	14 37	16 52	15 45	17 1	4	14	31	15	17	16	17
32	15 5	17 25	16 15	17 34	4	15	32	15	17	16	18
33	15 34	17 58	16 46	18 7	4	15	33	16	18	17	18
34	16 2	18 30	17 16	18 40	4	16	34	16	18	17	19
35	16 30	19 3	17 47	19 13	5	16	35	17	19	18	19
36	16 58	19 36	18 17	19 46	5	17	36	17	20	18	20
37	17 27	20 8	18 48	20 19	5	17	37	18	20	19	20
38	17 55	20 41	19 18	20 52	5	18	38	18	21	19	21
39	18 23	21 14	19 49	21 25	5	18	39	18	21	20	21
40	18 52	21 46	20 19	21 58	5	19	40	19	22	20	22
41	19 20	22 19	20 50	22 31	5	19	41	19	22	21	22
42	19 48	22 52	21 20	23 3	6	20	42	20	23	21	23
43	20 16	23 24	21 51	23 36	6	20	43	20	23	22	24
44	20 45	23 57	22 21	24 9	6	21	44	21	24	22	24
45	21 13	24 30	22 52	24 42	6	21	45	21	24	23	25
46	21 41	25 2	23 22	25 15	6	21	46	22	25	23	25
47	22 10	25 35	23 53	25 48	6	22	47	22	26	24	26
48	22 38	26 8	24 23	26 21	6	22	48	23	26	24	26
49	23 6	26 40	24 54	26 54	6	23	49	23	27	25	27
50	23 34	27 13	25 24	27 27	7	23	50	24	27	25	27
51	24 3	27 46	25 55	28 0	7	24	51	24	28	26	28
52	24 31	28 18	26 25	28 33	7	24	52	25	28	26	28
53	24 59	28 51	26 56	29 6	7	25	53	25	39	27	29
54	25 28	29 24	27 26	29 39	7	25	54	26	29	27	30
55	25 56	29 56	27 56	30 12	7	26	55	26	30	28	30
56	26 24	30 29	28 27	30 45	7	26	56	26	30	28	31
57	26 52	31 2	28 57	31 18	7	27	57	27	31	29	31
58	27 21	31 34	29 28	31 51	8	27	58	27	32	29	32
59	27 49	32 7	29 58	32 23	8	28	59	28	32	30	32
60	28 17	32 40	30 29	32 56	8	28	60	28	33	30	33

First Equation of Moon's Longitude. ARGUMENT 1.

Arg.	1	Diff.	Arg.	1	Diff.
0	12′ 40″	42″	5000	12′ 40″	40″
100	11 58	42	5100	13 20	41
200	11 16	42	5200	14 1	40
300	10 34	41	5300	14 41	39
400	9 53	41	5400	15 20	40
500	9 12	40	5500	16 0	38
600	8 32	38	5600	16 38	37
700	7 54	38	5700	17 15	37
800	7 16	36	5800	17 52	35
900	6 40	34	5900	18 27	34
1000	6 6	33	6000	19 1	32
1100	5 33	31	6100	19 33	31
1200	5 2	30	6200	20 4	29
1300	4 32	27	6300	20 33	28
1400	4 5	25	6400	21 1	26
1500	3 40	23	6500	21 27	23
1600	3 17	21	6600	21 50	22
1700	2 56	18	6700	22 12	19
1800	2 38	16	6800	22 31	17
1900	2 22	13	6900	22 48	15
2000	2 9	11	7000	23 3	12
2100	1 58	8	7100	23 15	10
2200	1 50	6	7200	23 25	7
2300	1 44	3	7300	23 32	5
2400	1 41	0	7400	23 37	2
2500	1 41	2	7500	23 39	0
2600	1 43	5	7600	23 39	3
2700	1 48	7	7700	23 36	6
2800	1 55	10	7800	23 30	8
2900	2 5	12	7900	23 22	11
3000	2 17	15	8000	23 11	13
3100	2 32	17	8100	22 58	16
3200	2 49	19	8200	22 42	18
3300	3 8	22	8300	22 24	21
3400	3 30	23	8400	22 3	23
3500	3 53	26	8500	21 40	25
3600	4 19	27	8600	21 15	27
3700	4 46	30	8700	20 48	30
3800	5 16	31	8800	20 18	31
3900	5 47	32	8900	19 47	33
4000	6 19	34	9000	19 14	34
4100	6 53	35	9100	18 40	36
4200	7 28	37	9200	18 4	38
4300	8 5	37	9300	17 26	38
4400	8 42	38	9400	16 48	40
4500	9 20	39	9500	16 8	41
4600	9 59	40	9600	15 27	41
4700	10 39	40	9700	14 46	42
4800	11 19	40	9800	14 4	42
4900	11 59	41	9900	13 22	42
5000	12 40		10000	12 40	

Equations 2 *to* 7 *of Moon's Longitude.* ARGUMENTS 2 to 7.

Arg.	2	3	4	5	6	7	Arg.
2500	4′ 57″	0′ 2″	6′ 30″	3′ 39″	0′ 6″	0′ 1″	2500
2600	4 57	0 2	6 30	3 39	0 6	0 1	2400
2700	4 56	0 3	6 29	3 38	0 7	0 1	2300
2800	4 55	0 3	6 27	3 37	0 8	0 2	2200
2900	4 53	0 4	6 24	3 36	0 9	0 3	2100
3000	4 50	0 5	6 21	3 34	0 10	0 4	2000
3100	4 47	0 6	6 17	3 32	0 12	0 5	1900
3200	4 43	0 8	6 12	3 29	0 14	0 6	1800
3300	4 39	0 9	6 7	3 26	0 17	0 8	1700
3400	4 34	0 11	6 1	3 22	0 19	0 10	1600
3500	4 29	0 13	5 54	3 18	0 22	0 12	1500
3600	4 23	0 15	5 47	3 14	0 25	0 14	1400
3700	4 17	0 18	5 39	3 10	0 29	0 17	1300
3800	4 11	0 20	5 30	3 5	0 33	0 19	1200
3900	4 4	0 23	5 21	3 0	0 37	0 22	1100
4000	3 57	0 26	5 12	2 54	0 41	0 25	1000
4100	3 49	0 29	5 2	2 49	0 45	0 28	900
4200	3 41	0 32	4 52	2 43	0 50	0 31	800
4300	3 33	0 35	4 41	2 37	0 54	0 35	700
4400	3 24	0 39	4 30	2 30	0 59	0 38	600
4500	3 15	0 42	4 19	2 24	1 4	0 42	500
4600	3 7	0 46	4 7	2 17	1 9	0 45	400
4700	2 58	0 49	3 56	2 10	1 14	0 49	300
4800	2 48	0 53	3 44	2 4	1 19	0 53	200
4900	2 39	0 56	3 32	1 57	1 25	0 56	100
5000	2 30	1 0	3 20	1 50	1 30	1 0	0000
5100	2 21	1 4	3 8	1 43	1 35	1 4	9900
5200	2 11	1 7	2 56	1 36	1 40	1 7	9800
5300	2 2	1 11	2 44	1 29	1 46	1 11	9700
5400	1 53	1 14	2 33	1 23	1 51	1 15	9600
5500	1 44	1 18	2 21	1 16	1 56	1 18	9500
5600	1 36	1 21	2 10	1 10	2 1	1 22	9400
5700	1 27	1 25	1 59	1 3	2 6	1 25	9300
5800	1 19	1 28	1 48	0 57	2 10	1 28	9200
5900	1 11	1 31	1 38	0 51	2 15	1 32	9100
6000	1 3	1 34	1 28	0 46	2 19	1 35	9000
6100	0 56	1 37	1 19	0 40	2 23	1 38	8900
6200	0 49	1 39	1 10	0 35	2 27	1 40	8800
6300	0 43	1 42	1 1	0 30	2 31	1 43	8700
6400	0 36	1 44	0 53	0 26	2 35	1 46	8600
6500	0 31	1 47	0 46	0 21	2 38	1 48	8500
6600	0 26	1 49	0 39	0 18	2 41	1 50	8400
6700	0 21	1 51	0 33	0 14	2 43	1 52	8300
6800	0 17	1 52	0 28	0 11	2 46	1 54	8200
6900	0 13	1 54	0 23	0 8	2 48	1 55	8100
7000	0 10	1 55	0 19	0 6	2 50	1 56	8000
7100	0 7	1 56	0 16	0 4	2 51	1 57	7900
7200	0 5	1 57	0 13	0 2	2 52	1 58	7800
7300	0 4	1 57	0 11	0 1	2 53	1 59	7700
7400	0 3	1 58	0 10	0 1	2 54	1 59	7600
7500	0 3	1 58	0 10	0 1	2 54	1 59	7500

TABLE XXXIX.

Equations 8 *and* 9.

Arg.	8	9	Arg.	8	9
0	1′ 20″	1′ 20″	5000	1′ 20″	1′ 20″
100	1 15	1 29	5100	1 24	1 36
200	1 11	1 37	5200	1 29	1 31
300	1 7	1 46	5300	1 33	1 37
400	1 2	1 54	5400	1 37	1 42
500	0 58	2 1	5500	1 42	1 47
600	0 54	2 8	5600	1 46	1 51
700	0 50	2 15	5700	1 50	1 55
800	0 46	2 20	5800	1 54	1 58
900	0 42	2 25	5900	1 58	2 0
1000	0 38	2 29	6000	2 1	2 1
1100	0 35	2 32	6100	2 5	2 2
1200	0 31	2 34	6200	2 8	2 2
1300	0 28	2 35	6300	2 11	2 1
1400	0 25	2 35	6400	2 14	1 59
1500	0 23	2 34	6500	2 17	1 56
1600	0 20	2 32	6600	2 19	1 52
1700	0 18	2 29	6700	2 22	1 48
1800	0 16	2 26	6800	2 24	1 43
1900	0 14	2 21	6900	2 25	1 38
2000	0 13	2 16	7000	2 27	1 32
2100	0 11	2 11	7100	2 28	1 25
2200	0 10	2 4	7200	2 29	1 18
2300	0 10	1 58	7300	2 30	1 11
2400	0 9	1 51	7400	2 30	1 4
2500	0 9	1 43	7500	2 31	0 56
2600	0 10	1 36	7600	2 30	0 49
2700	0 10	1 29	7700	2 30	0 42
2800	0 11	1 22	7800	2 29	0 36
2900	0 12	1 15	7900	2 28	0 29
3000	0 13	1 8	8000	2 27	0 24
3100	0 15	1 2	8100	2 26	0 18
3200	0 16	0 57	8200	2 24	0 14
3300	0 18	0 52	8300	2 22	0 10
3400	0 21	0 47	8400	2 20	0 8
3500	0 23	0 44	8500	2 17	0 6
3600	0 26	0 41	8600	2 15	0 5
3700	0 29	0 39	8700	2 12	0 5
3800	0 32	0 38	8800	2 9	0 6
3900	0 35	0 38	8900	2 5	0 8
4000	0 39	0 39	9000	2 2	0 11
4100	0 42	0 40	9100	1 58	0 15
4200	0 46	0 42	9200	1 54	0 20
4300	0 50	0 45	9300	1 50	0 25
4400	0 54	0 49	9400	1 46	0 32
4500	0 58	0 53	9500	1 42	0 39
4600	1 3	0 58	9600	1 38	0 46
4700	1 7	1 3	9700	1 33	0 54
4800	1 11	1 9	9800	1 29	1 3
4900	1 16	1 14	9900	1 24	1 11
5000	1 20	1 20	10000	1 20	1 20

TABLE XL.

Equations 10 *and* 11.

Arg.	10	11	Arg.	10	11
0	10″	10″	500	10″	10″
10	9	11	510	10	11
20	9	12	520	9	11
30	8	13	530	9	12
40	7	14	540	8	13
50	7	15	550	8	14
60	6	16	560	8	14
70	6	17	570	8	15
80	5	17	580	7	15
90	5	18	590	7	15
100	5	18	600	7	16
110	4	19	610	7	16
120	4	19	620	7	16
130	4	19	630	7	16
140	4	19	640	7	15
150	4	19	650	8	15
160	4	19	660	8	15
170	4	18	670	8	14
180	5	18	680	9	13
190	5	17	690	9	13
200	5	16	700	10	12
210	6	16	710	10	11
220	6	15	720	11	10
230	7	14	730	11	9
240	7	13	740	12	9
250	8	12	750	12	8
260	8	11	760	13	7
270	9	10	770	13	6
280	9	10	780	14	5
290	10	9	790	14	4
300	10	8	800	15	3
310	11	7	810	15	3
320	11	6	820	15	2
330	12	6	830	16	2
340	12	5	840	16	1
350	12	5	850	16	1
360	12	5	860	16	1
370	13	4	870	16	1
380	13	4	880	16	1
390	13	4	890	16	1
400	13	4	900	15	2
410	13	5	910	15	2
420	12	5	920	15	3
430	12	5	930	14	3
440	12	6	940	14	4
450	12	6	950	13	5
460	11	7	960	13	6
470	11	8	970	12	7
480	11	8	980	11	8
490	10	9	990	11	9
500	10	10	1000	10	10

TABLE XLI.

Equations 12 *to* 19.

Arg.	12	13	14	15	16	17	18	19	Arg.
250	2″	2′	8′	0″	34″	3″	17″	3″	250
260	2	2	8	0	34	3	17	3	240
270	2	2	8	0	34	3	17	3	230
280	3	2	8	0	33	3	17	3	220
290	3	2	8	0	33	4	16	3	210
300	3	2	8	0	33	4	16	3	200
310	3	3	9	1	33	4	16	3	190
320	4	3	9	1	32	4	16	4	180
330	4	4	9	1	32	4	16	4	170
340	5	4	10	2	32	4	16	4	160
350	6	5	10	2	31	5	15	4	150
360	6	6	11	2	31	5	15	5	140
370	7	7	11	3	30	5	15	5	130
380	8	7	12	3	29	5	15	5	120
390	9	8	12	4	29	6	14	6	110
400	10	9	13	4	28	6	14	6	100
410	10	10	13	5	27	6	14	6	90
420	11	11	14	5	27	7	13	7	80
430	12	12	15	6	26	7	13	7	70
440	13	13	15	6	25	8	12	7	60
450	14	14	16	7	24	8	12	8	50
460	16	15	17	7	23	8	12	8	40
470	17	16	18	8	23	9	11	9	30
480	18	18	18	9	22	9	11	9	20
490	19	19	19	9	21	10	10	10	10
500	20	20	20	10	20	10	10	10	000
510	21	21	21	11	19	10	10	10	990
520	22	22	21	11	18	11	9	11	980
530	23	23	22	12	17	11	9	11	970
540	24	25	23	12	17	12	8	12	960
550	25	26	24	13	16	12	8	12	950
560	26	27	24	14	15	12	7	13	940
570	27	28	25	14	14	13	7	13	930
580	28	29	26	15	13	13	7	13	920
590	29	30	26	15	13	13	6	14	910
600	30	31	27	16	12	14	6	14	900
610	31	32	28	16	11	14	6	14	890
620	32	33	28	17	11	14	5	15	880
630	33	33	29	17	10	15	5	15	870
640	34	34	29	18	9	15	5	15	860
650	34	35	30	18	9	15	5	16	850
660	35	36	30	18	8	16	4	16	840
670	35	36	31	19	8	16	4	16	830
680	36	37	31	19	8	16	4	16	820
690	36	37	31	19	7	16	4	17	810
700	37	37	32	19	7	16	4	17	800
710	37	38	32	20	7	16	3	17	790
720	37	38	32	20	6	16	3	17	780
730	38	38	32	20	6	16	3	17	770
740	38	38	32	20	6	17	3	17	760
750	38	38	32	20	6	17	3	17	750

TABLE XLII.

Equation 20.

Arg.	20	Arg.
0	10″	500
10	11	510
20	12	520
30	13	530
40	13	540
50	14	550
60	15	560
70	16	570
80	16	580
90	17	590
100	17	600
110	17	610
120	17	620
130	17	630
140	17	640
150	17	650
160	17	660
170	16	670
180	16	680
190	15	690
200	14	700
210	13	710
220	13	720
230	12	730
240	11	740
250	10	750
260	9	760
270	8	770
280	7	780
290	6	790
300	6	800
310	5	810
320	4	820
330	4	830
340	3	840
350	3	850
360	3	860
370	3	870
380	3	880
390	3	890
400	3	900
410	3	910
420	4	920
430	4	930
440	5	940
450	6	950
460	6	960
470	7	970
480	8	980
490	9	990
500	10	1000

Evection.

ARGUMENT. Evection, corrected.

	0s	Is	IIs	IIIs	IVs	Vs
0°	1°30′ 0″	2°10′ 43	2°40 10	2°50 25	2°39 8	2° 9 42
1	1 31 25	2 11 57	2 40 51	2 50 23	2 38 25	2 8 29
2	1 32 51	2 13 9	2 41 30	2 50 20	2 37 40	2 7 16
3	1 34 16	2 14 21	2 42 8	2 50 15	2 36 55	2 6 2
4	1 35 42	2 15 31	2 42 45	2 50 9	2 36 8	2 4 47
5	1 37 7	2 16 41	2 43 21	2 50 1	2 35 19	2 3 32
6	1 38 32	2 17 50	2 43 55	2 49 52	2 34 30	2 2 16
7	1 39 57	2 18 58	2 44 27	2 49 41	2 33 40	2 1 0
8	1 41 21	2 20 5	2 44 59	2 49 29	2 32 48	1 59 43
9	1 42 46	2 21 11	2 45 29	2 49 15	2 31 55	1 58 26
10	1 44 10	2 22 17	2 45 57	2 49 0	2 31 2	1 57 8
11	1 45 34	2 23 21	2 46 24	2 48 43	2 30 7	1 55 49
12	1 46 58	2 24 24	2 46 50	2 48 26	2 29 11	1 54 30
13	1 48 21	2 25 26	2 47 14	2 48 6	2 28 14	1 53 11
14	1 49 54	2 26 28	2 47 37	2 47 45	2 27 16	1 51 51
15	1 51 7	2 27 28	2 47 59	2 47 23	2 26 17	1 50 31
16	1 52 29	2 28 27	2 48 19	2 47 0	2 25 17	1 49 11
17	1 53 51	2 29 25	2 48 37	2 46 35	2 24 16	1 47 50
18	1 55 12	2 30 21	2 48 54	2 46 8	2 23 14	1 46 29
19	1 56 33	2 31 17	2 49 10	2 45 41	2 22 11	1 45 7
20	1 57 53	2 32 11	2 49 24	2 45 12	2 21 7	1 43 46
21	1 59 13	2 33 5	2 49 37	2 44 41	2 20 2	1 42 24
22	2 0 32	2 33 57	2 49 48	2 44 9	2 18 56	1 41 2
23	2 1 51	2 34 48	2 49 58	2 43 36	2 17 50	1 39 39
24	2 3 9	2 35 38	2 50 6	2 43 2	2 16 43	1 38 17
25	2 4 26	2 36 26	2 50 13	2 42 26	2 15 34	1 36 54
26	2 5 43	2 37 13	2 50 19	2 41 49	2 14 25	1 35 32
27	2 6 59	2 37 59	2 50 23	2 41 11	2 13 16	1 34 9
28	2 8 15	2 38 44	2 50 25	2 40 31	2 12 5	1 32 46
29	2 9 30	2 39 28	2 50 26	2 39 50	2 10 54	1 31 23
30	2 10 43	2 40 10	2 50 25	2 39 8	2 9 42	1 30 0

Evection.

ARGUMENT. Evection, corrected.

	VI^s	VII^s	VIII^s	IX^s	X^s	XI^s
0°	1°30′ 0″	0°50′ 18″	0°20′ 52″	0° 9′ 34″	0°19′ 50″	0°49′ 16″
1	1 28 37	0 49 6	0 20 10	0 9 34	0 20 32	0 50 30
2	1 27 14	0 47 55	0 19 29	0 9 35	0 21 16	0 51 45
3	1 25 51	0 46 44	0 18 49	0 9 37	0 22 1	0 53 1
4	1 24 28	0 45 34	0 18 11	0 9 41	0 22 47	0 54 17
5	1 23 6	0 44 26	0 17 34	0 9 47	0 23 34	0 55 33
6	1 21 43	0 43 17	0 16 58	0 9 54	0 24 22	0 56 51
7	1 20 20	0 42 10	0 16 24	0 10 2	0 25 12	0 58 9
8	1 18 58	0 41 4	0 15 50	0 10 12	0 26 3	0 59 28
9	1 17 36	0 39 58	0 15 19	0 10 23	0 26 55	1 0 47
10	1 16 14	0 38 53	0 14 48	0 10 36	0 27 48	1 2 7
11	1 14 52	0 37 49	0 14 19	0 10 50	0 28 43	1 3 27
12	1 13 31	0 36 46	0 13 51	0 11 5	0 29 39	1 4 48
13	1 12 10	0 35 44	0 13 25	0 11 23	0 30 35	1 6 9
14	1 10 49	0 34 43	0 13 0	0 11 41	0 31 33	1 7 31
15	1 9 29	0 33 43	0 12 37	0 12 1	0 32 32	1 8 53
16	1 8 9	0 32 44	0 12 14	0 12 23	0 33 32	1 10 16
17	1 6 49	0 31 46	0 11 54	0 12 45	0 34 34	1 11 39
18	1 5 30	0 30 49	0 11 34	0 13 10	0 35 36	1 13 2
19	1 4 11	0 29 53	0 11 16	0 13 35	0 36 39	1 14 26
20	1 2 52	0 28 58	0 11 0	0 14 3	0 37 43	1 15 50
21	1 1 34	0 28 5	0 10 45	0 14 31	0 38 48	1 17 14
22	1 0 17	0 27 12	0 10 31	0 15 1	0 39 55	1 18 39
23	0 59 0	0 26 20	0 10 19	0 15 33	0 41 2	1 20 3
24	0 57 44	0 25 30	0 10 8	0 16 5	0 42 10	1 21 28
25	0 56 28	0 24 40	0 9 59	0 16 39	0 43 19	1 22 53
26	0 55 13	0 23 52	0 9 51	0 17 15	0 44 29	1 24 18
27	0 53 58	0 23 5	0 9 45	0 17 52	0 45 39	1 25 44
28	0 52 44	0 22 20	0 9 40	0 18 30	0 46 51	1 27 9
29	0 51 31	0 21 35	0 9 36	0 19 9	0 48 3	1 28 34
30	0 50 18	0 20 52	0 9 34	0 19 50	0 49 16	1 30 0

Equation of Moon's Centre.

ARGUMENT. Anomaly, corrected.

	0s	Is	IIs	IIIs	IVs	Vs
0°	7° 0′ 0″	10°20′ 58″	12°38′ 44″	13°17′ 35″	12°16′ 21″	9°58′ 29″
1	7 7 5	10 26 52	12 41 43	13 17 5	12 12 48	9 52 58
2	7 14 10	10 32 42	12 44 35	13 16 28	12 9 11	9 47 24
3	7 21 15	10 38 27	12 47 20	13 15 44	12 5 29	9 41 48
4	7 28 19	10 44 8	12 49 59	13 14 53	12 1 41	9 36 10
5	7 35 23	10 49 43	12 52 30	13 13 56	11 57 49	9 30 29
6	7 42 26	10 55 14	12 54 55	13 12 52	11 53 52	9 24 46
7	7 49 28	11 0 39	12 57 12	13 11 41	11 49 50	9 19 1
8	7 56 28	11 6 0	12 59 23	13 10 24	11 45 44	9 13 13
9	8 3 28	11 11 15	13 1 26	13 9 1	11 41 33	9 7 24
10	8 10 26	11 16 24	13 3 23	13 7 31	11 37 17	9 1 32
11	8 17 22	11 21 29	13 5 12	13 5 54	11 32 57	8 55 39
12	8 24 17	11 26 27	13 6 55	13 4 12	11 28 33	8 49 44
13	8 31 10	11 31 20	13 8 30	13 2 23	11 24 5	8 43 47
14	8 38 1	11 36 8	13 9 59	13 0 27	11 19 32	8 37 49
15	8 44 50	11 40 49	13 11 20	12 58 26	11 14 55	8 31 49
16	8 51 36	11 45 25	13 12 34	12 56 18	11 10 14	8 25 48
17	8 58 20	11 49 54	13 13 41	12 54 5	11 5 30	8 19 46
18	9 5 1	11 54 18	13 14 41	12 51 45	11 0 41	8 13 42
19	9 11 39	11 58 35	13 15 34	12 49 19	10 55 49	8 7 38
20	9 18 15	12 2 47	13 16 20	12 46 47	10 50 53	8 1 32
21	9 24 47	12 6 52	13 16 59	12 44 10	10 45 53	7 55 26
22	9 31 16	12 10 50	13 17 31	12 41 27	10 40 50	7 49 18
23	9 37 42	12 14 42	13 17 56	12 38 38	10 35 43	7 43 10
24	9 44 4	12 18 28	13 18 14	12 35 43	10 30 33	7 37 1
25	9 50 23	12 22 7	13 18 24	12 32 43	10 25 20	7 30 52
26	9 56 38	12 25 40	13 18 28	12 29 37	10 20 4	7 24 42
27	10 2 49	12 29 6	13 18 25	12 26 26	10 14 45	7 18 32
28	10 8 56	12 32 25	13 18 16	12 23 10	10 9 22	7 12 21
29	10 14 59	12 35 38	13 17 59	12 19 48	10 3 57	7 6 11
30	10 20 58	12 38 44	13 17 35	12 16 21	9 58 29	7 0 0

Equation of Moon's Centre.

ARGUMENT. Anomaly, corrected.

	VI^s	VII^s	VIII^s	IX^s	X^s	XI^s
0°	7° 0′ 0″	4° 1′ 31″	1°43′ 39″	0°42′ 25″	1°21′ 16″	3°39′ 2″
1	6 53 49	3 56 3	1 40 12	0 42 1	1 24 22	3 45 1
2	6 47 39	3 50 38	1 36 50	0 41 44	1 27 35	3 51 4
3	6 41 28	3 45 15	1 33 34	0 41 35	1 30 54	3 57 11
4	6 35 18	3 39 56	1 30 23	0 41 32	1 34 20	4 3 22
5	6 29 8	3 34 40	1 27 17	0 41 36	1 37 53	4 9 37
6	6 22 59	3 29 26	1 24 17	0 41 46	1 41 32	4 15 55
7	6 16 50	3 24 17	1 21 22	0 42 4	1 45 18	4 22 18
8	6 10 42	3 19 10	1 18 33	0 42 29	1 49 10	4 28 44
9	6 4 34	3 14 7	1 15 50	0 43 1	1 53 8	4 35 13
10	5 58 28	3 9 7	1 13 12	0 43 40	1 57 13	4 41 45
11	5 52 22	3 4 11	1 10 41	0 44 26	2 1 24	4 48 21
12	5 46 17	2 59 19	1 8 15	0 45 19	2 5 42	4 54 59
13	5 40 14	2 54 30	1 5 55	0 46 19	2 10 5	5 1 40
14	5 34 12	2 49 46	1 3 42	0 47 26	2 14 35	5 8 24
15	5 28 11	2 45 5	1 1 34	0 48 40	2 19 11	5 15 10
16	5 22 11	2 40 28	0 59 33	0 50 1	2 23 52	5 21 59
17	5 16 13	2 35 55	0 57 37	0 51 30	2 28 39	5 28 50
18	5 10 16	2 31 27	0 55 48	0 53 5	2 33 32	5 35 43
19	5 4 21	2 27 3	0 54 6	0 54 47	2 38 31	5 42 37
20	4 58 28	2 22 43	0 52 29	0 56 37	2 43 35	5 49 34
21	4 52 36	2 18 27	0 50 59	0 58 33	2 48 45	5 56 32
22	4 46 47	2 14 16	0 49 36	1 0 37	2 54 0	6 3 31
23	4 40 59	2 10 10	0 48 19	1 2 48	2 59 21	6 10 32
24	4 35 14	2 6 8	0 47 8	1 5 5	3 4 46	6 17 34
25	4 29 31	2 2 11	0 46 4	1 7 30	3 10 17	6 24 37
26	4 23 50	1 58 19	0 45 7	1 10 1	3 15 52	6 31 41
27	4 18 11	1 54 31	0 44 16	1 12 40	3 21 33	6 38 45
28	4 12 35	1 50 49	0 43 32	1 15 25	3 27 18	6 45 50
29	4 7 2	1 47 11	0 42 55	1 18 17	3 33 8	6 52 55
30	4 1 31	1 43 39	0 42 25	1 21 16	3 39 2	7 0 0

TABLE XLV. *Variation.*

ARGUMENT. Variation, corrected.

	0s	Is	IIs	IIIs	IVs	Vs
0°	0°38′ 0″	1° 8′ 1″	1° 6′ 58″	0°35′ 54″	0° 5′ 29″	0° 6′ 2″
1	0 39 13	1 8 35	1 6 18	0 34 40	0 4 54	0 6 42
2	0 40 26	1 9 7	1 5 36	0 33 27	0 4 21	0 7 24
3	0 41 39	1 9 36	1 4 52	0 32 13	0 3 51	0 8 8
4	0 42 52	1 10 3	1 4 5	0 31 0	0 3 22	0 8 55
5	0 44 4	1 10 28	1 3 17	0 29 47	0 2 56	0 9 44
6	0 45 16	1 10 50	1 2 27	0 28 34	0 2 33	0 10 34
7	0 46 28	1 11 9	1 1 35	0 27 22	0 2 12	0 11 27
8	0 47 38	1 11 26	1 0 42	0 26 11	0 1 54	0 12 22
9	0 48 48	1 11 41	0 59 46	0 25 1	0 1 38	0 13 19
10	0 49 57	1 11 53	0 58 49	0 23 51	0 1 24	0 14 17
11	0 51 6	1 12 2	0 57 50	0 22 42	0 1 14	0 15 17
12	0 52 13	1 12 9	0 56 50	0 21 34	0 1 5	0 16 19
13	0 53 19	1 12 13	0 55 48	0 20 28	0 1 0	0 17 22
14	0 54 24	1 12 15	0 54 45	0 19 22	0 0 57	0 18 27
15	0 55 27	1 12 14	0 53 41	0 18 18	0 0 57	0 19 33
16	0 56 30	1 12 10	0 52 35	0 17 15	0 0 59	0 20 41
17	0 57 31	1 12 4	0 51 28	0 16 13	0 1 4	0 21 50
18	0 58 30	1 11 55	0 50 21	0 15 13	0 1 11	0 23 0
19	0 59 28	1 11 44	0 49 12	0 14 15	0 1 22	0 24 11
20	1 0 24	1 11 30	0 48 2	0 13 17	0 1 34	0 25 23
21	1 1 19	1 11 14	0 46 52	0 12 22	0 1 50	0 26 36
22	1 2 11	1 10 55	0 45 40	0 11 28	0 2 8	0 27 50
23	1 3 2	1 10 34	0 44 29	0 10 37	0 2 28	0 29 4
24	1 3 51	1 10 10	0 43 16	0 9 47	0 2 51	0 30 20
25	1 4 38	1 9 44	0 42 3	0 8 59	0 3 17	0 31 36
26	1 5 23	1 9 15	0 40 50	0 8 13	0 3 45	0 32 52
27	1 6 6	1 8 44	0 39 36	0 7 29	0 4 16	0 34 9
28	1 6 47	1 8 11	0 38 22	0 6 47	0 4 48	0 35 26
29	1 7 25	1 7 36	0 37 8	0 6 7	0 5 24	0 36 43
30	1 8 1	1 6 58	0 35 54	0 5 29	0 6 2	0 38 0

TABLE XLVI. *Reduction.*

ARGUMENT. Supplement of Node + Moon's Orb. Long.

	0s VIs	Is VIIs	IIs VIIIs	IIIs IXs	IVs Xs	Vs XIs
0°	7′ 0″	1′ 3″	1′ 3″	7′ 0″	12′ 57″	12′ 57″
1	6 46	0 56	1 10	7 14	13 4	12 50
2	6 31	0 49	1 18	7 29	13 10	12 42
3	6 17	0 43	1 26	7 43	13 17	12 33
4	6 3	0 38	1 35	7 57	13 22	12 25
5	5 48	0 33	1 44	8 12	13 27	12 16
6	5 34	0 28	1 54	8 26	13 32	12 6
7	5 20	0 24	2 3	8 40	13 36	11 56
8	5 6	0 20	2 14	8 54	13 40	11 46
9	4 53	0 17	2 24	9 7	13 43	11 36
10	4 39	0 14	2 35	9 21	13 46	11 25
11	4 26	0 12	2 46	9 34	13 48	11 14
12	4 12	0 10	2 58	9 48	13 50	11 2
13	3 59	0 9	3 9	10 1	13 51	10 50
14	3 46	0 8	3 22	10 13	13 52	10 38
15	3 34	0 8	3 34	10 26	13 52	10 26

TABLE XLV. *Variation.*

ARGUMENT. Variation, corrected.

	VI^s	VII^s	VIII^s	IX	X^s	XI^s
0°	0°38′ 0″	1° 9′ 58″	1°10′ 30″	0°40′ 6″	0° 9′ 2″	0° 7′ 58″
1	0 39 17	1 10 36	1 9 53	0 38 52	0 8 24	0 8 35
2	0 40 34	1 11 11	1 9 13	0 37 38	0 7 49	0 9 13
3	0 41 51	1 11 44	1 8 31	0 36 24	0 7 15	0 9 54
4	0 43 8	1 12 15	1 7 47	0 35 10	0 6 45	0 10 37
5	0 44 24	1 12 43	1 7 1	0 33 57	0 6 16	0 11 22
6	0 45 40	1 13 9	1 6 13	0 32 44	0 5 50	0 12 9
7	0 46 55	1 13 32	1 5 23	0 31 31	0 5 26	0 12 58
8	0 48 10	1 13 52	1 4 31	0 30 19	0 5 5	0 13 49
9	0 49 24	1 14 10	1 3 38	0 29 8	0 4 46	0 14 41
10	0 50 37	1 14 26	1 2 42	0 27 58	0 4 29	0 15 36
11	0 51 49	1 14 38	1 1 45	0 26 48	0 4 16	0 16 32
12	0 53 0	1 14 48	1 0 47	0 25 39	0 4 4	0 17 30
13	0 54 10	1 14 56	0 59 47	0 24 31	0 3 56	0 18 29
14	0 55 19	1 15 1	0 58 45	0 23 25	0 3 50	0 19 30
15	0 56 27	1 15 3	0 57 42	0 22 19	0 3 46	0 20 32
16	0 57 33	1 15 3	0 56 38	0 21 15	0 3 45	0 21 36
17	0 58 38	1 15 0	0 55 32	0 20 12	0 3 47	0 22 41
18	0 59 41	1 14 54	0 54 25	0 19 10	0 3 51	0 23 47
19	1 0 43	1 14 46	0 53 18	0 18 10	0 3 58	0 24 54
20	1 1 43	1 14 35	0 52 9	0 17 11	0 4 7	0 26 3
21	1 2 41	1 14 22	0 50 59	0 16 14	0 4 19	0 27 12
22	1 3 38	1 14 6	0 49 49	0 15 18	0 4 34	0 28 22
23	1 4 33	1 13 48	0 48 38	0 14 25	0 4 51	0 29 32
24	1 5 25	1 13 27	0 47 26	0 13 33	0 5 10	0 30 44
25	1 6 16	1 13 3	0 46 13	0 12 43	0 5 32	0 31 55
26	1 7 5	1 12 38	0 45 0	0 11 54	0 5 57	0 33 8
27	1 7 52	1 12 9	0 43 47	0 11 8	0 6 23	0 34 20
28	1 8 36	1 11 39	0 42 33	0 10 24	0 6 53	0 35 33
29	1 9 13	1 11 6	0 41 20	0 9 42	0 7 24	0 36 47
30	1 9 58	1 10 30	0 40 6	0 9 2	0 7 58	0 38 0

TABLE XLVI. *Reduction.*

ARGUMENT. Supplement of Node + Moon's Orb. Long.

	0^s VI^s	I^s VII^s	II^s VIII^s	III^s IX^s	IV^s X^s	V^s XI^s
15°	3′ 34″	0′ 8″	3′ 34″	10′ 26″	13′ 52	10′ 26″
16	3 22	0 8	3 46	10 38	13 52	10 13
17	3 9	0 9	3 59	10 50	13 51	10 1
18	2 58	0 10	4 12	11 2	13 50	9 48
19	2 46	0 12	4 26	11 14	13 48	9 34
20	2 35	0 14	4 39	11 25	13 46	9 21
21	2 24	0 17	4 53	11 36	13 43	9 7
22	2 14	0 20	5 6	11 46	13 40	8 54
23	2 3	0 24	5 20	11 56	13 36	8 40
24	1 54	0 28	5 34	12 6	13 32	8 26
25	1 44	0 33	5 48	12 16	13 27	8 12
26	1 35	0 38	6 3	12 25	13 22	7 57
27	1 26	0 43	6 17	12 33	13 17	7 43
28	1 18	0 49	6 31	12 42	13 10	7 29
29	1 10	0 56	6 46	12 50	13 4	7 14
30	1 3	1 3	7 0	12 57	12 57	7 0

Moon's Distance from the North Pole of the Ecliptic.

ARGUMENT. Suppl. of Node + Moon's Orbit Longitude.

	III^s	IV^s	V^s	VI^s	VII^s	VIII^s	
0°	84°39′ 16″	85°20′ 43″	87°13′ 47″	89°48′ 0″	92°22′ 13″	94°15′ 17″	30°
1	84 39 19	85 23 27	87 18 28	89 53 23	92 26 52	94 17 57	29
2	84 39 27	85 26 16	87 23 12	89 58 46	92 31 27	94 20 31	28
3	84 39 41	85 29 10	87 27 58	90 4 8	92 36 0	94 23 1	27
4	84 40 1	85 32 9	87 32 48	90 9 31	92 40 30	94 25 25	26
5	84 40 27	85 35 12	87 37 39	90 14 52	92 44 56	94 27 45	25
6	84 40 58	85 38 20	87 42 33	90 20 14	92 49 19	94 29 59	24
7	84 41 34	85 41 33	87 47 30	90 25 35	92 53 39	94 32 8	23
8	84 42 17	85 44 50	87 52 28	90 30 55	92 57 56	94 34 12	22
9	84 43 5	85 48 11	87 57 29	90 36 14	93 2 9	94 36 11	21
10	84 43 58	85 51 37	88 2 31	90 41 33	93 6 18	94 38 4	20
11	84 44 57	85 55 7	88 7 36	90 46 50	93 10 24	94 39 52	19
12	84 46 2	85 58 42	88 12 42	90 52 7	93 14 27	94 41 35	18
13	84 47 12	86 2 20	88 17 50	90 57 22	93 18 25	94 43 13	17
14	84 48 27	86 6 3	88 23 0	91 2 36	93 22 20	94 44 45	16
15	84 49 49	86 9 50	88 28 11	91 7 49	93 26 10	94 46 11	15
16	84 51 15	86 13 40	88 33 24	91 13 0	93 29 57	94 47 32	14
17	85 52 47	86 17 35	88 38 38	91 18 10	93 33 40	94 48 48	13
18	84 54 25	86 21 33	88 43 53	91 23 18	93 37 18	94 49 58	12
19	84 56 7	86 25 36	88 49 10	91 28 24	93 40 53	94 51 3	11
20	84 57 56	86 29 42	88 54 27	91 33 29	93 44 23	94 52 2	10
21	84 59 49	86 33 51	88 59 46	91 38 31	93 47 49	94 52 55	9
22	85 1 48	86 38 4	89 5 5	91 43 32	93 51 10	94 53 43	8
23	85 3 52	86 42 21	89 10 25	91 48 30	93 54 27	94 54 26	7
24	85 6 1	86 46 41	89 15 46	91 53 27	93 57 40	94 55 2	6
25	85 8 15	86 51 4	89 21 7	91 58 21	94 0 48	94 55 33	5
26	85 10 35	86 55 30	89 26 29	92 3 12	94 3 51	94 55 59	4
27	85 12 59	87 0 0	89 31 52	92 8 1	94 6 50	94 56 18	3
28	85 15 29	87 4 32	89 37 14	92 12 48	94 9 44	94 56 33	2
29	85 18 3	87 9 8	89 42 37	92 17 32	94 12 33	94 56 41	1
30	85 20 43	87 13 47	89 48 0	92 22 13	94 15 17	94 56 44	0
	II^s	I^s	0^s	XI^s	X^s	IX^s	

TABLE XLVIII.

Equation II. *of the Moon's Polar Distance.*

ARGUMENT II., corrected.

	III^s	IV^s	V^s	VI^s	VII^s	VIII^s	
0°	0 14″	1′24″	4′37″	9′ 0″	13′23″	16′36″	30°
1	0 14	1 29	4 45	9 9	13 31	16 40	29
2	0 14	1 34	4 53	9 18	13 39	16 45	28
3	0 14	1 39	5 1	9 27	13 47	16 49	27
4	0 15	1 44	5 9	9 37	13 54	16 53	26
5	0 16	1 49	5 18	9 46	14 2	16 57	25
6	0 17	1 54	5 26	9 55	14 9	17 1	24
7	0 18	2 0	5 34	10 4	14 17	17 4	23
8	0 19	2 5	5 43	10 13	14 24	17 8	22
9	0 20	2 11	5 51	10 22	14 31	17 11	21
10	0 22	2 17	6 0	10 31	14 38	17 14	20
11	0 23	2 23	6 9	10 40	14 45	17 17	19
12	0 25	2 29	6 17	10 49	14 52	17 20	18
13	0 27	2 35	6 26	10 58	14 59	17 23	17
14	0 29	2 41	6 35	11 7	15 5	17 26	16
15	0 32	2 48	6 44	11 16	15 12	17 28	15
16	0 34	2 54	6 53	11 25	15 18	17 31	14
17	0 37	3 1	7 2	11 34	15 25	17 33	13
18	0 40	3 8	7 11	11 43	15 31	17 35	12
19	0 42	3 15	7 20	11 51	15 37	17 36	11
20	0 45	3 22	7 29	12 0	15 43	17 38	10
21	0 49	3 29	7 38	12 9	15 49	17 40	9
22	0 52	3 36	7 47	12 17	15 55	17 41	8
23	0 56	3 43	7 56	12 26	16 0	17 42	7
24	0 59	3 51	8 5	12 34	16 6	17 43	6
25	1 3	3 58	8 14	12 42	16 11	17 44	5
26	1 7	4 6	8 23	12 51	16 16	17 45	4
27	1 11	4 13	8 32	12 59	16 21	17 45	3
28	1 15	4 21	8 42	13 7	16 26	17 46	2
29	1 20	4 29	8 51	13 15	16 31	17 46	1
30	1 24	4 37	9 0	13 23	16 36	17 46	0
	II^s	I^s	0^s	XI^s	X^s	IX^s	

TABLE XLIX.

Equation III. *of the Polar Distance.*

ARGUMENT. Moon's True Longitude.

	III^s	IV^s	V^s	VI	VII	VIII^s	
0°	16″	15″	12″	8″	4″	1″	30°
6	16	14	11	7	3	1	24
12	16	14	10	6	3	0	18
18	16	13	10	5	2	0	12
24	15	13	9	5	1	0	6
30	15	12	8	4	1	0	0
	II^s	I^s	0^s	XI^s	X^s	IX^s	

TABLE L.

To convert Degrees and Minutes into Decimal Parts.

Degrees and Minutes.	Dec. Parts.
1° 5′	.003
1 26	4
1 48	5
2 10	6
2 31	7
2 53	8
3 14	9
3 36	10
3 58	11
4 19	12
4 41	13
5 2	14
5 24	15
5 46	16
6 7	17
6 29	18
6 50	19
7 12	20
7 34	21
7 55	22
8 17	23
8 38	24
9 0	25
9 22	26
9 43	27
10 5	28
10 26	29
10 48	30
11 10	31
11 31	32
11 53	33
12 14	34
12 36	35
12 58	36
13 19	37
13 41	38
14 2	39
14 24	40
14 46	41
15 7	42
15 29	43
15 50	44
16 12	45
16 34	46
16 55	47
17 17	48
17 38	49
18 0	50
18 22	51
18 43	52
19 5	53

TABLE LI.

Equations of Polar Distance.

ARGUMENTS. 20 of Long.; V. to IX., corrected; and X., not corrected.

Arg.	20	V.	VI.	VII.	VIII.	IX.	X.	Arg.
250	0″	56″	6″	3″	25″	3″	11″	250
260	0	56	6	3	25	3	11	240
270	0	56	6	3	25	3	11	230
280	1	55	6	3	25	3	11	220
290	1	55	7	3	25	4	11	210
300	1	55	7	4	25	4	11	200
310	1	54	8	4	24	5	12	190
320	2	53	8	5	24	6	12	180
330	2	53	9	5	24	6	13	170
340	3	52	10	6	23	7	13	160
350	3	51	11	7	23	8	14	150
360	4	50	12	8	23	9	14	140
370	4	49	13	9	22	10	15	130
380	5	48	14	10	22	11	16	120
390	6	46	15	11	21	13	17	110
400	6	45	16	12	21	14	17	100
410	7	44	17	13	20	15	18	90
420	8	42	18	14	20	17	19	80
430	9	41	20	15	19	18	20	70
440	10	39	21	17	19	20	21	60
450	10	38	23	18	18	22	22	50
460	11	36	24	19	17	23	23	40
470	12	35	25	21	17	25	24	30
480	13	33	27	22	16	27	25	20
490	14	32	28	24	16	28	26	10
500	15	30	30	25	15	30	27	000
510	16	28	31	26	14	32	28	990
520	17	27	33	28	14	33	29	980
530	18	25	34	29	13	35	30	970
540	19	24	36	31	12	37	31	960
550	19	22	37	32	12	38	32	950
560	20	20	39	33	11	40	33	940
570	21	19	40	34	11	41	34	930
580	22	17	41	36	10	43	35	920
590	23	16	43	37	10	44	36	910
600	24	15	44	38	9	46	37	900
610	24	13	45	39	9	47	37	890
620	25	12	46	40	8	48	38	880
630	26	11	47	41	8	50	39	870
640	26	10	48	42	7	51	40	860
650	27	9	49	43	7	52	40	850
660	27	8	50	44	6	53	41	840
670	28	7	51	45	6	54	41	830
680	28	7	52	45	6	54	42	820
690	29	6	52	46	6	55	42	810
700	29	5	53	46	5	56	42	800
710	29	5	53	47	5	56	43	790
720	29	5	53	47	5	56	43	780
730	30	4	54	47	5	57	43	770
740	30	4	54	47	5	57	43	760
750	30	4	54	47	5	57	43	750

Moon's Equatorial Parallax.

ARGUMENT. Argument of the Evection.

	0s	Is	IIs	IIIs	IVs	Vs	
0°	1′28″	1′23″	1′ 9″	0′50″	0′32″	0′18″	30°
1	1 28	1 23	1 8	0 49	0 31	0 18	29
2	1 28	1 22	1 8	0 49	0 30	0 18	28
3	1 28	1 22	1 7	0 48	0 30	0 17	27
4	1 28	1 22	1 7	0 47	0 29	0 17	26
5	1 28	1 21	1 6	0 47	0 29	0 17	25
6	1 28	1 21	1 5	0 46	0 28	0 17	24
7	1 28	1 20	1 5	0 46	0 28	0 16	23
8	1 28	1 20	1 4	0 45	0 27	0 16	22
9	1 28	1 20	1 4	0 44	0 27	0 16	21
10	1 28	1 19	1 3	0 44	0 26	0 16	20
11	1 28	1 19	1 2	0 43	0 26	0 15	19
12	1 27	1 18	1 2	0 42	0 25	0 15	18
13	1 27	1 18	1 1	0 42	0 25	0 15	17
14	1 27	1 17	1 0	0 41	0 24	0 15	16
15	1 27	1 17	1 0	0 40	0 24	0 15	15
16	1 27	1 16	0 59	0 40	0 24	0 15	14
17	1 27	1 16	0 59	0 39	0 23	0 14	13
18	1 26	1 15	0 58	0 39	0 23	0 14	12
19	1 26	1 15	0 57	0 38	0 22	0 14	11
20	1 26	1 14	0 57	0 37	0 22	0 14	10
21	1 26	1 14	0 56	0 37	0 21	0 14	9
22	1 25	1 13	0 55	0 36	0 21	0 14	8
23	1 25	1 13	0 55	0 36	0 21	0 14	7
24	1 25	1 12	0 54	0 35	0 20	0 14	6
25	1 25	1 12	0 53	0 34	0 20	0 14	5
26	1 24	1 11	0 53	0 34	0 20	0 14	4
27	1 24	1 11	0 52	0 33	0 19	0 14	3
28	1 24	1 10	0 51	0 33	0 19	0 13	2
29	1 23	1 10	0 51	0 32	0 19	0 13	1
30	1 23	1 9	0 50	0 32	0 18	0 13	0
	XIs	Xs	IXs	VIIIs	VIIs	VIs	

Moon's Equatorial Parallax.

Argument. Anomaly.

	0s	Is	IIs	IIIs	IVs	Vs	
0°	58′58″	58′27″	57′ 8″	55′30″	54′ 2″	53′ 3″	30°
1	58 58	58 25	57 5	55 27	53 59	53 2	29
2	58 58	58 23	57 2	55 23	53 57	53 0	28
3	58 57	58 21	56 58	55 20	53 54	52 59	27
4	58 57	58 19	56 55	55 17	53 52	52 58	26
5	58 57	58 16	56 52	55 14	53 50	52 57	25
6	58 56	58 14	56 49	55 11	53 47	52 56	24
7	58 56	58 12	56 45	55 7	53 45	52 55	23
8	58 55	58 10	56 42	55 4	53 43	52 54	22
9	58 55	58 7	56 39	55 1	53 41	52 53	21
10	58 54	58 5	56 36	54 58	53 38	52 52	20
11	59 53	58 2	56 32	54 55	53 36	52 51	19
12	58 53	58 0	56 29	54 52	53 34	52 50	18
13	58 52	57 57	56 26	54 49	53 32	52 49	17
14	58 51	57 55	56 22	54 46	53 30	52 49	16
15	58 50	57 52	56 19	54 43	53 28	52 48	15
16	58 49	57 49	56 16	54 40	53 26	52 47	14
17	58 48	57 46	56 13	54 37	53 24	52 47	13
18	58 46	57 44	56 9	54 34	53 22	52 46	12
19	58 45	57 41	56 6	54 31	53 21	52 45	11
20	58 44	57 38	56 3	54 29	53 19	52 45	10
21	58 42	57 35	55 59	54 26	53 17	52 45	9
22	58 41	57 32	55 56	54 23	53 15	52 44	8
23	58 39	57 29	55 53	54 20	53 14	52 44	7
24	58 38	57 26	55 49	54 18	53 12	52 43	6
25	58 36	57 23	55 46	54 15	53 10	52 43	5
26	58 34	57 20	55 43	54 12	53 9	52 43	4
27	58 33	57 17	55 40	54 10	53 7	52 43	3
28	58 31	57 14	55 36	54 7	53 6	52 43	2
29	58 29	57 11	55 33	54 4	53 4	52 43	1
30	58 27	57 8	55 30	54 2	53 3	52 43	0
	XIs	Xs	IXs	VIIIs	VIIs	VIs	

TABLE LIV.

Moon's Equatorial Parallax.

Argument. Argument of the Variation.

	0s	Is	IIs	IIIs	IVs	Vs	
0°	56″	42″	16″	4″	18″	44″	30°
1	56	41	15	4	18	45	29
2	55	41	14	4	19	46	28
3	55	40	14	4	20	46	27
4	55	39	13	4	21	47	26
5	55	38	12	4	22	48	25
6	55	37	12	4	23	48	24
7	55	36	11	5	24	49	23
8	55	35	10	5	24	50	22
9	54	35	10	5	25	50	21
10	54	34	9	6	26	51	20
11	54	33	9	6	27	51	19
12	53	32	8	6	28	52	18
13	53	31	8	7	29	53	17
14	52	30	7	7	30	53	16
15	52	29	7	8	31	53	15
16	51	28	6	8	32	54	14
17	51	27	6	9	33	54	13
18	50	26	6	9	34	55	12
19	50	25	5	10	35	55	11
20	49	24	5	10	35	55	10
21	49	24	5	11	36	56	9
22	48	23	4	12	37	56	8
23	47	22	4	12	38	56	7
24	47	21	4	13	39	56	6
25	46	20	4	14	40	57	5
26	45	19	4	14	41	57	4
27	45	18	4	15	42	57	3
28	44	18	4	16	42	57	2
29	43	17	4	17	43	57	1
30	42	16	4	18	44	57	0
	XIs	Xs	IXs	VIIIs	VIIs	VIs	

TABLE LV.

Moon's Semidiameter.

Argument. Equatorial Parallax.

Eq. Par.	Semidi.
53′ 40″	14′ 37″
53 50	14 40
54 0	14 43
54 10	14 46
54 20	14 48
54 30	14 51
54 40	14 54
54 50	14 57
55 0	14 59
55 10	15 2
55 20	15 5
55 30	15 7
55 40	15 10
55 50	15 13
56 0	15 16
56 10	15 18
56 20	15 21
56 30	15 24
56 40	15 26
56 50	15 29
57 0	15 32
57 10	15 35
57 20	15 37
57 30	15 40
57 40	15 43
57 50	15 46
58 0	15 48
58 10	15 51
58 20	15 54
58 30	15 56
58 40	15 59
58 50	16 2
59 0	16 5
59 10	16 7
59 20	16 10
59 30	16 13
59 40	16 16
59 50	16 18
60 0	16 21
60 10	16 24
60 20	16 26
60 30	16 29
60 40	16 32
60 50	16 35
61 0	16 37
61 10	16 40
61 20	16 43
61 30	16 46
61 40	16 48

TABLE LVI.

Moon's Horary Motion in Longitude.

Arguments. 2, 3, 4, and 5 of Long.

Arg.	2	3	4	5	Arg.
0	6″	1″	3″	3″	100
5	5	2	3	3	95
10	5	2	3	3	90
15	4	2	3	3	85
20	4	3	2	2	80
25	3	3	2	2	75
30	2	3	2	2	70
35	2	4	1	1	65
40	1	4	1	1	60
45	1	4	1	1	55
50	0	5	1	1	50

Moon's Horary Motion in Longitude.

ARGUMENT. Argument of the Evection.

	0s	Is	IIs	IIIs	IVs	Vs	
0°	1′ 20″	1′ 15″	1′ 0″	0′ 39″	0′ 20″	0′ 6″	30°
1	1 20	1 14	0 59	0 39	0 19	0 6	29
2	1 20	1 14	0 58	0 38	0 19	0 5	28
3	1 20	1 14	0 58	0 37	0 18	0 5	27
4	1 20	1 13	0 57	0 37	0 18	0 5	26
5	1 20	1 13	0 56	0 36	0 17	0 4	25
6	1 20	1 12	0 56	0 35	0 16	0 4	24
7	1 20	1 12	0 55	0 35	0 16	0 4	23
8	1 20	1 11	0 54	0 34	0 15	0 4	22
9	1 20	1 11	0 54	0 33	0 15	0 3	21
10	1 20	1 11	0 53	0 33	0 14	0 3	20
11	1 20	1 10	0 52	0 32	0 14	0 3	19
12	1 19	1 10	0 52	0 31	0 13	0 3	18
13	1 19	1 9	0 51	0 31	0 13	0 3	17
14	1 19	1 9	0 50	0 30	0 12	0 2	16
15	1 19	1 8	0 50	0 29	0 12	0 2	15
16	1 19	1 8	0 49	0 29	0 11	0 2	14
17	1 18	1 7	0 48	0 28	0 11	0 2	13
18	1 18	1 7	0 48	0 27	0 11	0 2	12
19	1 18	1 6	0 47	0 27	0 10	0 2	11
20	1 18	1 5	0 46	0 26	0 10	0 1	10
21	1 18	1 5	0 46	0 25	0 9	0 1	9
22	1 17	1 4	0 45	0 25	0 9	0 1	8
23	1 17	1 4	0 44	0 24	0 8	0 1	7
24	1 17	1 3	0 44	0 23	0 8	0 1	6
25	1 16	1 3	0 43	0 23	0 8	0 1	5
26	1 16	1 2	0 42	0 22	0 7	0 1	4
27	1 16	1 1	0 41	0 22	0 7	0 1	3
28	1 15	1 1	0 41	0 21	0 7	0 1	2
29	1 15	1 0	0 40	0 20	0 6	0 1	1
30	1 15	1 0	0 39	0 20	0 6	0 1	0
	XIs	Xs	IXs	VIIIs	VIIs	VIs	

Moon's Horary Motion in Longitude.

ARGUMENTS. Sum of preceding Equations and Anomaly, corrected.

	0″	10″	20″	30″	40″	50″	60″	70″	80″	90″	100″	
0ˢ 0°	4′	5″	6″	8″	9″	10″	11″	12″	14″	15	16″	XIIˢ 0°
5	4	5	6	8	9	10	11	12	14	15	16	25
10	4	5	7	8	9	10	11	12	13	15	16	20
15	4	5	7	8	9	10	11	12	13	15	16	15
20	5	6	7	8	9	10	11	12	13	14	15	10
25	5	6	7	8	9	10	11	12	13	14	15	5
I 0	5	6	7	8	9	10	11	12	13	14	15	XI 0
5	5	6	7	8	9	10	11	12	13	14	15	25
10	6	7	7	8	9	10	11	12	13	13	14	20
15	6	7	8	8	9	10	11	12	12	13	14	15
20	7	7	8	9	9	10	11	11	12	13	13	10
25	7	8	8	9	9	10	11	11	12	12	13	5
II 0	7	8	8	9	9	10	11	11	12	12	13	X 0
5	8	8	9	9	10	10	10	11	11	12	12	25
10	8	9	9	9	10	10	10	11	11	11	12	20
15	9	9	9	10	10	10	10	10	11	11	11	15
20	9	10	10	10	10	10	10	10	10	10	11	10
25	10	10	10	10	10	10	10	10	10	10	10	5
III 0	10	10	10	10	10	10	10	10	10	10	10	IX 0
5	11	11	11	10	10	10	10	10	9	9	9	25
10	11	11	11	11	10	10	10	9	9	9	9	20
15	12	11	11	11	10	10	10	9	9	9	8	15
20	12	12	11	11	10	10	10	9	9	8	8	10
25	13	12	12	11	11	10	9	9	8	8	7	5
IV 0	13	12	12	11	11	10	9	9	8	8	7	VIII 0
5	13	13	12	11	11	10	9	9	8	7	7	25
10	14	13	12	11	11	10	9	9	8	7	6	20
15	14	13	12	12	11	10	9	8	8	7	6	15
20	14	13	12	12	11	10	9	8	8	7	6	10
25	14	13	13	12	11	10	9	8	7	7	6	5
V 0	15	14	13	12	11	10	9	8	7	6	5	VII 0
5	15	14	13	12	11	10	9	8	7	6	5	25
10	15	14	13	12	11	10	9	8	7	6	5	20
15	15	14	13	12	11	10	9	8	7	6	5	15
20	15	14	13	12	11	10	9	8	7	6	5	10
25	15	14	13	12	11	10	9	8	7	6	5	5
VI 0	15	14	13	12	11	10	9	8	7	6	5	VI 0
	0″	10″	20″	30″	40″	50″	60″	70″	80″	90″	100″	

Moon's Horary Motion in Longitude.

ARGUMENT. Anomaly, corrected.

	0s	Is	IIs	IIIs	IVs	Vs	
0°	34′ 51″	34′ 14″	32′ 39″	30′ 45″	29′ 6″	28′ 1″	30°
1	34 51	34 12	32 36	30 42	29 3	27 59	29
2	34 51	34 9	32 32	30 38	29 0	27 58	28
3	34 51	34 7	32 28	30 34	28 58	27 56	27
4	34 51	34 4	32 24	30 31	28 55	27 55	26
5	34 50	34 1	32 21	30 27	28 52	27 54	25
6	34 50	33 59	32 17	30 23	28 50	27 53	24
7	34 49	33 56	32 13	30 20	28 47	27 51	23
8	34 49	33 53	32 9	30 16	28 45	27 50	22
9	34 48	33 50	32 5	30 13	28 42	27 49	21
10	34 47	33 47	32 2	30 9	28 40	27 48	20
11	34 46	33 44	31 58	30 6	28 37	27 47	19
12	34 45	33 41	31 54	30 2	28 35	27 46	18
13	34 44	33 38	31 50	29 59	28 33	27 45	17
14	34 43	33 35	31 46	29 56	28 30	27 45	16
15	34 42	33 32	31 42	29 52	28 28	27 44	15
16	34 41	33 28	31 38	29 49	28 26	27 43	14
17	34 39	33 25	31 35	29 46	28 24	27 42	13
18	34 38	33 22	31 31	29 42	28 22	27 42	12
19	34 36	33 18	31 27	29 39	28 20	27 41	11
20	34 34	33 15	31 23	29 36	28 18	27 41	10
21	34 33	33 12	31 19	29 33	28 16	27 40	9
22	34 31	33 8	31 15	29 30	28 14	27 40	8
23	34 29	33 5	31 12	29 26	28 12	27 39	7
24	34 27	33 1	31 8	29 23	28 10	27 39	6
25	34 25	32 58	31 4	29 20	28 9	27 39	5
26	34 23	32 54	31 0	29 17	28 7	27 39	4
27	34 21	32 50	30 57	29 14	28 5	27 38	3
28	34 19	32 47	30 53	29 12	28 4	27 38	2
29	34 16	32 43	30 49	29 9	28 2	27 38	1
30	34 14	32 39	30 45	29 6	28 1	27 38	0
	XIs	Xs	IXs	VIIIs	VIIs	VIs	

Moon's Horary Motion in Longitude.

ARGUMENTS. Sum of preceding Equations and Arg. of Variation.

	27′	28′	29′	30′	31′	32′	33′	34′	35′	36′	37′	
0ˢ 0°	0″	1″	2″	4″	5″	6″	7′	8″	10″	11″	12″	XIIˢ 0°
5	0	1	2	4	5	6	7	8	10	11	12	25
10	0	1	3	4	5	6	7	8	9	11	12	20
15	1	2	3	4	5	6	7	8	9	10	11	15
20	1	2	3	4	5	6	7	8	9	10	11	10
25	2	3	4	4	5	6	7	8	8	9	10	5
I 0	3	4	4	5	5	6	7	7	8	8	9	XI 0
5	4	4	5	5	6	6	6	7	7	8	8	25
10	5	5	5	6	6	6	6	6	7	7	7	20
15	6	6	6	6	6	6	6	6	6	6	6	15
20	7	7	7	7	6	6	6	5	5	5	5	10
25	8	8	7	7	6	6	6	5	5	4	4	5
II 0	9	9	8	7	7	6	5	5	4	3	3	X 0
5	10	9	8	8	7	6	5	4	4	3	2	25
10	11	10	9	8	7	6	5	4	3	2	1	20
15	11	10	9	8	7	6	5	4	3	2	1	15
20	12	11	10	8	7	6	5	4	2	1	0	10
25	12	11	10	8	7	6	5	4	2	1	0	5
III 0	12	11	10	8	7	6	5	4	2	1	0	IX 0
5	12	11	10	8	7	6	5	4	2	1	0	25
10	12	11	10	8	7	6	5	4	2	1	0	20
15	11	10	9	8	7	6	5	4	3	2	1	15
20	11	10	9	8	7	6	5	4	3	2	1	10
25	10	9	8	8	7	6	5	4	4	3	2	5
IV 0	9	8	8	7	7	6	5	5	4	4	3	VIII 0
5	8	8	7	7	6	6	6	5	5	4	4	25
10	7	7	7	6	6	6	6	6	5	5	5	20
15	6	6	6	6	6	6	6	6	6	6	6	15
20	5	5	5	6	6	6	6	6	7	7	7	10
25	4	4	5	5	6	6	6	7	7	8	8	5
V 0	3	3	4	5	5	6	7	7	8	9	9	VII 0
5	2	3	3	4	5	6	7	8	9	9	10	25
10	1	2	3	4	5	6	7	8	9	10	11	20
15	0	2	3	4	5	6	7	8	9	10	12	15
20	0	1	2	4	5	6	7	8	10	11	12	10
25	0	1	2	3	5	6	7	9	10	11	12	5
VI 0	0	1	2	3	5	6	7	9	10	11	12	VI 0
	27′	28′	29′	30′	31′	32′	33′	34′	35′	36′	37′	

Moon's Horary Motion in Longitude.

ARGUMENT. Argument of the Variation.

	0s	Is	IIs	IIIs	IVs	Vs	
0°	1′ 17″	0′ 58″	0′ 20″	0′ 2″	0′ 22″	1′ 0″	30°
1	1 17	0 57	0 19	0 2	0 23	1 1	29
2	1 17	0 55	0 18	0 3	0 24	1 2	28
3	1 17	0 54	0 17	0 3	0 25	1 3	27
4	1 17	0 53	0 16	0 3	0 26	1 4	26
5	1 17	0 52	0 15	0 3	0 27	1 5	25
6	1 16	0 51	0 14	0 3	0 29	1 6	24
7	1 16	0 49	0 13	0 4	0 30	1 7	23
8	1 16	0 48	0 12	0 4	0 31	1 8	22
9	1 15	0 47	0 11	0 4	0 33	1 9	21
10	1 15	0 45	0 11	0 5	0 34	1 0	20
11	1 14	0 44	0 10	0 5	0 35	1 11	19
12	1 14	0 43	0 9	0 6	0 37	1 12	18
13	1 13	0 41	0 8	0 6	0 38	1 13	17
14	1 13	0 40	0 8	0 7	0 39	1 13	16
15	1 12	0 39	0 7	0 8	0 40	1 14	15
16	1 11	0 38	0 6	0 8	0 42	1 15	14
17	1 11	0 36	0 6	0 9	0 43	1 15	13
18	1 10	0 35	0 5	0 10	0 44	1 16	12
19	1 9	0 34	0 5	0 11	0 46	1 16	11
20	1 8	0 32	0 4	0 11	0 47	1 17	10
21	1 7	0 31	0 4	0 12	0 48	1 17	9
22	1 6	0 30	0 4	0 13	0 50	1 18	8
23	1 5	0 29	0 3	0 14	0 51	1 18	7
24	1 4	0 27	0 3	0 15	0 52	1 18	6
25	1 3	0 26	0 3	0 16	0 54	1 19	5
26	1 2	0 25	0 3	0 17	0 55	1 19	4
27	1 1	0 24	0 3	0 18	0 56	1 19	3
28	1 0	0 23	0 2	0 19	0 57	1 19	2
29	0 59	0 21	0 2	0 20	0 59	1 19	1
30	0 58	0 20	0 2	0 22	1 0	1 19	0
	XIs	Xs	IXs	VIIIs	VIIs	Vs	

Moon's Horary Motion in Longitude.

ARGUMENT. Argument of the Reduction.

	0s	Is	IIs	IIIs	IVs	Vs	
0°	2″	6″	14″	18″	14″	6′	30°
1	2	6	14	18	14	6	29
2	2	7	14	18	13	6	28
3	2	7	15	18	13	5	27
4	2	7	15	18	13	5	26
5	2	7	15	18	13	5	25
6	2	8	15	18	12	5	24
7	2	8	16	18	12	4	23
8	2	8	16	18	12	4	22
9	2	8	16	18	12	4	21
10	3	9	16	17	11	4	20
11	3	9	16	17	11	4	19
12	3	9	16	17	11	4	18
13	3	9	17	17	11	3	17
14	3	10	17	17	10	3	16
15	3	10	17	17	10	3	15
16	3	10	17	17	10	3	14
17	3	11	17	17	9	3	13
18	4	11	17	16	9	3	12
19	4	11	17	16	9	3	11
20	4	11	17	16	9	3	10
21	4	12	18	16	8	2	9
22	4	12	18	16	8	2	8
23	4	12	18	16	8	2	7
24	5	12	18	15	8	2	6
25	5	13	18	15	7	2	5
26	5	13	18	15	7	2	4
27	5	13	18	15	7	2	3
28	6	13	18	14	7	2	2
29	6	14	18	14	6	2	1
30	6	14	18	14	6	2	0
	XIs	Xs	IXs	VIIIs	VIIs	VIs	

Moon's Horary Motion in Latitude.

ARGUMENT. Arg. I. of Latitude.

	0ˢ+	Iˢ+	IIˢ+	IIIˢ—	IVˢ—	Vˢ—	
0°	2′ 58″	2′ 34″	1′ 29″	0′ 0″	1′ 29″	2′ 34″	30°
1	2 58	2 33	1 27	0 3	1 32	2 36	29
2	2 58	2 31	1 24	0 6	1 35	2 37	28
3	2 58	2 29	1 21	0 9	1 37	2 39	27
4	2 58	2 28	1 18	0 12	1 40	2 40	26
5	2 57	2 26	1 15	0 16	1 42	2 41	25
6	2 57	2 24	1 13	0 19	1 45	2 43	24
7	2 57	2 22	1 10	0 22	1 47	2 44	23
8	2 56	2 20	1 7	0 25	1 50	2 45	22
9	2 56	2 19	1 4	0 28	1 52	2 46	21
10	2 55	2 17	1 1	0 31	1 55	2 47	20
11	2 55	2 15	0 58	0 34	1 57	2 48	19
12	2 54	2 12	0 55	0 37	1 59	2 49	18
13	2 53	2 10	0 52	0 40	2 2	2 50	17
14	2 53	2 8	0 49	0 43	2 4	2 51	16
15	2 52	2 6	0 46	0 46	2 6	2 52	15
16	2 51	2 4	0 43	0 49	2 8	2 53	14
17	2 50	2 2	0 40	0 52	2 10	2 53	13
18	2 49	1 59	0 37	0 55	2 12	2 54	12
19	2 48	1 57	0 34	0 58	2 15	2 55	11
20	2 47	1 55	0 31	1 1	2 17	2 55	10
21	2 46	1 52	0 28	1 4	2 19	2 56	9
22	2 45	1 50	0 25	1 7	2 20	2 56	8
23	2 44	1 47	0 22	1 10	2 22	2 57	7
24	2 43	1 45	0 19	1 13	2 24	2 57	6
25	2 41	1 42	0 16	1 15	2 26	2 57	5
26	2 40	1 40	0 12	1 18	2 28	2 58	4
27	2 39	1 37	0 9	1 21	2 29	2 58	3
28	2 37	1 35	0 6	1 24	2 31	2 58	2
29	2 36	1 32	0 3	1 27	2 33	2 58	1
30	2 34	1 29	0 0	1 29	2 34	2 58	0
	XIˢ+	Xˢ+	IXˢ+	VIIIˢ—	VIIˢ—	VIˢ—	

TABLE LXIV.

Moon's Horary Motion in Latitude.

ARGUMENT. Arg. II. of Latitude.

	0ˢ+	Iˢ+	IIˢ+	IIIˢ—	IVˢ—	Vˢ—	
0	4″	4″	2″	0″	2″	4″	30°
6	4	3	2	0	3	4	24
12	4	3	1	1	3	4	18
18	4	3	1	1	3	4	12
24	4	3	0	2	3	4	6
30	4	2	0	2	4	4	0
	XIˢ+	Xˢ+	IXˢ+	VIIIˢ—	VIIˢ—	VIˢ—	

MAY, 1836.

AT APPARENT NOON.

Day of the Week.	Day of the Month.	THE SUN'S Right Ascension.	Diff. for 1 hour.	Declination.	Diff. for 1 hour.	Sidereal Time of the Semidiameter passing the Meridian.*	Equation of Time, *to be subtracted from Apparent Time.*	Diff. for 1 hour.
		h. m. s.	s.	° ′ ″	″	m. s.	m. s.	s.
Sun...	1	2 34 39 .57	9 .551	N.15 10 19 .0	44 .85	1 6 .00	3 6 .12	0 .305
Mon..	2	2 38 28 .79	9 .574	15 28 15 .4	44 .22	1 6 .08	3 13 .43	0 .283
Tues..	3	2 42 18 .56	9 .598	15 45 56 .6	43 .57	1 6 .16	3 20 .21	0 .259
Wed..	4	2 46 8 .90	9 .622	16 3 22 .2	42 .91	1 6 .24	3 26 .42	0 .234
Thur.	5	2 49 59 .82	9 .645	16 20 32 .1	42 .24	1 6 .32	3 32 .04	0 .211
Frid..	6	2 53 51 .31	9 .670	16 37 25 .8	41 .55	1 6 .40	3 37 .10	0 .186
Sat. ..	7	2 57 43 .38	9 .695	16 54 3 .1	40 .86	1 6 .48	3 41 .57	0 .161
Sun...	8	3 1 36 .05	9 .719	17 10 23 .7	40 .15	1 6 .56	3 45 .44	0 .137
Mon..	9	3 5 29 .31	9 .744	17 26 27 .2	39 .43	1 6 .64	3 48 .73	0 .112
Tues..	10	3 9 23 .16	9 .769	17 42 13 .4	38 .69	1 6 .72	3 51 .42	0 .088
Wed..	11	3 13 17 .61	9 .793	17 57 41 .9	37 .94	1 6 .81	3 53 .53	0 .063
Thur.	12	3 17 12 .64	9 .818	18 12 52 .5	37 .18	1 6 .89	3 55 .04	0 .038
Frid..	13	3 21 8 .27	9 .842	18 27 44 .9	36 .40	1 6 .97	3 55 .96	0 .014
Sat. ..	14	3 25 4 .48	9 .867	18 42 18 .6	35 .62	1 7 .05	3 56 .30	0 .010
Sun...	15	3 29 1 .28	9 .890	18 56 33 .6	34 .82	1 7 .13	3 56 .05	0 .034
Mon..	16	3 32 58 .65	9 .915	19 10 29 .4	34 .01	1 7 .21	3 55 .24	0 .058
Tues..	17	3 36 56 .61	9 .937	19 24 5 .6	33 .20	1 7 .29	3 53 .86	0 .080
Wed..	18	3 40 55 .10	9 .960	19 37 22 .3	32 .36	1 7 .37	3 51 .94	0 .103
Thur.	19	3 44 54 .14	9 .983	19 50 18 .9	31 .51	1 7 .45	3 49 .46	0 .126
Frid..	20	3 48 53 .73	10 .005	20 2 55 .1	30 .66	1 7 .53	3 46 .44	0 .148
Sat. ..	21	3 52 53 .85	10 .026	20 15 10 .9	29 .79	1 7 .60	3 42 .89	0 .169
Sun...	22	3 56 54 .47	10 .047	20 27 5 .8	28 .91	1 7 .68	3 38 .84	0 .190
Mon..	23	4 0 55 .60	10 .068	20 38 39 .7	28 .02	1 7 .75	3 34 .28	0 .211
Tues..	24	4 4 57 .22	10 .089	20 49 52 .2	27 .13	1 7 .83	3 29 .22	0 .232
Wed..	25	4 8 59 .35	10 .108	21 0 43 .3	26 .22	1 7 .90	3 23 .66	0 .252
Thur.	26	4 13 1 .95	10 .128	21 11 12 .6	25 .30	1 7 .96	3 17 .62	0 .270
Frid..	27	4 17 5 .02	10 .148	21 21 19 .9	24 .38	1 8 .03	3 11 .13	0 .290
Sat. ..	28	4 21 8 .56	10 .166	21 31 5 .1	23 .45	1 8 .10	3 4 .17	0 .309
Sun...	29	4 25 12 .55	10 .185	21 40 28 .0	22 .52	1 8 .16	2 56 .76	0 .328
Mon..	30	4 29 16 .99	10 .203	21 49 28 .4	21 .58	1 8 .22	2 48 .90	0 .346
Tues..	31	4 33 21 .86	10 .221	21 58 6 .2	20 .62	1 8 .28	2 40 .60	0 .363
Wed..	32	4 37 27 .17		N.22 6 21 .1		1 8 .34	2 31 .89	

* *Mean* Time of the Semidiameter passing may be found by subtracting 0s. 18 from the *Sidereal* Time.

MAY, 1836. II.

AT MEAN NOON.

Day of the Week.	Day of the Month.	THE SUN'S Right Ascension.	Declination.	Semidiam.*	Equation of Time, *to be added to Mean Time.*	Sidereal Time.
		h. m. s.	° ′ ″	′ ″	m. s.	h. m. s.
Sun...	1	2 34 40 .06	N.15 10 21 .3	15 52 .9	3 6 .14	2 37 46 .20
Mon..	2	2 38 29 .30	15 28 17 .8	15 52 .7	3 13 .45	2 41 42 .75
Tues..	3	2 42 19 .09	15 45 59 .0	15 52 .5	3 20 .22	2 45 39 .31
Wed..	4	2 46 9 .45	16 3 24 .7	15 52 .2	3 26 .43	2 49 35 .88
Thur.	5	2 50 0 .39	16 20 34 .6	15 52 .0	3 32 .05	2 53 32 .44
Frid..	6	2 53 51 .89	16 37 28 .3	15 51 .8	3 37 .11	2 57 29 .00
Sat. ..	7	2 57 43 .98	16 54 5 .6	15 51 .5	3 41 .58	3 1 25 .56
Sun...	8	3 1 36 .66	17 10 26 .2	15 51 .3	3 45 .45	3 5 22 .11
Mon..	9	3 5 29 .93	17 26 29 .7	15 51 .1	3 48 .74	3 9 18 .67
Tues..	10	3 9 23 .79	17 42 15 .9	15 50 .9	3 51 .43	3 13 15 .22
Wed..	11	3 13 18 .24	17 57 44 .4	15 50 .7	3 53 .53	3 17 11 .77
Thur.	12	3 17 13 .28	18 12 55 .0	15 50 .5	3 55 .04	3 21 8 .32
Frid..	13	3 21 8 .91	18 27 47 .3	15 50 .3	3 55 .96	3 25 4 .87
Sat. ..	14	3 25 5 .13	18 42 21 .0	15 50 .1	3 56 .30	3 29 1 .43
Sun...	15	3 29 1 .93	18 56 35 .9	15 49 .9	3 56 .05	3 32 57 .98
Mon..	16	3 32 59 .30	19 10 31 .6	15 49 .7	3 55 .24	3 36 54 .54
Tues..	17	3 36 57 .25	19 24 7 .8	15 49 .5	3 53 .86	3 40 51 .11
Wed..	18	3 40 55 .74	19 37 24 .4	15 49 .3	3 51 .93	3 44 47 .67
Thur.	19	3 44 54 .78	19 50 20 .9	15 49 .1	3 49 .45	3 48 44 .23
Frid..	20	3 48 54 .36	20 2 57 .1	15 48 .9	3 46 .43	3 52 40 .79
Sat. ..	21	3 52 54 .47	20 15 12 .8	15 48 .8	3 42 .88	3 56 37 .35
Sun...	22	3 56 55 .08	20 27 7 .6	15 48 .6	3 38 .83	4 0 33 .91
Mon..	23	4 0 56 .20	20 38 41 .4	15 48 .5	3 34 .27	4 4 30 .47
Tues..	24	4 4 57 .81	20 49 53 .8	15 48 .3	3 29 .21	4 8 27 .02
Wed..	25	4 8 59 .92	21 0 44 .8	15 48 .1	3 23 .65	4 12 23 .57
Thur.	26	4 13 2 .51	21 11 14 .0	15 48 .0	3 17 .61	4 16 20 .12
Frid..	27	4 17 5 .56	21 21 21 .2	15 47 .8	3 11 .12	4 20 16 .68
Sat. ..	28	4 21 9 .08	21 31 6 .3	15 47 .7	3 4 .15	4 24 13 .23
Sun...	29	4 25 13 .05	21 40 29 .1	15 47 .5	2 56 .74	4 28 9 .79
Mon..	30	4 29 17 .47	21 49 29 .4	15 47 .4	2 48 .88	4 32 6 .35
Tues..	31	4 33 22 .32	21 58 7 .1	15 47 .3	2 40 .59	4 36 2 .91
Wed..	32	4 37 27 .60	N.22 6 21 .9	15 47 .1	2 31 .88	4 39 59 .48

* The Semidiameter for *Apparent* Noon may be assumed the same as that for *Mean* Noon.

III. MAY, 1836.

MEAN TIME.

Day of the Month.	THE SUN'S Longitude.	THE SUN'S Latitude.	Logarithm of the Radius Vector of the Earth.	THE MOON'S Semidiameter.		THE MOON'S Horizontal Parallax.	
	Noon.	Noon.	Noon.	Noon.	Midnight.	Noon.	Midnight.
	° ′ ″	″		′ ″	′ ″	′ ″	′ ″
1	41 5 56 .6	N.0 .40	0 .0036278	16 27 .2	16 29 .8	60 22 .7	60 32 .2
2	42 4 3 .3	0 .27	0 .0037345	16 31 .3	16 31 .8	60 37 .9	60 39 .5
3	43 2 8 .4	0 .14	0 .0038404	16 31 .2	16 29 .5	60 37 .3	60 31 .2
4	44 0 12 .0	N.0 .02	0 .0039458	16 26 .9	16 23 .5	60 21 .7	60 9 .1
5	44 58 14 .3	S.0 .08	0 .0040506	16 19 .4	16 14 .7	59 54 .1	59 36 .8
6	45 56 15 .0	0 .15	0 .0041544	16 9 .5	16 4 .1	59 17 .8	58 57 .9
7	46 54 14 .4	0 .20	0 .0042572	15 58 .5	15 52 .8	58 37 .6	58 16 .5
8	47 52 12 .4	0 .23	0 .0043587	15 47 .2	15 41 .6	57 55 .8	57 35 .4
9	48 50 9 .2	0 .22	0 .0044588	15 36 .2	15 31 .0	57 15 .7	56 56 .6
10	49 48 4 .7	0 .18	0 .0045575	15 26 .0	15 21 .3	56 38 .2	56 20 .8
11	50 45 58 .8	0 .10	0 .0046545	15 16 .8	15 12 .5	56 4 .4	55 48 .6
12	51 43 51 .7	S.0 .01	0 .0047499	15 8 .5	15 4 .7	55 33 .9	55 20 .1
13	52 41 43 .3	N.0 .10	0 .0048433	15 1 .2	14 57 .9	55 7 .3	54 55 .2
14	53 39 33 .6	0 .23	0 .0049349	14 55 .0	14 52 .3	54 44 .3	54 34 .5
15	54 37 22 .6	0 .37	0 .0050244	14 49 .9	14 47 .7	54 25 .6	54 17 .7
16	55 35 10 .2	0 .50	0 .0051119	14 45 .9	14 44 .5	54 11 .1	54 5 .7
17	56 32 56 .5	0 .63	0 .0051972	14 43 .4	14 42 .7	54 1 .8	53 59 .4
18	57 30 41 .2	0 .74	0 .0052805	14 42 .5	14 42 .7	53 58 .4	53 59 .4
19	58 28 24 .5	0 .83	0 .0053616	14 43 .5	14 44 .7	54 2 .0	54 6 .6
20	49 26 6 .2	0 .90	0 .0054407	14 46 .5	14 48 .9	54 13 .2	54 22 .1
21	60 23 46 .4	0 .94	0 .0055181	14 52 .0	14 55 .7	54 33 .4	54 46 .8
22	61 21 25 .0	0 .95	0 .0055936	14 59 .9	15 4 .9	55 2 .4	55 20 .6
23	62 19 2 .0	0 .93	0 .0056674	15 10 .4	15 16 .5	55 41 .0	56 3 .4
24	63 16 37 .4	0 .88	0 .0057396	15 23 .2	15 30 .3	56 27 .8	56 54 .0
25	64 14 11 .5	0 .80	0 .0058103	15 37 .8	15 45 .5	57 21 .5	57 49 .9
26	65 11 44 .1	0 .70	0 .0058798	15 53 .4	16 1 .2	58 18 .7	58 47 .3
27	66 9 15 .4	0 .58	0 .0059480	16 8 .8	16 15 .9	59 15 .2	59 41 .4
28	67 6 45 .4	0 .45	0 .0060147	16 22 .6	16 28 .4	60 5 .7	60 27 .1
29	68 4 14 .3	0 .32	0 .0060804	16 33 .3	16 37 .1	60 45 .1	60 59 .1
30	69 1 42 .2	0 .18	0 .0061448	16 39 .7	16 41 .0	61 8 .7	61 13 .5
31	69 59 9 .1	N.0 .06	0 .0062079	16 41 .1	16 39 .9	61 13 .8	61 9 .3
32	70 56 35 .1	S.0 .05	0 .0062700	16 37 .4	16 33 .8	61 0 .0	60 46 .9

MAY, 1836. IV.

MEAN TIME.

Day of the Week.	Day of the Month.	THE MOON'S Longitude. Noon.	Longitude. Midnight.	Latitude. Noon.	Latitude. Midnight.	Age. Noon.	Meridian Passage.
		° ′ ″	° ′ ″	° ′ ″	° ′ ″	d.	h. m.
Sun...	1	223 24 44 .6	230 48 24 .7	N.0 36 44 .1	S 0 4 13 .5	15 .5	12 35 .0
Mon..	2	238 14 0 .4	245 40 32 .3	S. 0 45 18 .0	1 25 41 .7	16 .5	13 34 .1
Tues..	3	253 7 3 .7	260 32 39 .3	2 4 37 .8	2 41 22 .2	17 .5	14 37 .1
Wed..	4	267 56 26 .2	275 17 38 .7	3 15 14 .8	3 45 41 .1	18 .5	15 41 .7
Thur.	5	282 35 36 .1	289 49 43 .9	4 12 12 .5	4 34 27 .4	19 .5	16 45 .1
Frid..	6	296 59 36 .8	304 4 54 .5	4 52 10 .9	5 5 14 .4	20 .5	17 44 .8
Sat. ..	7	311 5 24 .5	318 1 1 .4	5 13 34 .8	5 17 14 .5	21 .5	18 39 .5
Sun...	8	324 51 42 .1	331 37 31 .7	5 16 20 .2	5 11 2 .0	22 .5	19 29 .6
Mon..	9	338 18 35 .0	344 55 4 .2	5 1 33 .1	4 48 10 .1	23 .5	20 15 .8
Tues..	10	351 27 10 .2	357 55 6 .3	4 31 10 .3	4 10 53 .2	24 .5	20 59 .5
Wed..	11	4 19 6 .3	10 39 26 .4	3 47 39 .3	3 21 50 .0	25 .5	21 41 .8
Thur.	12	16 56 20 .2	23 10 2 .7	2 53 47 .4	2 23 54 .4	26 .5	22 24 .0
Frid..	13	29 20 48 .0	35 28 50 .3	1 52 33 .8	1 20 8 .3	27 .5	23 7 .0
Sat. ..	14	41 34 23 .5	47 37 41 .1	S. 0 47 0 .9	S. 0 13 33 .9	28 .5	23 51 .5
Sun...	15	53 38 56 .6	59 38 24 .2	N.0 19 50 .8	N.0 52 51 .9	29 .5	☌
Mon..	16	65 36 18 .6	71 32 55 .4	1 25 8 .9	1 56 23 .0	0 .9	0 38 .0
Tues..	17	77 28 30 .4	83 23 21 .6	2 26 15 .0	2 54 28 .7	1 .9	1 26 .6
Wed..	18	89 17 47 .4	95 12 9 .2	3 20 47 .4	3 44 56 .7	2 .9	2 16 .7
Thur.	19	101 6 48 .1	107 2 8 .7	4 6 42 .4	4 25 52 .1	3 .9	3 7 .5
Frid..	20	112 58 35 .7	118 56 36 .5	4 42 13 .3	4 55 36 .2	4 .9	3 58 .0
Sat. ..	21	124 56 39 .4	130 59 14 .3	5 5 49 .6	5 12 45 .1	5 .9	4 47 .3
Sun...	22	137 4 51 .3	143 14 2 .9	5 16 13 .2	5 16 7 .2	6 .9	5 35 .0
Mon..	23	149 27 19 .5	155 45 12 .4	5 12 19 .7	5 4 46 .3	7 .9	6 21 .1
Tues..	24	162 8 10 .8	168 36 42 .7	4 53 22 .1	4 38 5 .5	8 .9	7 6 .2
Wed..	25	175 11 12 .1	181 52 0 .5	4 18 57 .4	3 56 2 .2	9 .9	7 51 .1
Thur.	26	188 39 21 .7	195 33 24 .6	3 29 26 .4	2 59 23 .7	10 .9	8 37 .2
Frid..	27	202 34 10 .3	209 41 29 .8	2 26 11 .1	1 50 12 .6	11 .9	9 25 .6
Sat. ..	28	216 55 6 .2	224 14 29 .8	N.1 11 57 .4	N.0 32 2 .0	12 .9	10 17 .8
Sun...	29	231 39 3 .1	239 7 55 .3	S. 0 8 53 .2	S. 0 50 1 .5	13 9	11 14 .9
Mon..	30	246 40 8 .5	254 14 37 .0	1 30 34 .1	2 9 41 .7	14 .9	12 17 .2
Tues..	31	261 50 9 .7	269 25 32 .6	2 46 35 .6	3 20 29 .8	15 .9	13 23 .1
Wed..	32	276 59 33 .8	284 31 2 .3	S. 3 50 45 .6	S. 4 16 49 .7	16 .9	14 29 .7

MEAN TIME.

THE MOON'S RIGHT ASCENSION AND DECLINATION.

Hour.	Right Ascension.	Declination.	Diff. Dec. for 10m.	Hour.	Right Ascension.	Declination.	Diff. Dec. for 10m.
	FRIDAY 13.				SUNDAY 15.		
	h. m. s.	° ′ ″	″		h. m. s.	° ′ ″	″
0	1 51 49 .41	N. 9 29 48 .4	132 .95	0	3 24 41 .69	N.19 1 24 .7	101 .38
1	1 53 42 .44	9 43 6 .1	132 .47	1	3 26 41 .88	19 11 33 .0	100 .52
2	1 55 35 .55	9 56 20 .9	132 .02	2	3 28 42 .26	19 21 36 .1	99 .65
3	1 57 28 .75	10 9 33 .0	131 .53	3	3 30 42 .84	19 31 34 .0	98 .77
4	1 59 22 .05	10 22 42 .2	131 .05	4	3 32 43 .62	19 41 26 .6	97 .88
5	2 1 15 .44	10 35 48 .5	130 .55	5	3 34 44 .60	19 51 13 .9	97 .00
6	2 3 8 .93	10 48 51 .8	130 .05	6	3 36 45 .77	20 0 55 .9	96 .10
7	2 5 2 .53	11 1 52 .1	129 .53	7	3 38 47 .14	20 10 32 .5	95 .18
8	2 6 56 .23	11 14 49 .3	129 .02	8	3 40 48 .72	20 20 3 .6	94 .27
9	2 8 50 .04	11 27 43 .4	128 .48	9	3 42 50 .49	20 29 29 .2	93 .35
10	2 10 43 .96	11 40 34 .3	127 .95	10	3 44 52 .46	20 38 49 .3	92 .42
11	2 12 37 .99	11 53 22 .0	127 .42	11	3 46 54 .64	20 48 3 .8	91 .48
12	2 14 32 .14	12 6 6 .5	126 .85	12	3 48 57 .01	20 57 12 .7	90 .53
13	2 16 26 .42	12 18 47 .6	126 .28	13	3 50 59 .59	21 6 15 .9	89 .58
14	2 18 20 .81	12 31 25 .3	125 .73	14	3 53 2 .37	21 15 13 .4	88 .62
15	2 20 15 .34	12 43 59 .7	125 .13	15	3 55 5 .34	21 24 5 .1	87 .63
16	2 22 9 .99	12 56 30 .5	124 .55	16	3 57 8 .52	21 32 50 .9	86 .68
17	2 24 4 .78	13 8 57 .8	123 .97	17	3 59 11 .91	21 41 31 .0	85 .68
18	2 25 59 .70	13 21 21 .6	123 .35	18	4 1 15 .49	21 50 5 .1	84 .70
19	2 27 54 .76	13 33 41 .7	122 .75	19	4 3 19 .27	21 58 33 .3	83 .68
20	2 29 49 .97	13 45 58 .2	122 .12	20	4 5 23 .25	22 6 55 .4	82 .70
21	2 31 45 .31	13 58 10 .9	121 .48	21	4 7 27 .44	22 15 11 .6	81 .67
22	2 33 40 .80	14 10 19 .8	120 .87	22	4 9 31 .82	22 23 21 .6	80 .67
23	2 35 36 .44	N.14 22 25 .0	120 .20	23	4 11 36 .40	N.22 31 25 .6	79 .63
	SATURDAY 14.				MONDAY 16.		
0	2 37 32 .22	N.14 34 26 .2	119 .57	0	4 13 41 .17	N.22 39 23 .4	78 .58
1	2 39 28 .16	14 46 23 .6	118 .88	1	4 15 46 .15	22 47 14 .9	77 .57
2	2 41 24 .25	14 58 16 .9	118 .23	2	4 17 51 .32	22 55 0 .3	76 .50
3	2 43 20 .51	15 10 6 .3	117 .53	3	4 19 56 .69	23 2 39 .3	75 .45
4	2 45 16 .92	15 21 51 .5	116 .87	4	4 22 2 .25	23 10 12 .0	74 .38
5	2 47 13 .49	15 33 32 .7	116 .15	5	4 24 8 .01	23 17 38 .3	73 .32
6	2 49 10 .23	15 45 9 .6	115 .47	6	4 26 13 .96	23 24 58 .2	72 .25
7	2 51 7 .14	15 56 42 .4	114 .75	7	4 28 20 .10	23 32 11 .7	71 .15
8	2 53 4 .21	16 8 10 .9	114 .02	8	4 30 26 .43	23 39 18 .6	70 .07
9	2 55 1 .45	16 19 35 .0	113 .30	9	4 32 32 .95	23 46 19 .0	68 .97
10	2 56 58 .87	16 30 54 .8	112 .55	10	4 34 39 .65	23 53 12 .8	67 .87
11	2 58 56 .46	16 42 10 .1	111 .82	11	4 36 46 .55	24 0 0 .0	66 .75
12	3 0 54 .22	16 53 21 .0	111 .07	12	4 38 53 .62	24 6 40 .5	65 .63
13	3 2 52 .16	17 4 27 .4	110 .30	13	4 41 0 .88	24 13 14 .3	64 .52
14	3 4 50 .28	17 15 29 .2	109 .52	14	4 43 8 .32	24 19 41 .4	63 .38
15	3 6 48 .58	17 26 26 .3	108 .75	15	4 45 15 .93	24 26 1 .7	62 .25
16	3 8 47 .07	17 37 18 .8	107 .97	16	4 47 23 .72	24 32 15 .2	61 .10
17	3 10 45 .73	17 48 6 .6	107 .17	17	4 49 31 .69	24 38 21 .8	59 .95
18	3 12 44 .59	17 58 49 .6	106 .37	18	4 51 39 .83	24 44 21 .5	58 .82
19	3 14 43 .63	18 9 27 .8	105 .55	19	4 53 48 .14	24 50 14 .4	57 .65
20	3 16 42 .86	18 20 1 .1	104 .73	20	4 55 56 .62	24 56 0 .3	56 48
21	3 18 42 .28	18 30 29 .5	103 .90	21	4 58 5 .27	25 1 39 .2	55 .32
22	3 20 41 .89	18 40 52 .9	103 .08	22	5 0 14 .07	25 7 11 .1	54 .13
23	3 22 41 .69	18 51 11 .4	102 .22	23	5 2 23 .04	25 12 35 .9	52 .97
24	3 24 41 .69	N.19 1 24 .7		24	5 4 32 .17	N.25 17 53 .7	

Second Differences.

Hours & Min.		1′	2′	3′	4′	5′	6′	7′	8′	9′	10′	11′
h. m.	h. m.	″	″	″	″	″	″	″	″	″	″	″
0 0	12 0	0.0	0.0	0.0	0.0	0.0	0.0	0.0	0.0	0.0	0.0	0.0
0 10	11 50	0.4	0.8	1.2	1.6	2.0	2.4	2.9	3.3	3.7	4.1	4.5
0 20	11 40	0.8	1.6	2.4	3.2	4.1	4.9	5.7	6.5	7.3	8.1	8.9
0 30	11 30	1.2	2.4	3.6	4.8	6.0	7.2	8.4	9.6	10.8	12.0	13.2
0 40	11 20	1.6	3.1	4.7	6.3	7.9	9.4	11.0	12.6	14.2	15.7	17.3
0 50	11 10	1.9	3.9	5.8	7.8	9.7	11.6	13.6	15.5	17.4	19.4	21.4
1 0	11 0	2.3	4.6	6.9	9.2	11.5	13.8	16.0	18.3	20.6	22.9	25.2
1 10	10 50	2.6	5.3	7.9	10.5	13.2	15.8	18.4	21.1	23.7	26.3	29.0
1 20	10 40	3.0	5.9	8.9	11.9	14.8	17.8	20.7	23.7	26.7	29.6	32.6
1 30	10 30	3.3	6.6	9.8	13.1	16.4	19.7	23.0	26.3	29.5	32.8	36.1
1 40	10 20	3.6	7.2	10.8	14.4	17.9	21.5	25.1	28.7	32.3	35.9	39.5
1 50	10 10	3.9	7.8	11.6	15.5	19.4	23.3	27.2	31.0	34.9	38.8	42.7
2 0	10 0	4.2	8.3	12.5	16.7	20.8	25.0	29.2	33.3	37.5	41.7	45.8
2 10	9 50	4.4	8.9	13.3	17.8	22.2	26.6	31.1	35.5	40.0	44.4	48.8
2 20	9 40	4.7	9.4	14.1	18.8	23.5	28.2	32.9	37.6	42.3	47.0	51.7
2 30	9 30	4.9	9.9	14.8	19.8	24.7	29.7	34.6	39.6	44.5	49.5	54.4
2 40	9 20	5.2	10.4	15.6	20.7	25.9	31.1	36.3	41.5	46.7	51.9	57.0
2 50	9 10	5.4	10.8	16.2	21.6	27.1	32.5	37.9	43.3	48.7	54.1	59.5
3 0	9 0	5.6	11.3	16.9	22.5	28.1	33.8	39.4	45.0	50.6	56.3	61.9
3 10	8 50	5.8	11.7	17.5	23.3	29.1	35.0	40.8	46.6	52.4	58.3	64.1
3 20	8 40	6.0	12.0	18.1	24.1	30.1	36.1	42.1	48.1	54.2	60.2	66.2
3 30	8 30	6.2	12.4	18.6	24.8	31.0	37.2	43.4	49.6	55.8	62.0	68.2
3 40	8 20	6.4	12.7	19.1	25.5	31.8	38.2	44.6	50.9	57.3	63.7	70.0
3 50	8 10	6.5	13.0	19.6	26.1	32.6	39.1	45.7	52.2	58.7	65.2	71.7
4 0	8 0	6.7	13.3	20.0	26.7	33.3	40.0	46.7	53.3	60.0	66.7	73.3
4 10	7 50	6.8	13.6	20.4	27.2	34.0	40.8	47.6	54.4	61.2	68.0	74.8
4 20	7 40	6.9	13.8	20.8	27.7	34.6	41.5	48.4	55.4	62.3	69.2	76.1
4 30	7 30	7.0	14.1	21.1	28.1	35.2	42.2	49.2	56.2	63.3	70.3	77.3
4 40	7 20	7.1	14.3	21.4	28.5	35.6	42.8	49.9	57.0	64.2	71.3	78.4
4 50	7 10	7.2	14.4	21.6	28.0	36.1	43.3	50.5	57.7	64.9	72.2	79.4
5 0	7 0	7.3	14.6	21.9	29.2	36.5	43.8	51.0	58.3	65.6	72.9	80.2
5 10	6 50	7.4	14.7	22.1	29.4	36.8	44.1	51.5	58.8	66.2	73.6	80.9
5 20	6 40	7.4	14.8	22.2	29.6	37.0	44.4	51.9	59.3	66.7	74.1	81.5
5 30	6 30	7.4	14.9	22.3	29.8	37.2	44.7	52.1	59.6	67.0	74.5	81.9
5 40	6 20	7.5	15.0	22.4	29.9	37.4	44.9	52.3	59.8	67.3	74.8	82.2
5 50	6 10	7.5	15.0	22.5	30.0	37.5	45.0	52.5	60.0	67.4	74.9	82.4
6 0	6 0	7.5	15.0	22.5	30.0	37.5	45.0	52.5	60.0	67.5	75.0	82.5

Second Differences.

Hours & Min.		10″	20″	30″	40″	50″	1″	2″	3″	4″	5″	6″	7″	8″	9″
h. m.	h. m.	″	″	″	″	″	″	″	″	″	″	″	″	″	″
0 0	12 0	0.0	0.0	0.0	0.0	0.0	0.0	0.0	0.0	0.0	0.0	0.0	0.0	0.0	0.0
0 10	11 50	0.1	0.1	0.2	0.3	0.3	0.0	0.0	0.0	0.0	0.0	0.0	0.0	0.1	0.1
0 20	11 40	0.1	0.3	0.4	0.5	0.7	0.0	0.0	0.0	0.1	0.1	0.1	0.1	0.1	0.1
0 30	11 30	0.2	0.4	0.6	0.8	1.0	0.0	0.0	0.1	0.1	0.1	0.1	0.1	0.2	0.2
0 40	11 20	0.3	0.5	0.8	1.0	1.3	0.0	0.1	0.1	0.1	0.1	0.2	0.2	0.2	0.2
0 50	11 10	0.3	0.6	1.0	1.3	1.6	0.0	0.1	0.1	0.1	0.2	0.2	0.2	0.3	0.3
1 0	11 0	0.4	0.8	1.1	1.5	1.9	0.0	0.1	0.1	0.2	0.2	0.2	0.3	0.3	0.3
1 10	10 50	0.4	0.9	1.3	1.8	2.2	0.0	0.1	0.1	0.2	0.2	0.3	0.3	0.4	0.4
1 20	10 40	0.5	1.0	1.5	2.0	2.5	0.0	0.1	0.1	0.2	0.2	0.3	0.3	0.4	0.4
1 30	10 30	0.5	1.1	1.6	2.2	2.7	0.1	0.1	0.2	0.2	0.3	0.3	0.4	0.4	0.5
1 40	10 20	0.6	1.2	1.8	2.4	3.0	0.1	0.1	0.2	0.2	0.3	0.4	0.4	0.5	0.5
1 50	10 10	0.6	1.3	1.9	2.6	3.2	0.1	0.1	0.2	0.3	0.3	0.4	0.5	0.5	0.6
2 0	10 0	0.7	1.4	2.1	2.8	3.5	0.1	0.1	0.2	0.3	0.3	0.4	0.5	0.6	0.6
2 10	9 50	0.7	1.5	2.2	3.0	3.7	0.1	0.1	0.2	0.3	0.4	0.4	0.5	0.6	0.7
2 20	9 40	0.8	1.6	2.3	3.1	3.9	0.1	0.2	0.2	0.3	0.4	0.5	0.5	0.6	0.7
2 30	9 30	0.8	1.6	2.5	3.3	4.1	0.1	0.2	0.2	0.3	0.4	0.5	0.6	0.7	0.7
2 40	9 20	0.9	1.7	2.6	3.5	4.3	0.1	0.2	0.3	0.3	0.4	0.5	0.6	0.7	0.8
2 50	9 10	0.9	1.8	2.7	3.6	4.5	0.1	0.2	0.3	0.4	0.5	0.5	0.6	0.7	0.8
3 0	9 0	0.9	1.9	2.8	3.8	4.7	0.1	0.2	0.3	0.4	0.5	0.6	0.7	0.7	0.8
3 10	8 50	1.0	1.9	2.9	3.9	4.9	0.1	0.2	0.3	0.4	0.5	0.6	0.7	0.8	0.9
3 20	8 40	1.0	2.0	3.0	4.0	5.0	0.1	0.2	0.3	0.4	0.5	0.6	0.7	0.8	0.9
3 30	8 30	1.0	2.1	3.1	4.1	5.2	0.1	0.2	0.3	0.4	0.5	0.6	0.7	0.8	0.9
3 40	8 20	1.1	2.1	3.2	4.2	5.3	0.1	0.2	0.3	0.4	0.5	0.6	0.7	0.8	1.0
3 50	8 10	1.1	2.2	3.3	4.3	5.4	0.1	0.2	0.3	0.4	0.5	0.7	0.8	0.9	1.0
4 0	8 0	1.1	2.2	3.3	4.4	5.6	0.1	0.2	0.3	0.4	0.6	0.7	0.8	0.9	1.0
4 10	7 50	1.1	2.3	3.4	4.5	5.7	0.1	0.2	0.3	0.5	0.6	0.7	0.8	0.9	1.0
4 20	7 40	1.2	2.3	3.5	4.6	5.8	0.1	0.2	0.3	0.5	0.6	0.7	0.8	0.9	1.0
4 30	7 30	1.2	2.3	3.5	4.7	5.9	0.1	0.2	0.4	0.5	0.6	0.7	0.8	0.9	1.1
4 40	7 20	1.2	2.4	3.6	4.8	5.9	0.1	0.2	0.4	0.5	0.6	0.7	0.8	1.0	1.1
4 50	7 10	1.2	2.4	3.6	4.8	6.0	0.1	0.2	0.4	0.5	0.6	0.7	0.8	1.0	1.1
5 0	7 0	1.2	2.4	3.6	4.9	6.1	0.1	0.2	0.4	0.5	0.6	0.7	0.9	1.0	1.1
5 10	6 50	1.2	2.5	3.7	4.9	6.1	0.1	0.2	0.4	0.5	0.6	0.7	0.9	1.0	1.1
5 20	6 40	1.2	2.5	3.7	4.9	6.1	0.1	0.2	0.4	0.5	0.6	0.7	0.9	1.0	1.1
5 30	6 30	1.2	2.5	3.7	5.0	6.2	0.1	0.2	0.4	0.5	0.6	0.7	0.9	1.0	1.1
5 40	6 20	1.2	2.5	3.7	5.0	6.2	0.1	0.2	0.4	0.5	0.6	0.7	0.9	1.0	1.1
5 50	6 10	1.2	2.5	3.7	5.0	6.2	0.1	0.2	0.4	0.5	0.6	0.7	0.9	1.0	1.1
6 0	6 0	1.3	2.6	3.8	5.0	6.3	0.1	0.2	0.4	0.5	0.6	0.7	0.9	1.0	1.1

Third Differences.

Time after noon or midnight.	10″	20″	30″	40″	50″	1′	2′	3′	4′	5′	Time after noon or midnight.
+	″	″	″	″	″	″	″	″	″	″	—
0h. 0m.	0.0	0.0	0.0	0.0	0.0	0.0	0.0	0.0	0.0	0.0	12h. 0m.
0 30	0.0	0.1	0.1	0.1	0.2	0.2	0.4	0.5	0.7	0.9	11 30
1 0	0.1	0.1	0.2	0.2	0.3	0.3	0.6	1.0	1.3	1.5	11 0
1 30	0.1	0.1	0.2	0.3	0.3	0.4	0.8	1.2	1.6	2.1	10 30
2 0	0.1	0.2	0.2	0.3	0.4	0.5	0.9	1.4	1.9	2.3	10 0
2 30	0.1	0.2	0.2	0.3	0.4	0.5	1.0	1.4	1.9	2.4	9 30
3 0	0.1	0.2	0.2	0.3	0.4	0.5	0.9	1.4	1.9	2.3	9 0
3 30	0.1	0.1	0.2	0.3	0.4	0.4	0.9	1.3	1.7	2.2	8 30
4 0	0.1	0.1	0.2	0.2	0.3	0.4	0.7	1.1	1.5	1.9	8 0
4 30	0.0	0.1	0.1	0.2	0.2	0.3	0.6	0.9	1.2	1.5	7 30
5 0	0.0	0.1	0.1	0.1	0.2	0.2	0.4	0.6	0.8	1.0	7 0
5 30	0.0	0.0	0.1	0.1	0.1	0.1	0.2	0.3	0.4	0.5	6 30
6 0	0.0	0.0	0.0	0.0	0.0	0.0	0.0	0.0	0.0	0.0	6 0
+											—

TABLE LXVIII.

Fourth Differences.

Time after noon or midnight.	10″	20″	30″	40″	50″	1′	2′	3′	Time after noon or midnight.
h. m.	″	″	″	″	″	″	″	″	h. m.
0 0	0.0	0.0	0.0	0.0	0.0	0.0	0.0	0.0	12 0
0 30	0.0	0.1	0.1	0.1	0.2	0.2	0.4	0.6	11 30
1 0	0.1	0.1	0.2	0.3	0.3	0.4	0.8	1.2	11 0
1 30	0.1	0.2	0.3	0.4	0.5	0.6	1.2	1.7	10 30
2 0	0.1	0.2	0.4	0.5	0.6	0.7	1.5	2.2	10 0
2 30	0.1	0.3	0.4	0.6	0.7	0.9	1.8	2.7	9 30
3 0	0.2	0.3	0.5	0.7	0.9	1.0	2.1	3.1	9 0
3 30	0.2	0.4	0.6	0.8	0.9	1.1	2.3	3.4	8 30
4 0	0.2	0.4	0.6	0.8	1.0	1.2	2.5	3.7	8 0
4 30	0.2	0.4	0.7	0.9	1.1	1.3	2.6	3.9	7 30
5 0	0.2	0.5	0.7	0.9	1.1	1.4	2.7	4.1	7 0
5 30	0.2	0.5	0.7	0.9	1.2	1.4	2.8	4.2	6 30
6 0	0.2	0.5	0.7	0.9	1.2	1.4	2.8	4.2	6 0

TABLE LXIX.

Mercury's Epochs.

	M. Longitude.				Aphelion.				Node.				II.	III.
	s	°	′	″	s	°	′	″	s	°	′	″		
1835	7	11	3	34	8	14	52	47	1	16	21	44	11	1559
1836 B.	9	8	52	11	8	14	53	43	1	16	22	26	377	1925
1837	11	2	35	14	8	14	54	39	1	16	23	8	336	228
1838	0	26	18	18	8	14	55	35	1	16	23	50	296	592
1839	2	20	1	21	8	14	56	31	1	16	24	32	255	955
1840 B.	4	17	49	58	8	14	57	27	1	16	25	15	217	1320
1841	6	11	33	1	8	14	58	23	1	16	25	57	176	1682
1842	8	5	16	5	8	14	59	19	1	16	26	39	135	2047
1843	9	28	59	8	8	15	0	15	1	16	27	21	95	351
1844 B.	11	26	47	45	8	15	1	11	1	16	28	3	56	717
1845	1	20	30	48	8	15	2	7	1	16	28	45	15	1083
1846	3	14	13	52	8	15	3	3	1	16	29	27	380	1450
1847	5	7	56	55	8	15	3	59	1	16	30	10	340	1816
1848 B.	7	5	45	32	8	15	4	55	1	16	30	52	300	113
1849	8	29	28	35	8	15	5	51	1	16	31	34	260	478
1850	10	23	11	39	8	15	6	47	1	16	32	16	220	844
1851	0	16	54	42	8	15	7	43	1	16	32	58	180	1209
1852 B.	2	14	43	19	8	15	8	39	1	16	33	40	140	1575
1853	4	8	26	22	8	15	9	35	1	16	34	23	100	1940
1854	6	2	9	26	8	15	10	31	1	16	35	5	59	235
1855	7	25	52	29	8	15	11	27	1	16	35	47	19	600
1856 B.	9	23	41	6	8	15	12	23	1	16	36	29	385	966
1857	11	17	24	9	8	15	13	19	1	16	37	11	345	1331
1858	1	11	7	13	8	15	14	15	1	16	37	53	304	1696
1859	3	4	50	16	8	15	15	11	1	16	38	35	264	2061
1860 B.	5	2	38	53	8	15	16	7	1	16	39	18	224	357
1861	6	26	21	56	8	15	17	4	1	16	40	0	184	722
1862	8	20	5	0	8	15	18	0	1	16	40	42	144	1087
1863	10	13	48	3	8	15	18	56	1	16	41	24	103	1452
1864 B.	0	11	36	40	8	15	19	52	1	16	42	6	64	1818
1865	2	5	19	43	8	15	20	48	1	16	42	48	24	113
1866	3	29	2	47	8	15	21	44	1	16	43	30	389	478
1867	5	22	45	50	8	15	22	40	1	16	44	13	348	843
1868 B.	7	20	34	27	8	15	23	36	1	16	44	55	309	1209
1869	9	14	17	30	8	15	24	32	1	16	45	37	268	1574
1870	11	8	0	34	8	15	25	28	1	16	46	19	228	1939
1871	1	1	43	37	8	15	26	24	1	16	47	1	188	234
1872 B.	2	29	32	14	8	15	27	20	1	16	47	43	148	599
1873	4	23	15	17	8	15	28	16	1	16	48	26	108	965
1874	6	16	58	21	8	15	29	12	1	16	49	8	67	1330
1875	8	10	41	24	8	15	30	8	1	16	49	50	27	1695
1876 B.	10	8	30	1	8	15	31	4	1	16	50	32	393	2061
1877	0	2	13	4	8	15	32	0	1	16	51	14	353	356
1878	1	25	56	8	8	15	32	56	1	16	51	56	312	720
1879	3	19	39	11	8	15	33	52	1	16	52	38	272	1086
1880 B.	5	17	27	48	8	15	34	48	1	16	53	21	232	1452
1881	7	11	10	51	8	15	35	44	1	16	54	3	192	1817
1882	9	4	53	55	8	15	36	40	1	16	54	45	152	112
1883	10	28	36	58	8	15	37	36	1	16	55	27	111	477
1884 B.	0	26	25	35	8	15	38	32	1	16	56	9	72	843
1885	2	20	8	38	8	15	39	28	1	16	56	51	32	1208

TABLE LXX.

Mercury's Motions for Months.

Months.		D.	Longitude.				Aph.	Nod.
			s	°	′	″	″	″
Jan...	Bis..	0	0	0	0	0	0	0
	Com.	1	0	4	5	33	0	0
Feb...	Bis..	31	4	6	51	49	5	4
	Com.	32	4	10	57	22	5	4
March........		60	8	5	32	33	9	7
April..........		91	0	12	24	23	14	11
May		121	4	15	10	39	19	14
June...........		152	8	22	2	29	23	18
July		182	0	24	48	46	28	21
August........		213	5	1	40	35	33	25
September...		244	9	8	32	24	38	28
October		274	1	11	18	41	42	32
November....		305	5	18	10	30	47	35
December....		335	9	20	56	47	52	39

TABLE LXXI.

Motions for Days and Hours.

Days.	Longitude.				Ap.	No.	Hours.	Longitude.		
	s	°	′	″	″	″		°	′	″
1	0	0	0	0	0	0	1	0	10	14
2	0	4	5	33	0	0	2	0	20	28
3	0	8	11	5	0	0	3	0	30	42
4	0	12	16	38	1	0	4	0	40	55
5	0	16	22	10	1	0	5	0	51	9
6	0	20	27	43	1	1	6	1	1	23
7	0	24	33	15	1	1	7	1	11	37
8	0	28	38	48	1	1	8	1	21	51
9	1	2	44	20	1	1	9	1	32	5
10	1	6	49	53	2	1	10	1	42	19
11	1	10	55	26	2	1	11	1	52	32
12	1	15	0	58	2	1	12	2	2	46
13	1	19	6	31	2	1	13	2	13	0
14	1	23	12	3	2	2	14	2	23	14
15	1	27	17	36	2	2	15	2	33	28
16	2	1	23	8	2	2	16	2	43	42
17	2	5	28	41	3	2	17	2	53	56
18	2	9	34	13	3	2	18	3	4	9
19	2	13	39	46	3	2	19	3	14	23
20	2	17	45	19	3	2	20	3	24	37
21	2	21	50	51	3	2	21	3	34	51
22	2	25	56	24	3	2	22	3	45	5
23	3	0	1	56	4	3	23	3	55	19
24	3	4	7	29	4	3	24	4	5	33
25	3	8	13	1	4	3				
26	3	12	18	34	4	3				
27	3	16	24	6	4	3				
28	3	20	29	39	4	3				
29	3	24	35	12	4	3				
30	3	28	40	44	4	3				
31	4	2	46	17	5	3				

TABLE LXXII.

Mots. Min. and Sec.

Min.	Long.		Sec.	Long.
	′	″		″
1	0	10	1	0
2	0	20	2	0
3	0	31	3	1
4	0	41	4	1
5	0	51	5	1
6	1	1	6	1
7	1	12	7	1
8	1	22	8	1
9	1	32	9	2
10	1	42	10	2
11	1	53	11	2
12	2	3	12	2
13	2	13	13	2
14	2	23	14	2
15	2	33	15	3
16	2	44	16	3
17	2	54	17	3
18	3	4	18	3
19	2	14	19	3
20	3	25	20	3
21	3	35	21	4
22	3	45	22	4
23	3	55	23	4
24	4	6	24	4
25	4	16	25	4
26	4	26	26	4
27	4	36	27	5
28	4	46	28	5
29	4	57	29	5
30	5	7	30	5
31	5	17	31	5
32	5	27	32	5
33	5	38	33	6
34	5	48	34	6
35	5	58	35	6
36	6	8	36	6
37	6	18	37	6
38	6	29	38	6
39	6	39	39	7
40	6	49	40	7
41	6	59	41	7
42	7	10	42	7
43	7	20	43	7
44	7	30	44	8
45	7	40	45	8
46	7	51	46	8
47	8	1	47	8
48	8	11	48	8
49	8	21	49	8
50	8	31	50	9
51	8	42	51	9
52	8	52	52	9
53	9	2	53	9
54	9	12	54	9
55	9	23	55	9
56	9	33	56	10
57	9	43	57	10
58	9	53	58	10
59	10	4	59	10
60	10	15	60	10

TABLE LXXIII.

Equation of Mercury's Centre.

ARGUMENT. Mean Anomaly.

	0s—	Is—	IIs—	IIIs—	IVs—	Vs—	
°	° ′ ″	° ′ ″	° ′ ″	° ′ ″	° ′ ″	° ′ ″	°
0	0 0 0	9 34 57	17 47 21	22 56 6	22 46 14	14 55 46	30
1	0 19 36	9 53 11	18 1 15	23 1 46	22 38 36	14 31 30	29
2	0 39 12	10 11 19	18 14 57	23 7 3	22 30 26	14 6 42	28
3	0 58 48	10 29 22	18 28 26	23 11 59	22 21 44	13 41 25	27
4	1 18 23	10 47 18	18 41 42	23 16 33	22 12 29	13 15 39	26
5	1 37 57	11 5 9	18 54 45	23 20 45	22 2 42	12 49 23	25
6	1 57 31	11 22 53	19 7 35	23 24 33	21 52 23	12 22 39	24
7	2 17 3	11 40 30	19 20 11	23 27 59	21 41 30	11 55 27	23
8	2 36 35	11 58 1	19 32 33	23 31 1	21 30 4	11 27 48	22
9	2 56 5	12 15 24	19 44 41	23 33 39	21 18 5	10 59 44	21
10	3 15 33	12 32 41	19 56 34	23 35 52	21 5 31	10 31 14	20
11	3 34 59	12 49 50	20 8 13	23 37 41	20 52 25	10 2 20	19
12	3 54 24	13 6 51	20 19 37	23 39 5	20 38 44	9 33 2	18
13	4 13 47	13 23 45	20 30 45	23 40 4	20 24 29	9 3 22	17
14	4 33 7	13 40 30	20 41 38	23 40 37	20 9 40	8 33 20	16
15	4 52 25	13 57 8	20 52 15	23 40 44	19 54 17	8 2 58	15
16	5 11 40	14 13 37	21 2 36	23 40 24	19 38 20	7 32 16	14
17	5 30 53	14 29 57	21 12 40	23 39 38	19 21 48	7 1 17	13
18	5 50 2	14 46 8	21 22 27	23 38 24	19 4 43	6 30 0	12
19	6 9 8	15 2 10	21 31 57	23 36 43	18 47 3	5 58 27	11
20	6 28 11	15 18 3	21 41 10	23 34 34	18 28 49	5 26 40	10
21	6 47 11	15 33 46	21 50 5	23 31 57	18 10 0	4 54 39	9
22	7 6 7	15 49 19	21 58 42	23 28 51	17 50 38	4 22 26	8
23	7 24 58	16 4 42	22 7 0	23 25 16	17 30 42	3 50 1	7
24	7 43 46	16 19 55	22 15 0	23 21 12	17 10 13	3 17 28	6
25	8 2 30	16 34 57	22 22 41	23 16 38	16 49 10	2 44 46	5
26	8 21 9	16 49 48	22 30 2	23 11 34	16 27 34	2 11 57	4
27	8 39 43	17 4 29	22 37 3	23 6 0	16 5 25	1 39 3	3
28	8 58 13	17 18 58	22 43 45	22 59 56	15 42 44	1 6 4	2
29	9 16 37	17 33 16	22 50 6	22 53 21	15 19 31	0 33 3	1
30	9 34 57	17 47 21	22 56 6	22 46 14	14 55 46	0 0 0	0
	XIs+	Xs+	IXs+	VIIIs+	VIIs+	VIs+	

TABLE LXXIV.

Equations 2 *and* 3. *Always Affirmative.*

Arg.	2	Arg.	2	Arg.	2	Arg.	3	Arg.	3	Arg.	3	Arg.	3
0	9″	260	1″	520	5″	0	5″	650	5″	1300	19″	1950	8″
20	9	280	2	540	4	50	4	700	6	1350	19	2000	7
40	9	300	4	560	3	100	3	750	7	1400	19	2050	6
60	8	320	5	580	2	150	2	800	8	1450	19	2100	5
80	7	340	6	600	1	200	2	850	10	1500	19	2150	4
120	6	360	7	620	1	250	1	900	11	1550	18	2200	3
140	5	380	8	640	1	300	1	950	12	1600	17	2250	2
160	4	400	9	660	1	350	1	1000	14	1650	16	2300	1
180	2	420	9	680	2	400	1	1050	15	1700	15	2350	1
200	1	440	9	700	3	450	1	1100	16	1750	14	2400	1
220	1	460	9	720	5	500	2	1150	17	1800	13	2450	1
240	1	480	8	740	6	550	3	1200	18	1850	11		
260	1	500	7	760	7	600	4	1250	18	1900	10		
280	1	520	5	780	8	650	5	1300	19	1950	8		

Mercury's Radius Vector.

ARGUMENT. Mean Anomaly.

	0ˢ	Iˢ	IIˢ	IIIˢ	IVˢ	Vˢ	
0°	0.46669	0.45923	0.43734	0.40303	0.36138	0.32362	30°
1	0.46668	0.45872	0.43638	0.40171	0.35997	0.32263	29
2	0.46666	0.45820	0.43540	0.40039	0.35857	0.32167	28
3	0.46662	0.45767	0.43441	0.39907	0.35717	0.32073	27
4	0.46656	0.45712	0.43341	0.39773	0.35578	0.31982	26
5	0.46649	0.45655	0.43240	0.39639	0.35440	0.31893	25
6	0.46639	0.45597	0.43137	0.39504	0.35302	0.31807	24
7	0.46628	0.45537	0.43032	0.39368	0.35164	0.31725	23
8	0.46616	0.45475	0.42927	0.39231	0.35028	0.31645	22
9	0.46602	0.45412	0.42820	0.39094	0.34892	0.31568	21
10	0.46586	0.45347	0.42712	0.38956	0.34757	0.31495	20
11	0.46569	0.45281	0.42602	0.38818	0.34624	0.31424	19
12	0.46549	0.45213	0.42492	0.38679	0.34491	0.31357	18
13	0.46529	0.45144	0.42380	0.38540	0.34359	0.31293	17
14	0.46506	0.45073	0.42266	0.38400	0.34229	0.31232	16
15	0.46482	0.45001	0.42152	0.38260	0.34099	0.31175	15
16	0.46456	0.44927	0.42036	0.38120	0.33972	0.31121	14
17	0.46429	0.44851	0.41919	0.37979	0.33845	0.31071	13
18	0.46400	0.44774	0.41802	0.37838	0.33720	0.31024	12
19	0.46369	0.44696	0.41682	0.37696	0.33597	0.30981	11
20	0.46337	0.44616	0.41562	0.37555	0.33475	0.30941	10
21	0.46302	0.44534	0.41441	0.37413	0.33355	0.30905	9
22	0.46267	0.44451	0.41318	0.37271	0.33236	0.30873	8
23	0.46229	0.44366	0.41195	0.37129	0.33120	0.30844	7
24	0.46190	0.44280	0.41070	0.36987	0.33005	0.30819	6
25	0.46150	0.44193	0.40945	0.36845	0.32893	0.30798	5
26	0.46108	0.44104	0.40818	0.36704	0.32782	0.30781	4
27	0.46064	0.44014	0.40691	0.36562	0.32674	0.30768	3
28	0.46018	0.43922	0.40562	0.36420	0.32567	0.30758	2
29	0.45971	0.43829	0.40433	0.36279	0.32464	0.30752	1
30	0.45923	0.43734	0.40303	0.36138	0.32362	0.30750	0
	XIˢ	Xˢ	IXˢ	VIIIˢ	VIIˢ	VIˢ	

TABLE LXXVI.

Reduction to the Ecliptic.

ARGUMENT. Orbit Long. of Mercury—Long. of Node.

	0s , VIs	Is , VIIs	IIs , VIIIs	IIIs , IXs	IVs , Xs	Vs , XIs
0°	— 0′ 15″	— 11′ 22″	— 11′ 25″	— 0′ 15″	+ 10′ 55″	+ 10′ 52″
1	0 42	11 35	11 11	+ 0 12	11 8	10 38
2	1 9	11 48	10 56	0 39	11 19	10 24
3	1 35	11 59	10 41	1 6	11 31	10 8
4	2 2	12 10	10 25	1 33	11 42	9 52
5	2 29	12 20	10 8	1 59	11 51	9 35
6	— 2 55	— 12 28	— 9 50	+ 2 26	+ 12 0	+ 9 17
7	3 21	12 36	9 32	2 52	12 8	8 59
8	3 47	12 43	9 13	3 18	12 15	8 40
9	4 13	12 49	8 53	3 44	12 21	8 20
10	4 38	12 55	8 33	4 10	12 26	8 0
11	— 5 3	— 12 59	— 8 12	+ 4 35	+ 12 30	+ 7 39
12	5 28	13 2	7 50	5 0	12 33	7 17
13	5 52	13 5	7 28	5 24	12 35	6 55
14	6 16	13 6	7 5	5 49	12 37	6 33
15	6 40	13 7	6 42	6 12	12 37	6 10
16	— 7 3	— 13 7	— 6 19	+ 6 35	+ 12 36	+ 5 46
17	7 25	13 5	5 54	6 58	12 35	5 22
18	7 47	13 3	5 30	7 20	12 32	4 58
19	8 9	13 0	5 5	7 42	12 29	4 33
20	8 30	12 56	4 40	8 3	12 25	4 8
21	— 8 50	— 12 51	— 4 14	+ 8 23	+ 12 19	+ 3 43
22	9 10	12 45	3 48	8 43	12 13	3 17
23	9 29	12 38	3 22	9 2	12 6	2 51
24	9 47	12 30	2 56	9 20	11 58	2 25
25	10 5	12 21	2 29	9 38	11 49	1 59
26	— 10 22	— 12 12	— 2 3	+ 9 55	+ 11 40	+ 1 32
27	10 38	12 1	1 36	10 11	11 29	0 55
28	10 54	11 49	1 9	10 26	11 18	0 38
29	11 8	11 38	0 42	10 41	11 5	+ 0 12
30	— 11 22	— 11 25	— 0 15	+ 10 55	+ 10 52	— 0 15

TABLE LXXVII.

Heliocentric Longitudes, &c., of the Planet Venus, at the times of the next two Transits over the Sun's Disc.

Times.	Hel. Long. from true Equinox.	Hel. Lat.	Rad. Vector.
1874, Dec. 8th, at 12h.	76° 41′ 36.6″	4′ 6.3″N.	0.7203632
16h.	76 57 44.1	5 3.5	0.7203449
20h.	77 13 51.5	6 1.0	0.7203268
1882, Dec. 6th, at noon.	74 12 55.7	4 58.1S.	0.7205502
4h.	74 29 2.5	4 0.8	0.7205315
8h.	74 45 9.7	3 3.5	0.7205127

NOTE.—The Aberration in Longitude at the time of each Transit, is 3.4″; to be *added* to the true geocentric longitude, to obtain the apparent longitude.

TABLE LXXVIII.

Mercury's Heliocentric Latitude for the year 1800, *with the Secular Variation.*

ARGUMENT. Orbit Longitude of Mercury—Long. of Node.

	Latitude. 0ˢ N. VIˢ S.	Sec. Var. +	Log. Cos. Lat.	Latitude. Iˢ N. VIIˢ S.	Sec. Var. +	Log. Cos. Lat.	Latitude. IIˢ N. VIIIˢ S.	Sec. Var. +	Log. Cos. Lat.	
°	° ′ ″	″		° ′ ″	″		° ′ ″	″		°
0	0 0 0	0	10.00000	3 29 39	9	9.99919	6 3 35	16	9.99757	30
1	0 7 19	0	10.00000	3 35 58	9	9.99914	6 7 13	16	9.99752	29
2	0 14 37	1	10.00000	3 42 13	10	9.99909	6 10 43	16	9.99747	28
3	0 21 56	1	9.99999	3 48 24	10	9.99904	6 14 7	16	9.99742	27
4	0 29 14	1	9.99998	3 54 31	10	9.99899	6 17 24	16	9.99738	26
5	0 36 31	2	9.99998	4 0 33	10	9.99894	6 20 34	17	9.99733	25
6	0 43 48	2	9.99996	4 6 31	11	9.99888	6 23 37	17	9.99729	24
7	0 51 4	2	9.99995	4 12 25	11	9.99883	6 26 33	17	9.99725	23
8	0 58 19	2	9.99994	4 18 14	11	9.99877	6 29 22	17	9.99721	22
9	1 5 33	3	9.99992	4 23 59	11	9.99872	6 32 4	17	9.99717	21
10	1 12 46	3	9.99990	4 29 38	12	9.99866	6 34 39	17	9.99713	20
11	1 19 58	3	9.99988	4 35 13	12	9.99861	6 37 6	17	9.99710	19
12	1 27 8	4	9.99986	4 40 43	12	9.99855	6 39 27	17	9.99706	18
13	1 34 17	4	9.99984	4 46 8	12	9.99849	6 41 39	17	9.99703	17
14	1 41 24	4	9.99981	4 51 27	13	9.99844	6 43 45	18	9.99700	16
15	1 48 29	5	9.99978	4 56 41	13	9.99838	6 45 43	18	9.99697	15
16	1 55 32	5	9.99975	5 1 50	13	9.99832	6 47 34	18	9.99694	14
17	2 2 33	5	9.99972	5 6 53	13	9.99827	6 49 17	18	9.99691	13
18	2 9 31	6	9.99969	5 11 51	14	9.99821	6 50 52	18	9.99689	12
19	2 16 28	6	9.99966	5 16 43	14	9.99815	6 52 21	18	9.99687	11
20	2 23 22	6	9.99962	5 21 29	14	9.99810	6 53 41	18	9.99685	10
21	2 30 13	6	9.99959	5 26 9	14	9.99804	6 54 54	18	9.99683	9
22	2 37 2	7	9.99955	5 30 44	14	9.99799	6 55 59	18	9.99681	8
23	2 43 48	7	9.99951	5 35 12	15	9.99793	6 56 57	18	9.99680	7
24	2 50 31	7	9.99947	5 39 34	15	9.99788	6 57 47	18	9.99678	6
25	2 57 11	8	9.99942	5 43 51	15	9.99782	6 58 29	18	9.99677	5
26	3 3 47	8	9.99938	5 48 0	15	9.99777	6 59 4	18	9.99676	4
27	3 10 21	8	9.99933	5 52 4	15	9.99772	6 59 31	18	9.99676	3
28	3 16 50	9	9.99929	5 56 1	15	9.99767	6 59 51	18	9.99675	2
29	3 23 17	9	9.99924	5 59 51	16	9.99762	7 0 2	18	9.99675	1
30	3 29 39	9	9.99919	6 3 35	16	9.99757	7 0 6	18	9.99675	0
	XIˢ S. Vˢ N.	+		Xˢ S. IVˢ N.	+		IXˢ S. IIIˢ N.	+		

Multiplier for Aber. in Lat. Arg. Geoc. Lat.

Arg.	Mult.
0°	0.00
1	0.02
2	0.03
3	0.05
4	0.07
5	0.09
6	0.10
7	0.12

Aberration in Latitude. Part IV.

ARGUMENT. Arg. of Latitude.

Arg.	Aber.	Arg.	Arg.	Aber.	Arg.
0°	— 5″	360°	90°	— 1″	270°
10	5	350	100	0	260
20	5	340	110	+ 1	250
30	4	330	120	1	240
40	4	320	130	2	230
50	3	310	140	2	220
60	3	300	150	3	210
70	2	290	160	3	200
80	1	280	170	3	190
90	— 1	270	180	+ 3	180

Aberration of Mercury in Longitude.

Part I. Arg. Elong.			Part II. Arg. Ann. Par.			Part III. Arg. Geoc. Long.		
Elon.	Aberrat.	Elon.	Ann. Par.	Aberrat.	Ann. Par.	Geoc. Long.	Aberrat.	Geoc. Long.
°	″	°	°	″	°	°	″	°
0	− 20 −	360	0	− 33 −	360	0	− 2 +	180
4	20	356	4	33	356	4	2	184
8	20	352	8	32	352	8	3	188
12	20	348	12	32	348	12	3	192
16	19	344	16	31	344	16	4	196
20	19	340	20	31	340	20	4	200
24	18	336	24	30	336	24	4	204
28	18	332	28	29	332	28	5	208
32	17	328	32	28	328	32	5	212
36	16	324	36	26	324	36	5	216
40	15	320	40	25	320	40	6	220
44	14	316	44	24	316	44	6	224
48	13	312	48	22	312	48	6	228
52	12	308	52	20	308	52	6	232
56	11	304	56	18	304	56	7	236
60	10	300	60	16	300	60	7	240
64	9	296	64	14	296	64	7	244
68	8	292	68	12	292	68	7	248
72	6	288	72	10	288	72	7	252
76	5	284	76	8	284	76	7	256
80	3	280	80	6	280	80	7	260
84	2	276	84	3	276	84	7	264
88	− 1 −	272	88	− 1 −	272	88	7	268
92	+ 1 +	268	92	+ 1 +	268	92	7	272
96	2	264	96	3	264	96	7	276
100	3	260	100	6	260	100	6	280
104	5	256	104	8	256	104	6	284
108	6	252	108	10	252	108	6	288
112	8	248	112	12	248	112	5	292
116	9	244	116	14	244	116	5	296
120	10	240	120	16	240	120	5	300
124	11	236	124	18	236	124	4	304
128	12	232	128	20	232	128	4	308
132	13	228	132	22	228	132	4	312
136	14	224	136	24	224	136	3	316
140	15	220	140	25	220	140	3	320
144	16	216	144	26	216	144	2	324
148	17	212	148	28	212	148	2	328
152	18	208	152	29	208	152	1	332
156	18	204	156	30	204	156	1	336
160	19	200	160	31	200	160	− 0 +	340
164	19	196	164	31	196	164	+ 0 −	344
168	20	192	168	32	192	168	0	348
172	20	188	172	32	188	172	1	352
176	20	184	176	33	184	176	1	356
180	+ 20 +	180	180	+ 33 +	180	180	+ 2 −	360

TABLE LXXX.

Logistical Logarithms.

	0′	1′	2′	3′	4′	5′	6′	7′
0″	00000	17782	14771	13010	11761	10792	10000	9331
1	35563	17710	14735	12986	11743	10777	9988	9320
2	32553	17639	14699	12962	11725	10763	9976	9310
3	30792	17570	14664	12939	11707	10749	9964	9300
4	29542	17501	14629	12915	11689	10734	9952	9289
5	28573	17434	14594	12891	11671	10720	9940	9279
6	27782	17368	14559	12868	11654	10706	9928	9269
7	27112	17302	14525	12845	11636	10692	9916	9259
8	26532	17238	14491	12821	11619	10678	9905	9249
9	26021	17175	14457	12798	11601	10663	9893	9238
10	25563	17112	14424	12775	11584	10649	9881	9228
11	25149	17050	14390	12753	11566	10635	9869	9218
12	24771	16990	44357	12730	11549	10621	9858	9208
13	24424	16930	14325	12707	11532	10608	9846	9198
14	24102	16871	14292	12685	11515	10594	9834	9188
15	23802	16812	14260	12663	11498	10580	9823	9178
16	23522	16755	14228	12640	11481	10566	9811	9168
17	23259	16698	14196	12618	11464	10552	9800	9158
18	23010	16642	14165	12596	11447	10539	9788	9148
19	22775	16587	14133	12574	11430	10525	9777	9138
20	22553	16532	14102	12553	11413	10512	9765	9128
21	22341	16478	14071	12531	11397	10498	9754	9119
22	22139	16425	14040	12510	11380	10484	9742	9109
23	21946	16372	14010	12488	11363	10471	9731	9099
24	21761	16320	13979	12467	11347	10458	9720	9089
25	21584	16269	13949	12445	11331	10444	9708	9079
26	21413	16218	13919	12424	11314	10431	9697	9070
27	21249	16168	13890	12403	11298	10418	9686	9060
28	21091	16118	13860	12382	11282	10404	9675	9050
29	20939	16069	13831	12362	11266	10391	9664	9041
30	20792	16021	13802	12341	11249	10378	9652	9031
31	20649	15973	13773	12320	11233	10365	9641	9021
32	20512	15925	13745	12300	11217	10352	9630	9012
33	20378	15878	13716	12279	11201	10339	9619	9002
34	20248	15832	13688	12259	11186	10326	9608	8992
35	20122	15786	13660	12239	11170	10313	9597	8983
36	20000	15740	13632	12218	11154	10300	9586	8973
37	19881	15695	13604	12198	11138	10287	9575	8964
38	19765	15651	13576	12178	11123	10274	9564	8954
39	19652	15607	13549	12159	11107	10261	9553	8945
40	19542	15563	13522	12139	11091	10248	9542	8935
41	19435	15520	13495	12119	11076	10235	9532	8926
42	19331	15477	13468	12099	11061	10223	9521	8917
43	19228	15435	13441	12080	11045	10210	9510	8907
44	19128	15393	13415	12061	11030	10197	9499	8898
45	19031	15351	13388	12041	11015	10185	9488	8888
46	18935	15310	13362	12022	10999	10172	9478	8879
47	18842	15269	13336	12003	10984	10160	9467	8870
48	18751	15229	13310	11984	10969	10147	9456	8861
49	18661	15189	13284	11965	10954	10135	9446	8851
50	18573	15149	13259	11946	10939	10122	9435	8842
51	18487	15110	13233	11927	10924	10110	9425	8833
52	18403	15071	13208	11908	10909	10098	9414	8824
53	18320	15032	13183	11889	10894	10085	9404	8814
54	18239	14994	13158	11871	10880	10073	9393	8805
55	18159	14956	13133	11852	10865	10061	9383	8796
56	18081	14918	13108	11834	10850	10049	9372	8787
57	18004	14881	13083	11816	10835	10036	9362	8778
58	17929	14844	13059	11797	10821	10024	9351	8769
59	17855	14808	13034	11779	10806	10012	9341	8760
60	17782	14771	13010	11761	10792	10000	9331	8751

TABLE LXXX.

Logistical Logarithms.

	8′	9′	10′	11′	12′	13′	14′	15′	16′
0″	8751	8239	7782	7368	6990	6642	6320	6021	5740
1	8742	8231	7774	7361	6984	6637	6315	6016	5736
2	8733	8223	7767	7354	6978	6631	6310	6011	5731
3	8724	8215	7760	7348	6972	6625	6305	6006	5727
4	8715	8207	7753	7341	6966	6620	6300	6001	5722
5	8706	8199	7745	7335	6960	6614	6294	5997	5718
6	8697	8191	7738	7328	6954	6609	6289	5992	5713
7	8688	8183	7731	7322	6948	6603	6284	5987	5709
8	8679	8175	7724	7315	6942	6598	6279	5982	5704
9	8670	8167	7717	7309	6936	6592	6274	5977	5700
10	8661	8159	7710	7302	6930	6587	6269	5973	5695
11	8652	8152	7703	7296	6924	6581	6264	5968	5691
12	8643	8144	7696	7289	6918	6576	6259	5963	5686
13	8635	8136	7688	7283	6912	6570	6254	5958	5682
14	8626	8128	7681	7276	6906	6565	6248	5954	5677
15	8617	8120	7674	7270	6900	6559	6243	5949	5673
16	8608	8112	7667	7264	6894	6554	6238	5944	5669
17	8599	8104	7660	7257	6888	6548	6233	5939	5664
18	8591	8097	7653	7251	6882	6543	6228	5935	5660
19	8582	8089	7646	7244	6877	6538	6223	5930	5655
20	8573	8081	7639	7238	6871	6532	6218	5925	5651
21	8565	8073	7632	7232	6865	6527	6213	5920	5646
22	8556	8066	7625	7225	6859	6521	6208	5916	5642
23	8547	8058	7618	7219	6853	6516	6203	5911	5637
24	8539	8050	7611	7212	6847	6510	6198	5906	5633
25	8530	8043	7604	7206	6841	6505	6193	5902	5629
26	8522	8035	7597	7200	6836	6500	6188	5897	5624
27	8513	8027	7590	7193	6830	6494	6183	5892	5620
28	8504	8020	7583	7187	6824	6489	6178	5888	5615
29	8496	8012	7577	7181	6818	6484	6173	5883	5611
30	8487	8004	7570	7175	6812	6478	6168	5878	5607
31	8479	7997	7563	7168	6807	6473	6163	5874	5602
32	8470	7989	7556	7162	6801	6467	6158	5869	5598
33	8462	7981	7549	7156	6795	6462	6153	5864	5594
34	8453	7974	7542	7149	6789	6457	6148	5860	5589
35	8445	7966	7535	7143	6784	6451	6143	5855	5585
36	8437	7959	7528	7137	6778	6446	6138	5850	5580
37	8428	7951	7522	7131	6772	6441	6133	5846	5576
38	8420	7944	7515	7124	6766	6435	6128	5841	5572
39	8411	7936	7508	7118	6761	6430	6123	5836	5567
40	8403	7929	7501	7112	6755	6425	6118	5832	5563
41	8395	7921	7494	7106	6749	6420	6113	5827	5559
42	8386	7914	7488	7100	6743	6414	6108	5823	5554
43	8378	7906	7481	7093	6738	6409	6103	5818	5550
44	8370	7899	7474	7087	6732	6404	6099	5813	5546
45	8361	7891	7467	7081	6726	6398	6094	5809	5541
46	8353	7884	7461	7075	6721	6393	6089	5804	5537
47	8345	7877	7454	7069	6715	6388	6084	5800	5533
48	8337	7869	7447	7063	6709	6383	6079	5795	5528
49	8328	7862	7441	7057	6704	6377	6074	5790	5524
50	8320	7855	7434	7050	6698	6372	6069	5786	5520
51	8312	7847	7427	7044	6692	6367	6064	5781	5516
52	8304	7840	7421	7038	6687	6362	6059	5777	5511
53	8296	7832	7414	7032	6681	6357	6055	5772	5507
54	8288	7825	7407	7026	6676	6351	6050	5768	5503
55	8279	7818	7401	7020	6670	6346	6045	5763	5498
56	8271	7811	7394	7014	6664	6341	6040	5758	5494
57	8263	7803	7387	7008	6659	6336	6035	5754	5490
58	8255	7796	7381	7002	6653	6331	6030	5749	5486
59	8247	7789	7374	6996	6648	6325	6025	5745	5481
60	8239	7782	7368	6990	6642	6320	6021	5740	5477

Logistical Logarithms.

	17′	18′	19′	20′	21′	22′	23′	24′	25′
0″	5477	5229	4994	4771	4559	4357	4164	3979	3802
1	5473	5225	4990	4768	4556	4354	4161	3976	3799
2	5469	5221	4986	4764	4552	4351	4158	3973	3796
3	5464	5217	4983	4760	4549	4347	4155	3970	3793
4	5460	5213	4979	4757	4546	4344	4152	3967	3791
5	5456	5209	4975	4753	4542	4341	4149	3964	3788
6	5452	5205	4971	4750	4539	4338	4145	3961	3785
7	5447	5201	4967	4746	4535	4334	4142	3958	3782
8	5443	5197	4964	4742	4532	4331	4139	3955	3779
9	5439	5193	4960	4739	4528	4328	4136	3952	3776
10	5435	5189	4956	4735	4525	4325	4133	3949	3773
11	5430	5185	4952	4732	4522	4321	4130	3946	3770
12	5426	5181	4949	4728	4518	4318	4127	3943	3768
13	5422	5177	4945	4724	4515	4315	4124	3940	3765
14	5418	5173	4941	4721	4511	4311	4120	3937	3762
15	5414	5169	4937	4717	4508	4308	4117	3934	3759
16	5409	5165	4933	4714	4505	4305	4114	3931	3756
17	5405	5161	4930	4710	4501	4302	4111	3928	3753
18	5401	5157	4926	4707	4498	4298	4108	3925	3750
19	5397	5153	4922	4703	4494	4295	4105	3922	3747
20	5393	5149	4918	4699	4491	4292	4102	3919	3745
21	5389	5145	4915	4696	4488	4289	4099	3917	3742
22	5384	5141	4911	4692	4484	4285	4096	3914	3739
23	5380	5137	4907	4689	4481	4282	4092	3911	3736
24	5376	5133	4903	4685	4477	4279	4089	3908	3733
25	5372	5129	4900	4682	4474	4276	4086	3905	3730
26	5368	5125	4896	4678	4471	4273	4083	3902	3727
27	5364	5122	4892	4675	4467	4269	4080	3899	3725
28	5359	5118	4889	4671	4464	4266	4077	3896	3722
29	5355	5114	4885	4668	4460	4263	4074	3893	3719
30	5351	5110	4881	4664	4457	4260	4071	3890	3716
31	5347	5106	4877	4660	4454	4256	4068	3887	3713
32	5343	5102	4874	4657	4450	4253	4065	3884	3710
33	5339	5098	4870	4653	4447	4250	4062	3881	3708
34	5335	5094	4866	4650	4444	4247	4059	3878	3705
35	5331	5090	4863	4646	4440	4244	4055	3875	3702
36	5326	5086	4859	4643	4437	4240	4052	3872	3699
37	5322	5082	4855	4639	4434	4237	4049	3869	3696
38	5318	5079	4852	4636	4430	4234	4046	3866	3693
39	5314	5075	4848	4632	4427	4231	4043	3863	3691
40	5310	5071	4844	4629	4424	4228	4040	3860	3688
41	5306	5067	4841	4625	4420	4224	4037	3857	3685
42	5302	5063	4837	4622	4417	4221	4034	3855	3682
43	5298	5059	4833	4618	4414	4218	4031	3852	3679
44	5294	5055	4830	4615	4410	4215	4028	3849	3677
45	5290	5051	4826	4611	4407	4212	4025	3846	3674
46	5285	5048	4822	4608	4404	4209	4022	3843	3671
47	5281	5044	4819	4604	4400	4205	4019	3840	3668
48	5277	5040	4815	4601	4397	4202	4016	3837	3665
49	5273	5036	4811	4597	4394	4199	4013	3834	3663
50	5269	5032	4808	4594	4390	4196	4010	3831	3660
51	5265	5028	4804	4590	4387	4193	4007	3828	3657
52	5261	5025	4800	4587	4384	4189	4004	3825	3654
53	5257	5021	4797	4584	4380	4186	4001	3822	3651
54	5253	5017	4793	4580	4377	4183	3998	3820	3649
55	5249	5013	4789	4577	4374	4180	3995	3817	3646
56	5245	5009	4786	4573	4370	4177	3991	3814	3643
57	5241	5005	4782	4570	4367	4174	3988	3811	3640
58	5237	5002	4778	4566	4364	4171	3985	3808	3637
59	5233	4998	4775	4563	4361	4167	3982	3805	3635
60	5229	4994	4771	4559	4357	4164	3979	3802	3632

TABLE LXXX.

Logistical Logarithms.

	26′	27′	28′	29′	30′	31′	32′	33′	34′
0″	3632	3468	3310	3158	3010	2868	2730	2596	2467
1	3629	3465	3307	3155	3008	2866	2728	2594	2465
2	3626	3463	3305	3153	3005	2863	2725	2592	2462
3	3623	3460	3302	3150	3003	2861	2723	2590	2460
4	3621	3457	3300	3148	3001	2859	2721	2588	2458
5	3618	3454	3297	3145	2998	2856	2719	2585	2456
6	3615	3452	3294	3143	2996	2854	2716	2583	2454
7	3612	3449	3292	3140	2993	2852	2714	2581	2452
8	3610	3446	3289	3138	2991	2849	2712	2579	2450
9	3607	3444	3287	3135	2989	2847	2710	2577	2448
10	3604	3441	3284	3133	2986	2845	2707	2574	2445
11	3601	3438	3282	3130	2984	2842	2705	2572	2443
12	3598	3436	3279	3128	2981	2840	2703	2570	2441
13	3596	3433	3276	3125	2979	2838	2701	2568	2439
14	3593	3431	3274	3123	2977	2835	2698	2566	2437
15	3590	3428	3271	3120	2974	2833	2696	2564	2435
16	3587	3425	3269	3118	2972	2831	2694	2561	2433
17	3585	3423	3266	3115	2969	2828	2692	2559	2431
18	3582	3420	3264	3113	2967	2826	2689	2557	2429
19	3579	3417	3261	3110	2965	2824	2687	2555	2426
20	3576	3415	3259	3108	2962	2821	2685	2553	2424
21	3574	3412	3256	3105	2960	2819	2683	2551	2422
22	3571	3409	3253	3103	2958	2817	2681	2548	2420
23	3568	3407	3251	3101	2955	2815	2678	2546	2418
24	3565	3404	3248	3098	2953	2812	2676	2544	2416
25	3563	3401	3246	3096	2950	2810	2674	2542	2414
26	3560	3399	3243	3093	2948	2808	2672	2540	2412
27	3557	3396	3241	3091	2946	2805	2669	2538	2410
28	3555	3393	3238	3088	2943	2803	2667	2535	2408
29	3552	3391	3236	3086	2941	2801	2665	2533	2405
30	3549	3388	3233	3083	2939	2798	2663	2531	2403
31	3546	3386	3231	3081	2936	2796	2660	2529	2401
32	3544	3383	3228	3078	2934	2794	2658	2527	2399
33	4541	3380	3225	3076	2931	2792	2656	2525	2397
34	3538	3378	3223	3073	2929	2789	2654	2522	2395
35	3535	3375	3220	3071	2927	2787	2652	2520	2393
36	3533	3372	3218	3069	2924	2785	2649	2518	2391
37	3530	3370	3215	3066	2922	2782	2647	2516	2389
38	3527	3367	3213	3064	2920	2780	2645	2514	2387
39	3525	3365	3210	3061	2917	2778	2643	2512	2384
40	3522	3362	3208	3059	2915	2775	2640	2510	2382
41	3519	3359	3205	3056	2912	2773	2638	2507	2380
42	3516	3357	3203	3054	2910	2771	2636	2505	2378
43	3514	3354	3200	3052	2908	2769	2634	2503	2376
44	3511	3351	3198	3049	2905	2766	2632	2501	2374
45	3508	3349	3195	3047	2903	2764	2629	2499	2372
46	3506	3346	3193	3044	2901	2762	2627	2497	2370
47	3503	3344	3190	3042	2898	2760	2625	2494	2368
48	3500	3341	3188	3039	2896	2757	2623	2492	2366
49	3497	3338	3185	3037	2894	2755	2621	2490	2364
50	3495	3336	3183	3034	2891	2753	2618	2488	2362
51	3492	3333	3180	3032	2889	2750	2616	2486	2359
52	3489	3331	3178	3030	2887	2748	2614	2484	2357
53	3487	3328	3175	3027	2884	2746	2612	2482	2355
54	3484	3325	3173	3025	2882	2744	2610	2480	2353
55	3481	3323	3170	3022	2880	2741	2607	2477	2351
56	3479	3320	3168	3020	2877	2739	2605	2475	2349
57	3476	3318	3165	3018	2875	2737	2603	2473	2347
58	3473	3315	3163	3015	2873	2735	2601	2471	2345
59	3471	3313	3160	3013	2870	2732	2599	2469	2343
60	3468	3310	3158	3010	2868	2730	2596	2467	2341

TABLE LXXX.

Logistical Logarithms.

	35′	36′	37′	38′	39′	40′	41′	42′	43′
0″	2341	2218	2099	1984	1871	1761	1654	1549	1447
1	2339	2216	2098	1982	1869	1759	1652	1547	1445
2	2337	2214	2096	1980	1867	1757	1650	1546	1443
3	2335	2212	2094	1978	1865	1755	1648	1544	1442
4	2333	2210	2092	1976	1863	1754	1647	1542	1440
5	2331	2208	2090	1974	1862	1752	1645	1540	1438
6	2328	2206	2088	1972	1860	1750	1643	1539	1437
7	2326	2204	2086	1970	1858	1748	1641	1537	1435
8	2324	2202	2084	1968	1856	1746	1640	1535	1433
9	2322	2200	2082	1967	1854	1745	1638	1534	1432
10	2320	2198	2080	1965	1852	1743	1636	1532	1430
11	2318	2196	2078	1963	1850	1741	1634	1530	1428
12	2316	2194	2076	1961	1849	1739	1633	1528	1427
13	2314	2192	2074	1959	1847	1737	1631	1527	1425
14	2312	2190	2072	1957	1845	1736	1629	1525	1423
15	2310	2188	2070	1955	1843	1734	1627	1523	1422
16	2308	2186	2068	1953	1841	1732	1626	1522	1420
17	2306	2184	2066	1951	1839	1730	1624	1520	1418
18	2304	2182	2064	1950	1838	1728	1622	1518	1417
19	2302	2180	2062	1948	1836	1727	1620	1516	1415
20	2300	2178	2061	1946	1834	1725	1619	1515	1413
21	2298	2176	2059	1944	1832	1723	1617	1513	1412
22	2296	2174	2057	1942	1830	1721	1615	1511	1410
23	2294	2172	2055	1940	1828	1719	1613	1510	1408
24	2291	2170	2053	1938	1827	1718	1612	1508	1407
25	2289	2169	2051	1936	1825	1716	1610	1506	1405
26	2287	2167	2049	1934	1823	1714	1608	1504	1403
27	2285	2165	2047	1933	1821	1712	1606	1503	1402
28	2283	2163	2045	1931	1819	1711	1605	1501	1400
29	2281	2161	2043	1929	1817	1709	1603	1499	1398
30	2279	2159	2041	1927	1816	1707	1601	1498	1397
31	2277	2157	2039	1925	1814	1705	1599	1496	1395
32	2275	2155	2037	1923	1812	1703	1598	1494	1393
33	2273	2153	2035	1921	1810	1702	1596	1493	1392
34	2271	2151	2033	1919	1808	1700	1594	1491	1390
35	2269	2149	2032	1918	1806	1698	1592	1489	1388
36	2267	2147	2030	1916	1805	1696	1591	1487	1387
37	2265	2145	2028	1914	1803	1694	1589	1486	1385
38	2263	2143	2026	1912	1801	1693	1587	1484	1383
39	2261	2141	2024	1910	1799	1691	1585	1482	1382
40	2259	2139	2022	1908	1797	1689	1584	1481	1380
41	2257	2137	2020	1906	1795	1687	1582	1479	1378
42	2255	2135	2018	1904	1794	1686	1580	1477	1377
43	2253	2133	2016	1903	1792	1684	1578	1476	1375
44	2251	2131	2014	1901	1790	1682	1577	1474	1373
45	2249	2129	2012	1899	1788	1680	1575	1472	1372
46	2247	2127	2010	1897	1786	1678	1573	1470	1370
47	2245	2125	2009	1895	1785	1677	1571	1469	1368
48	2243	2123	2007	1893	1783	1675	1570	1467	1367
49	2241	2121	2005	1891	1781	1673	1568	1465	1365
50	2239	2119	2003	1889	1779	1671	1566	1464	1363
51	2237	2117	2001	1888	1777	1670	1565	1462	1362
52	2235	2115	1999	1886	1775	1668	1563	1460	1360
53	2233	2113	1997	1884	1774	1666	1561	1459	1359
54	2231	2111	1995	1882	1772	1664	1559	1457	1357
55	2229	2109	1993	1880	1770	1663	1558	1455	1355
56	2227	2107	1991	1878	1768	1661	1556	1454	1354
57	2225	2105	1989	1876	1766	1659	1554	1452	1352
58	2223	2103	1987	1875	1765	1657	1552	1450	1350
59	2220	2101	1986	1873	1763	1655	1551	1449	1349
60	2218	2099	1984	1871	1761	1654	1549	1447	1347

Logistical Logarithms.

	44′	45′	46′	47′	48′	49′	50′	51′	52′
0″	1347	1249	1154	1061	969	880	792	706	621
1	1345	1248	1152	1059	968	878	790	704	620
2	1344	1246	1151	1057	966	877	789	703	619
3	1342	1245	1149	1056	965	875	787	702	617
4	1340	1243	1148	1054	963	874	786	700	616
5	1339	1241	1146	1053	962	872	785	699	615
6	1337	1240	1145	1051	960	871	783	697	613
7	1335	1238	1143	1050	959	869	782	696	612
8	1334	1237	1141	1048	957	868	780	694	610
9	1332	1235	1140	1047	956	866	779	693	609
10	1331	1233	1138	1045	954	865	777	692	608
11	1329	1232	1137	1044	953	863	776	690	606
12	1327	1230	1135	1042	951	862	774	689	605
13	1326	1229	1134	1041	950	860	773	687	603
14	1324	1227	1132	1039	948	859	772	686	602
15	1322	1225	1130	1037	947	857	770	685	601
16	1321	1224	1129	1036	945	856	769	683	599
17	1319	1222	1127	1034	944	855	767	682	598
18	1317	1221	1126	1033	942	853	766	680	596
19	1316	1219	1124	1031	941	852	764	679	595
20	1314	1217	1123	1030	939	850	763	678	594
21	1313	1216	1121	1028	938	849	762	676	592
22	1311	1214	1119	1027	936	847	760	675	591
23	1309	1213	1118	1025	935	846	759	673	590
24	1308	1211	1116	1024	933	844	757	672	588
25	1306	1209	1115	1022	932	843	756	670	587
26	1304	1208	1113	1021	930	841	754	669	585
27	1303	1206	1112	1019	929	840	753	668	584
28	1301	1205	1110	1018	927	838	751	666	583
29	1300	1203	1109	1016	926	837	750	665	581
30	1298	1201	1107	1015	924	835	749	663	580
31	1296	1200	1105	1013	923	834	747	662	579
32	1295	1198	1104	1012	921	833	746	661	577
33	1293	1197	1102	1010	920	831	744	659	576
34	1291	1195	1101	1008	918	830	743	658	574
35	1290	1193	1099	1007	917	828	741	656	573
36	1288	1192	1098	1005	915	827	740	655	572
37	1287	1190	1096	1004	914	825	739	654	570
38	1285	1189	1095	1002	912	824	737	652	569
39	1283	1187	1093	1001	911	822	736	651	568
40	1282	1186	1091	999	909	821	734	649	566
41	1280	1184	1090	998	908	819	733	648	565
42	1278	1182	1088	996	906	818	731	647	563
43	1277	1181	1087	995	905	816	730	645	562
44	1275	1179	1085	993	903	815	729	644	561
45	1274	1178	1084	992	902	814	727	642	559
46	1272	1176	1082	990	900	812	726	641	558
47	1270	1174	1081	989	899	811	724	640	557
48	1269	1173	1079	987	897	809	723	638	555
49	1267	1171	1078	986	896	808	721	637	554
50	1266	1170	1076	984	894	806	720	635	552
51	1264	1168	1074	983	893	805	719	634	551
52	1262	1167	1073	981	891	803	717	633	550
53	1261	1165	1071	980	890	802	716	631	548
54	1259	1163	1070	978	888	801	714	630	547
55	1257	1162	1068	977	887	799	713	628	546
56	1256	1160	1067	975	885	798	711	627	544
57	1254	1159	1065	974	884	796	710	626	543
58	1253	1157	1064	972	883	795	709	624	541
59	1251	1156	1062	971	881	793	707	623	540
60	1249	1154	1061	969	880	792	706	621	539

Logistical Logarithms.

	53′	54′	55′	56′	57′	58′	59′
0″	539	458	378	300	223	147	73
1	537	456	377	298	221	146	72
2	536	455	375	297	220	145	71
3	535	454	374	296	219	143	69
4	533	452	373	294	218	142	68
5	532	451	371	293	216	141	67
6	531	450	370	292	215	140	66
7	529	448	369	291	214	139	64
8	528	447	367	289	213	137	63
9	526	446	366	288	211	136	62
10	525	444	365	287	210	135	61
11	524	443	363	285	209	134	60
12	522	442	362	284	208	132	58
13	521	440	361	283	206	131	57
14	520	439	359	282	205	130	56
15	518	438	358	280	204	129	55
16	517	436	357	279	202	127	53
17	516	435	356	278	201	126	52
18	514	434	354	276	200	125	51
19	513	432	353	275	199	124	50
20	512	431	352	274	197	122	49
21	510	430	350	273	196	121	47
22	509	428	349	271	195	120	46
23	507	427	348	270	194	119	45
24	506	426	346	269	192	117	44
25	505	424	345	267	191	116	42
26	503	423	344	266	190	115	41
27	502	422	342	265	189	114	40
28	501	420	341	264	187	112	39
29	499	419	340	262	186	111	38
30	498	418	339	261	185	110	36
31	497	416	337	260	184	109	35
32	495	415	336	258	182	107	34
33	494	414	335	257	181	106	33
34	403	412	333	256	180	105	31
35	491	411	332	255	179	104	30
36	490	410	331	253	177	103	29
37	489	408	329	252	176	101	28
38	487	407	328	251	175	100	27
39	486	406	327	250	174	99	25
40	484	404	326	248	172	98	24
41	483	403	324	247	171	96	23
42	482	402	323	246	170	95	22
43	480	400	322	244	169	94	21
44	479	399	320	243	167	93	19
45	478	398	319	242	166	91	18
46	476	396	318	241	165	90	17
47	475	395	316	239	163	89	16
48	474	394	315	238	162	88	15
49	472	392	314	237	161	87	13
50	471	391	313	235	160	85	12
51	470	390	312	234	158	84	11
52	468	388	310	233	157	83	10
53	467	387	309	232	156	82	8
54	466	386	307	230	155	80	7
55	464	384	306	229	153	79	6
56	463	383	305	228	152	78	5
57	462	382	304	227	151	77	4
58	460	381	302	225	150	75	2
59	459	379	301	224	148	74	1
60	458	378	300	223	147	73	0

TABLE LXXXI.

Reduction to the Meridian.

	m.									
Sec.	0m	1m	2m	3m	4m	5m	6m	7m	8m	9m
	″	″	″	″	″	″	″	″	″	″
0	0.00	1.96	7.85	17.67	31.41	49.09	70.68	96.20	125.65	159.02
1	0.00	2.03	7.99	17.87	31.68	49.41	71.07	96.66	126.17	159.61
2	0.00	2.10	8.12	18.09	31.94	49.74	71.47	97.12	126.70	160.20
3	0.00	2.16	8.25	18.27	32.21	50.07	71.86	97.58	127.23	160.79
4	0.01	2.23	8.39	18.47	32.47	50.40	72.26	98.04	127.75	161.39
5	0.01	2.30	8.52	18.67	32.74	50.74	72.66	98.51	128.28	161.98
6	0.02	2.38	8.66	18.87	33.01	51.07	73.06	98.97	128.81	162.58
7	0.03	2.45	8.80	19.07	33.27	51.40	73.46	99.44	129.34	163.17
8	0.03	2.52	8.94	19.28	33.54	51.74	73.86	99.90	129.87	163.77
9	0.04	2.60	9.08	19.48	33.82	52.07	74.26	100.37	130.41	164.37
10	0.05	2.67	9.22	19.69	34.09	52.41	74.66	100.85	130.94	164.97
11	0.07	2.75	9.36	19.90	34.36	52.75	75.07	101.31	131.48	165.57
12	0.08	2.83	9.50	20.11	34.64	53.09	75.47	101.78	132.01	166.17
13	0.09	2.91	9.65	20.32	34.91	53.43	75.88	102.25	132.55	166.77
14	0.11	2.99	9.79	20.53	35.19	53.77	76.29	102.72	133.09	167.37
15	0.12	3.07	9.94	20.74	35.46	54.12	76.70	103.20	133.63	167.98
16	0.14	3.15	10.09	20.95	35.74	54.46	77.10	103.67	134.17	168.58
17	0.16	3.23	10.24	21.17	36.02	54.81	77.51	104.15	134.71	169.19
18	0.18	3.32	10.39	21.38	36.30	55.15	77.92	104.63	135.25	169.80
19	0.20	3.40	10.54	21.60	36.59	55.50	78.34	105.10	135.79	170.41
20	0.22	3.49	10.69	21.82	36.87	55.85	78.75	105.58	136.34	171.02
21	0.24	3.58	10.84	22.03	37.15	56.20	79.17	106.06	136.88	171.63
22	0.26	3.67	11.00	22.25	37.44	56.55	79.58	106.55	137.43	172.24
23	0.29	3.76	11.15	22.48	37.72	56.90	80.00	107.03	137.98	172.86
24	0.31	3.85	11.31	22.70	38.01	57.25	80.42	107.51	138.53	173.47
25	0.34	3.94	11.47	22.92	38.30	57.61	80.84	108.00	139.08	174.09
26	0.37	4.03	11.63	23.14	38.59	57.96	81.26	108.48	139.63	174.70
27	0.40	4.13	11.79	23.37	38.88	58.32	81.68	108.97	140.18	175.32
28	0.43	4.22	11.95	23.60	39.17	58.68	82.10	109.46	140.74	175.94
29	0.46	4.32	12.11	23.82	39.47	59.03	82.53	109.95	141.29	176.56
30	0.49	4.42	12.27	24.05	39.76	59.39	82.95	110.14	141.85	177.18
31	0.52	4.52	12.44	24.28	40.05	59.76	83.38	110.93	142.40	177.80
32	0.56	4.62	12.60	24.51	40.35	60.15	83.81	111.42	142.96	178.42
33	0.59	4.72	12.77	24.74	40.65	60.48	84.23	111.91	143.52	179.05
34	0.63	4.82	12.94	24.98	40.95	60.84	84.66	112.41	144.08	179.68
35	0.67	4.92	13.10	25.21	41.25	61.21	85.09	112.90	144.64	180.30
36	0.71	5.03	13.27	25.45	41.55	61.57	85.52	113.40	145.20	180.93
37	0.75	5.13	13.44	25.68	41.85	61.94	85.96	113.90	145.77	181.56
38	0.79	5.24	13.62	25.92	42.15	62.31	86.39	114.40	146.34	182.19
39	0.83	5.35	13.79	26.16	42.45	62.68	86.82	114.90	146.90	182.82
40	0.87	5.45	13.96	26.40	42.76	63.05	87.26	115.40	147.46	183.45
41	0.92	5.56	14.14	26.64	43.06	63.42	87.70	115.90	148.03	184.09
42	0.96	5.67	14.31	26.88	43.37	63.79	88.13	116.40	148.60	184.72
43	1.01	5.79	14.49	27.12	43.68	64.16	88.57	116.91	149.17	185.35
44	1.06	5.90	14.67	27.37	43.99	64.54	89.01	117.41	149.74	185.99
45	1.10	6.01	14.85	27.61	44.30	64.91	89.46	117.92	150.31	186.63
46	1.15	6.13	15.03	27.86	44.61	65.29	89.90	118.43	150.88	187.27
47	1.20	6.24	15.21	28.10	44.92	65.67	90.34	118.94	151.46	187.91
48	1.26	6.36	15.39	28.35	45.24	66.05	90.79	119.45	152.03	188.55
49	1.31	6.48	15.58	28.60	45.55	66.43	91.23	119.96	152.61	189.19
50	1.36	6.60	15.76	28.85	45.87	66.81	91.68	120.47	153.19	189.83
51	1.42	6.72	15.95	29.10	46.18	67.19	92.13	120.98	153.77	190.47
52	1.48	6.84	16.14	29.36	46.50	67.58	92.57	121.50	154.35	191.12
53	1.53	6.96	16.32	29.61	46.82	67.96	93.02	122.01	154.93	191.76
54	1.59	7.09	16.51	29.86	47.14	68.35	93.47	122.53	155.51	192.41
55	1.65	7.21	16.70	30.12	47.46	68.73	93.93	123.05	156.09	193.06
56	1.71	7.34	16.89	30.38	47.79	69.12	94.38	123.57	156.68	193.71
57	1.77	7.47	17.08	30.64	48.11	69.51	94.83	124.09	157.26	194.36
58	1.83	7.59	17.28	30.89	48.43	69.90	95.29	124.61	157.85	195.02
59	1.90	7.72	17.48	31.15	48.76	70.29	95.75	125.13	158.43	195.67

TABLE LXXXI.

Reduction to the Meridian.

	m.							*n.*	
Sec.	10^m	11^m	12^m	13^m	14^m	15^m	16^m		
	″	″	″	″	″	″	″	m. s.	″
0	196.32	237.54	282.68	331.75	384.74	441.63	502.45	5 0	0.01
1	196.97	238.26	283.46	332.61	385.66	442.62	503.50	6 0	0.01
2	197.63	238.98	284.25	333.46	386.58	443.61	504.55	6 30	0.02
3	198.29	239.70	285.04	334.32	387.50	444.59	505.60	7 0	0.02
4	198.94	240.42	285.83	335.17	388.42	445.58	506.66	7 30	0.03
5	199.60	241.15	286.62	336.03	389.34	446.56	507.71	8 0	0.04
6	200.26	241.87	287.41	336.89	390.26	447.55	508.76	10	0.04
7	200.93	242.60	288.20	337.74	391.18	448.54	509.81	20	0.05
8	201.59	243.33	288.99	338.60	392.10	449.52	510.86	30	0.05
9	202.25	244.06	289.79	339.45	393.02	450.51	511.92	40	0.05
10	202.92	244.79	290.58	340.31	393.94	451.49	512.97	50	0.06
11	203.58	245.52	291.38	341.17	394.87	452.49	514.03	9 0	0.06
12	204.25	246.26	292.18	342.04	395.80	453.49	515.09	10	0.07
13	204.92	246.99	292.98	342.91	396.73	454.49	516.16	20	0.07
14	205.59	247.72	293.78	343.77	397.66	455.48	517.22	30	0.08
15	206.26	248.46	294.58	344.64	398.59	456.48	518.28	40	0.08
16	206.93	249.19	295.38	345.51	399.52	457.48	519.34	50	0.09
17	207.60	249.93	296.18	346.38	400.45	458.48	520.41	10 0	0.09
18	208.27	250.67	296.99	347.24	401.39	459.48	521.47	10	0.10
19	208.95	251.41	297.79	348.11	402.32	460.48	522.53	20	0.11
20	209.62	252.15	298.60	348.98	403.25	461.48	523.60	30	0.11
21	210.30	252.89	299.40	349.85	404.19	462.48	524.67	40	0.12
22	210.98	253.63	300.21	350.73	405.13	463.48	525.74	50	0.13
23	211.66	254.38	301.02	351.61	406.07	464.49	526.81	11 0	0.14
24	212.34	255.12	301.83	352.49	407.02	465.50	527.89	10	0.15
25	213.02	255.87	302.65	353.36	407.96	466.51	528.96	20	0.16
26	213.70	256.62	303.46	354.24	408.90	467.52	530.04	30	0.16
27	214.38	257.37	304.27	355.12	409.84	468.53	531.11	40	0.17
28	215.07	258.12	305.09	356.00	410.79	469.54	532.18	50	0.18
29	215.75	258.87	305.90	356.87	411.73	470.55	533.26	12 0	0.19
30	216.44	259.62	306.72	357.75	412.67	471.56	534.34	10	0.20
31	217.12	260.37	307.54	358.64	413.62	472.57	535.42	20	0.22
32	217.81	261.12	308.36	359.53	414.58	473.59	536.50	30	0.23
33	218.50	261.88	309.18	360.42	415.53	474.61	537.58	40	0.24
34	219.19	262.64	310.00	361.30	416.49	475.63	538.67	50	0.25
35	219.89	263.39	310.82	362.19	417.44	476.65	539.75	13 0	0.27
36	220.58	264.15	311.65	363.08	418.40	477 66	540.84	10	0.28
37	221.27	264.91	312.47	363.97	419.35	478.68	541.92	20	0.30
38	221.97	265.67	313.30	364.86	420.31	479.70	543.01	30	0.31
39	222.66	266.43	314.12	365.75	421.26	480.72	544.09	40	0.33
40	223.36	267.20	314.95	366.64	422.22	481.74	545.18	50	0.34
41	224.06	267.96	315.78	367.54	423.18	482.77	546.27	14 0	0.36
42	224.76	268.72	316.61	368.43	424.14	483.80	547.36	10	0.38
43	225.46	269.49	317.44	369.33	425.11	484.83	548.45	20	0.39
44	226.16	270.26	318.27	370.23	426.07	485.86	549.55	30	0.41
45	226.86	271.03	319.11	371.13	427.04	486.89	550.64	40	0.43
46	227.57	271.79	319.94	372.03	428.00	487.92	551.74	50	0.45
47	228.27	272.57	320.78	372.93	428.97	488.95	552.83	15 0	0.47
48	228.98	273.34	321.62	373.83	429.94	489.98	553.93	10	0.49
49	229.69	274.11	322.45	374.73	430.90	491.01	555.02	20	0.52
50	230.40	274.88	323.29	375.64	431.87	492.04	555.12	30	0.54
51	231.11	275.66	324.13	376.54	432.85	493.08	558.22	40	0.56
52	231.81	276.43	324.97	377.45	433.82	494.12	559.33	50	0.59
53	232.53	277.21	325.82	378.36	434.80	495.16	560.44	16 0	0.61
54	233.24	277.99	326.66	379.27	435.78	496.20	561.54	10	0.64
55	233.95	278.77	327.50	380.18	436.75	497.25	562.65	20	0.67
56	234.67	279.55	328.35	381.09	437.73	498.29	563.76	30	0.69
57	235.38	280.33	329.20	382.00	438.70	499.33	564.87	40	0.72
58	236.10	281.11	330.04	382.91	439.68	500.37	565.97	50	0.75
59	236.82	281.89	330.89	383.82	440.66	501.41	567.08	17 0	0.78

www.ingramcontent.com/pod-product-compliance
Lightning Source LLC
LaVergne TN
LVHW020925110826
845150LV00004B/774

* 9 7 8 1 4 2 5 5 5 3 6 0 9 *